Automotive Steels

Related titles

Lightweight Composite Structures in Transport: Design, Manufacturing, Analysis and Performance
(ISBN 978-1-78242-325-6)

Future Development of Thermal Spray Coatings: Types, Designs, Manufacture and Applications
(ISBN 978-0-85709-769-9)

Pekguleryuz, Kainer and Kaya, *Fundamentals of Magnesium Alloy Metallurgy*
(ISBN 978-0-85709-088-1)

Woodhead Publishing Series in Metals
and Surface Engineering

Automotive Steels

Design, Metallurgy, Processing and Applications

Edited by

Radhakanta Rana

Shiv Brat Singh

ELSEVIER

AMSTERDAM • BOSTON • HEIDELBERG • LONDON
NEW YORK • OXFORD • PARIS • SAN DIEGO
SAN FRANCISCO • SINGAPORE • SYDNEY • TOKYO
Woodhead Publishing is an imprint of Elsevier

WP
WOODHEAD
PUBLISHING

Woodhead Publishing is an imprint of Elsevier
The Officers' Mess Business Centre, Royston Road, Duxford, CB22 4QH, United Kingdom
50 Hampshire Street, 5th Floor, Cambridge, MA 02139, United States
The Boulevard, Langford Lane, Kidlington, OX5 1GB, United Kingdom

Notices

Knowledge and best practice in this field are constantly changing. As new research and experience broaden our understanding, changes in research methods, professional practices, or medical treatment may become necessary.

Practitioners and researchers must always rely on their own experience and knowledge in evaluating and using any information, methods, compounds, or experiments described herein. In using such information or methods they should be mindful of their own safety and the safety of others, including parties for whom they have a professional responsibility.

To the fullest extent of the law, neither the Publisher nor the authors, contributors, or editors, assume any liability for any injury and/or damage to persons or property as a matter of products liability, negligence or otherwise, or from any use or operation of any methods, products, instructions, or ideas contained in the material herein.

British Library Cataloguing-in-Publication Data
A catalogue record for this book is available from the British Library

Library of Congress Cataloging-in-Publication Data
A catalog record for this book is available from the Library of Congress

ISBN: 978-0-08-100638-2 (print)
ISBN: 978-0-08-100653-5 (online)

Publisher: Matthew Deans
Acquisition Editor: Gwen Jones
Editorial Project Manager: Charlotte Cockle
Production Project Manager: Debasish Ghosh
Cover Designer: Christian J. Bilbow

Typeset by MPS Limited, Chennai, India

Contents

List of contributors

E.H. Atzema Tata Steel, IJmuiden, The Netherlands

E. Billur Billur Metal Form Ltd., Bursa, Turkey; Atılım University, Ankara, Turkey

B.C. De Cooman Graduate Institute of Ferrous Technology, POSTECH, Pohang, South Korea

E. De Moor Colorado School of Mines, Golden, CO, United States

J.R. Fekete National Institute of Standards and Technology, Boulder, CO, United States

N. Fonstein ArcelorMittal Global R&D, East Chicago Labs, United States

C.I. Garcia University of Pittsburgh, Pittsburgh, PA, United States

P. Ghosh Tata Steel, Jamshedpur, India

J.N. Hall Steel Market Development Institute, Southfield, MI, United States

G. Krauss Colorado School of Mines, Golden, CO, United States

O.N. Mohanty RSB Group, Pune, India

M. Mukherjee Tata Steel Ltd., Jamshedpur, Jharkhand, India

E. Pereloma University of Wollongong, Wollongong, NSW, Australia

R.K. Ray Indian Institute of Engineering Science and Technology, Shibpur, West Bengal, India

J.G. Speer Colorado School of Mines, Golden, CO, United States

K. Sugimoto Shinshu University, Wakasato, Nagano, Japan

I. Timokhina Deakin University, Geelong, VIC, Australia

Design of auto body: materials perspective

1

J.R. Fekete[1] *and J.N. Hall*[2]
[1]National Institute of Standards and Technology, Boulder, CO, United States,
[2]Steel Market Development Institute, Southfield, MI, United States

1.1 History of steel usage in vehicle body structures and closures

Steel has been an important material for body construction of motor vehicles in North America since the early 1900s. At that time, steel competed with aluminum and wood for predominance in body construction, but by the 1920s it was the material of choice. Its low cost, coupled with its ability to be pressed into complex shapes, and easily joined through welding processes, led to this position in the industry. From these early days, the auto industry depended on secure supplies of sheet steel, and the steel industry responded by developing a strong capability for thin, wide steel sheets to support one of its major customers. However, starting in the 1960s, the automotive industry faced significant new challenges that would fundamentally change vehicle structural requirements. These challenges included regulatory demands for safer, cleaner, and more fuel efficient vehicles, as well as increased competition from new materials entrants in the North American market and customer demands for higher performance, comfort, and reliability. The responses to these challenges required the development of new steel products with higher strength and improved manufacturability.

1.2 Significant events in history impacting steel application in vehicle design

The 20th century, particularly its second half, was a time of rapid development of both the steel and auto industries. The amazing improvements in the ability of people and goods to be moved across great distances resulted in rapid growth of the transportation industry. This came with a price, though, as injury and deaths resulting from accidents skyrocketed, and skies darkened with the emissions of the expanding numbers of internal combustion engines. At the same time, customers came to expect an ever increasing level of comfort and speed in their vehicles. The experience of the United States in the latter half of the 20th century serves as a

Automotive Steels. DOI: http://dx.doi.org/10.1016/B978-0-08-100638-2.00001-8

relevant example of how the steel and auto industries worked together to meet these emerging needs.

The post-World War II economic expansion in the U.S. resulted in rapid growth of the automotive industry in the 1950s and 1960s. With this success came increasing public pressure to improve the safety and environmental performance of this growing industry. The U.S. government responded to these events through several legislative actions. The Federal Clean Air Act was passed in 1970. This act established the regulatory framework for monitoring and reducing emissions of air pollutants, and created the Environmental Protection Agency (EPA), whose mandate included reducing pollution from motor vehicles. In the same year, the Highway Safety Act was passed, creating the National Highway and Traffic Safety Administration (NHTSA), charged with establishing safety requirements for both motor vehicles and the roads on which they traveled. Examples of these new requirements include implementation of energy absorbing bumpers, three-point restraint systems, and improved structural requirements for frontal and side impact energy absorption.

At the same time, the Arab oil embargo of 1973 resulted in disruptions in the supply of gasoline for motor vehicle usage. The price of gasoline increased dramatically and became very unstable. One consequence of these events was increasing demand for smaller, more fuel efficient vehicles. At this time, small cars constituted a relatively small part of the U.S. market, as the domestic manufacturers responded to the demand from their customers for larger, more luxurious vehicles. However, small cars had been exported to the U.S. market for many years by a number of overseas suppliers (in relatively small numbers). These vehicles included the Volkswagen Beetle, Honda Civic, and Toyota Corolla. The "gas shocks" helped boost the demand for these vehicles in the U.S. market, a demand that has increased over time. These events also resulted in public pressure for political solutions to the need for improved fuel economy in motor vehicles. The result was the implementation of CAFE (Corporate Average Fuel Economy) standards by the EPA.

It quickly became clear to automotive engineers that these new regulatory and consumer demands would necessitate significant vehicle mass reduction. Reducing mass resulted in higher fuel economy, lower vehicle emissions, and helped engineers meet new safety requirements. Vehicle downsizing and migration from body-on-frame (BOF) to body-frame-integral (BFI) structures were two early initiatives used to accomplish the mass reduction. Fig. 1.1 demonstrates the dramatic mass reductions that were accomplished by the domestic automakers, and the improvement in fuel mileage that followed.

This focus on mass reduction led to demonstrations of the improvement in structural efficiency made possible when the strength-to-weight ratio of the materials of construction is increased. An example of this work in the late 1970s was the development of the "Charger XL" by Chrysler Corporation, where application of both higher strength steel and aluminum resulted in a 286 kg reduction in vehicle mass with no impact on vehicle quality or performance [1,2]. This work was an early demonstration of the potential of high-strength steel.

In the early days of automotive high-strength steel development, many different concepts were investigated. At this time, ingot casting and rolling were still the most widely used processes for producing slabs. The so-called "rimmed"

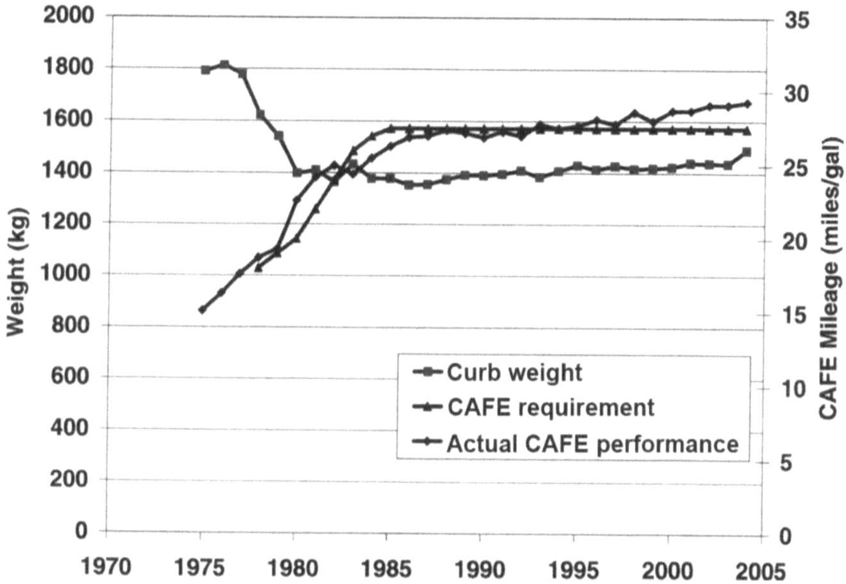

Figure 1.1 History of vehicle curb weight, CAFE mileage requirements and actual CAFE performance for the U.S. fleet [11].

steels (named for the "rimming" action—the boiling caused by dissolved oxygen reacting with carbon in the mold to create CO and CO_2) were commonly used for automotive applications because of their superior surface quality, cleanliness, and ductility. Nitrogen and carbon remained in solid solution in rimmed steel, and metallurgists could take advantage of this characteristic to increase the strength of steel parts through strain aging. The strain was induced during the forming processes and the subsequent aging occurred during a post forming heat treatment, which sometimes involved the paint bake cycle. Nitrogen could be added to these materials to make even higher yield strength grades, up to 500 MPa [3]. These steels were the precursors to the bake hardenable grades described below. However, there were two problems with this approach. First, the materials were susceptible to stretcher strains or "Lüders lines," an objectionable surface condition, especially for exposed quality material. Second, and most important, the industry at this time was moving rapidly toward continuous casting of slabs, a much more efficient process than the traditional casting of ingots and subsequent production of slabs through rolling. The continuous casting process requires "killed" steel, the opposite of "rimmed" steel. Aluminum is added to "kill" the oxidation of carbon in these steels by replacing the carbon in the oxidation reaction. It also combines with nitrogen and, to a lesser extent, carbon itself, removing them from solution. Thus, the strain aging was significantly reduced, and the high strength levels of rimmed steel could not be reached with killed steels. There were few applications of strain-aging high strength steel at this time, and the onset of continuous cast, killed steel quickly ended the use of these materials in automotive applications.

So-called "ultra-high strength steels," with tensile strength levels above 600 MPa, were also in development at this time. These included martensitic steels [4,5] which were produced in continuous annealing lines, and recovery annealed steels, which were cold rolled to very high strength levels, then annealed below the recrystallization temperature to recover enough ductility to survive rudimentary forming processes [6]. Both of these materials found niches in the marketplace, mainly in roll-formed parts such as bumpers and beams where formability requirements were not as difficult. Initial development of dual phase (DP) steels also occurred during this time [7,8]. These materials were processed to produce microstructures of martensite and/or bainite islands in a ferrite matrix through careful intercritical annealing and subsequent fast cooling. The potential of these products was successfully demonstrated, but it was difficult to produce a uniform product with the available process control technology. Also, the relatively low cooling capabilities of steel processing lines demanded higher alloy contents to achieve the needed hardenability. This resulted in products that were difficult to weld. It would be another 20 years before DP steel could be developed into an important structural material in the automotive industry.

The high-strength steel products that would become most widely used at this time were the microalloyed high strength, low alloy (HSLA) steels [9]. Automotive steel makers used a combination of alloying with carbo-nitride formers, such as Nb, V, Ti, and Zr, and careful thermomechanical processing to produce fine grained, precipitation strengthened steels. The final products had yield strength levels of 280–550 MPa and relatively high ductility. Additions of rare earth elements such as Ca or Zr were found to transform sulfide inclusions from long "stringers" to a more globular morphology, and the resulting improved transverse ductility was critical to the successful early application of HSLA steels [10]. However, as with the DP steels, the processing requirements of these products tested the process control capabilities of steel mills and early versions of these products had much larger ranges of mechanical properties than the commonly used mild steels. This fact, along with the reduced formability and higher springback after stamping, made early applications difficult to produce through stamping. The feedback from the press shops caused product engineers to slow down their application of high-strength steel. However, the need for more efficient structures was not going away, which forced both the automotive and steel industries to improve their processes to successfully produce parts with these steels and to utilize their capability to reduce vehicle mass.

The regulatory pressure steadily increased during the decade of the 1980s. The frontal and side impact requirements conceived and proposed earlier were now fully implemented. Additional requirements for pole impacts and bumper integrity were also implemented. As shown in Fig. 1.1, the CAFE requirements for cars steadily increased from 18 mpg (miles per gallon) at the beginning of the decade to 27.5 mpg by the end. The California Air Resources Board and EPA also continued to drive reductions in vehicle emissions through regulatory actions.

During the 1980s, the pressure to improve fuel efficiency to reduce weight caused the majority of car platforms in the United States to convert from BOF to

BFI (also known as monocoque). New vehicles were also substantially downsized in order to improve fuel efficiency. High-strength steel continued to make inroads to improve the structural performance of these vehicles, but forming and joining challenges slowed the progress. The increasing pressure to reduce mass and cost coupled with the difficulties in implementing high-strength steel to this point led the auto manufacturers to begin seriously considering alternatives to steel for construction of their cars, especially plastics. There was a sense in the industry that steel had no more to offer, and alternatives were needed to achieve the needed mass and cost reductions. The launch of the plastic-skinned Pontiac Fiero in 1984 and plans to further replace steel with plastic on new lines of sport coupes and minivans sent a shudder through the steel industry [12]. The growing threat of alternative materials inspired steel producers to approach the domestic automotive manufacturers and begin a dialog about how they could work more closely together to the benefit of both.

One result of this dialog was the formation of the Auto/Steel Partnership in 1987 [13]. According to its Vision Statement, published on its website (www.a-sp.org), its purpose is to leverage the resources of the automotive and steel industries, “ensuring that steel is the ‘competitive material of choice’ in a changing automotive market, using inter-company and inter-industry cooperative programs to ensure the success of the member companies, and proactively resolving governmental regulatory agency requirements and customer needs.” The Partnership ushered in a new era of cooperation between the steel and auto industries. A major part of the work of the Partnership was to further the development of high-strength steel and the application technology required for its successful usage.

The threat of plastics to the steel industry was not confined to perceived mass and cost benefits; plastics also had the advantage of being resistant to corrosion. The steel industry response was development of automotive quality galvanized products, both hot dip galvanized and electrolytically coated. The increased application of zinc-coated steels resulted in an increase in the complexity of resistance spot welding of these structures. These issues needed to be addressed before high-strength steel applications could grow, and car industry engineers were hard pressed to devote enough time to resolve them. They needed more support from the steel manufacturers.

The steel industry responded to the challenge of high-strength steel with deployment of customer-focused technical staffs. National Steel Corporation opened its Product Application Center in Detroit in 1983, the first steel company technical center devoted entirely to automotive customer technical support. Others followed, usually with technical support personnel located close to customer facilities coordinating the activities of the mill’s central research and development centers. In conjunction with development work spearheaded by the Auto/Steel Partnership, these activities supported the engineering resources of the car companies and helped achieve rapid advancements in the forming, joining, coating, and corrosion resistance of vehicle structures and related components. The application of high-strength steel, which had slowed at the beginning of the 1980s, grew in the second half of the decade. HSLA steels were the most popular form of steel used. Rails, rocker panels, door beams, and bumpers were the most common applications of HSLA steels. Application of martensitic sheet products continued to grow in bumper beams, and they were also applied to door impact beams, a part required to meet side impact requirements. Dual phase steels, produced in continuous annealing lines, started being applied at this time [14,15].

Another factor in the development of the market for high-strength steel was the beginning of large-scale vehicle manufacturing in the United States by several Japanese automakers. The Accord and the smaller Civic models entered the U.S. market in the 1970s, and were successful in part because of their fuel efficiency in the face of the oil shortages. Their market success, coupled with the uncertain trade situation of the time, convinced Honda that manufacturing in North America was a necessary and feasible strategy. In 1982, Honda began manufacturing the Accord in Marysville, Ohio. The Honda plant was followed in 1983 by a Nissan plant in Smyrna, Tennessee to manufacture small pickup trucks. In 1984, Toyota began manufacturing in the United States through a joint venture with General Motors, named New United Motor Manufacturing Inc. (NUMMI) located in Fremont, California. In 1988, Toyota opened their first wholly-owned manufacturing plant in Georgetown, Kentucky.

These so-called "new domestic" manufacturers were familiar with steel products developed in their home country and demanded these products from the North American steel industry [16]. Their demands were paralleled by a substantial investment by the Japanese steel industry in North America. The domestic steel industry was going through a period of low profitability, and some companies needed infusions of cash and technology to improve processes and develop new products. Several companies took equity positions with North American steelmakers (e.g., NKK with National Steel, Kawasaki Steel with Armco to form AK Steel, etc.) and others developed joint ventures (e.g., Nippon Steel and Inland Steel formed cold rolling and coating facilities I/N Tek and I/N Kote). A particularly useful set of products developed at this time were formable, high tensile strength steels, based on interstitial-free (IF) steelmaking practices and alloyed with solid solution strengtheners, such as manganese, silicon or phosphorus [17]. These met requirements of the Japanese automakers, but were also made available to the U.S. manufacturers, who began to consider them for applications requiring high strength and high formability. New applications were now candidates for higher strength steels, parts with formability requirements that would preclude the use of the less formable HSLA steels.

This decade also marked the development and first applications of bake hardenable steels for exposed body panels, new products that harkened back to the strain-aging steels of old. Through improved processes and alloy development, a new generation of strain aging products was born that strengthened through the strain produced in the stamping process and subsequent aging during the paint bake cycle. Unlike those former products, these new materials were able to suppress the stretcher strains that resulted in unsightly Lüders lines and could, therefore, be used on exposed body panels. These materials could be produced either by cold rolling and continuous annealing of conventional low-carbon ($\simeq$0.05 wt.%

C) steel products, or by cold rolling and batch annealing of extra-low-carbon (≃0.01 wt.% C) products. In both cases, the key was to leave sufficient carbon dissolved in the steel to develop the strain-aging response, but not so much as to result in aging under ambient conditions prior to stamping [18]. Bake hardenable steels could also be produced with partially stabilized ultra-low-carbon chemistries (≃0.005 wt.% C) using a continuous annealing process [19]. Application of bake hardenable steels resulted in increased outer body panel dent resistance, improving performance and reducing vehicle mass [20]. Further, in many materials, the aging started immediately after press forming. These panels were more resistant to handling damage during transportation from the press shop to the body shop [21] resulting in significant cost savings from lower scrap rates. The application of bake hardenable and other medium strength steels, such as the high strength IF steels, has grown to where nearly all steel exterior body panels are manufactured from one of these materials.

The last decade of the 20th century was a period of heightened competitiveness and challenge for the automotive industry. The CAFE targets mandated by regulation were stable during this time. However, demand for fuel increased, in part because of increasing vehicle size. Not only were new car models larger than their predecessors but truck-based, full-frame sport utility vehicles (SUVs) began capturing a large share of the market. These vehicles had lower CAFE targets than cars (20.7 mpg in 1996, vs 27.5 mpg), and were less fuel efficient. However, they were very successful in the market place. Thus, improving (or even maintaining) fuel economy in the face of the market demand for larger, higher content vehicles was a severe challenge. In addition, continuing improvements in passive safety, based in part on new regulations such as from offset crash and side pole impact tests, resulted in additional structural requirements (and vehicle mass). There also was an increased awareness by the public of the results of these tests, which increased pressure on automakers to build more crash resistant vehicles. Thus the need for improvement in the performance, cost effectiveness, and efficiency of vehicle structures continued to increase.

HSLA steels entered this decade with great promise for vehicle structural applications and ended it as the material of choice. Its mass reduction potential was demonstrated both in the literature [22] and in practical application. The formability of HSLA steels, though lower than that of mild steels, was shown to be predictable using the existing formability evaluation tools [23]. However, the higher level of springback (and springback variation) in these parts remained an obstacle to efficient die engineering and productive press shop operation. Higher variability in mechanical properties of HSLA steels (compared to mild steel) was one cause of increased dimensional variability in final parts. Implementation of improved process control technology and careful operating practices by the steel industry resulted in improvements in uniformity of mechanical properties of high-strength steels, as documented by the Auto/Steel Partnership [24]. Process and product design recommendations to improve the dimensional performance of HSLA steel parts were also described [25], which helped accelerate applications, and provided foundations for developments of future processes and applications.

The need for structural efficiency to meet both regulatory requirements and customer demands has not relented, and the steel industry has continued to innovate to face the challenge. Most recently, improved strength and ductility in advanced steel products such as DP, transformation-induced plasticity and complex phase (or multi-phase) steels are enabling application of high-strength steel to an ever increasing number of parts. The result is expected to be a significant increase in the fraction of high-strength steel in a vehicle body, and an increase in the strength level of the steel grades that are used. Discussions of these products, their metallurgy, manufacturing characteristics, and their application will form the main part of this book.

1.3 Breakdown in vehicle by material mass and application

Material trends have largely been driven by both regulation and customer desire for vehicles that are safer, more energy efficient and environmentally friendly. In large part because of a constant string of innovations, steel continues to hold an important share of the vehicle materials market, although alternative materials, especially aluminum, have made inroads in recent years.

A recent report by Ducker Analysis [26] outlined the usage of steel in vehicles. It showed the average 2013 vehicle weighs 3821 lbs, 57% was all forms of steel and iron. The ongoing replacement of mild steel with varying grades of high-strength steels was also noted. In 2013, the average steel content of a North American light vehicle was 1615 lbs. When compared to previous years, mild steel usage was down, whereas all categories of higher strength steels were up, as shown in Fig. 1.2. The replacement of high-strength steels with mild steels is a persistent

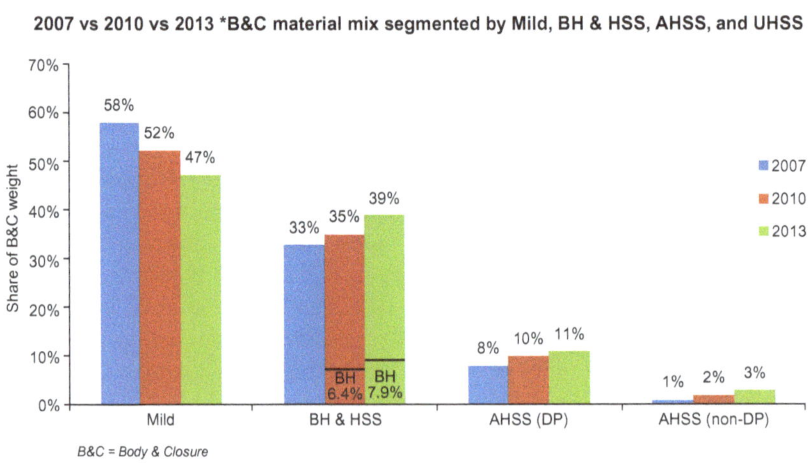

Figure 1.2 Comparison of steel grade changes from 2007 to 2013 [26].

trend that started in the 1980s, gained momentum in the 1990s and continues to this day, resulting from the combination of increasing customer and regulatory demands coupled with innovations in steel processing and product development aimed at maintaining its historic dominance in vehicle applications.

1.4 Improved safety and fuel economy: current regulations

Two of the biggest drivers influencing material selection for auto body applications are safety and fuel economy regulations. These have significantly impacted vehicle mass and performance over the past several decades. Safety regulations were first introduced in the 1960s starting with seat belts. In the early 1990s they began to increase rapidly to further enhance protection of passengers in crash events by requiring anti-intrusion metrics based on specific crash scenarios. CAFE was introduced in 1975. While safety was driving increased mass to protect passengers, fuel economy standards were driving the need for innovations which decreased the mass of the vehicle while maintaining the same performance. In addition to all of this, innovations in passenger convenience, such as navigation and infotainment systems, motorized windows, lift gate, side door, etc., were accelerating and led to significant increases in vehicle mass. A look at the impact of these competing technologies on mass is illustrated in Fig. 1.3. To add more challenge to the industry, powertrain performance demands were on the rise, as shown by acceleration time from 0 to 60 miles per hour. Automakers have worked diligently to meet both government requirements and customer expectations. Details of steel's role in supporting these challenges are discussed in the following sections.

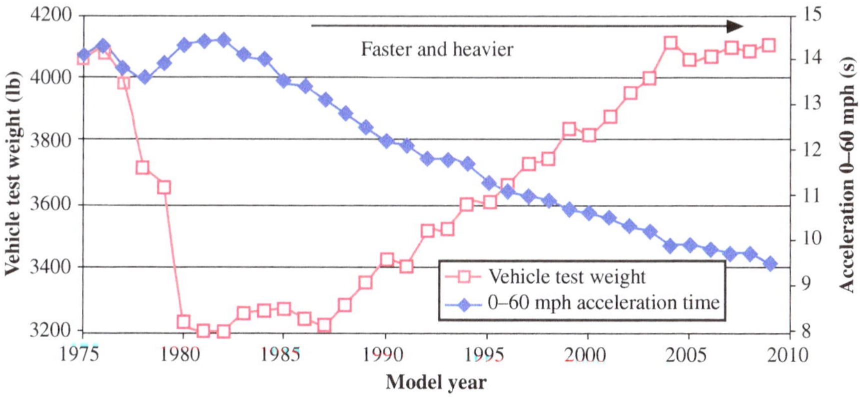

Figure 1.3 Change in vehicle weight and acceleration time since the introduction of CAFE [27].

1.4.1 Safety regulations

In the 1960s, NHTSA was given a legislative mandate for motor vehicle safety to issue Federal Motor Vehicle Safety Standards and regulations. The purpose is to protect the public against unreasonable risk of crashes occurring as a result of design, construction or performance of motor vehicles, and to also protect against unreasonable risk of death or injury in the event of a crash occurring. The regulations are classified by:

Crash Avoidance—regulates technology to help the driver avoid a collision.
Crash Worthiness—focuses on the vehicle response in a collision.
Post-crash standards—covers vehicle response to containing fuel after a crash including the fuel system integrity and flammability of materials.
Other—includes a wide range of standards from fuel economy to theft prevention and manufacturer identification, and many more not covered by crash related regulations.

In 1967 the first safety standard came into effect for vehicles manufactured on and after January 1, 1968. Since then, several new standards and amendments have been published in the Federal Register. Fig. 1.4 shows examples of several of these and includes a few recommendations from the International Institute for Highway Safety.

Crashworthiness regulations placed a big demand on automakers to design vehicles to meet these requirements. The vehicle body structure provides the bulk of the performance. Since the structures are predominately made from steel,

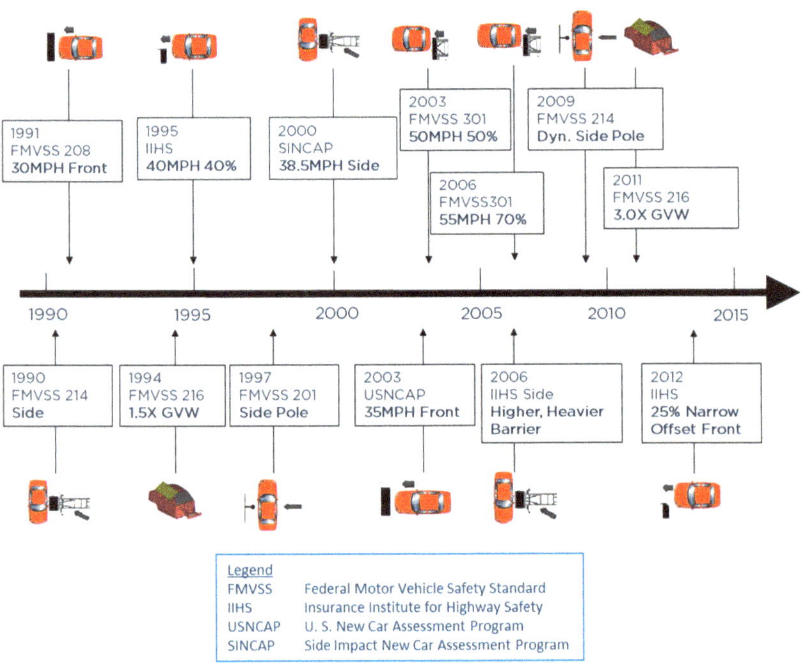

Figure 1.4 Introductions of crash worthiness regulations, 1990–2012.

the steel industry worked collaboratively with the automakers to design grades of steel to help deliver the desired performance for various sections of the body as needed. A review of these grades and how they are applied is covered in Chapter 3, Formability of Auto Components.

1.4.2 CAFE and relationship to CO_2 emissions

CAFE standards were first introduced by Congress in 1975 to help reduce the country's dependence on foreign oil. The regulations at first applied only to passenger cars in 1978, then included light duty trucks up to 6000 pounds in 1980, and finally increased to all vehicles up to 8500 pounds the next year. Regulations varied during the 1980s for both cars and trucks before reaching a steady target for cars in 1990 through 2010, with trucks moderately increasing during the period from 20 to 21 mpg through 2005, then reaching 23.5 mpg by 2010. NHTSA was later assigned responsibility by the Department of Transportation to enforce these standards.

In December 2007, the Energy Independence and Security Act was signed which increased the fuel economy standards by 40% to further reduce dependence on oil. This was the first legislative change to CAFE since its creation. This regulation set a goal for fuel economy standards to reach 35 miles per gallon for the fleet average of cars and trucks by 2020. NHTSA proposed a plan in April 2008 for model year (MY) 2011–15 vehicles to reach the 2020 goal. The new rules also introduced a fuel economy to be assigned to cars and trucks by their footprint. This helps the manufacturer meet regulations based on what they manufacture and sell, as opposed to the volume mix the government expects the public to buy. This removes the concern with regulations driving automakers to manufacture vehicles the consumer is not interested in purchasing.

In 2009, the Obama administration proposed a change in the program for 2012–16 to get the fleet average to 35.5 mpg by 2016. Then, in 2011, President Obama announced a plan to get to 54.5 mpg by 2025. An agreement between automakers and the administration was reached and CAFE for 2017–25 vehicles was announced. The agreement finalized MY 2017–21 targets and proposed targets for 2022–25 to be reviewed and finalized by April 1, 2018. This review could result in changes for fuel economy more or less stringent than originally proposed.

Fig. 1.5 shows the changes in the fuel economy standards by year for passenger cars and light trucks. The dashed lines for MY 2017 through 2025 represent the expected increases needed to reach 54.5 mpg equivalent (mpge) fleet-wide average by 2025. "Equivalent" refers to the combined NHTSA fuel economy regulation and EPA tailpipe emissions converted to mpg. EPA joined forces with NHTSA to include a greenhouse gas (GHG) emissions component to fuel economy to simplify the metric for the industry. As a result of the direct tie between tailpipe emissions and the amount of fuel burned during use of the vehicle, this is a logical conclusion. A detailed discussion on the importance of measuring GHG emissions through the life of a vehicle is covered in Chapter 2, Steels for autobodies: a general overview.

The automakers approach and strategy on how to meet fuel economy regulations are different depending on type and mix of vehicles, manufacturing capabilities,

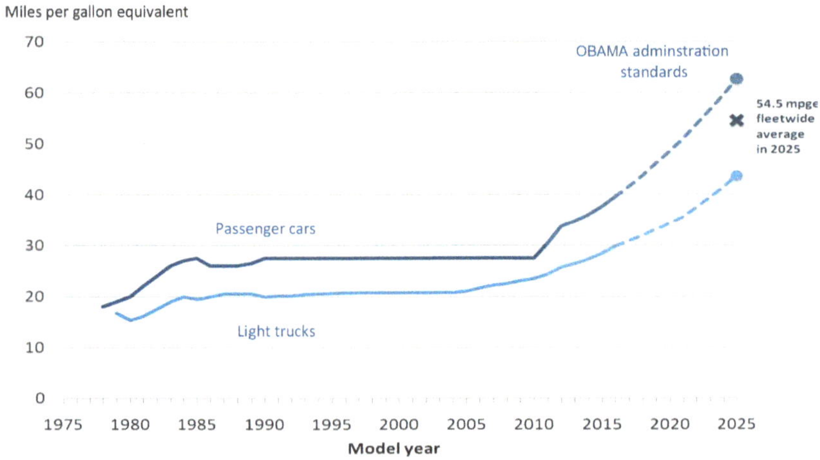

Figure 1.5 History of fuel economy regulation [28].

and other criteria. However, they all must meet the regulations and make vehicles which are affordable to the consumer. A review of technologies used to improve fuel economy is covered in the next section.

1.5 Vehicle energy losses and contribution to fuel economy through mass reduction

The engine and transmission components of the vehicle powertrain, make up more than half of the energy loss on a vehicle. Aerodynamics and tire rolling resistance also account for significant energy loss; however, mass reduction is an important contributor to energy loss because major decreases in mass will result in powertrain downsizing.

Several areas contribute to energy loss in automobiles. An example in a typical mid-sized sedan is shown in Fig. 1.6. This breakdown represents combined driving conditions of city and highway. The amount of impact does change depending on the type of vehicle (car vs truck vs SUV) and speed. For example, weight is a much larger factor in city driving conditions than highway, whereas aerodynamics becomes much larger on the highway. The other areas of energy loss remain about the same regardless of speed.

Although vehicle mass is fifth on the list for energy loss in a sedan, it gets a lot of attention because it can have significant effects on other areas. A lighter vehicle requires a smaller engine and brakes for equivalent performance. Therefore, in addition to directly influencing fuel economy, vehicle mass has a compounding effect and will help reduce mass in these related areas. Because the focus of this book is on automotive steel, lightweighting is clearly the largest area of impact for improving fuel economy.

Figure 1.6 Example of energy losses in a mid-size sedan [29].

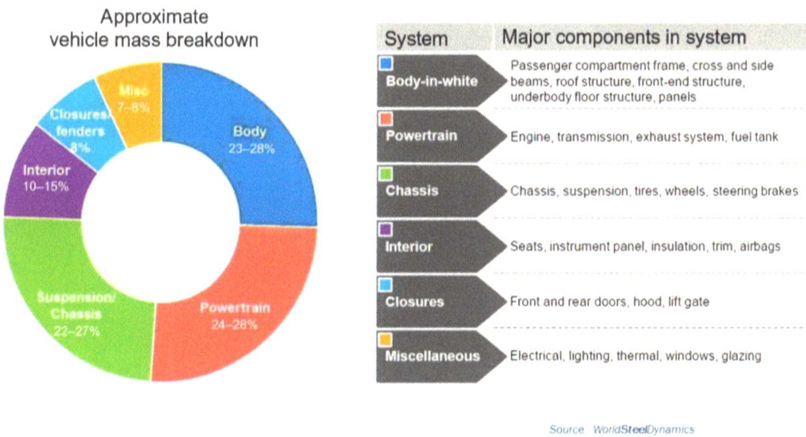

Figure 1.7 Vehicle mass breakdown by major sub-system [30].

An example of the distribution of mass by major vehicle component systems is shown in Fig. 1.7. This figure shows approximately one quarter of the vehicle mass is made up of the body structure (not including the closure panels, i.e., doors, hoods, fenders, and deck lids). Another quarter of the vehicle is the suspension and chassis, and another quarter is the powertrain system. The final quarter of the vehicle is made up of the interior, closures, and miscellaneous (such as glass and tires).

Competition among materials for lightweighting helps automakers achieve their fuel economy targets while also maintaining performance and safety requirements. In addition to steel, the most commonly used materials for body structure and closure applications include aluminum and magnesium alloys, as well as reinforced polymer composites, typically carbon fiber reinforced plastics for structural parts. It is very important to note that vehicle performance, including safety, is designed into the vehicle using selected materials. The materials themselves don't provide improved handling, safety, etc. However, their properties provide the tools and boundaries within which an engineer can design high performance vehicle components. All of these materials offer mass reduction opportunities, but they also come with several challenges including cost and manufacturability. The section below explores the advantages and boundaries each group of materials provide the engineer.

1.5.1 Aluminum

The importance of reducing weight was recognized early in the development of the automobile and aluminum was one of the materials identified for its lower density and good strength. The density of aluminum is about one third the density of steel (2.7 vs 7.8 g. cm^{-3}). Typical automotive aluminum sheet grades include 5000 and 6000 series and have tensile strength ranges on the order of 190–300 MPa. This strength range places aluminum sheet in the same relative range as mild steel. However, elongation ranges are lower than mild steel, and aluminum sheet has been shown to exhibit lower formability.

The main challenges for aluminum applications in the automotive market include designing for aluminum, improved formability or forming processes, improved joining processes, avoidance of stress corrosion, and lower cost of material production. As the industry is forecasting more than 90% of its growth in sheet, they are developing and implementing several new technologies to support their forecasted proliferation in closure and structure applications.

Aluminum alloy development is similar to that in the steel industry. Individual aluminum suppliers obtain feedback from the automakers as to what alloy improvements are needed. Most of this feedback is in the form of asking for improved properties to enhance formability and strength. New grades are delivered as quickly as possible to meet these demands on an ongoing basis. Future aluminum grades currently under development are targeting tensile strengths of 500 MPa (compared to 250–300 MPa now) in order to compete with high-strength steel grades. In 2009, Novelis introduced a multi-layer, multi-alloy sheet aimed at both higher strength and higher formability. Currently, this alloy has only been deployed on limited luxury class, low-volume vehicles.

Innovations in joining technology have been recently deployed to increase aluminum sheet usage through a pretreatment coating applied at the rolling mill by Alcoa and Novelis. This coating improves adhesive bonding and durability with a thin profile to facilitate spot welding and riveting.

In addition to alloy development for improved formability, Original Equipment Manufacturers (OEM) and aluminum companies have teamed up to develop other forming process aids. "Warm forming" allows the aluminum sheet to be drawn to deeper depths and formed to tighter radii without splitting. Warm forming is commercially available technology, however there is no known application at this time. General Motors had a hot forming process which is no longer in production because of cost and quality issues.

A heat treating and forming process combination is a recent technology patented by Ford Motor Company. This process first heat-treats the sheet to improve formability, and then it is cooled and immediately (within 30 minutes) stamped. The part is then post heat treated to regain strength. Ford reported at a Society of Automotive Engineers (SAE) meeting that 26% of the F-150 body structure consists of postformed heat treated 6000-series aluminum [31].

One other major area of development in aluminum is the continuous casting/sheet production process. Alcoa announced a new "micromill," to produce aluminum sheet from molten aluminum for a variety of products, including containers and automotive sheet. The benefits of the micromill are claimed to be:

- 40% greater formability
- 30% stronger
- 20 minutes instead of 20 days
- 50% lower energy
- ¼ plant floor footprint

Additional announcements indicate commercial production will not be available for several years. Details are closely held by Alcoa regarding capacity, gage, width, surface quality, consumables, etc. Although process time and cost are expected to be significantly reduced, Alcoa stated it intends to charge a premium for these grades because of their increased strength and formability.

1.5.2 Magnesium

The main attraction of magnesium for the automotive industry is its very low density (1.8 g. cm^{-3}). The strength range of magnesium sheet alloys is on the order of 180–300 MPa, which is also in the range of mild steel. However, ductility values are even lower than aluminum, on the order of 5–15% total elongations. Magnesium sheet alloys available today for automotive applications are not formable at room temperature and have severe corrosion issues. In addition, improvements are needed in alloy cost and processing quality.

United States Automotive Materials Partnership LLC (USAMP), a consortium of the United States Center for Automotive Research, has a magnesium development project to address challenges for new automotive applications. They summarized the technology development needs in a 2006 report called "Magnesium Vision 2020" summarized by these areas:

- Enhancing corrosion resistance through alloy development, coatings and joint designs
- Improving die casting quality and ductility by reducing porosity
- Developing warm forming sheet processes
- Creating low-cost wrought alloys for stamping, extrusion and forging

This project is still actively working on advanced development for magnesium and involves suppliers, OEMs and academia.

1.5.3 Carbon fiber reinforced polymers

Carbon fiber reinforced polymers are composite materials which rely on the carbon fiber to provide the strength and stiffness while the polymer provides a cohesive matrix to protect and hold the fibers together and provides some toughness. Carbon fibers provide highly directional properties much different than the metals most commonly used for these automotive applications. They can be engineered to achieve mass reductions not achievable by the metals. Since these are artificially composited materials their properties and performance can be tailored to the application through varying strength, length, directionality and amount of the reinforcing fibers and in the selection of the polymer matrix. The largest drawbacks are the high cost in producing the fibers and the low throughput rates at which components can be manufactured. The cumulative time to place the fibers in a mold, inject the polymer and allow the part to set is in the order of a few minutes.

Carbon fiber reinforced polymers are gaining popularity in the luxury, sport segment for mass reduction. These materials, which may also be reinforced with glass or other fibers, have high-price tags and are more suited for lower volume manufacturing as a result of molding cycle times. New technology in reducing fiber cost and panel processing has been in the news in the past few years and the increased implementation is a good indicator progress is being made. More announcements of improvements and applications over the next several years are expected. However, application to higher volume, lower-cost vehicle segments is not expected any time soon.

1.6 Summary

Steel has a long history in the automotive industry. The mass of the average vehicle has been well over 50% steel for almost 100 years. The introduction of new regulations stemming from the implementation of the Highway Safety Act and Clean Air Act (which instituted Corporate Average Fuel Economy) has challenged the steel industry to develop innovative solutions to meet new requirements. The introduction of new steel grades with higher strength enabled designers to achieve greater performance in crash without adding mass. In fact, vehicle bodies are now both safer and lighter as a result of designs with advanced high-strength steel grades. Although aluminum, magnesium, and carbon fiber reinforced polymers compete with steel for many applications, steel offers automakers a robust, economical and environmentally responsible way to meet their needs for body structure and closure applications, along with many suspension/chassis and powertrain components. More discussion on current automotive steel applications using advanced high-strength steels is included in Chapter 2, Steels for auto bodies: a general overview.

References

[1] D.G. Adams, J.A. DiCello, C. Hoppe, A.S. Kasper, A.N. Keisoglou, W.W. McVinnie, High strength materials and vehicle weight reduction analysis, Soc. Automot. Eng. (1975), Paper No. 750221.

[2] D.G. Adams, S. Dinda, R.A. George, R.W. Karry, A.S. Kasper, J. Pogorel, et al., Charger XL: A lightweight materials development vehicle, Soc. Automot. Eng. (1976), Paper No. 76020.

[3] R.L. Pascorek, A. Sipler, Strain aging properties of high-strength hot-rolled steels, SAE Trans. 86 (1978) 740.
[4] W.H. McFarland, Mechanical properties of low carbon- alloy free martensites, Trans. AIME 233 (1965) 2028.
[5] W.H. McFarland, H.L. Taylor, Properties and applications of low carbon martensitic steel sheets, Soc. Automot. Eng. (1969), Paper No. 690263.
[6] P.B. Lake, J.I. Grenawalt, Partially annealed cold rolled steel sheet, SAE Trans. 86 (1978) 718.
[7] K. Araki, Y. Takada, K. Nakaoka, Work Hardening of Continuously Annealed Dual Phase Steels, Technical Report, Technical Research Center, Nippon Kokan K.K., Japan, 1976.
[8] M.S. Rashid, GM 980X- A unique high strength sheet steel with superior formability, Soc. Automot. Eng. (1976), Paper No. 760206.
[9] R.P. Krupitzer, R.E. Miner, P.J. VanderArend, F. Reis, J.A. Slane, J.K. Abraham, et al., Progress in HSLA steels in automotive applications, SAE Trans. 86 (1978) 708.
[10] L. Luyckx, J.R. Bell, A. McLean, M. Korchynsky, AIME Metallurgical Transactions 1 (1970) 3341.
[11] U.S. National Highway and Transportation Safety Administration.
[12] GM to use plastic body on Camaro: also on Firebirds; Pontiac site set, American Metal Market, 1985.
[13] G.S. Vasilash, How the Pontiac Fiero helped save the North American steel industry, Automot. Des. Prod. (1999).
[14] B.S. Levy, Design and manufacturing guidelines for ultra high strength steel bumper reinforcement beams, Soc. Automot. Eng. (1979), Paper No. 790333.
[15] T.E. Fine, S. Dinda, Development of lightweight door intrusion beams utilizing an ultra high strength steel, Soc. Automot. Eng. (1975), Paper No. 750222.
[16] K. Osawa, T. Shimomura, M. Kinoshita, K. Matsudo, Development of high-strength cold-rolled steel sheets for automotive use by continuous annealing, high strength steel for automotive use P-124, Soc. Automot. Eng. (1983) 109.
[17] J.R. Fekete, D.C. Strugala, Z. Yao, Advanced sheet steels for automotive applications, JOM 44 (1) (1992) 17.
[18] R. Pradhan, Dent-resistant bake hardening steels for automotive outer-body applications, Soc. Automot. Eng. (1991), Paper No. 910290.
[19] M. Kuroshawa, S. Satoh, T. Obara, K. Tsumoyama, T. Irie, Age hardening behavior of deep drawable and bake hardenable steel sheet produced by high temperature continuous annealing, Int. J. Mater. Prod. Technol. 4 (1989) 244.
[20] T. Hayashida, M. Oda, T. Yamada, Y. Matsukawa, J. Tanaka, Development and applications of continuous- annealed low-carbon Al-Killed BH Steel Sheets, Proceedings of Symposium On High Strength Steels for Automotive (1994) 135, R. Pradhan, ed.
[21] I. Kovch, T. Owens, M. Bala, R. Thompson, Use of continuously annealed bake hardenable steels for automobile outer panels, Soc. Automot. Eng. (1990), Paper No. 900715.
[22] General Motors J-Car High Strength Steel Development Study, High Strength Steel Awareness Bulletin No. 16, Auto/Steel Partnership, 1996.
[23] S.P. Keeler, W.G. Brazier, Relationship Between, Laboratory Material Characterization and Press Shop Formability, Proceedings of International Symposium On High Strength Low Alloy Steels, Microalloying 75, (1977).
[24] Material Uniformity of High Strength Steels, Vol. 2, Auto/Steel Partnership, (1998).
[25] High Strength Steel Stamping Design Manual, Auto/ Steel Partnership, (2000).

[26] A. Abraham, Metallic material trends in the North American light vehicle, Great Designs in Steel Seminar, Livonia MI, May 13, (2015).
[27] National Highway Transportation Safety Administration.
[28] U.S. Department of Energy.
[29] General Motors internal report.
[30] P.F. Marcus, AutoBody Warfare: Aluminum Attack, WorldSteelDynamics Report, October (2014).
[31] AAM Research Downgrades Aluminum's Penetration into the Global Car Market, *American Metal Market Research Whitepapers*, (2015).

Steels for auto bodies: a general overview

2

J.N. Hall[1] *and J.R. Fekete*[2]
[1]Steel Market Development Institute, Southfield, MI, United States, [2]National Institute of Standards and Technology, Boulder, CO, United States

2.1 Steel grades and design strategy for auto body applications

Through over 100 years of development and application, steel has proven itself to be a versatile and effective material for vehicle body structures. During this time, the evolving requirements for safety, durability and economy have driven steel producers and vehicle engineers to work together developing new grades aimed at the unique requirements of various areas of the vehicle. Through careful control of chemistry and processing, steel can be tailored to provide optimum performance for specific applications. The variety of steel grades can be seen in Fig. 2.1, which also shows the relationship between strength and ductility for automotive steel grades. This relationship impacts the application of steel grades and will be further described below including the characteristics desired for typical applications and the types of steels that meet these requirements most effectively.

2.1.1 Critical safety structure applications

Steel selection for critical safety applications follows two general guidelines. These guidelines are founded on the principle that in a sudden deceleration resulting from a crash, the energy must be dissipated in a controlled manner so that the deceleration of the occupants does not exceed certain thresholds, maximizing survivability and minimizing chances for injury [2]. The first of these guidelines is the creation of zones in the structure whose role in a crash event is to absorb the vehicle's kinetic energy and provide for a rapid but controlled deceleration. The objective of the structural designer in this case is to use material as efficiently as possible. This requires (1) a given structural member involve as much of the material as possible during the crash event and (2) the material itself should absorb the maximum amount of energy per unit of material used. Involving the maximum amount of material requires careful structural design, the details of which are outside the scope of this book. But an example can be seen in Fig. 2.2, depicting a structural beam which folds like an accordion during an axial crush event [2]. Note almost all of the material in this structure experiences some deformation. Compare this to a member

Automotive Steels. DOI: http://dx.doi.org/10.1016/B978-0-08-100638-2.00002-X

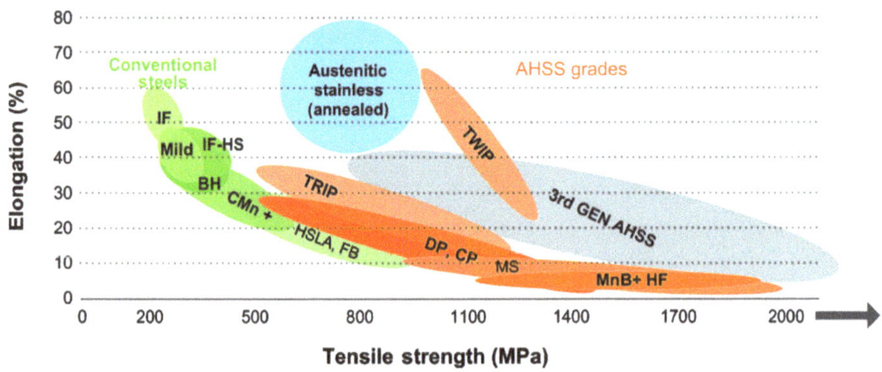

Figure 2.1 Illustration of automotive sheet steel grades based on strength and ductility [1].

Figure 2.2 Example of structural beam after drop tower testing simulating an axial crush event [3].

that is so slender it only kinks at one location. In this case, most of the material does not contribute to energy absorption, and this is a suboptimal design.

The amount of energy absorbed by the material is strongly influenced by the stress–strain behavior of the material which itself is controlled by the microstructure of the material. The absorbed energy of a given element of material is described by the area under the stress–strain curve.[1] Thus, the best materials will have high flow stresses at a given strain, and will be able to continue deforming over large ranges of strain, maximizing the amounts of absorbed energy. Vehicle

[1] The integral of the stress–strain curve over the range of strain to which the material is subjected.

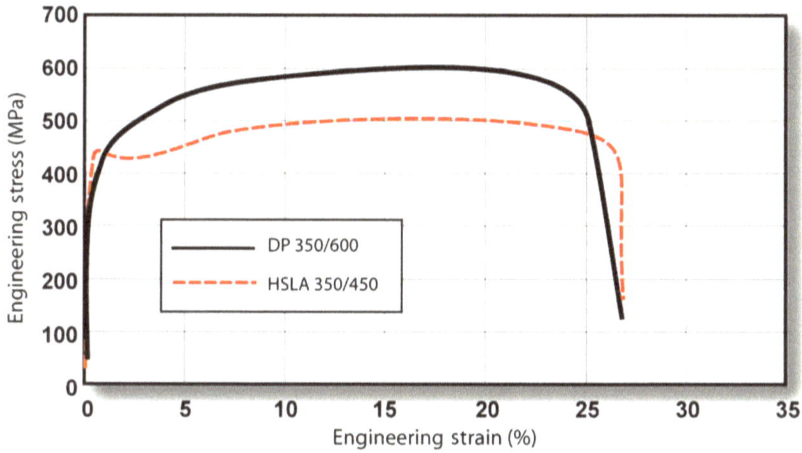

Figure 2.3 Comparison of stress–strain curves for HSLA-340 MPa YS and DP-600 MPa TS steels [4].

engineers have exploited steel's combination of strength and ductility for this purpose for many years, but most recently dual phase (DP) steels have emerged as the material of choice when maximum absorbed crash energy is required. The unique microstructure of dual phase steels, combining ductile ferrite with high strength martensite and/or bainite, results in significantly higher work hardening rates than conventional mild or high strength low alloy (HSLA) steels. An example is shown in Fig. 2.3, which compares the stress–strain curves for an HSLA steel with a minimum yield strength (YS) of 340 MPa, to a dual phase steel with a minimum tensile strength (TS) of 600 MPa. These materials have similar yield strength and ductility but the dual phase material work hardens to a higher tensile strength, which results in increased energy absorption for a given increment of strain. Although HSLA and even mild steels are still applied in these applications, depending on the structural requirements, dual phase steels are expanding rapidly into applications that require the material to exhibit the maximum ability to absorb energy.

The other guideline for safe vehicle structure design is that the so called "safety cell" which contains the occupants must resist intrusion, and maintain its integrity. This allows engineers to design the interior of the car with some assurance of the position and velocities of the occupants during crash events.[2] They can then provide needed space for the occupants, design energy absorptive features in the interior, and design the restraint systems to provide maximum safety.

Structural design in this case requires materials and components whose main purpose is to transfer the applied forces stemming from crash events. Structural members must resist deformation, or in some cases provide a very small and well controlled amount of deformation in order to manage the crash loads. They also must be effectively joined to mating components to ensure that loads can be

[2] Seat belts and air bags are also critical elements of this design philosophy.

transferred as designed and not concentrated in an unwanted location resulting from the failure of a connection or joint. Steel grades designed for these applications should have the highest possible yield strength, have good ductility (though not necessarily over a high level of strain), be capable of forming strong welds to itself and to other high strength materials, and be compatible with other manufacturing processes, such as forming, painting, and assembly (e.g., sealing and coating for corrosion mitigation).

Steels with a martensitic microstructure provide the highest possible yield strengths, upwards of 2 GPa, though more typically at the 1.2–1.5 GPa level.[3] This makes them ideal candidates for use in constructing the passenger safety cell. Typical applications would include the B-pillar structure, important for controlling side impact intrusion, and A-pillar/roof rail components, critical for controlling intrusion during a rollover. However, as seen in Fig. 2.1, these materials have limited ductility compared to other high strength grades. Although this level of ductility is sufficient for the intended application, it does limit the available processes for forming the required parts from martensitic sheet products. Stamping can be used for certain simple parts, and roll forming can be used to create somewhat more complex shapes, though the process requires these parts to have a constant section. Hot stamping is an alternative technology that enables forming of complex martensitic parts, and its usage is rapidly expanding. In this process, the incoming sheet is an annealed material, with high ductility. It is heated to austenitizing temperatures in a furnace, and then pressed and simultaneously quenched in special water-cooled tools. Thus complex shapes can be formed, and the rapid cooling causes the final part to have a martensitic microstructure.

Dual phase steels are also used in anti-intrusion applications in the passenger safety cell. Though the yield strengths of dual phase steels are typically below that of martensitic grades, they can be similar to or higher than conventional high-strength steels, and their higher level of formability compared to martensitic grades makes them attractive alternatives for the use of lower cost conventional stamping processes for complex parts that still require high strength. Conventional high strength low alloy steels, which have yield strengths that typically top out at 550 MPa, are also used in certain applications where they make sense as a lower cost alternative to dual phase or martensitic steels.

2.1.2 Other body applications

The beautiful and durable bodies of today's vehicles are enabled by steel's strength, formability, and surface quality. Today's available grades give vehicle engineers the tools to continue to design high quality, corrosion-resistant vehicle bodies with dramatic design details and glossy painted surfaces. A discussion of the available materials and their usage follows below.

[3] For more detail on steel microstructures and their relationship to properties, see "Steels: Processing, Structure and Performance," G. Krauss, ASM International, 2015.

For unexposed applications, the key considerations are formability and location on the vehicle. An engineer will select from several grades of steel, offering increasing levels of formability with commensurate reductions in yield strength. These applications are often driven by stiffness requirements rather than strength, which make material strength a secondary requirement to formability. Highly formable grades can be exploited to enable shapes that would otherwise be impossible to form reliably in high volumes. The increased capability can also be used to consolidate parts, reducing the number of tool sets required and reducing the amount of welding. Although many manufacturers carry their own specifications for these grades, they can also be found in a relevant Society of Automotive Engineers (SAE) specification [5].

Many of these applications reside in areas of the vehicle exposed to road debris and salt. In these cases, zinc coatings are specified to protect the steel from corrosion. In recent years, industry has settled on two broad categories of zinc coating. The first is pure zinc, which can be applied either through a hot-dip galvanizing process or via electrogalvanizing. Hot-dip galvanizing draws a preheated strip of material through a molten zinc bath. The strip then passes through an air knife, which blows excess molten zinc off the surface, leaving behind the precise thickness of zinc coating specified. The coating thickness is important, because enough is needed to ensure corrosion protection, but too much makes the material difficult to weld. The electrogalvanizing process uses electric current to plate zinc onto steel in a series of electrolytic cells. The alternative process is known as galvannealing. This process uses a hot-dip galvanizing line, but after the steel is coated it proceeds through another furnace which causes the zinc to alloy with the iron on the surface of the steel, resulting in intermetallic zinc–iron compounds. This process results in a material that is easier to weld than pure zinc coated steel.

Exposed applications have additional requirements, including resistance to superficial denting damage and high surface quality to ensure a glossy, highly-reflective painted surface. The ability to resist denting has been shown to be related to the geometry of the surface, the elastic modulus and the yield strength of the material [6]. The surface geometry is a feature of the product, and is generally not negotiable. However the steel industry has developed steel products with the excellent formability required for complex surfaces, but with higher yield strengths to enhance the dent resistance of the finished panels. There are two types of products used in these applications. Grades with enhanced dent resistance are generally alloyed with a solid solution strengthener, such as phosphorus. These grades have higher yield strength and work hardenability than mild steel, both of which enhance the yield strength of the final formed panel. As an example, mild steel for automotive applications has a minimum yield strength of 140 MPa, whereas the dent resistant grades have yield strengths ranging from 180 to 300 MPa (with commensurate reductions in formability, which must be considered).[4]

[4] See, e.g., "Categorization and Properties of Dent Resistant, High Strength and Ultra High Strength Automotive Sheet Steel," SAE Surface Vehicle Recommended Practice J2340_199910, 1999.

For further dent resistance, "bake hardenable" grades have been developed. In addition to the increased incoming strength described above, these grades take advantage of the metallurgical trait of strain-aging to gain even higher strength in the finished part. In this case, strain comes by virtue of the panel forming process, and the aging takes place in the ovens used to cure the paint applied to the panel during the assembly process (typically around 170°C). The combination of straining and subsequent aging can add 10 MPa or more to the yield strength of the finished panel and, perhaps more importantly, causes the return of a yield point in the finished material, which also improves the dent resistance of the finished panel by increasing the "apparent" stiffness of the material. Although the elastic modulus of steel does not change, in the absence of the yield point, steel can start to exhibit deformation at the micro level prior to reaching its macro yield strength. This deformation can cause the material to act like a material with a lower elastic modulus, which in this application can result in poorer dent resistance compared to bake hardenable steels.

Painted surface appearance is a critical requirement for exposed surfaces (also known as "Class 1" applications). Steel requirements to ensure excellent painted surfaces include tight requirements for surface texture (specified as average surface roughness (R_a) and peaks per inch or cm [7]). Also, an absence of surface defects is critical. These include dents, dirt, strain (Lüder's lines, resulting from straining an already strain-aged steel) and stains. Vehicle manufacturers have their own specifications and requirements for these items, based on the characteristics of their own processing lines. There have been attempts to modify steel surfaces to improve painted surface appearance (e.g., laser or electron beam texturing of work rolls used to create the steel surface) and to develop and correlate other surface quality parameters to painted surface appearance (e.g., surface waviness (W_{ca}), a measure of long wavelength surface undulations) but none have yet been accepted widely by the industry.

The development of robust processes for laser butt welding of steel sheets has enabled the application of blanks fabricated from multiple steel sheets to be pressed into stamped parts. These are known as "tailored blanks" or "tailor-welded blanks," and can be used in a number of ways to optimize vehicle structures. Since multiple materials can be used, the thickness, strength, and coatings can be varied, permitting cost and mass savings through the use of multiple thicknesses in a part, blank size reduction (through more efficient blank nesting), part consolidation and more precise application of corrosion-resistant coatings. An example of applying tailored blank technology to a vehicle door ring is shown in Fig. 2.4. In this example, multiple thicknesses and multiple steel strength grades are used to enable a 20% mass reduction in the component.

Steel products have also been developed which help improve vehicle noise, vibration, and harshness (NVH) performance. These products are laminates, in which a noise damping polymeric material is encased between two thin steel sheets. The resulting laminate can be formed in conventional presses, and can also be resistance spot welded if the laminated layer contains some electrically conductive material. A common application for these materials is the floor pan and the firewall of the vehicle, helping to insulate the vehicle from road noise. These materials are more expensive than conventional cold rolled steel, but their application can eliminate the need for other noise damping treatments, lowering the overall cost of the vehicle.

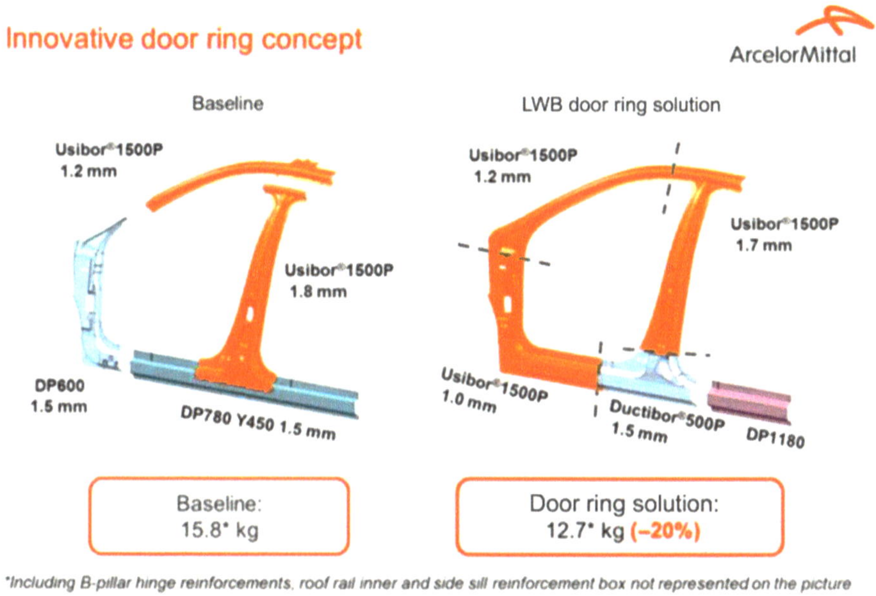

Figure 2.4 Use of tailor-welded blanks for innovative door ring concept [8].

2.2 Steel's contribution to fuel economy through mass reduction

As a result of the significant increases in steel grade strength, less steel is needed to provide the same load carrying performance than for lower strength grades. This allows steel to be considered an effective material for lightweighting and thus contributing to improved fuel economy. Several studies on reducing vehicle mass with high strength and advanced high-strength steel were completed in the 2009–11 timeframe [9–12]. These studies resulted in mass reductions from 13% to 29% as compared to baseline vehicles. The variation in mass reduction depends primarily on the type of optimization method used and will be discussed below.

2.2.1 2-G grade and gauge optimization

Often vehicle mass reduction starts with looking at lower density "alternative" materials such as aluminum, magnesium or carbon fiber reinforced polymers as described in Chapter 1, Design of Auto Body: Materials Perspective. The focus of this section is achieving vehicle mass reduction using steel in the body structure and closures.

A relatively simple way to reduce mass in a vehicle using steel is to replace each strength dominant part with one made from a higher strength steel grade. Mass reduction occurs because a smaller thickness part of a higher strength material can carry more load. A Lotus Engineering study on the 2009 Toyota Venza funded by the U.S. Energy Foundation evaluated mass reduction opportunities for a typical vehicle in the crossover utility segment. The baseline 2009 Venza body was predominately mild steel with less than 10% high-strength steel. The study evaluated two architectural changes classified as "low" or "high" development. The "low development" vehicle targeted a 20% mass reduction through the use of technologies feasible by 2014 for inclusion in a 2017 vehicle. This vehicle used mainly high-strength steel grades with very few nonferrous applications. The "high development" vehicle targeted a 40% mass reduction for 2020 production with technology feasible by 2017. This version focused on using alternative materials such as aluminum, magnesium, and polymer composites.

By incorporating new advanced high-strength steel (AHSS) grades into the 2009 Toyota Venza, Lotus engineering was able to reduce the body-in-white (BIW) mass by approximately 16% which was consistent with other vehicle models already using moderate amounts of high-strength steel. Previous studies by WorldAutoSteel on the UltraLight Family of Research [8] showed 25% mass reduction is achievable when replacing mild steel with AHSS. The "high development" vehicle using lower density, alternative materials only showed mass savings of 29%.

In this study, the life cycle emissions of the "low" and "high" development vehicles are also compared. Even with the large mass savings differences and thus difference in tailpipe emissions based on improved fuel economy, the vehicles total emissions are about the same. The disparity lies in the material production phase where alternative materials emit greater greenhouse gas (GHG) during production thus negating their advantages during the driving phase of its life cycle. A more detailed explanation on GHG is covered at the end of this chapter.

A phase two study of the Toyota Venza was completed by FEV in 2010 [7]. In this phase crash analysis capability was added along with computer-aided engineering analysis for NVH, durability, and stiffness. In addition, more rigorous cost analysis methodology was included. The body mass savings for this study showed nearly 13% reduction but used less high-strength steel. The mass reduction resulted in an overall vehicle mass savings of over 18%, with a net cost savings for the vehicle.

ArcelorMittal performed a series of studies on mass reduction in the BIW and chassis systems called S-in Motion. These studies started with the evaluation of a medium-sized (C-segment) sedan in 2010 and later included a pick-up truck, battery electric vehicle (BEV), and others. Each study incorporated the use of AHSS both cold and hot stamped as well as laser-welded blanks, long products and tubes. The mass reduction for the sedan achieved 19% by moving from 36% AHSS applications to 54% and significantly increasing steel strength in each category of grade. This mass reduction was enabled by currently available technology and with no additional cost to the vehicle. The pickup study using the same principles achieved a 23% mass reduction when compared with an equivalent baseline vehicle in 2014. An additional 4% reduction was anticipated with application of grades under development during the time of the study.

2.2.2 3-G grade, gauge, and geometry optimization

In 2011, the WorldAutoSteel organization completed an extensive project to develop a fully-engineered, steel-intensive electric vehicle with reduced GHG emissions over its entire life cycle [9]. This FutureSteelVehicle (FSV) achieved over 35% mass reduction over a 2009 benchmark vehicle using an AHSS-intensive body structure and reduced its GHG by nearly 70%. The redesigned vehicle met all global crash and durability requirements, enabling a five-star safety rating while avoiding high cost penalties for mass reduction.

The mass reduction in the body structure was achieved through the combination of new AHSS grades, thus reducing part thicknesses as discussed above, but also with reevaluating the load paths and thus optimizing the geometry of the parts based on the added strength provided. The results of the project included the development of 19 new steel grades to provide the most effective grades to meet the load path requirements while still being manufacturable. In addition, several new or improved manufacturing methods were used to ensure the best grade could be used for each individual component application.

As demonstrated by FSV, the S-in Motion pick-up, and others shown in Fig. 2.1 combining 3-G methodology for mass reduction along with emerging steel grades, significant mass savings of around 22% and greater are achievable with steel as compared to baseline vehicles with mainly mild and high strength steel grades. The question remains: is this enough mass reduction to realize future CAFE (Corporate Average Fuel Economy) requirements?

2.2.3 NHTSA[5] Volpe model for CAFE compliance and effects

In order to help assess the automakers progress or potential in achieving CAFE Compliance, the Department of Transportation developed a model referred to as the Volpe Model to support NHTSA's CAFE rulemaking [13]. The model can be used as a tool to estimate manufacturer's compliance to anticipated future vehicle fleets. The model includes calculating costs, effect and benefits of technologies to meet CAFE standards including Monte Carlo simulation. Fig. 2.5 shows the NHTSA Volpe Model results for the full 2025 U.S. vehicle fleet capability of meeting 54.5 mpg performance based on EPA's projections of sales as a function of powertrain performance and mass reduction achieved in the BIW. The chart shows that if the anticipated powertrain improvements are met in their entirety, no additional mass reduction is necessary to meet the fuel economy. However, if there is any amount of shortfall in these improvements, mass reduction becomes essential. As steel has shown mass reduction potential of approximately 25%, the model results show steel bodies will help offset powertrain shortfalls around 10%. However, if powertrain improvements under deliver by around 15% or greater, additional mass reduction technologies will be needed.

[5] National Highway Traffic and Safety Administration, an agency of the United States Department of Transportation.

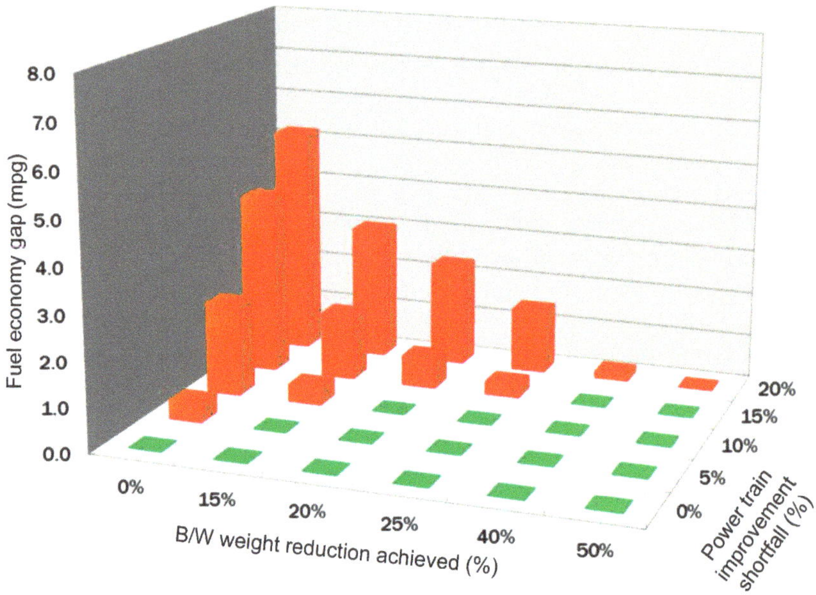

Figure 2.5 AHSS mass reduction potential using NHTSA Volpe model on the forecasted 2025 fleet.[6]

The results of these studies show very promising outcome of steel to support the industry in meeting future CAFE regulations. The next section will discuss how these studies have impacted actual vehicle designs to validate the models.

2.3 Recent body structure & closures production applications

The FSV project results were shared with all major automakers in 2011. During the launch of new vehicle designs for model year 2015 it became evident that many of the new grades and improved manufacturing techniques had been implemented into several of these new vehicles. This rapid adoption of new technology showed clearly the need for lightweighting, and the desire to lightweight with steel.

Because of the number of steel grades and manufacturing techniques available, automakers have several choices of strategies they can use to design the required performance into each segment of the body. The following examples will demonstrate some of these strategies.

The first 2015 model year introduction highlighting the application of AHSS was the Chevrolet Colorado shown in Fig. 2.6. The cab structure of this vehicle is over 72% high-strength or AHSS as shown by the pie chart illustrating the amount

[6] ©2013 ArcelorMittal USA LLC, All rights reserved in all countries.

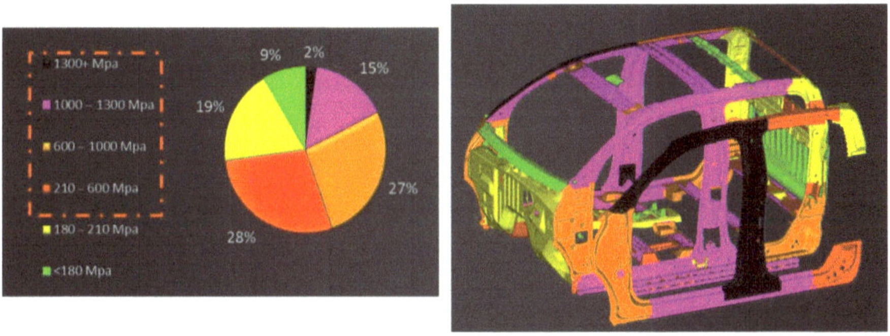

Figure 2.6 2015 Chevrolet Colorado body-in-white cab structure [14].

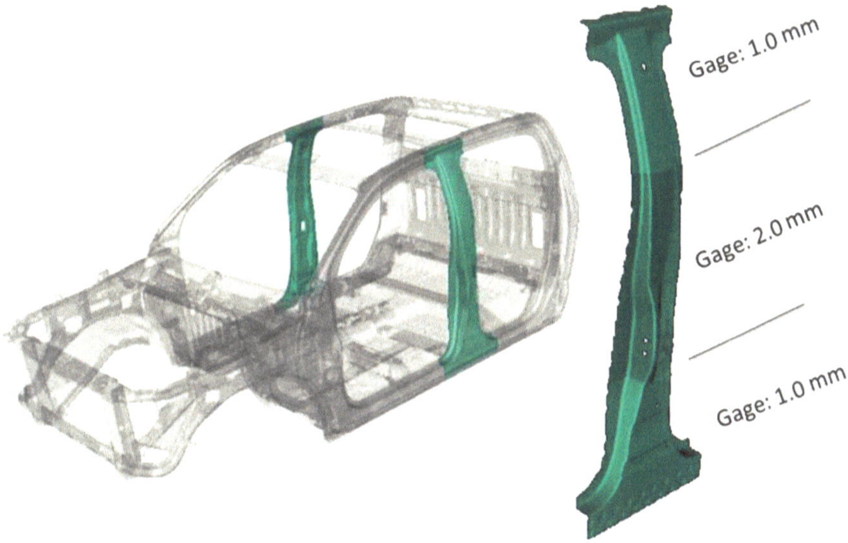

Figure 2.7 Chevrolet Colorado B-pillar hot stamped with variable thickness blank [14].

of various grades of steel by their tensile strengths. The red portion showing 28% represents the high-strength steel portion of the structure and is mainly made up of HSLA grades. The remaining portions of the pie chart included in the dashed square outline are various grades of AHSS ranging in tensile strength levels from 600 MPa to over 1300 MPa. These grades are mainly dual phase grades of various strength levels; roll-formed martensite and hot-stamped steels. The application of these steels allowed the engineers and designers to achieve mass reduction while increasing the stiffness of the vehicle.

In addition to the efficient use of AHSS grades, a variable-thickness blank was used on the press hardened B-pillar, Fig. 2.7. A variable thickness blank is a single blank with thicknesses varying along the length. This allows the part to be designed and

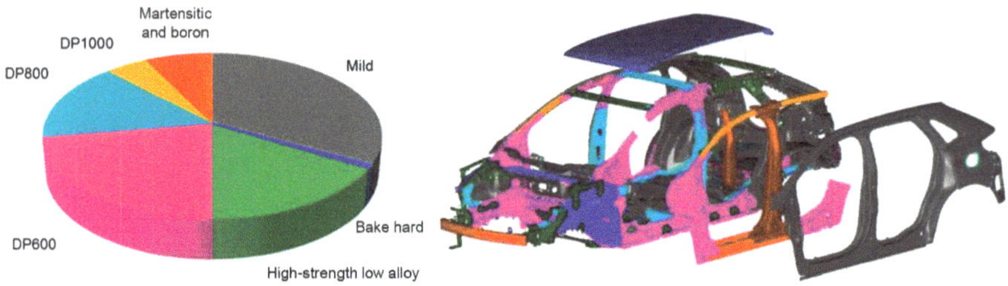

Figure 2.8 2015 Ford edge body-in-white [15].

stamped with the appropriate thickness of material only where it is needed thus providing additional mass reduction. The B-pillar has the added benefit of not using a laser welded blank to vary thickness which results in weak heat-affected zones which negatively impact performance. This part was hot stamped to achieve the desired shape.

The 2015 Ford Edge also is a great example of efficient design with various steel grades for optimizing both performance and mass. The material distribution shown in Fig. 2.8 is made up of 50% AHSS with about 20% high-strength steel. The AHSS grades used for the Edge are also mainly dual phase, martensitic or hot-stamped steels as shown. However, Ford engineers chose a different strategy to meet the requirements for the upper structure. A hydro-formed tube made of DP1000 grade was used for the A-pillar and roof rail, shown in orange in Fig. 2.8, which was first introduced on the 2014 Ford Mustang. In addition, hydro-formed components were also introduced in the lower section of the D-pillar. These components significantly add to the stiffness and rigidity needed for an efficient mass design. In addition, while several automakers have chosen to use aluminum for hoods, all closures on the Edge are steel.

Honda designed the 2015 Acura TLX with 52% high-strength and AHSS grades to help achieve the high rigidity to promote a smooth and quiet ride, crisp steering and handling, safety performance, and long-term durability. The enhanced performance was achieved without an increase in mass. This design also features a one-piece hot stamped door ring, shown in Fig. 2.9. This technology was also applied to the 2016 Honda Pilot to achieve mass reduction and increased rigidity.

Nissan introduced a multiphase 1180 grade steel for their solution to meet the performance requirements in a stamped A-pillar and roof rail. This grade, first used on the 2015 Murano, along with other AHSS grades were applied to help reduce vehicle mass while maintaining performance.

These examples demonstrate there are numerous ways to achieve the desired performance in a critical area of the passenger compartment. Clearly there is no one right way to accomplish this. It is highly dependent on the selection of material, the vehicle design and the general philosophy of the automaker around assembly and dimensional control. One additional major consideration is the manufacturing footprint to support all of this. Major shifts in materials, design and assembly process will significantly impact the tooling required in manufacturing. With the hundreds of parts making up the body-in-white, the options are unlimited. Although steel's

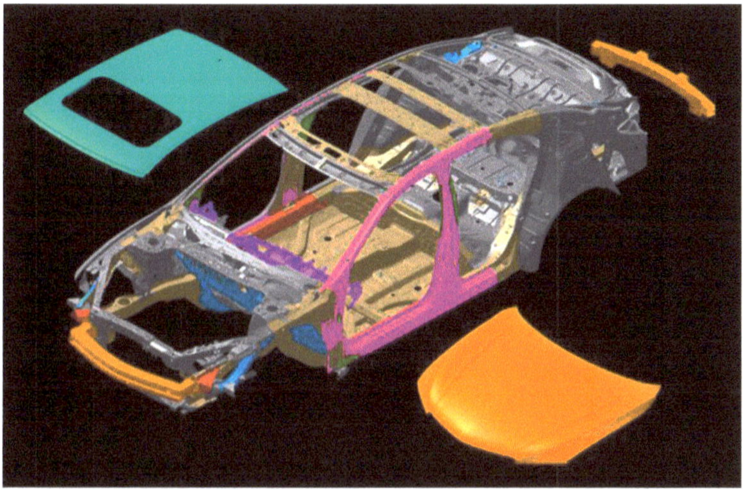

Figure 2.9 Example of a one-piece hot stamped door ring, shown in purple, in the 2015 Acura TLX [16].

innovations offer such a wide variety of strength grades and many manufacturing possibilities, increasing the strength of steel to support lightweighting does come with some manufacturing challenges.

2.4 Manufacturing concerns

In vehicle manufacturing, there are three main processes for which the compatibility of the steel must be ensured to guarantee robust, high quality manufacturing. These are forming, welding/joining, and painting. Each are considered below.

2.4.1 Forming

The most commonly used forming process in vehicle manufacturing is stamping. The following comments will focus on stamping, but are applicable to other processes (e.g., roll forming, stretch-bending, hydroforming, etc.). The goal of stamping processes is to provide a part with the correct geometry, required surface quality (varies depending on application), desired features (e.g., holes of the correct size and location, properly oriented flanges, etc.) with no splits or objectionable wrinkles. The last two of these, splits and wrinkles, are strongly influenced by the selection of material. Wrinkles are caused by excess material in a specific location of the stamping. Some wrinkles are acceptable, considering their location and severity. Splits (or fractures) are caused when the strain capacity of the material is exceeded. They are never acceptable.

Stamping process design includes procedures to ensure that when material of the specified formability is used, splits do not occur. The foundation of these

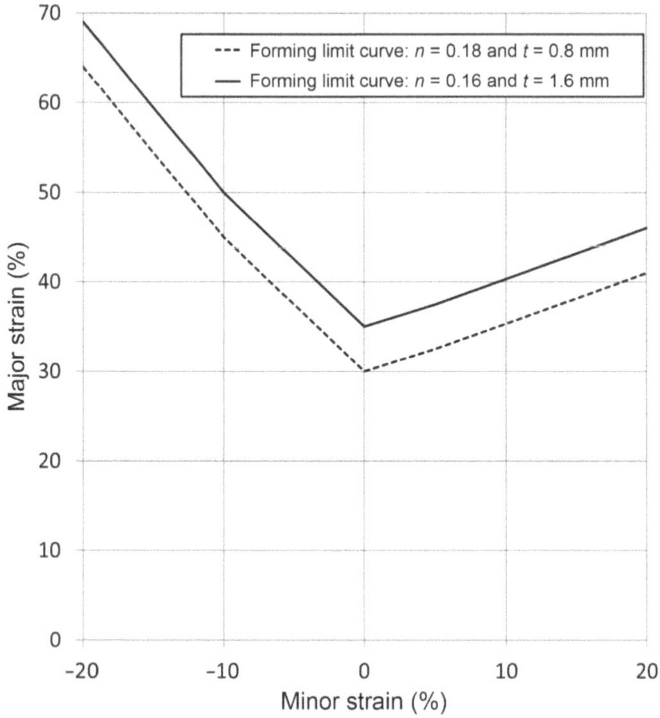

Figure 2.10 Example of forming limit diagram.

procedures is the forming limit diagram (FLD) [17]. The FLD is a graph with axes of minor strain on the *x*-axis and major strain on the *y*-axis,[7] as shown in Fig. 2.10. Upon this graph, a line exists which separates the areas of strain space where no splits will occur (safe), from where splits are expected.[8] The shape of this line has been experimentally shown to be identical for steels, but its position on the graph is determined by the point that describes the failure strain in plane strain (no minor strain) conditions. This value is known as FLD_0. To use this diagram, first the strains of a stamping are measured (or calculated, this process is also used during mathematical analysis of stamping designs). This can be done through circle grid analysis [18] or by ultrasonically measuring thickness strains and using them to estimate surface strains [19]. The strains are then plotted on this graph, and the values compared to the forming limit values. If all strains fall below the forming limit line, failures are not expected and the stamping is "safe." If any values fall above the line, either the tooling must be modified to reduce the strains, or the

[7] At any point on a stamping, the major strain is the highest measurable strain (as a function of direction of measurement) and the minor strain is the lowest measurable strain (and can be negative).

[8] In some cases a "marginal" zone is also described where splits are possible if the process gets slightly out of control.

material must be changed to increase FLD_0. A detailed description of FLDs is provided in Chapter 3, Formability of Auto Components.

FLD_0 is a function of the formability of the material and its thickness. Formability is characterized by the n-value of the material, which is a parameter measured during tensile testing and represents the work hardening capacity of the material. A higher n-value indicates a more formable material. FLD_0 is calculated by the following equation [17]:

$$FLD_0 = (23.3 + 360 \times t) \times \left(\frac{n}{0.21}\right)$$

This equation shows that increasing thickness (t) or n-value can increase FLD_0, making the forming process more robust. Obviously, increasing thickness increases mass, which is almost always undesirable, so the steel industry has made improving formability a priority.

Reducing friction during stamping can also enhance formability, and this can be accomplished by the use of lubricants, some of which can be supplied on the steel coil itself. This is often a desirable situation, as the application of lubricants in a stamping plant, while effective, can create quite severe housekeeping problems. The most common lubricant supplied on a steel coil is "mill oil," whose purpose is mainly corrosion protection. If reduced friction is desired, a purchaser can specify a special pre-lube with enhanced tribological characteristics along with corrosion protection. Finally, dry film lubricants can provide the lowest possible friction; however these are usually difficult to remove prior to painting, resulting in the need for a separate cleaning operation prior to assembly and painting.

2.4.2 Welding/joining

The most common welding process used in vehicle manufacturing is resistance spot welding. In this process, sheets of steel are held between two water-cooled copper electrodes. Thousands of amperes are conducted through the sheets, causing melting that results in a weld nugget holding the two (or more) sheets together. The characteristics of this nugget control the performance of the welded joint.

The size of the nugget is important to the quality of a spot weld, and its minimum size is usually specified (typically around 4 mm diameter). The welding process is designed to ensure this minimum size, by controlling both the amount and duration of the applied current. The nugget is also expected to exceed the strength of the surrounding metal (both the base metal and the heat-affected metal). Destructive tests such as the lap-shear and cross tension tests [20] are used to demonstrate this characteristic. Given the complexity of the welding process itself, as well as the composition, thickness and number of materials being welded (this is known as the "stack-up"), each new combination is tested in the laboratory before being implemented into production. The ability of a steel to meet these requirements is generally known as "weldability."

Alloying increases the strength level of steel, but alloying also can make steels more difficult to weld. The desire to have weldable high-strength steels drove the initial development of HSLA steels. The amount of alloy in these materials is low compared to more traditional C–Mn steels at the same strength level. A parameter used as a general guide to weldability of a steel is known as the "carbon equivalent" (C_{eq}). This parameter is a function of the amounts of alloying elements, and represents the equivalent level of carbon that would result in the same weldability [21].

Some manufacturers specify a maximum C_{eq} value for various grades. Other manufacturers use this value as a trigger for more comprehensive testing of stack-ups containing high C_{eq} steels prior to production release. Note that the strength level of a steel does not dictate its weldability; chemical composition and microstructure also influence welding characteristics. Most advanced high-strength steels do have high C_{eq} values, but there are some that have been specifically designed to have low C_{eq} values. These are known as "low carbon equivalent" or LCE steels.

The use of zinc coatings also makes steels more difficult to weld. The coating weight control in early versions of zinc coated automotive steels was inadequate to allow development of robust production weld processes, as typical weld shop equipment couldn't cope with wide swings in coating weight. However, the new generation of hot-dip coating lines were built with improved coating weight control systems and this problem has largely disappeared. Typical coated steel today has a coating thickness of 7 μm, with an allowable range of slightly more than ±1 μm.

Laser processes including welding and trimming are implemented in modern vehicle body shops today. Laser welding has many advantages, including smaller heat affected zones (enabling smaller weld flanges compared to spot welding), energy efficiency and speed. The main disadvantage to laser welding in the body shop is the requirement for very precise fit between the panels being welded. In spot welding, the weld gun itself can be used to clamp the parts together prior to applying current, but this is not possible with laser processes. Thus, additional fixturing must be used to fit the parts together prior to welding. Laser processes are less sensitive to composition than spot weld processes, and a wide variety of different steel grades have been successfully welded together with lasers [22].

Other welding processes, such as gas metal arc (GMAW), metal/tungsten inert gas (MIG/TIG), are uncommon in the body shop and will not be considered here. For more information on GMAW as used for chassis applications, this reference [23] will be helpful.

Structural adhesives are becoming more prevalent in body shops, though they comprise a small fraction of joining methods in a typical shop. All steel grades are compatible with these adhesives. One of the advantages of adhesives is the creation of a continuous joint, rather than the discrete joint formed by spot welding (this is can be an advantage for laser welding as well). This can improve the stiffness of a component, which enables mass reduction through down-gaging.

2.4.3 Painting

Properly processed automotive steels do not cause any problems in painting. Care must be taken in selecting mill-applied lubricants and other coatings because they can be difficult to remove prior to painting. Vehicle manufacturers have lists of approved materials that can be used on incoming steel coils without further qualification. If alternative materials are to be used, they must pass an extensive qualification process to ensure compatibility with paint shop cleaners and other chemicals. Alternatively, if, e.g., a product such as a dry film lube is needed to form a difficult part, it can be washed and recoated with an approved material prior to shipment to the customer.

2.5 Future steel technology

Steel grade development over the past several decades has successfully provided increasing strength levels to the automotive industry. These higher strength grades have significantly lower elongation and are more difficult to form at room temperature. Roll forming is a common process to make rail-type geometries through bending a continuous narrow coil of material. This process is cost effective, however, limited in types of parts meeting the geometry limitations, such as rocker beams or roof rails. More complex geometries of typically stamped parts require hot stamping to achieve deeper draws and smaller radii. Because of the energy intensity and slow cycle times of the hot stamping process, the cost to produce these parts is significantly higher than stamped components. The added cost puts hot stamped parts in contention with alternative materials for the same mass savings.

Second generation AHSS (2nd Gen AHSS) were developed to offer high strength in combination with high formability and are shown in Fig. 2.1 identified as austenitic stainless steel and TWIP steel. These grades require high amounts of alloying elements such as chromium, nickel, and manganese to stabilize austenite at room temperature to give the excellent formability while twinning or transformation to martensite under strain yields high strength levels. Unfortunately, these grades proved to be limiting in value to the automotive industry because of problems associated with the high alloying content; cracking issues with traditional spot welding and cost.

Examining the steel strength-ductility diagram show in Fig. 2.1, a gap can be observed between the 1st Gen AHSS (represented by TRIP, DP, CP, and MS grades, shown in orange) and 2nd Gen AHSS (austenitic stainless and TWIP grades, shown in blue). This gap is where automakers were looking for the steel grades that would give them the best combination of strength, formability, and weldability at a reasonable cost. This gap represents the target area where researchers and steel makers are working to develop new grades with strength levels at or above 1000 MPa and ductility around 20% or above. This new category of AHSS is referred to as the third generation of AHSS (3rd Gen AHSS).

Significant research has been completed in this area with several processing/alloying paths to achieve this combination of properties [24]. Although this research has shown several ways to get to the desired performance in this category, it can be described by the type of microstructure necessary to achieve this. Fig. 2.11 shows through the use of simple rule of mixture of microstructures that a combination of stable austenite and martensite would make up the bulk of the microstructure. Mild

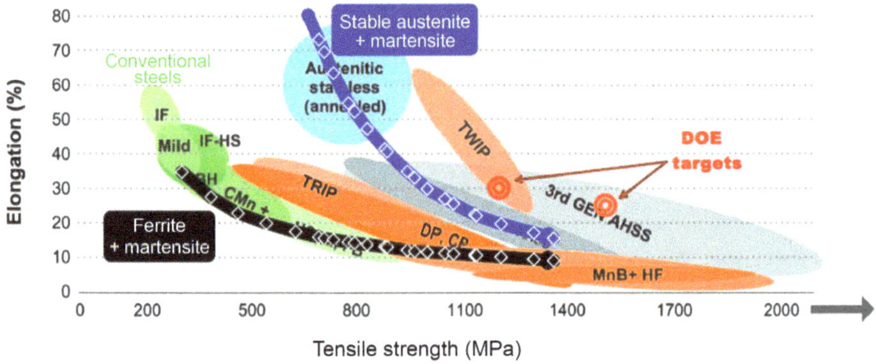

Figure 2.11 Microstructural rule of mixtures used to demonstrate steel grade strength and ductility [18].

steel and high-strength steel grades have ferritic microstructures giving the best formability. The evolution of 1st Gen AHSS added martensite to the matrix which added high strength without sacrificing too much ductility provided from the ferrite. Retained austenite phase is the basis of the 2nd Gen AHSS and provides ductility from its phase structure and strength by transforming to martensite under strain as described above. To achieve the combination of the high strength with high elongations it seems clear a combination of austenite and martensite would be needed to achieve the performance of the 3rd Gen AHSS. However, as illustrated in Fig. 2.11, these two phases alone will not achieve the desired outcome. Additional phases will be needed to augment the relatively low ductility of the martensitic phase.

Achieving these microstructures is not an easy path. The thermo-mechanical processing must be carefully controlled to consistently produce such complex microstructures. In addition, higher alloying content will be needed to support the stability of the austenite phase. The details of this technology are covered in Chapter 11, High Mn TWIP Steel and Medium Mn Steel and Chapter 12, Hot Formed Steels.

Another innovative project exploring the development of 3rd Gen AHSS is the Integrated Computational Materials Engineering (ICME) of 3rd Gen AHSS sponsored by the Department of Energy (DOE). This was a 4-year project proposed by the Auto/Steel Partnership (A/SP) through the United States Center for Automotive Research (USCAR) organization. The project developed models for making and predicting material behavior from the microstructural level through part formability and vehicle crash simulation. There were two grade targets provided by the DOE, a 1200 MPa tensile strength material with 30% elongation and another with 1500 MPa tensile strength and 25% elongation. Laboratory scale heats of the two grades were produced and used to validate the ICME models. The results of this work showed a potential for reducing mass of a body-side system on a mid-size sedan by up to 34% [25]. The success of this project provided the research and development foundation for the steel industry to move forward in commercializing these grades.

In addition to 3rd Gen AHSS, research and development to improve hot stamped steels to even higher strengths is well underway and tensile strengths of 2000 MPa are expected from this grade/processing combination. More information on the development of hot formed steels is included in Chapter 12, Hot Formed Steels. It is exciting to see so much innovation in the steel industry to meet the challenges in the automotive sector.

2.6 Sustainability/life cycle assessment

Currently, sustainability and life cycle assessment are not part of the engineering process when designing a vehicle. Product engineers have many requirements to fulfill when choosing materials and designing the vehicle. The main considerations include vehicle performance, mass and cost. In the United States, this does not include anything related to the environmental impact of the vehicle before or after its driving phase. However, the importance of the materials production phase and its end-of-life reuse or recyclability has been shown to significantly impact the environment such that it should be considered during the upfront design and engineering [26].

2.6.1 Sustainability

The first thought that comes to mind when sustainability is mentioned is end-of-life and most typically recyclability. In this regard, steel has a very good story compared with competing body materials. Steel is easily recycled. Because of its magnetic properties, steel is easily sorted and reclaimed after a vehicle is shredded. Its physical properties also allow steel to be recycled into any other steel grade without loss of quality whereas other materials, such as aluminum, must be recycled to the same grade or a lower-quality product. This is why stamping plants with higher volumes of aluminum products have implemented costly scrap sorting systems. This allows the stamper to recoup optimum cost from the aluminum offal.

Nearly all automotive steel is collected and recycled. There are more than 80 million tons of recycled steel available per year that can be manufactured into new steel products [27].

Since sustainability is used in reference to maintaining the ecological balance without destroying the environment, another major factor to be considered is the upfront materials production phase. There is no shortage of raw materials to make these popular body materials. However, it is a matter of where these natural resources are located and how they get processed into the usable form of the metal or composite.

It is fairly well known within the scientific community that the process to turn bauxite into aluminum is energy intensive, on the order of seven times more energy to process than steel (kilogram for kilogram), and the bauxite-to-aluminum process is also a “dirty” process. Several aluminum smelting facilities have shut down in North America because of economics and profitability.

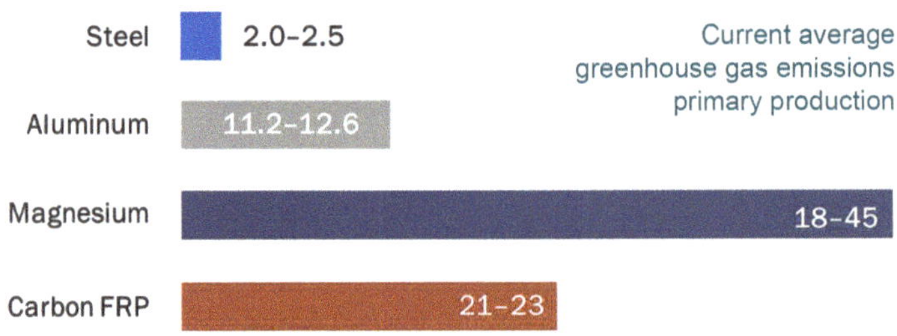

Figure 2.12 Average greenhouse gas emissions during material production (in kg CO_2e[9]/kg material) [28].

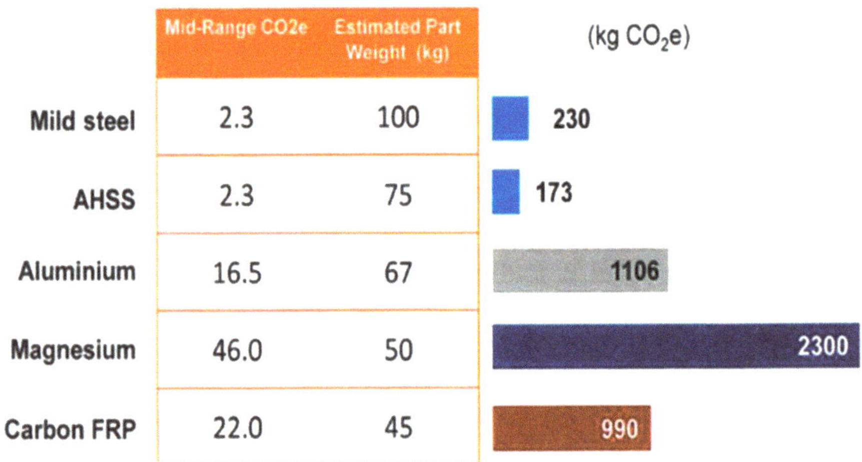

Figure 2.13 Material production greenhouse gas emissions normalized by average mass reduction potential [29].

A good way to compare these materials and their impact on the atmosphere is to look at the relative GHG emissions produced during the material production phase. Fig. 2.12 shows the amount of CO_2 gases emitted during production of one kilogram of each material. Steel is the cleanest material compared to each of the others by at least a factor of 4-5. It is argued however that steel does not achieve the same amount of mass reduction as the other materials and therefore these figures are unbalanced. Fig. 2.13 takes a more balanced look at material production emissions by accounting for the average amount of mass reduction for each. The numbers are reduced for all materials relative to a mild steel baseline with AHSS still showing a significant advantage over all other body materials.

So if sustainability is looked at from a big picture perspective, how do these factors of recyclability and material production play into the whole life of a vehicle?

[9] CO_2e is the term used to define all gas emissions relative to carbon dioxide.

2.6.2 Life cycle assessment

As discussed in Chapter 1, Design of Auto Body: Materials Perspective, fuel economy regulations have been an effective mechanism for improving vehicle fuel economy and reducing GHG emissions associated with fuel combustion. While the focus of federal regulations has typically been on the vehicle use phase (tailpipe emissions), the true GHG profile of a vehicle is only evident by considering the entire life cycle. A vehicle's life cycle has three parts (or phases): production, use (driving) and end-of-life (recycling and/or disposal). Life cycle assessment (LCA) is an established methodology of measuring the environmental footprint of a product over its entire life, including all of these phases.

The production phase can account for a significant portion of the overall life cycle emissions. WorldAutoSteel and the Steel Market Development Institute (SMDI) have studied several cases looking at lightweighting options for automotive bodies [30]. For the SMDI cases, the production phase comprised nearly 20% of total GHG emissions for internal combustion engine vehicles, and as much as 47% for BEVs. WorldAutoSteel found even greater values for the ratio of production emissions to total life cycle emissions. As overall fuel economy improves over time, production emissions will become even more important. If alternate powertrain vehicles such as BEVs increase in share, the importance of production emissions is increased even further. These production emissions result in environmental impacts before a vehicle is ever driven and they are not accounted for in current fuel economy regulations or factored into most automotive design practices. Once emitted, GHGs immediately begin absorbing energy from the sun leading to warming of the atmosphere. Major GHGs remain in the atmosphere for a range of decades (CH_4) to centuries (CO_2) after being released [31]. Therefore, the timing of GHG emissions is a critical consideration.

The University of California Santa Barbara Automotive Materials GHG Comparison Model V4 (UCSB Model) was used to model these various automotive lightweight scenarios. The UCSB Model, developed by Dr. Roland Geyer initially in 2007, calculates GHG emissions and energy over the entire life cycle of a vehicle. It has been peer reviewed and has gone through three update cycles since it was first developed. The model is fully transparent and publicly available at http://www.worldautosteel.org/projects/vehicle-lca-study/auto-materials-ghg-comparison-model/. The potential unintended consequences of a focus solely on tailpipe emissions was demonstrated through several approaches using the UCSB Model and can be found on the WorldAutoSteel and SMDI websites, worldautosteel.org and autosteel.org, respectively.

An example demonstrating an unintended consequence is a comparison of a mid-sized sedan made with an AHSS-intensive body to one made with an aluminum-intensive body. The default vehicle in the model (based on material composition, mass and fuel efficiency representative of model year 2007 vehicles) was compared to the two lightweight contender vehicles. The contender vehicles are created within the UCSB Model by replacing the steel body structure of the baseline vehicle with either AHSS or aluminum. The UCSB Model compares a baseline vehicle to one or more contender vehicles, based on a comprehensive set of input parameters. These parameters include, but are not limited to:

- Lifetime driving distance
- Material replacement coefficients
- Secondary mass savings
- Material production GHG emissions, including recycled content
- Degree of powertrain optimization/fuel reduction values
- Forming (stamping) yields

The results from this scenario, comparing an AHSS-intensive contender vehicle to an aluminum-intensive option, are shown in Table 2.1. In this simple case, the aluminum-intensive vehicle results in slightly higher (0.9%) life cycle GHG emissions, although the emissions from the two contender vehicles are close enough they can be said to be essentially equal from a life cycle perspective. However, the aluminum-intensive option results in nearly 30% higher GHG emissions at the materials production phase.

In addition to default parameters, the UCSB Model includes several default vehicles as baseline vehicles, including an SUV, pick-up truck, a HBEV and a BEV. The results for these model runs are very similar to sedan: the full life cycle emissions are higher in each case for the aluminum-intensive vehicle, and the aluminum-intensive options exhibit significantly greater production phase emissions (31%, 31%, 30%, and 19% greater, respectively) versus the AHSS-intensive options. A summary of these results is shown in Table 2.2.

The baseline vehicles were then updated to reflect curb weights and fuel efficiencies of model year 2013 vehicles. These updated vehicles are then used as baseline vehicles in the UCSB Model and updated GHG scenarios are again calculated for AHSS-intensive and aluminum-intensive contender vehicles. The results from the updated mid-size sedan scenario are shown in Table 2.3. The results are very similar to the earlier case compared to a 2007 baseline. Additional results tables for the other models can be found in Table 2.2. All results for these updated baseline vehicle scenarios (generally lighter curb weight and better fuel efficiency) are similar to the corresponding results from scenarios using the default (older model) baseline vehicles.

Table 2.1 Greenhouse gas emissions of the life cycle for a mid-size sedan with a 2007 baseline

	Scenario MS1 (kg CO_2eq/vehicle)			
	Production	**Use**	**End of life**	**Total**
Baseline	8760	50,048	−2218	56,590
Contender 1 (C1): AHSS Body	8260	48,284	−1985	54,558
Contender 2 (C2): Alum. Body	10,703	47,578	−3217	55,064
Difference C2−C1	2444	−706	−1232	506
Percent increase Alum. versus AHSS	**29.6%**	−1.5%	62.1%	**0.9%**

Table 2.2 **Summary of life cycle emissions for example vehicles**

Scenario	Description	GHG increase by selecting alum. Over AHSS-intensive vehicle	
		Full life cycle (%)	Production-phase only (%)
MS1	UCSB default baseline sedan	0.9	29.6
SUV1	UCSB default baseline SUV	1.0	31.3
TR1	UCSB default baseline pickup truck	0.6	31.0
HEV1	UCSB default baseline HEV	1.8	29.5
BEV1	UCSB default baseline BEV	1.8	19.1
MS2	Updated (2013) baseline sedan	0.9	29.7
SUV2	Updated (2013) baseline SUV	0.8	30.5
TR2	Updated (2013) baseline pickup truck	0.6	31.6
HEV2	Updated (2013) baseline HEV	1.7	29.4
BEV2	Updated (2013) baseline BEV	1.7	17.8
TR3	Future (2025) example—pickup truck	2.5	39.3
TR3a	Future (2025) example—w/ cutoff method	6.8	40.3
TR3b	Future (2025) example—recycled content sensitivity	2.3	37.9

Table 2.3 **Greenhouse gas emissions of the life cycle for a mid-size sedan with a 2013 baseline**

	Scenario MS2 (kg CO_2eq/vehicle)			
	Production	Use	End of life	Total
Baseline	8066	47,723	−1992	53,798
Contender 1 (C1): AHSS Body	7604	46,091	−1777	51,919
Contender 2 (C2): Alum. Body	9864	45,439	−2916	52,386
Difference C2−C1	2260	−653	−1140	468
Percent increase Alum. versus AHSS	**29.7%**	−1.4%	64.2%	**0.9%**

Finally, a scenario representative of a future vehicle was modeled, when fuel efficiency requirements (in compliance with CAFE regulations) will have improved and will be very similar within identical vehicle footprints. All of the previously described modeling relies on the data and formulas built into the UCSB Model for calculating the fuel efficiencies of baseline and contender vehicles. However, if this life cycle thinking approach is projected to a typical model year 2025 vehicle, the modeling must acknowledge the realities of CAFE regulations, in that all automakers are striving to meet very difficult MPG targets through 2025. Since cars are designed many years in advance, engineers are already designing to meet these

Table 2.4 Greenhouse gas emissions of the life cycle for a 2025 pick-up truck

	Scenario TR3 (kg CO_2eq/vehicle)			
	Production	**Use**	**End of life**	**Total**
Baseline	13,700	90,393	−4174	99,920
Contender 1 (C1): AHSS Body	13,199	62,953	−3915	72,237
Contender 2 (C2): Alum. Body	18,386	62,955	−7274	74,068
Difference C2−C1	5188	3	−3359	1831
Percent increase Alum. versus AHSS	**39.3%**	0.0%	85.8%	**2.5%**

specific future targets. As this design process moves forward, it becomes clear production phase GHG emissions will become even more significant, since use-phase emissions within vehicle footprints will, by necessity, become significantly lower and essentially equal. For this phase of the GHG modeling, the fuel efficiencies of both the contender vehicles (AHSS-intensive and aluminum-intensive) are set equal to the CAFE standard targets for the modeled vehicle, and thus the use-phase emissions are equal. This scenario is based on the fact automakers are designing vehicles to meet, not exceed, regulatory targets. Automakers will be challenged to meet CAFE-required minimum MPG within each vehicle footprint and will be unlikely to invest the additional funds necessary to take fuel efficiency beyond the required minimums.

The scenario shown in Table 2.4 provides an example of this approach. For this model run, the baseline vehicle was modeled after the curb weight and material composition of an actual 2014 model year pickup truck, and the use-phase emissions were set equal for the two contender vehicles. The primary aluminum ingot GHG profile, aluminum sheet recycled content, and end-of-life recycling rates for steel and aluminum were also updated to reflect future conditions. All other parameters were the same as in previous scenarios. The results from this assessment show an even greater life cycle GHG penalty for the use of high-production-emissions materials.

In this example, since the use-phase emissions are equal, the differences in the two contenders arise only from the production and end-of-life phases. The aluminum-intensive option results in 39% higher production phase GHG emissions and 2.5% higher life cycle emissions when the end-of-life phase is included. These differences are substantial, and it is again important to note the production phase emissions occur before the vehicles are ever driven.

A common criticism of the type of modeling described above is the results can be skewed in one direction or another by selecting input parameters favorable to one material. While the parameters used in the above modeling are credible and well documented, this criticism has been addressed by conducting several Monte Carlo simulations for an AHSS-intensive versus aluminum-intensive vehicle comparison. In this context, the Monte Carlo approach involves 5000 simulations of randomly selecting a value within a specified range of possible values for each of

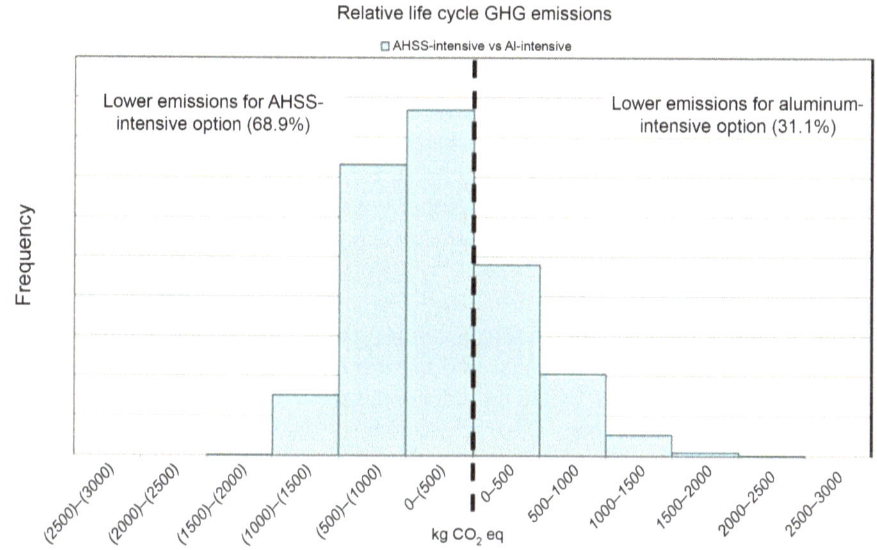

Figure 2.14 Monte Carlo analysis of the life cycle emissions for a mid-size sedan.

several parameters and performing GHG calculations within the UCSB model. The parameter ranges have been chosen to include all of the reasonable and credible values for each variable. The results are displayed in a histogram showing the percentage of the cases where one contender vehicle produces greater life cycle CO_2 emissions versus the other contender. By this approach, the entire range of credible parameter values can be assessed by the model, thereby addressing the sensitivity of the parameters and reducing uncertainty.

An example of the mid-sized sedan results is shown in Fig. 2.14. Here almost 70% of the 5000 scenarios showed the AHSS-intensive sedan to have more favorable life cycle GHG emissions over the aluminum-intensive vehicle. Similar results were found for the SUV, pick-up, HBEV, and BEV.

As automakers move to lightweighting as a significant component of their strategy to meet increasingly stringent CAFE regulations, it becomes critically important to look beyond the tailpipe, and instead consider the full life cycle emissions of vehicles. Because of its lower material production emissions and favorable recyclability steel offers the best total package for sustainability.

2.7 Summary

Controlling steel chemistry and processing allows grades to be developed for specific applications. This is how more than 200 individual grades of steel are available today. Individual vehicle designs are likely to only include 20–30 unique grades, but the innovation in sheet steel making has resulted in automakers having far more flexibility in their designs. Steel producers have worked together with vehicle engineers to develop new grades targeted to meet unique requirements for different applications on a variety of vehicles. These applications have ranged from increased strength for performance and mass reduction to zinc coatings for improved corrosion resistance to bake hardenable grades for closure panel dent resistance. Steel is versatile in providing great formability for styling body panels to great strength for crash performance. Not only has steel become stronger to resist crushing in the passenger compartment, it also can provide energy absorption in the front and rear crumple zones.

Steel strength increases through the last few decades have also allowed automakers to reduce vehicle mass through thinner gauges. Innovative manufacturing processes such as laser welded blanks and tailor rolled blanks offer additional mass reduction in individual parts by having thicker areas where more strength is needed and thinner where it is not. Other development and innovation have helped manufacturing engineers form and join new steel components for minimal disruption to vehicle assembly. Research and development continues to expand opportunities for steel applications, currently through the introduction of 3rd Generation AHSS to help meet the increasing needs of automakers.

In addition to steel's ability to meet performance needs in the vehicle, it also is a good choice for the environment. The mass-reduction capability improves fuel economy and thus lowers GHG emissions during use. Even before a car is ever driven, steel provides the lowest production emissions of all viable choices for auto bodies. At the end of life, steel is easily sorted from other vehicle materials as a result of its magnetic properties and is re-melted into new sheet products of any grade needed. It is clear to see why steel has long been, and is expected to continue to be, a significant part of automotive history.

References

[1] Steel Market Development Institute, Southfield, MI, 2015.

[2] P. DuBois, C.C. Chou, B.B. Fileta, T.B. Khalil, A.I. King, H.F. Mahmood, et al., Vehicle Crashworthiness and Occupant Protection, American Iron and Steel Institute, Southfield, MI, 2004.

[3] Michael Worswick, Professor, <worswick@lagavulin.uwaterloo.ca, http://www.formingandimpact.uwaterloo.ca/Research/crash.html >.

[4] Advanced High-Strength Steel Applications Guidelines Version 5.0, WorldAutoSteel, 2014.

[5] Categorization and Properties of Low-Carbon Automotive Sheet Steels, *SAE J2329*, 201504.

[6] J. Dicello, R. George, Design Criteria for the Dent Resistance of Auto Body Panels, SAE Technical Paper No. 740081, 1974.

[7] Surface Roughness and Peak Count Measurement of Cold-Rolled Sheet Steel, SAE J911, 1998.

[8] http://automotive.arcelormittal.com/Solutions_lightweighting/S_in_motion/SinMotionGeneral.

[9] http://www.worldautosteel.org/life-cycle-thinking/case-studies/lotus-engineering-2009-toyota-venza-mass-reduction-programme/.
[10] http://www.fev.com/press/always-informed/news-article/article/fev-inc-completes-vehicle-mass-reduction-study-for-epa.html.
[11] http://www.autosteel.org/Programs/ULSAB.aspx.
[12] http://www.worldautosteel.org/projects/future-steel-vehicle/phase-2-results/.
[13] http://www.nhtsa.gov/Laws+&+Regulations/CAFE+-+Fuel+Economy/cafe-volpe-model.
[14] W. Malone, The All New 2015 Chevrolet Colorado and GMC Canyon Cab Structure, *Great Designs in Steel*, Livonia MI, 13 May 2015.
[15] J. Reed, Advanced High-Strength Steel Technologies in the 2015 Ford Edge, *Great Designs in Steel*, Livonia MI, 13 May 2015.
[16] A. Madsen, 2015 Acura TLX Body Structure Review, *Great Designs in Steel*, Livonia MI, 13 May 2015.
[17] R.H. Wagoner, S. Chan Kwai, S. Chan, S.P. Keeler, Forming limit diagrams: concepts, methods, and applications, Commonwealth Dr, Warrendale, Pennsylvania 15086, USA The Minerals, Metals & Materials Society, 420, 1989.
[18] S. Dinda, How to Use Circle Grid Analysis for Die Tryout, ASM Intl, 1981.
[19] D.J. Schaeffler, E.J. Vineberg, Troubleshooting formability problems using strain analysis, Metalworking: Sheet Forming (ASM Handbook Volume 14 B), 2006 pp. 697–706.
[20] ANSI/AWS/SAE/D8.9–97, Recommended Practices for Test Methods for Evaluating the Resistance Spot Welding Behavior of Automotive Sheet Steel Materials, *American Welding Society*, Miami 1997.
[21] H. Oikawa, et al., Resistance Spot Weldability of High Strength Steel Sheets for Automobiles, Nippon Steel Technical Report No. 95, January 2007.
[22] B. Breakiron, J.R. Fekete, Formability analysis of high strength steel laser welded blanks, SAE Technical Paper 2005.
[23] GMAW Weld Design Guidelines for Chassis Structures, Final Project Report, Auto/Steel Partnership, November 2007.
[24] G. Thomas, ICME 3GAHSS Lab Heat Results Supporting DOE Targets, *Great Designs in Steel*, Livonia MI, 13 May 2015.
[25] Integrated Computational Materials Engineering Approach to Development of Lightweight 3GAHSS Vehicle Assembly, DOE#: DE-EE0005976AMP-802, Auto/Steel Partnership Project Review Meeting –Task 5 EDAG May 23rd, 2016.
[26] http://www.worldautosteel.org/life-cycle-thinking/case-studies/.
[27] http://www.smdisteel.org/sustainability.aspx.
[28] World Steel Association, "Steel in the Circular Economy," 2015.
[29] WorldAutoSteel Internal report, 2015.
[30] Steel Market Development Institute, The Importance of the Production Phase in Vehicle Life Cycle Emissions, 1 March 2016.
[31] https://www3.epa.gov/climatechange/ghgemissions/gases.html.

Formability of auto components

3

E.H. Atzema
Tata Steel, IJmuiden, The Netherlands

3.1 Introduction

Increasing demands on CO_2 emission and fuel efficiency drive for lower weight whilst a main selling point of cars is their safety. These aspects combine to an increased use of stronger steels in car bodies [1,2], as discussed in some detail in Chapter 1, Design of auto body: materials perspective and Chapter 2, Steels for auto bodies: a general overview.

The entire process chain in the automotive industry needs to be viewed holistically. Blanks are cut from coil, and subsequently formed. Although "formability" is used as a concept to indicate how much change of shape a steel sheet can endure before problems occur, there are many different forming processes which have different limits and more importantly place different demands on the steel product. On the one hand there are processes with uniform deformation over the thickness, such as stretching and deep drawing often referred to simultaneously as stamping. And on the other hand there are processes with a strong gradient over thickness such as bending and flanging. But other classifications are possible, for instance based on what demands are placed on material properties. A flavor of available processes will be given along with specific tests (if applicable) that are in widespread use.

Outer panels are generally made of mild low carbon steels or interstitial free (IF) steels for premium formability. A main requirement here is the surface quality, but this falls outside of the scope of this chapter. For these ductile materials decades of experience exist, hence the basic concepts of formability will be explained for these materials.

Structural parts are more often made from high strength steel (HSS), the group of single phase steels which are stronger due to precipitation hardening and solid solution hardening, for instance. The increased strength comes at the expense of elongation, with a repercussion on formability; albeit mostly in stamping. Flangeability and bendability are generally excellent, certainly when taking the strength into account.

For even higher strength, multiphase steels known collectively as advanced high strength steels (AHSS) are used. The composite nature of dual phase (DP) steels can enhance stretchability whereas complex phase (CP) steels are better in bending and flangeability. The DP steels, while limited in stretch formability and deep drawability with respect to mild steels, mostly show problems in practice with bendability and flangeability. Whereas conventional HSS (such as high strength low alloy

Automotive Steels. DOI: http://dx.doi.org/10.1016/B978-0-08-100638-2.00003-1

(HSLA) steels) can be bent flat onto themselves (by 180°) without problems, DP steels generally exhibit a lower bending angle before fracture starts. Conversely, the CP steels exhibit bad stretchability.

But even in AHSS, failure during forming is not always governed by aspects other than those in conventional steels. In this chapter, therefore, first the (widely accepted) general concepts of formability will be treated, then some lesser known advanced concepts, and finally the implications for the different steels in forming processes will be treated. To enable the latter, a brief section on forming processes is incorporated as well. The different sections will be shortly introduced here to enable the reader to retain an overview.

3.2 Basic concepts

A first characterization of material properties is generally done with a tensile test. This gives us an engineering stress-strain curve. However, forming analysts prefer to use true stress and strain over engineering stress and strain. Although an insight into the material behavior is obtained from uniaxial tensile tests, real forming processes exhibit multiaxial stress states. The yield locus concept provides the link between multiaxial and uniaxial stress. It also models anisotropy, both normal anisotropy and planar anisotropy. Whereas the yield locus is most important in deep drawing, stretching operations are mainly governed by hardening. AHSS exhibit relatively strong hardening behavior and obey different laws than HSS, for instance to enable modeling the Bauschinger effect.

Current AHSS have low planar anisotropy limiting the deep drawability, although research is underway into how to change this. Moreover, the hardening may initially be very high but it levels off with increasing strains, more than in HSS or mild steels, which is a drawback in terms of stretchability.

Finally, the Forming Limit Curve (FLC) concept is used to judge how safe obtained strain distributions are in a real stamping or in the simulation of one. Although the FLC is the onset of necking, it is generally at higher strains than in a tensile test because in the latter a diffuse neck develops which cannot occur in sheet metal forming. In sheet metal forming only local necking occurs.

Many processes are (implicitly) designed to avoid the (local) necking instability mentioned above. In such processes other limits are encountered, much less researched, but still some observations will be given [3,4].

3.3 Advanced process analysis

Most topics that will be treated in the basic concepts section can be found in many textbooks. This section will deal with advanced insights that are not commonplace yet. Material characterization will generally be by tensile test data. This data is (near) adiabatic. When sufficient contact with the stamping tools is made, the process will be (near) isothermal. For stronger materials this effect is larger.

Steel also exhibits strain rate sensitivity (SRS): if deformation is to occur faster this takes more stress. This may not appear to be of importance to (relatively slow) forming processes but it will be explained that it stabilizes the necking and thus increases formability. SRS is modeled as an addition to the stress according to Ref. [5] rather than multiplication, which is more commonly used. Also based on Ref. [5] and in Tata Steel R&D experience the effect is roughly similar in most steels and thus relatively smaller the higher the strength.

The weight gained by using stronger steels is obtained by down gauging. As springback increases with strength, and decreases with thickness, the dimensional stability becomes a concern. Drawing over a radius first produces bending and then unbending introducing the Bauschinger effect, which is stronger for AHSS than for other types of steel. Design guidelines usually advise to use smaller draw radii [6], which helps in controlling springback but may induce (brittle) fracture problems. The latter is not currently understood.

Although standard theory assumes the yield locus to be of constant shape and the same hardening law to apply regardless of stress state, differences have been found in recent investigations and an evolving yield loci may be necessary for ultimate accuracy [7–9].

The FLC is widely used to judge if strains measured in a stamping or simulated by FEA are safe in terms of process. There are however multiple ways to determine an FLC and there is still debate over which is the best. The newest method uses the time derivative in addition to the spatial derivative [10], but this introduces new issues. Alternatively, new empirical models appear to be able to predict the FLC with good accuracy from tensile data alone [11].

Besides these theoretical issues there are practical limitations to the FLC as well. Stamping processes typically exhibit change of strain path during forming, for which the FLC is not directly applicable. Various approaches to account for nonlinear strains paths are found in the literature [12].

A final aspect to consider is the impact of forming stage on inservice performance. All design is currently done by Computer Aided Engineering (CAE) methods and for ultimate accuracy the material in the car body cannot be considered virgin. The stamping process introduces work hardening and thinning of the material, and most AHSS also show some bake hardening, which has to be taken into account.

After treating the basic and advanced concepts a brief overview of the most important forming processes are given. First, the sheets need to be cut off of the coil.

Then the distinction between stretching and drawing is important mostly from a point of view of which material parameters matter most. But it will be pointed out that the distinction is often academic, as in a given (complex) geometry there is not much freedom of choice.

Bending is a special case where there is a strong gradient of deformation over the thickness, not otherwise encountered in sheet metal forming. And where for conventional steels bending never was much of a problem, in AHSS problems do occur.

The same holds more or less for flangeability. Although it is also limited in mild steels and HSS it never was much of a problem because the limits were often beyond the demands. For the stronger HSLAs and especially for AHSS it is of paramount importance to realize the limitations in the process design phase.

The roll forming process places entirely different demands on the material than do other forming processes. It is therefore better suited to other materials than those suitable for stamping.

In the last section, an attempt will be made to link the concepts and the forming processes to enable understanding of opportunities and problems that can rise from the combination of material and process.

3.4 Basic concepts

In this section the basic concepts will be discussed starting with strain and stress, because forming analysts favor a different metric than that usually employed by metallurgists. Subsequently, the implications of the hardening phenomenon on formability will be treated and then the discussion will be extended to multiaxial stress states. Finally, the FLC approach will be described.

3.4.1 Strain measures

The elongation of any object in a certain direction can be characterized by the ratio of change in length (Δl) over initial length (l_0), this is called engineering strain e:

$$e = \frac{\Delta l}{l_0} \tag{3.1}$$

Engineering strains are usually (though not necessarily) expressed as a percentage.

Whereas this is a very common and easy to understand definition, the true strain metric is much more valuable in forming technology. This is because engineering strains do not add up for subsequent processes and elongation versus compression shows an asymmetry in engineering strains. In other words, for comparable forming severity the numbers are different (in magnitude) for elongation and compression. A more convenient measure is the true strain ε. The definition of true strain is by natural logarithm of total length l over initial length l_0 (it is also referred to as logarithmic strain):

$$\varepsilon = \ln\left(\frac{l}{l_0}\right) \tag{3.2}$$

To illustrate this once more: If two forming processes executed in series add 50% engineering strain in each case then together these processes will add 125%.

$$\begin{aligned} &l_1 = 1.5 * l_0;\ l_2 = 1.5 * l_1;\ e_{0-1} = \frac{1.5l_0 - l_0}{l_0} = \frac{0.5l_0}{l_0} = 50\%;\ e_{1-2} = 50\% \\ &\text{But: } e_{0-2} = \frac{1.5 \cdot 1.5l_0 - l_0}{l_0} = 125\%; \end{aligned} \tag{3.3}$$

In true strains these numbers would be 0.405 and 0.810.

If in a process a sheet gets twice as long and half as wide (see Fig. 3.1), the engineering strains in length and width direction would be +100% and −50%, and the thickness would be unchanged. However, true strains would be of the same magnitude but opposite in sign (±0.693), reflecting the forming severity much better. To further illustrate this, compressive engineering strain is limited to −100% at the largest, whereas tensile elongation is limitless. We will use true strains from here on.

During plastic deformation of metals, any change of volume is negligible and hence preservation of volume V is assumed. In terms of true strains this means the sum of principal strains is zero:

$$0 = \ln\left(\frac{V}{V_0}\right) = \ln\left(\frac{l \cdot w \cdot t}{l_0 \cdot w_0 \cdot t_0}\right) = \ln\left(\frac{l}{l_0}\right) + \ln\left(\frac{w}{w_0}\right) + \ln\left(\frac{t}{t_0}\right) = \varepsilon_l + \varepsilon_w + \varepsilon_t \tag{3.4}$$

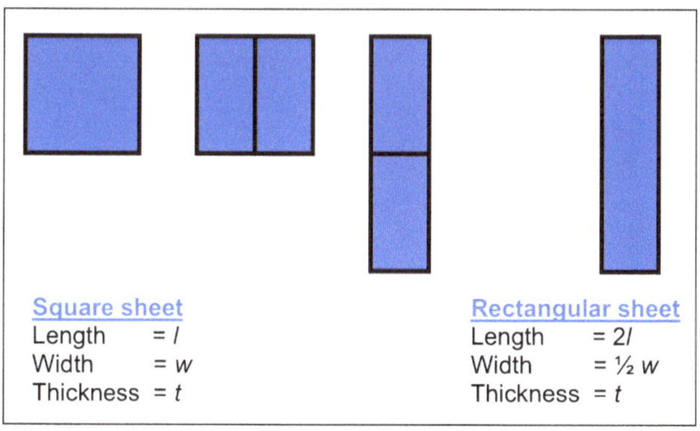

Figure 3.1 Strain illustration.
Source: Picture from Performance-Product-Process. Internal Tata Steel R&D course [13].

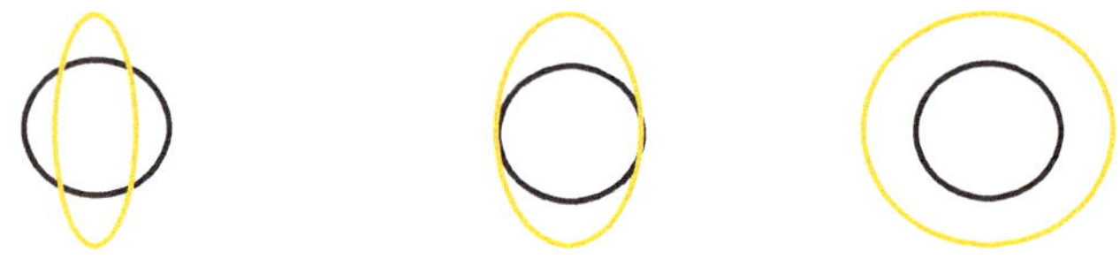

Figure 3.2 Deformed ellipses illustrating major (vertical) and minor (horizontal) strains.

So far, implicitly principal strains have been assumed. Without further proof it will be stipulated here that there are always directions (the so-called principal directions) in a material frame where all shear strains are zero and only normal strains are present. If we use (as researchers did initially when analyzing forming processes) circles scribed on the underformed sheet, these will be transformed to ellipses in the forming process. The main axes of these ellipses will be the principal directions. The largest axis will be termed major strain, the smaller one minor strain. In Fig. 3.2 the strains are aligned just like in an FLD (see Section 3.4.5, Forming Limit Curve): major strain is vertical, minor strain horizontal, and ranges from negative (left) via zero to positive (right).

The ratio of the width strain to the thickness strain is also of importance. In an isotropic material this would be 1.0, as the length strain is accommodated by both thickness and width equally. This ratio is known as r or Lankfords coefficient (Eq. 3.5). Although the symbol is simply r, forming technologists often refer to it in written text as r-value, as will be done in this chapter.

$$r = \frac{\varepsilon_2}{\varepsilon_1} \tag{3.5}$$

As will be shown later in the subsection "limits of forming," the final failure in forming will be by progressive thinning. So if length strain is accommodated more by width and less by thickness strain this would be beneficial. This implies that an r-value larger than 1.0 is preferred. The r-value can be different in different directions within the sheet. Usually three directions are measured: rolling direction (RD), transverse direction (TD), and diagonal (DD). The rolling direction is the reference so RD is also known as 0 degree, TD as 90 degree, and DD as 45 degree. For general assessment of the deep drawability a mean r-value ($\overline{r}$) is usually taken, defined as average over the directions like in Eq. (3.6). As can be imagined, a large difference in r-value over the directions is not wanted, hence definition of Δr (Eq. 3.7) that preferably is low.

$$\overline{r} = \frac{1}{4}(r_0 + r_{90} + 2 \cdot r_{45}) \tag{3.6}$$

$$\Delta r = \frac{1}{4}(r_0 + r_{90} - 2 \cdot r_{45}) \tag{3.7}$$

3.4.2 Stress measures

Obviously, a thick steel sheet can carry more load than a thin one. So the load F is not an adequate measure. Dividing the load by initial cross section area A_0 gives us engineering stress S.

$$S = \frac{F}{A_0} \tag{3.8}$$

Whilst this is convenient, it is not fair, as the cross section decreases when straining the tensile specimen. It would therefore be much fairer if momentary or instantaneous cross section area A is considered.

Necking can then be analyzed in terms of a competition between the decrease in cross section and the hardening. This will also be illustrated in the next section on hardening when comparing Fig. 3.4 with Fig. 3.5. The load divided by actual cross section gives us true stress σ:

$$\sigma = \frac{F}{A} \tag{3.9}$$

The product of true strain and true stress also equals the energy dissipated, something not true of engineering stress strain.

It can thus be concluded that true strain and true stress measures may be less intuitive but are more convenient for process analysis.

3.4.3 Hardening

The first test anyone would do when characterizing a metal would be a tensile test. It is assumed the reader is familiar with how such a test is performed. What can be observed is that at first the response is elastic: if unloaded, the specimen will return to its original length. In this part the response is very steep, i.e., the stress increases very rapidly when strain is increased. Then, when the so-called yield stress (YS) is reached, the response becomes plastic: the change in length is (largely) permanent. This is of course the useful part, if a metal part is to be shaped. The onset of yielding is commonly referred to as R_p (from the French *résistance plastique*) and is not easily found. Therefore, for practical purposes the stress at 0.2% plastic strain (designated $R_{p0.2}$) is usually used as the yield stress, Fig. 3.3. The red dotted line is the tangent to the initial part of the curve and assumed to be the elastic modulus. Often the subscript indicating the reference strain is omitted and reference is made simply to R_p.

After the yield stress it can be observed that the stress keeps on increasing with increasing strain. This is known as work hardening. Upon unloading the response is elastic, i.e., steep, and upon immediate reloading the yielding begins at the same stress level that was reached just before unloading. So the hardening is permanent.

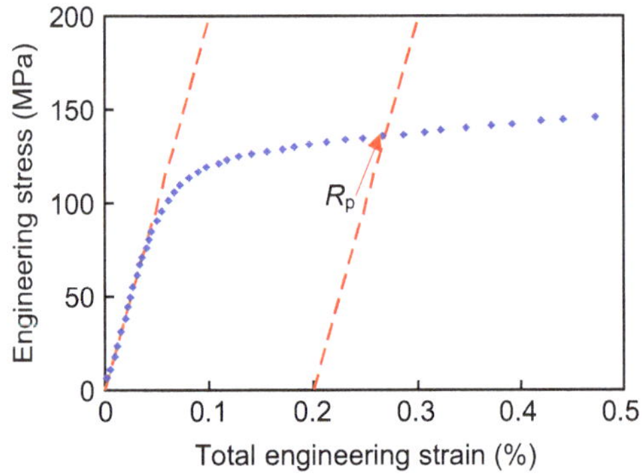

Figure 3.3 Practical definition of yield stress R_p (at 0.2% plastic strain).

In the traditional engineering stress-strain plot the curve goes through maximum strength, known as the ultimate tensile strength (UTS) or R_m (French: *résistance matériaux*) The strain at this point is known as uniform strain, A_g, because after this point it is observed that the specimen starts to neck. Locally the strains will be higher than elsewhere in the specimen. And *apparently* the hardening decreases since the force drops. However, when transforming tensile test results to true stress-true strain it can be seen that the hardening is not zero at the point of (diffuse) necking, but rather equal in magnitude to the decrease in area.

After necking has started the measured strain does not represent the actual local strain in the neck as we measure an average over a certain gauge length. For steel sheets this is typically 80 or 50 mm depending on the standard employed. This implies no data beyond A_g is available, although strains larger than A_g do occur in forming so data needs to be constructed to enable simulations to be run. This extrapolation of measured tensile data is most often done by fitting a hardening law relating yield stress σ_y to plastic strain ε_p. This "law" is fitted to the uniform strained part of the curve and used for arbitrarily large strains. For this fit, most often a power law is used [14,15]. This law has a hardening coefficient C and a hardening exponent n. As with r-value forming technologists often refer to n as n-value in written text.

$$\sigma_y = C \cdot (\varepsilon_p)^n \tag{3.10}$$

By the standard applicable to tensile tests (e.g., Ref. [16]), the n-value is determined by fitting the data between $e = 10\%$ and the smallest of A_g and 20%. The hardening coefficient can subsequently be determined from ultimate tensile strength (R_m) as being:

$$C = R_m \cdot \frac{n^n}{\exp(n)} \tag{3.11}$$

Plotting the true stress as a function of true strain for different hardening exponents (Fig. 3.4) and comparing to reduction in area (Fig. 3.5) gives us some insight into why necking initiates at higher strains for higher n-value.

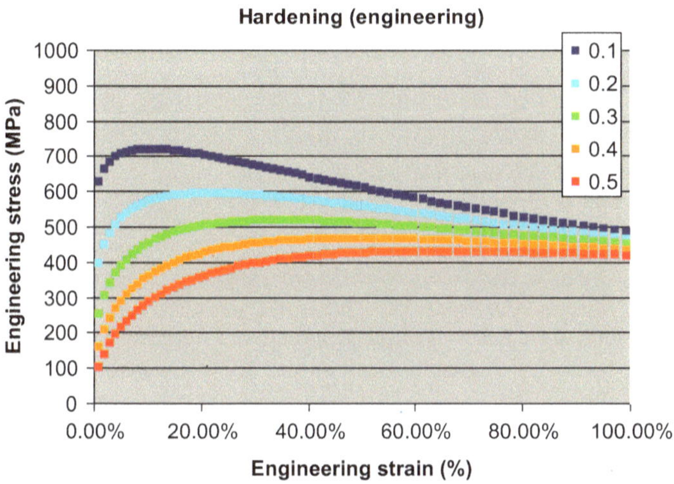

Figure 3.4 True stress-strain (i.e., hardening) for different hardening exponents.

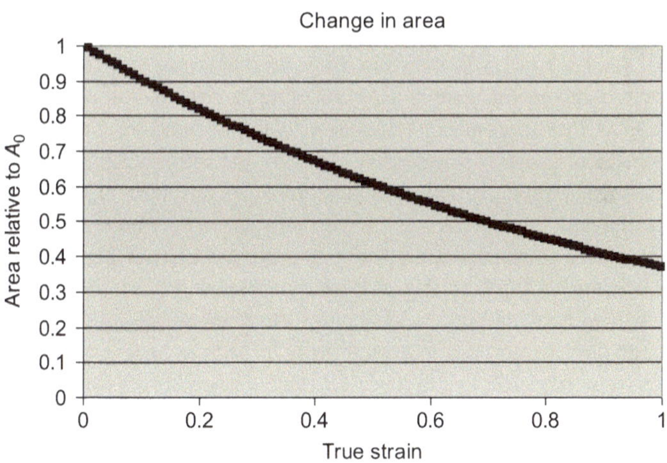

Figure 3.5 Reduction in area in a tensile test.

In this very simplified hardening law the stress is zero for zero plastic strain, so no yield stress is modeled. To overcome this, the prestrain ε_0 can be defined and related to yield stress (R_p).

$$\sigma_y = C \cdot (\varepsilon_0 + \varepsilon_p)^n; \varepsilon_0 = \left(\frac{R_p}{C}\right)^{\frac{1}{n}} \tag{3.12}$$

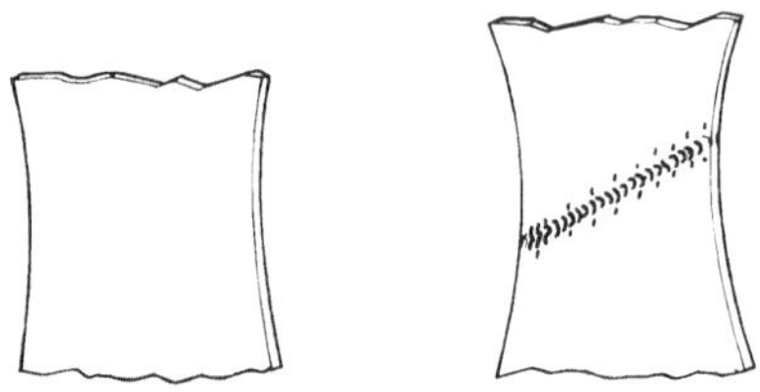

Figure 3.6 Diffuse neck (left) and local neck (right).
Source: R. Pearce, Sheet Metal Forming, Adam Hilger, Bristol, Philadelphia and New York, 1991, ISBN 0-7503-0101-5. With permission of Springer.

3.4.4 Limits of forming

Necking is the term used when a sheet progressively deforms locally whilst in the direct neighborhood strains develop much slower. Before necking, any gradients in strain will generally be very benign. In a tensile test the strains in the first stage will be the same over the entire length, i.e., uniform. Then, typically, a sheet steel will first exhibit a diffuse neck: locally the width of the sheet will become less than in the rest of the specimen. After some more strain a local neck forms. That is when locally the thickness of the sheet becomes progressively less than in the surrounding area. Finally, when all strain concentrates in this area, a fracture will occur (Fig. 3.6).

As explained above, in an engineering stress-strain plot of the tensile test the onset of the diffuse neck is clearly seen: it is the maximum of the curve. The local necking, however, cannot be distinguished easily. But in sheet metal forming practice this local necking point is the most important feature, as the diffuse neck cannot be formed. This is because during forming a part out of a metal sheet, the material is not as free to contract transverse to the applied load as it is in a tensile test. So the practical limit of formability is set by the occurrence of local necking.

Local necking is unacceptable because it gives an ugly appearance to outer panels and will result in a weak spot for structural parts. There are the odd exceptions, though, where in noncritical structural parts a small neck may be accepted if all attempts to improve the forming process have failed.

First, the diffuse necking process is analyzed mathematically. That the tensile force (F) goes through a maximum can mathematically be seen as having a horizontal tangent, i.e., its derivative with respect to strain is zero. An equation for the force is derived, being true stress multiplied by area, of which the derivative is taken and set at equal to zero:

$$\begin{aligned} &F=\sigma\cdot A;\ A\cdot l=V=V_0=A_0\cdot l_0\\ &A=A_0\cdot\frac{l_0}{l}=A_0\cdot e^{-\varepsilon};\ \frac{dA}{d\varepsilon}=-A_0\cdot e^{-\varepsilon}\\ &\frac{dF}{d\varepsilon}=0\Rightarrow\frac{d\sigma}{d\varepsilon}\cdot A+\sigma\cdot\frac{dA}{d\varepsilon}=0\\ &\qquad\Leftrightarrow\frac{d\sigma}{d\varepsilon}\cdot A_0\cdot e^{-\varepsilon}-\sigma\cdot A_0\cdot e^{-\varepsilon}=0\\ &\qquad\Leftrightarrow\frac{d\sigma}{d\varepsilon}=\sigma \end{aligned}\tag{3.13}$$

The result is a necking criterion. No assumptions have been made so far; the criterion is independent of the material investigated. This criterion can be stated as: "at the point of necking the derivative of true stress with respect to true strain is equal to the stress itself." For sheet metal it may be assumed that hardening follows power law and the derivative can be computed:

$$\begin{aligned}&\sigma = C \cdot (\varepsilon)^n \\ &\frac{d\sigma}{d\varepsilon} = nC \cdot (\varepsilon)^{n-1}\end{aligned} \tag{3.14}$$

And substitute it in the necking criterion:

$$\begin{aligned}&\frac{d\sigma}{d\varepsilon} = \sigma \\ &\Leftrightarrow nC \cdot (\varepsilon)^{n-1} = C \cdot (\varepsilon)^n \\ &\Rightarrow \varepsilon_{\text{diffuse}} = n\end{aligned} \tag{3.15}$$

So, in terms of true strain, this is equal to the n-value at the point of necking. However, this still applies to diffuse necking. A similar analysis can be done for local necking [15]. This employs the observation that once local necking sets in there is no additional deformation along the neck but all strain is perpendicular to the neck. Hence local necking takes place in plane strain mode. This is the reason why the plane strain point on the yield locus (explained in the next section) is so important. Then criterion for local necking can be deduced as [15]:

$$\varepsilon_{\text{local}} = 2n \tag{3.16}$$

Incidentally, although at first sight the local neck seems to be at 45 degree, this is actually not the case. Using Mohr's circles we can find the direction to be 54.7 degree with respect to the tensile direction [15]. The derivation is for isotropic material, but normal anisotropy does not change the number enough to warrant the extra effort required for the analysis.

By extrapolating the hardening beyond A_g through fitting a phenomenological law, a measure for the practical forming limit of strain in a tensile test has been obtained. Unfortunately, most forming processes do not resemble the uniaxial tensile test.

3.4.5 Forming limit curve

In sheet metal forming both in-plane stresses play a role, and the uniaxial stress situation is but one extreme of all possible conditions. The onset of local necking needs to be established for all encountered stress states. It was initially observed by Keeler [17] for stretching and later by Goodwin [18] for drawing that failure in sheet metal forming occurred at the same major and minor strain combination regardless of process. The curve connecting these points has become known as FLC.

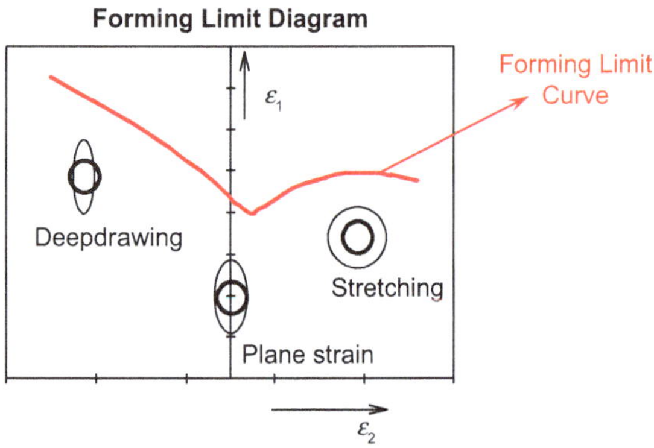

Figure 3.7 Forming limit diagram with forming limit curve [13].

The FLC concept is of key importance in practical formability analysis. In Fig. 3.7 a so-called forming limit diagram (FLD) is shown. The diagram uses the major and minor strain as the two axes and shows the major and minor strain combinations. The strains measured on a formed part can be plotted on such a diagram, as will be done much later in Section 3.8.3, Fig. 3.29. In this FLD the FLC can be plotted indicating the limits of the material, i.e., the onset of (local) necking. For ε_1 above the FLC failure is expected, whilst below the FLC the sheet remains intact.

3.4.6 Yield locus

Besides the point of (diffuse) necking the stress at which plastic deformation starts, i.e., yield stress, is also obtained from the tensile test. Once again, these apply to a uniaxial stress situation only. The yield stress changes with different stress situations as encountered in forming processes. The yield locus can be drawn to picture the yield stress in a biaxial stress state. In a two-dimensional graph with both in-plane stresses on the two axes, the yield locus is a curve connecting all stress combinations at which yielding starts. Obviously, this locus crosses the axis at the uniaxial yield stress. The concept is illustrated in Fig. 3.8, along with photographs of the tests as employed by Tata Steel R&D to determine individual stress points, as well as the tangent to the yield locus [20]. In the uniaxial stress the tangent is related to the r-value.

Again, like with strains, implicitly the principal stresses have been used here. Generally, in a given frame of reference there will be two normal stresses in the sheet plane plus a shear stress. There is, however, always a reference frame where the shear stress is zero and only normal stresses exist. In such a frame a two dimensional representation of the yield locus as seen in Fig. 3.8 can be made.

In the previous paragraph only the in-plane stresses were mentioned. In sheet metal forming the through thickness stress is generally regarded as negligible in comparison to in-plane stresses. Typically, a blank holder (also known as binder),

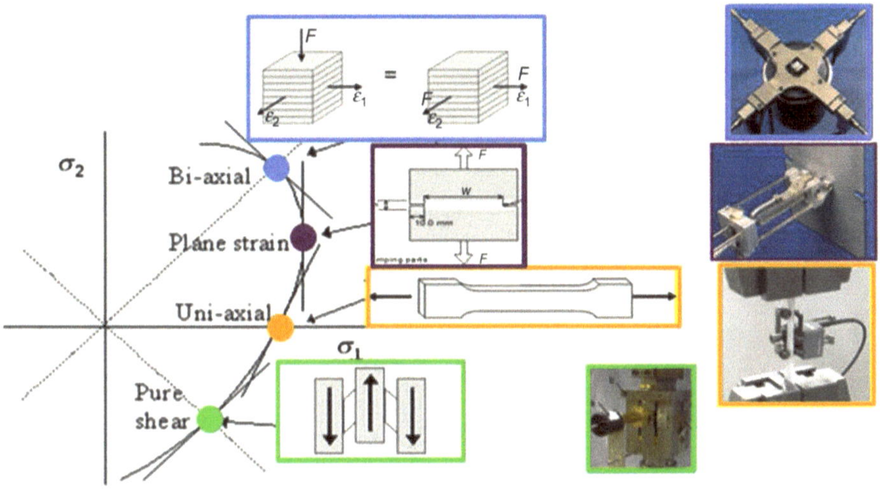

Figure 3.8 Yield locus with tests to determine individual stress points [19].

which serves to keep the flange of a cup flat during drawing (see Section 3.8.3), will exert a pressure of a few MPa on the surface of the sheet. Compare this to a yield stress of 140 MPa for the weakest of steels, and up to 1200 MPa or so for martensitic sheet steels and it is easy to understand why the through thickness stress is usually neglected. To be fair, on tight die radii or in draw beads the surface pressure can be significant.

The main use of the yield locus is in analyzing how different stress states compare in terms of yield stress. Consider the drawing of a hemispherical cup. In the bottom of the cup a biaxial stress state exists. In the flange of the cup the circumferential stress is negative whereas the axial stress is positive. If both stresses are of equal magnitude and opposite sign, the stress state is known as pure shear. It is now clear why the distribution of strains over the cup strongly depends upon the shape of the yield locus. For example, if for materials A and B (that have the same yield stress in a tensile test) the biaxial point is much larger for A, and at the same time the shear point is lower than for B, then material A will have less strain in the bottom and more in the flange than material B.

The yield locus thus determines the distribution of strains in a stretched or drawn product. In Fig. 3.9 the different stress states are indicated with a colored dot. The same dots can be seen on the pie-slice of a drawn cup in Fig. 3.10. Although initially, the cup wall is in shear deformation (at the start it is part of the flange), in the critical stage of the process the circumferential direction has no strain any more and the situation has become plane strain: red point.

In Fig. 3.11 it can be seen that the point where such a drawn cup fails is indeed in the wall of the cup. The force of the punch is brought to the flange via the wall of the cup. So the stress resistance needs to be high here. A high plane strain point is beneficial. The useful deformation is in the flange, as the deformation

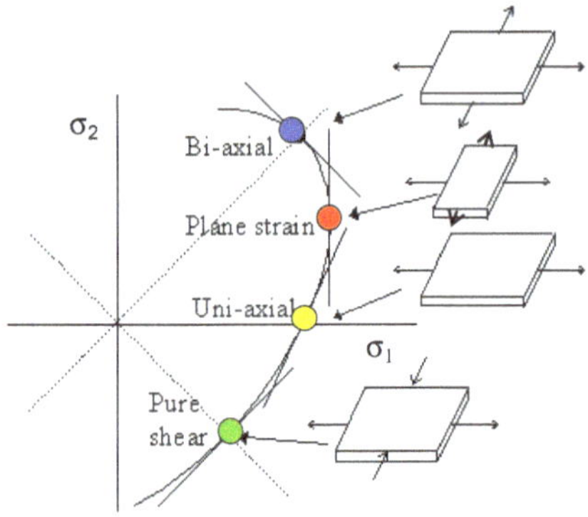

Figure 3.9 Yield locus with distinct stress states where tests are available.

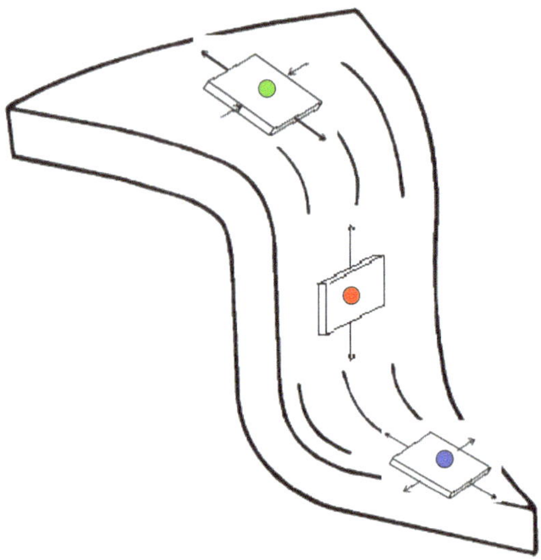

Figure 3.10 Stress states in a drawn cup.
Source: R. Pearce, Sheet Metal Forming, Adam Hilger, Bristol, Philadelphia and New York, 1991, ISBN 0-7503-0101-5. With permission of Springer.

Figure 3.11 Failure location on a deep drawn cup.

of the flange is what creates the cup wall. This is in shear mode, because the circumferential strain is compressive but the axial strain is tensile: green point. If the shear stress is low this is beneficial for the process. Combining these two it can be stated that the ratio of first principal stress in the plane strain (σ_{PS}) point over that in the shear point (σ_{SH}) determines the deep drawability of a material. To clarify, for the points as drawn in Fig. 3.9, the first principal stress is σ_1.

This ratio thus defined is referred to as Deep Drawing Ratio (DDR) [21]. If we use the yield locus model of Hill dating from 1948 [22] and assume planar isotropy we can formulate DDR in terms of r-value:

$$\mathrm{DDR} = \frac{\sigma_{PS}}{\sigma_{SH}} = \frac{\sigma_y \dfrac{r+1}{\sqrt{2r+1}}}{\sigma_y \sqrt{\dfrac{r+1}{2(2r+1)}}} = \sqrt{2(r+1)} \tag{3.17}$$

In Fig. 3.12 the correlation of the DDR with r-value is illustrated graphically. The α indicates the ratio of second principal stress over first principal stress and this will not concern us here. But the curve for $\alpha = r/(1+r)$ indicates the plane strain points. It can be seen that with increasing r, the first principal stress in the plane strain point increases whereas the shear stress ($\alpha = -1$) decreases. So, assuming the Hill model holds; the DDR increases with increasing r.

This corresponds with conventional wisdom that deep drawability is governed by r-value, with the added nuance that it is an indirect effect. This distinction is

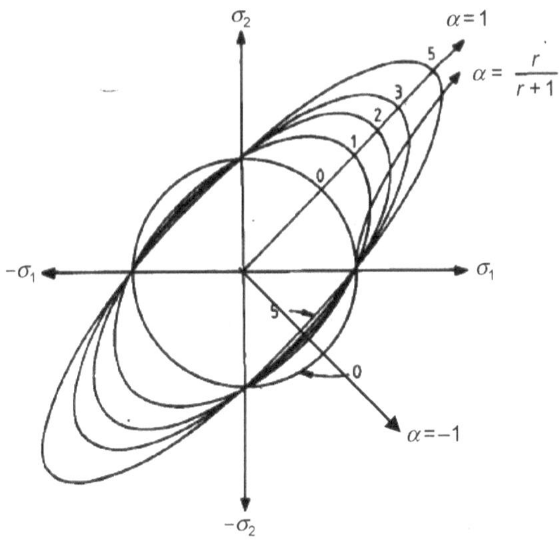

Figure 3.12 yield locus shape as function of r-value.
Source: R. Pearce, Sheet Metal Forming, Adam Hilger, Bristol, Philadelphia and New York, 1991, ISBN 0-7503-0101-5. With permission of Springer.

important because highly anisotropic materials often have a DDR slightly lower, and almost isotropic material often a higher DDR than expected based on r-value.

The shape of the yield locus is determined by the crystallographic texture of the material. The Lankford coefficient, or r-value, is an expression of crystallographic texture in the tensile test. It determines the tangent to the yield locus at the uniaxial points. And although a measured r-value at the uniaxial point correlates clearly with the plane strain stress, it can be improved upon by more advanced yield loci, for instance those that use the measured plane strain points directly [19].

A similar reasoning to the above can be held for (biaxial) stretching. The useful deformation is in biaxial strain, and again the fracture finally occurs in plane strain mode. So the plane strain point must again be high, and this time the biaxial point needs to be low. This can be expressed by the stretching ratio (STR) [21]. Again assuming the Hill model, it can be expressed as a function of r-value:

$$\mathrm{STR} = \frac{\sigma_{\mathrm{PS}}}{\sigma_{\mathrm{bi}}} = \frac{\sigma_{\mathrm{y}} \frac{r+1}{\sqrt{2r+1}}}{\sigma_{\mathrm{y}} \sqrt{\frac{r+1}{2}}} = \sqrt{\frac{2(r+1)}{2r+1}} \tag{3.18}$$

The stretchability can be seen to decrease with increasing r-value. Again here it is clear formability is not a single aspect but often a trade-off: what is good for drawability is bad for stretchability. The trend in deep drawing ratio is much stronger though, and for stretching the hardening plays an important role as well. A good compromise would be high anisotropy for good drawability and high hardening exponent to make up for stretchability.

Fitting the different models to the same material may show only small differences, Fig. 3.13, but a 1% difference in the stress of the plane strain point can make a noticeable difference in a simulation in terms of strain distribution,

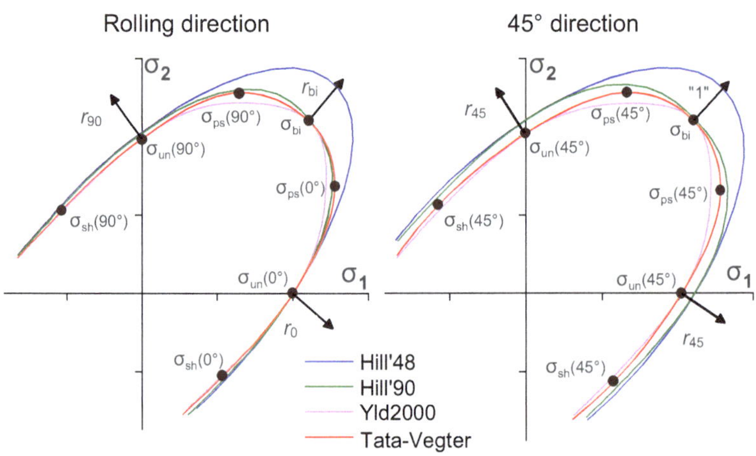

Figure 3.13 Different yield locus models fitted to the same material.

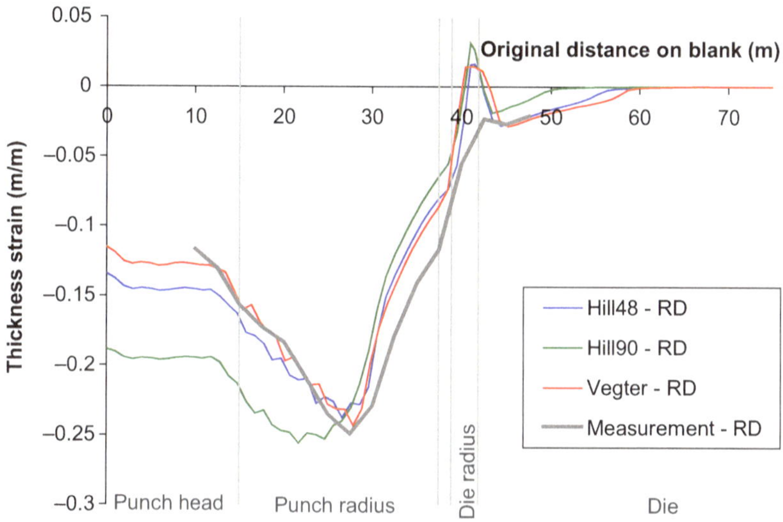

Figure 3.14 Thickness distribution in a stretch test: different yield locus models.

as can be seen in Fig. 3.14. It is therefore of prime importance to have an accurate yield locus model, and the only model available that is able to fit all measured points exactly is the Vegter criterion [19]. Alternatively, good results are obtained by reverse engineering [23]: fitting parameters of more flexible yield locus models than Hill (e.g., the BBC2005 model, [24]) to optimize predictions of actual forming tests.

3.5 Advanced process analysis

So much for the basic concepts of sheet metal forming that can be found in many textbooks. We will now briefly discuss the shortcomings of, and amendments to, these simple models.

3.5.1 Strain rate hardening

Steel exhibits so-called strain *rate* hardening (at ambient temperature). This means if it is to be deformed faster more stress needs to be applied. This is not the case for all metals, not for aluminum for instance. It is obvious this is a beneficial property for crash resistance since accidents happen at high strain rate. However, it is less obvious to see how it is beneficial for forming. This is because the limits of forming are given by local necking. A neck can only develop when there is more strain in the neck than outside the neck, so the strain rate in a neck must be higher. But if higher strain rate requires more stress this will dampen the strain rate difference and thus hinder the formation of a neck. Because a higher stress in the neck needs to be provided by a higher stress next to the neck (equilibrium of force), therefore this region is forced to still deform as well to harden to the stress needed. In other words, the strain rate dependence stabilizes the material, as it enhances formability by postponing the necking till higher strains. Before adding the strain rate sensitivity (SRS) to the hardening curve, however, the hardening curve will be examined in more detail.

Although the power law for hardening is used very often (indeed almost exclusively), it has no physical basis; it is simply a phenomenological fit. And there is reason to doubt if extrapolation will go well, although some firmly believe it does [25]. Alternatively, a hardening law can be based on the underlying physical phenomena. This is precisely what Bergström [26] did, and what van Liempt later amended [27]. The hardening stems from the fact that deformation requires dislocations to move, and when they do they run into obstacles in the microstructure. When they cannot move anymore this is called immobilization.

These obstacles can be interstitial or substitutional atoms, this is known as solid solution hardening. They can be precipitates, known as precipitation hardening. But they can also be other dislocations that act as obstacles to dislocation motion. This case is referred to as work hardening. If dislocations are immobilized, others need to be formed to keep the deformation going. For this an increase in stress is needed. At higher stress, some immobilized dislocations can be re-mobilized. Bergström arrived at a set of differential equations describing the statistical behavior of dislocations. This was picked up by van Liempt for application to rolling processes, where high deformations occur (hundreds of percent). For these high deformations the formation of dislocation cell substructures is an important effect and van Liempt formulated a way to take this into account.

Later on, Vegter [28] derived a closed form approximation of the hardening law from the set of differential equations that is easier to use for most engineers. In this form an additional term to take account of the strain rate sensitivity was also added, based on the work of Krabiell and Dahl [5]. The final equation looks daunting but many parameters turn out to be constants for steel.

$$\sigma_y = \underbrace{\sigma_0}_{\substack{\text{Basic}\\ \text{yield}\\ \text{stress}}} + \underbrace{\Delta\sigma_m(\beta(\varepsilon_p + \varepsilon_0) + [1 - e^{-\Omega\,(\varepsilon_p+\varepsilon_0)}])^{n'}}_{\text{Strain hardening}} + \underbrace{\sigma_0^*\left(1 + \frac{k \cdot T}{\Delta G_0} \cdot \ln\left(\frac{\dot{\varepsilon}}{\dot{\varepsilon}_0}\right)\right)^{m'}}_{\text{Strain rate hardening}} \tag{3.19}$$

In Eq. (3.19) the actual yield stress, σ_y, is given as a function of plastic strain ε_p and strain rate ε. While some of the parameters need to be fitted, all others are constants of nature (e.g., k is Boltzmann's constant) or nearly constant within the family of low carbon steels. The initial resistance σ_0, a saturation stress $\Delta\sigma_m$ to account for hardening and a dynamic stress σ_0^* taking care of the stress added by

SRS need to be found by fitting. Also to be fitted is the Ω parameter, which determines the curvature of hardening for lower strains, in other words how quickly the hardening saturates. Finally m' is a scaling factor for SRS that also needs to be determined.

Other parameters for low carbon steels can be assumed constant. The initial strain ε_0 in the material (for instance from temper rolling) and exponent n' do not vary appreciably. The β parameter determines the slope of hardening for higher strains (which would otherwise be 0) and is given. Finally, in the SRS term ΔG_0 an activation stress for dislocation movement and the reference strain rate (a maximum rate obtainable with dislocation movement) $\dot{\varepsilon}_0$ are also assumed constant.

For practical purposes even the two strain rate hardening parameters ($\sigma_0{}^*$ and m') can be seen to vary very little over the materials and hence a reasonably good description around the rate of deformation as found in forming can be obtained with fixed values for these parameters. The implication of nearly constant strain rate sensitivity parameters for all steels is that the higher the strength, the smaller the relative effect would be. As an order of magnitude it can be indicated here that increase of strain rate by a factor of 10 will add around 30 MPa to the stress [28]. The SRS will also influence the FLC as discussed later in this chapter.

If one is familiar with strain rate hardening models it will be noticed that this description has an additive term where most have a multiplicative term of the power law shape, see, e.g., Ref. [14]. Where the multiplicative models can only be fitted to a small strain rate range, one or two decades, the Krabiell and Dahl equation holds over the five decades of strain rates encountered from tensile tests to full vehicle crash.

A final step made in the modeling of hardening with a physical base to inspire more confidence in extrapolation was to add the composite hardening effect found in dual phase steels. It was again Vegter that made this step [29].

3.5.2 Yield locus

The yield locus in principle applies to the yield stress of the materials, whereas the shape of the yield locus after some forming has taken place is of greater interest. Therefore more advanced models are needed if the shape does not remain constant, when looked at closely [30]. Up until now a reasonable solution has been to determine the shape of the yield locus over a range of strains, not just the onset of plasticity (see Ref. [31] and everyday practice Tata Steel R&D using Ref. [19]). However, as demands on accuracy increase it will become necessary to take into account the changing shape of the yield locus with increasing plastic strain [7].

Measurements of hardening in the biaxial state are shown in Fig. 3.15, reproduced from Ref. [9]. All three tests for hardening in equibiaxial strain show the same results, after correction for different strain rates and temperature increase in the different tests. Though large differences between different equibiaxial tests have sometimes been reported in the literature (e.g., Fig. 4 in Ref. [25]), it is usually due

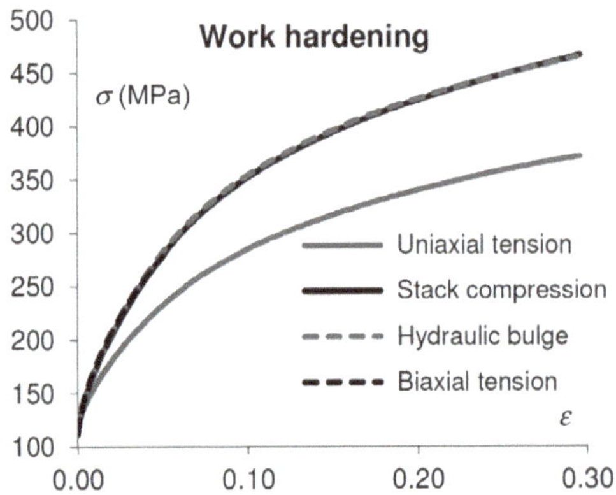

Figure 3.15 Hardening in biaxial state, multiple tests compared.
Source: P. Hora, P. Peters, N. Manopulo, M. Gorji, Challenges in the accurate modelling of sheet metal forming processes, in: 8th Forming Technology Forum 2015, Zurich, Switzerland, 29 & 30 June 2015, p. 5.

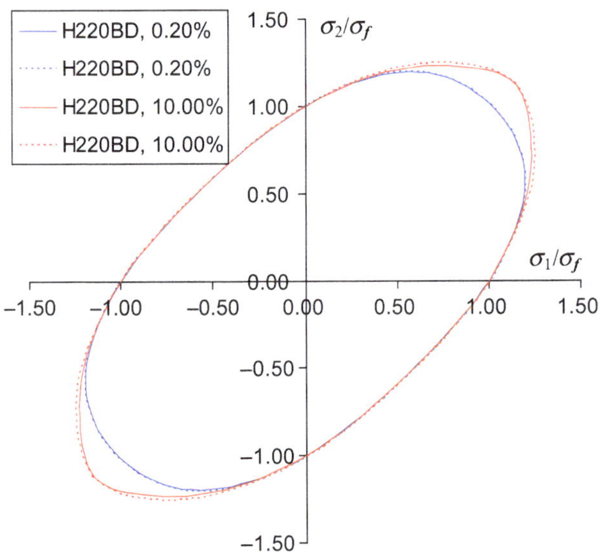

Figure 3.16 Initial yield locus versus that at 10% plastic strain (for bake hardening HSS, steel).
Source: J. Mulder, H. Vegter, Int. J. Mater. Form. 3 (2010) 191. http://dx.doi:10.1007/s12289-010-0739-7 [8].

to omission of these corrections. For example, the effect of temperature is not appreciated by many researchers. And whereas most tests can be described by (Eq. 3.19) some results in literature show a larger influence (Fig. 7 in Ref. [32]) indicating that (Eq. 3.19) still needs further work. The implication of a different hardening rate in the equibiaxial state is that the shape of the yield locus changes with evolving plastic strain. An example of this (from Ref. [9]) can be seen in Fig. 3.16.

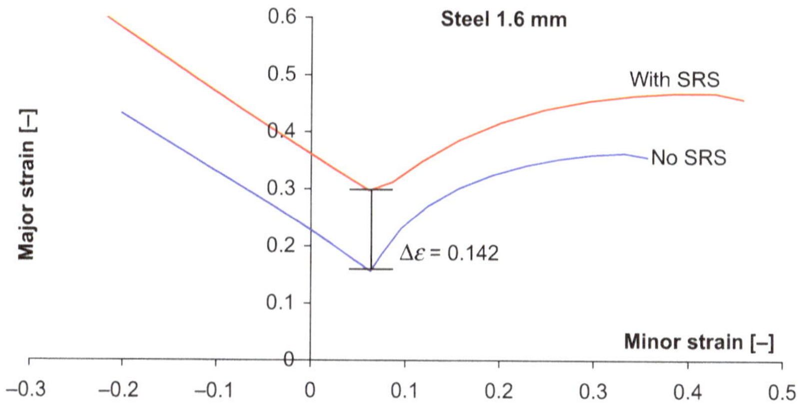

Figure 3.17 Influence of strain rate sensitivity on FLC (MK-type model).

3.5.3 Forming limit curves

The SRS is one reason why for the same shape of the hardening curve (i.e., for the same *n*-value) a steel will show higher FLC than an aluminum. This is illustrated by a model result for a steel switching SRS on and off, see Fig. 3.17. A more in-depth analysis can be found in [33] or in Ref. [34], treating the contribution of Vegter et al. to the benchmarks of Numisheet 2008.

After the initial introduction of the FLC by Keeler [17] and Goodwin [18], much research has been done on how to determine the FLC experimentally. In most of those the importance of proportional loading (or linear strain path) is stressed. This implicitly means that for nonlinear strain paths the concept does not work well. And indeed this is the case [35].

Sklad et al. [12] have shown that if after strain path change sufficient strain can be endured before failure occurs the change of strain path is not detrimental, but in general this limitation must be considered. Again Sklad, but also Stoughton [36] (and many others as well) use the stress based FLC which does not compare measured strains to a limit strain but rather a "measured" stress to a limit stress. We have put measured in quotes as the stress cannot easily be measured and usually the strains are measured and converted to stress. Although this appears to make the method strongly dependent on stress strain law this is not actually the case, as both "measured" stress and limit stress are obtained with the same law and hence errors in the constitutive law are not grave. In fact Stoughton has proposed an alternative approach in terms of strain, with the same benefits as the stress based approach exploiting the equivalent strain principle [36]. Other approaches are more phenomenological in nature, trying to combine a series of FLCs after different prestrains [37,38].

As the measurements are cumbersome and time consuming, Tata Steel researchers have done an analysis on FLC test data from the past four decades. They came up with an empirical set of relationships [11] with which the FLC can be predicted quite well, based solely on thickness and tensile data (total elongation according to Euronorm A_{80} and *r*-value).

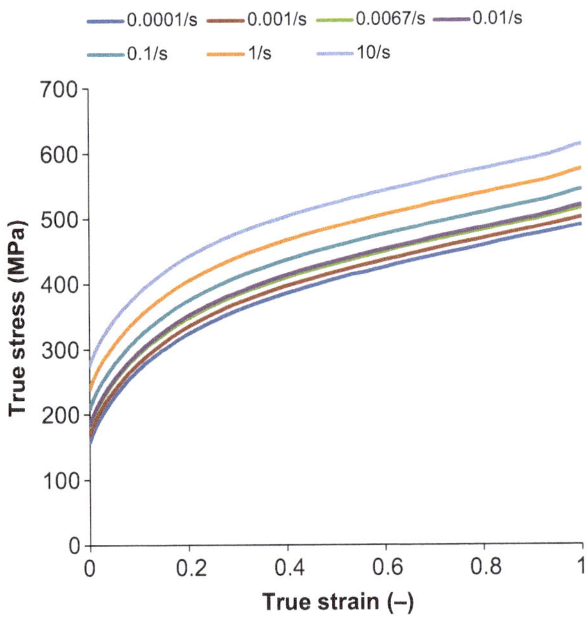

Figure 3.18 Isothermal hardening, different strain rates.

3.5.4 Temperature changes

Even in cold forming the temperature is not constant; it does not remain at ambient temperature when 15 parts are stamped every minute. The temperature can reach 100 C and while this may still seem cold it is enough to be noted in the hardening behavior, especially when (like in tensile testing) the steel heats up during the test. When the temperature effect overlays the isothermal hardening it is easy to draw the wrong conclusions. Compare Fig. 3.18 to Fig. 3.19 to get a feel for the difference. It may be surprising but the tensile test is nearly adiabatic, the heat simply cannot be conducted fast enough even though the strain rate is low (order of magnitude $7 \times 10^{-3}\ s^{-1}$).

3.5.5 Hardening model

Apart from the idea of evolving yield loci we have looked at hardening and yield locus separately so far. But they have to be connected somehow. If the yield locus models the onset of yielding than what does it look like after some strain, unloading and reloading in a tensile test? The uniaxial point must then be shifted to the new stress, but how about the rest of the yield locus? The simplest model is the isotropic hardening model: this lets the whole yield locus grow with the hardening curve. Schematically, shown schematically in Fig. 3.20.

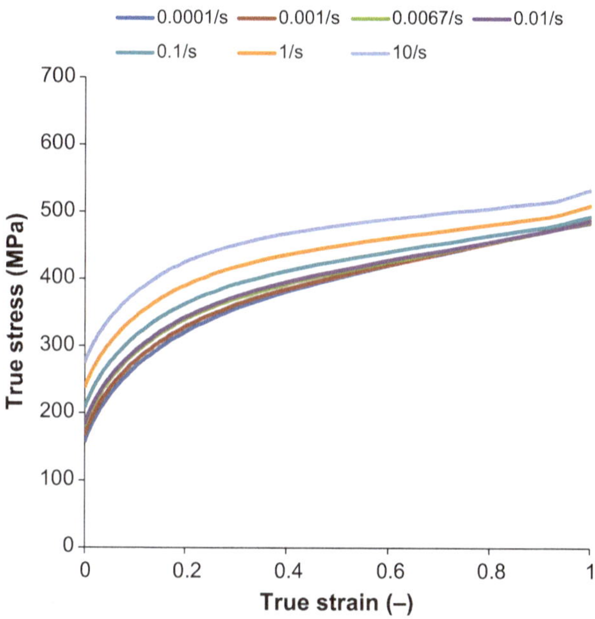

Figure 3.19 Adiabatic hardening, different strain rates.

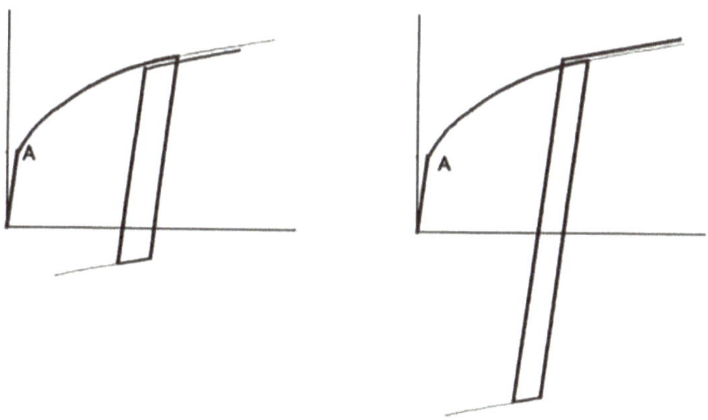

Figure 3.20 Kinematic (left) vs isotropic (right) hardening model.
Source: Reproduced from P.T. Vreede, A Finite Element Method for Simulations of 3 Dimensional Simulations of Sheet Metal Forming, PhD thesis, University of Twente, Enschede, the Netherlands, 1992, ISBN 90-9004754-9 [39].

An interesting aspect of material testing is the Bauschinger effect. If after plastic loading the load is reversed the onset of plasticity does not take place at the same (equivalent) stress (like it does in the isotropic hardening model, Fig. 3.20, right) but mostly at a lower stress.

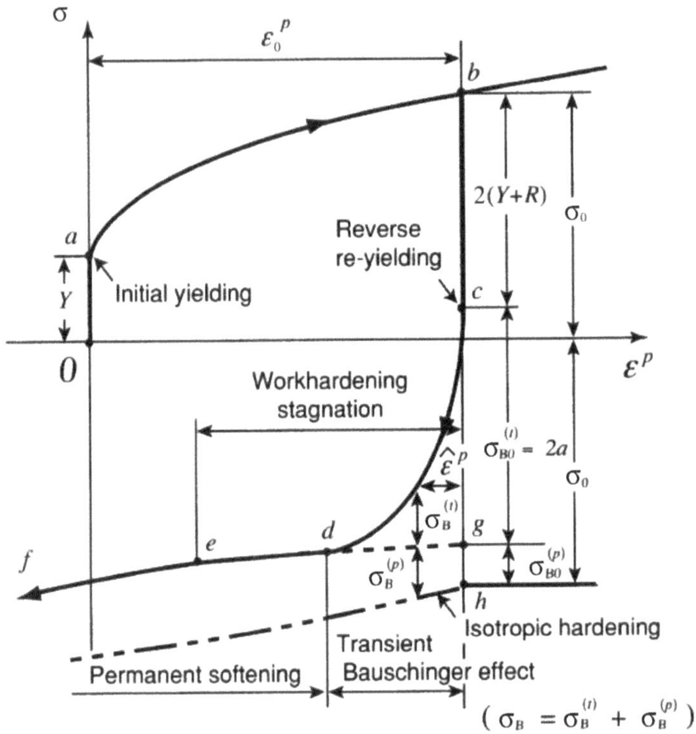

Figure 3.21 Advanced modeling of Bauschinger effect and more.

A hardening model capable of capturing this is the kinematic hardening model, see Fig. 3.20, left. This shifts the yield surface (so the size remains the same) to account for hardening in a tensile test, and consequently the yield stress in the opposite loading direction goes down rather than up. This represents another extreme also not encountered in practice, mixed models are usually needed to accurately describe the material behavior. The model of Ref. [40] is particularly versatile and seeing ever more use. Determining the parameters for such models is a nontrivial task though. Fig. 3.21 is reproduced from Ref. [40] but will not be explained in detail here. The purpose here is solely to indicate that such models are available in literature.

3.6 Forming processes

In this section a very short and introduction to forming processes will be given. The intention is to highlight which parts of the concepts discussed earlier in this chapter apply to each process.

Most formability analyses are concerned with stamping and thus the FLC and in more general terms plastic instability are of prime importance. There are many forming processes however, that do not trigger the plastic instability and hence much higher strains can be achieved than would be thought possible when looking at the FLC. Bending processes are a case in point.

In hydro forming of tubes for example, end feeding is employed to establish such deformation in the tube that the thickness remains approximately constant. Obviously this does causes catastrophic thinning (i.e., local necking) to be avoided and hence formability to be increased. Another way of looking at this is to plot the strain path in an FLC and realize that the strain and the FLC are parallel and thus will never cross.

In practice of course there are other limits to the deformation of the material, which will then become apparent. Because these are much more important in AHSS steels they will be treated there.

3.6.1 Cutting

As strip steel is coiled after production the first thing to be done with it is decoil and cut into blanks. Generally shaped blanks or developed blanks are employed for complex deep drawn parts. By cutting the blank to a shape more or less resembling the outline of the final product the forming severity can be reduced (Fig. 3.22).

An important aspect of the process is the cutting clearance, Fig. 3.23, which is generally chosen between 10–15% of the sheet thickness. For HSS and especially AHSS this is critical to the quality of the cut edge. While for steels 12% is traditionally viewed as the maximum sensible clearance, for AHSS 15% seems to offer better performance [42]. Even more important than the clearance is the condition of the tools: sharp cutting knives will produces better quality edges. In Ref. [43], the influence of worn tooling on HEC is clearly demonstrated.

After cutting, the blanks are stretched or drawn. These are very similar processes geometrically and differ only by boundary conditions. We will treat stretching first, followed by drawing.

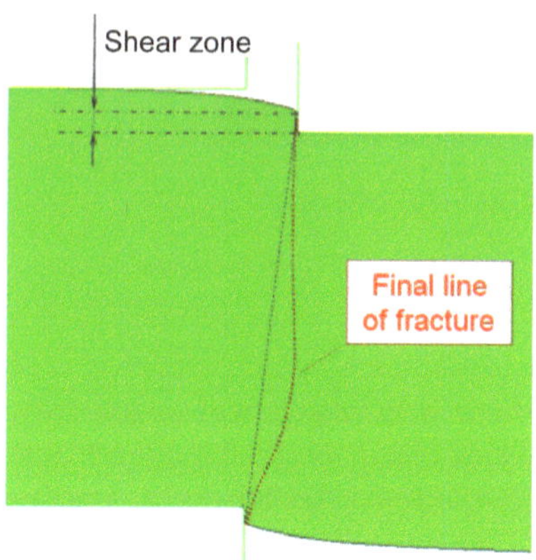

Figure 3.22 Shearing process in detail [41].

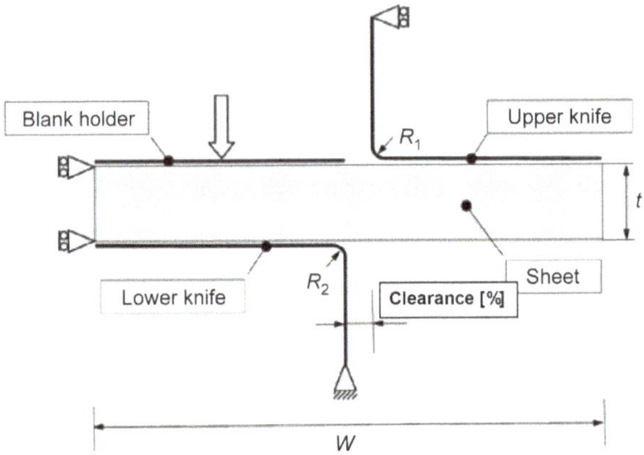

Figure 3.23 Cutting clearance [41].

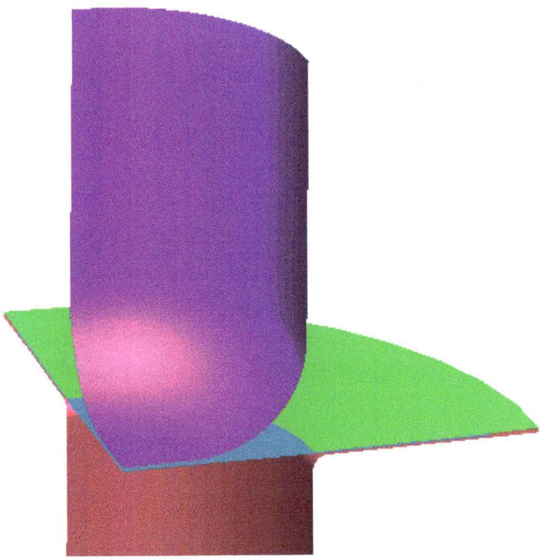

Figure 3.24 Stretching process schematically.

3.6.2 Stretching

Stretching refers to processes whereby all deformation is obtained at the expense of thickness. In a simple shape a hemispherical punch (purple) can be pushed into a metal sheet that is fully clamped, Fig. 3.24, by a blank holder (green) onto the dye (red). Typically under the blank holder no deformation takes place and initial thickness is maintained, blue zone, Fig. 3.25. But over the punch severe thinning will occur. The location of the thinnest spot depends not only on the material, but quite strongly on the friction. However, material properties do interact with the

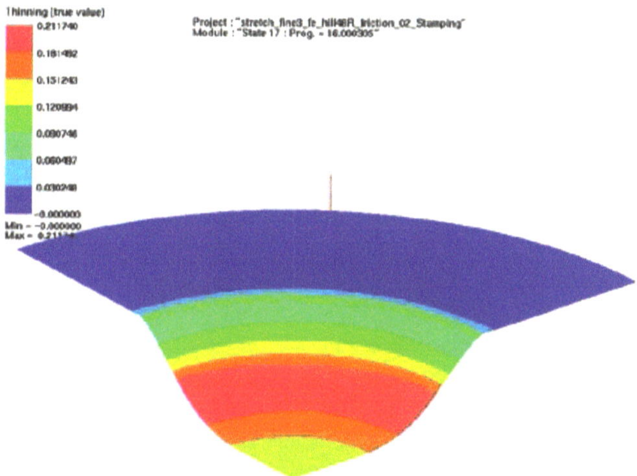

Figure 3.25 Stretching process deformations.

friction and therefore the yield locus needs to be modeled carefully to ensure accurate predictions. This was illustrated in an earlier section, Fig. 3.14, and a useful metric is the STR (Eq. 3.18).

That said, the most important parameter is not the yield locus but the hardening. The higher the hardening the better the imposed thinning can be withstood. So when looking for stretchable material one should look for n-value first, and then STR. Strain rate sensitivity once again stabilizes the material and therefore should be taken into account when simulating the stretching process. Usually due to tool contact the process is expected to stay isothermal and therefore the effect of temperature can generally be neglected. But, like in most forming processes, the strains are higher than the uniform strain in tensile testing, and an advanced hardening law like (Eq. 3.19) is beneficial for ultimate accuracy in simulations.

A picture of a test specimen (painted with black on white speckle pattern for measurement purposes) can be seen in Fig. 3.26, note that the flange stays undeformed!

3.6.3 Deep drawing

In deep drawing, essentially the same tools are used but now the flange is allowed to flow in, or draw-in, and is transformed to a cylindrical wall. This way much deeper products can be made. This is evident from Fig. 3.27 where three experimental cups can be seen, ranging from low blank holder force (left, drawn) to high blank holder force (right, stretched).

The process does have a lower limit on blank holder force though, because otherwise wrinkles will be formed, some examples of which are shown in Fig. 3.28. The formation of wrinkles is directly related to blank holder force. The final

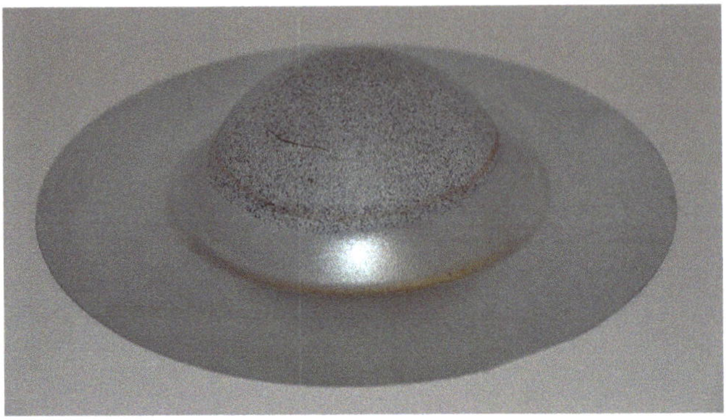

Figure 3.26 Experimental example of stretching [44].

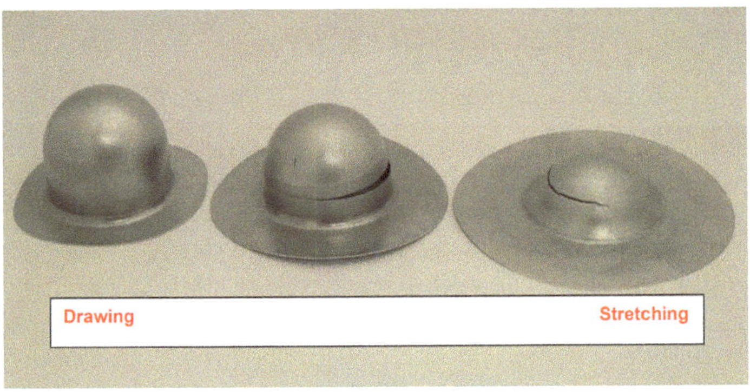

Figure 3.27 Varying blank holder force to change from drawing to stretching.

Figure 3.28 Series of experiments with varying blank holder force showing wrinkling and fracture trade-off [45].

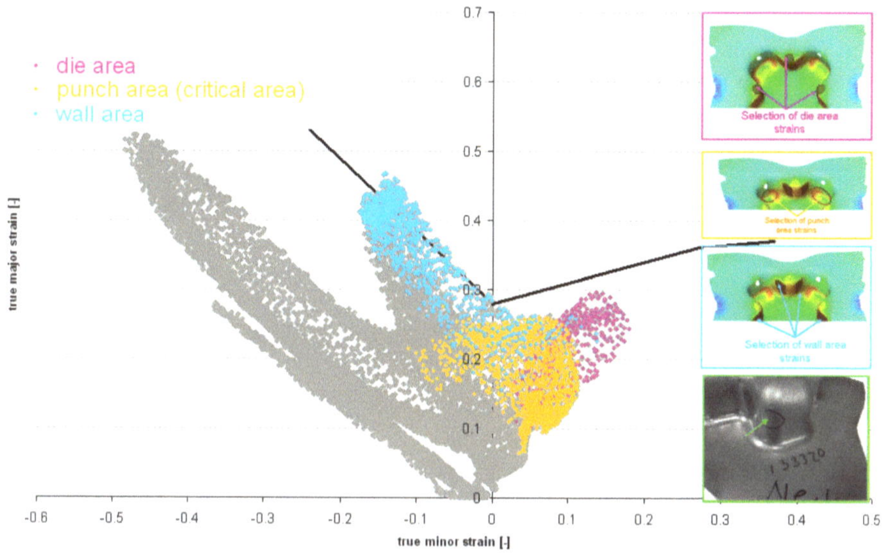

Figure 3.29 Comparison of stretched and drawn areas in Cross-Die test geometry. *Source*: Figure reproduced from M. Roelofsen, E.H. Atzema, A. Schouws, G. Botman, Ranking Formability of Materials by Means of the Cross Die – value, Internal Tata Steel R&D Report No. 155759, 2012 [46].

fracture is caused by the restraining force generated by blank holder through friction with the sheet. This implies that low friction is beneficial for drawability.

As most of the deformation is in the flange and this is useful deformation, i.e., the stress should be as low as possible, high hardening is not actually beneficial for stamping. Some hardening is needed, as early in the process some stretching occurs before the sheets starts "sliding" into the die opening. But high hardening (in the simplest view: high n-values) will be bad for deep drawability. The influence of hardening is not as strong as the yield locus so one should look for high r-value, or better still: good DDR first and limited hardening second.

In a typical stamping the complex geometry causes some parts to be stretched and other parts to be drawn. In areas that are (in mathematical terms) not developable, stretching will have to occur to enable the desired shape. Other areas may shift more to drawing depending on blank holder force. In Fig. 3.29 the yellow and magenta points indicate stretched parts of the geometry. The blue parts are deformed near plane strain mode. And finally, the gray points are the remainder but mainly represent the flange, which shows typical drawing behavior.

3.6.4 Bending

Many household products made of sheet metal are produced by bending. But in automotive products often a flange or a tab is bent as well, and this may cause formability problems.

Bending is mainly characterized by the gradient of strain over the thickness. With a Kirchhoff assumption a simple estimate of the strains in bending can be made. The Kirchhoff assumption states that flat planes perpendicular to the mid-plane of the sheet remain flat and perpendicular during forming. Under this assumption a sheet folded onto itself has an outer fiber strain of 100% (or 0.69 in true strain). For even the most formable steels this is above the FLC. However, the FLC is not applicable here. It is also more than the total elongation, but it is not a tensile test so this does not apply either (Figs. 3.30–3.33).

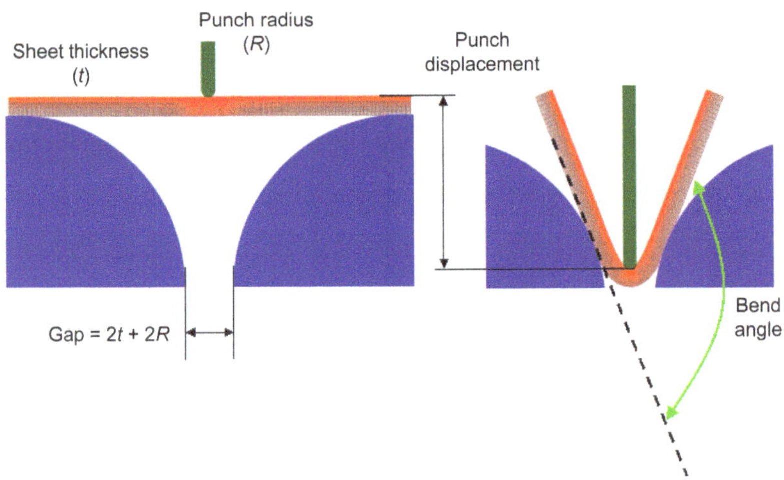

Figure 3.30 VDA 238 schematic [4].

Figure 3.31 Photograph of VDA238 test equipment [4].

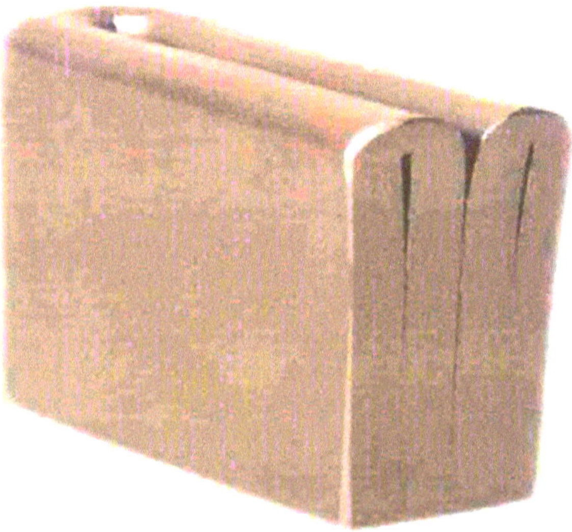

Figure 3.32 Doubly folded HSLA grade steel, taken from product brochure of Tata Steel Europe.

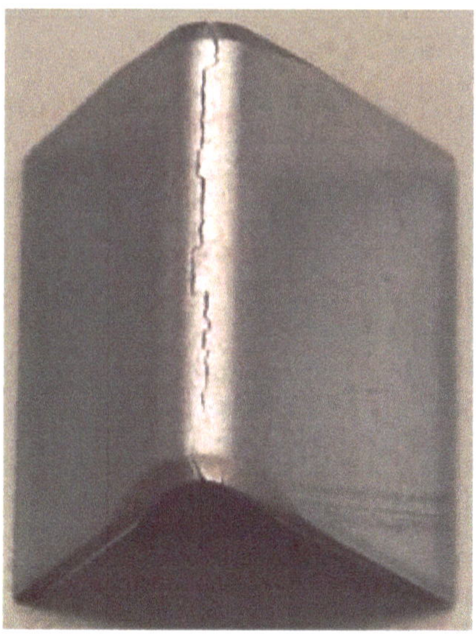

Figure 3.33 Picture of a DP steel test sample of the VDA 238 bend test [4].

Most steels can be folded flat without a problem, but the higher the strength the larger the risk of problems. With the advent of AHSS the need for a test arose and the German automotive industry developed a standard: the VDA 238-100. This test has become more or less the de facto standard for the entire European automotive industry. A very sharp bending knife with a radius 0.3 mm is used. The discussion in this section follows [4].

During the test a force displacement diagram can be recorded and a typical example is shown in Fig. 3.34. The VDA standard states that the point of load drop should be identified. However, by geometrical effects alone (as can be seen in the red simulated line) there is a load maximum, even without any fracture occurring. Therefore the load maximum is not an accurate indicator of damage occurring. For DP600 the bendability is still excellent and then this definition issue is encountered, and the test result then is essentially a test limit not a material limit. For higher strength DPs however, the bendability is less and the test protocol is applicable.

In Fig. 3.34 a higher strength DP is used and the point of fracture can clearly be seen, as that is where the experimental curves start to deviate from the FE simulations. It can also be seen that the point of the maximum load now is a good criterion as most of the experimental curves show a force maximum almost as soon as they start to deviate from the FE results.

3.6.5 Flanging

Flanging a concave edge (right in Fig. 3.35) will result in the edge of the blank to elongate. If this is a cut edge (as they usually are) this is then sensitive to fracture.

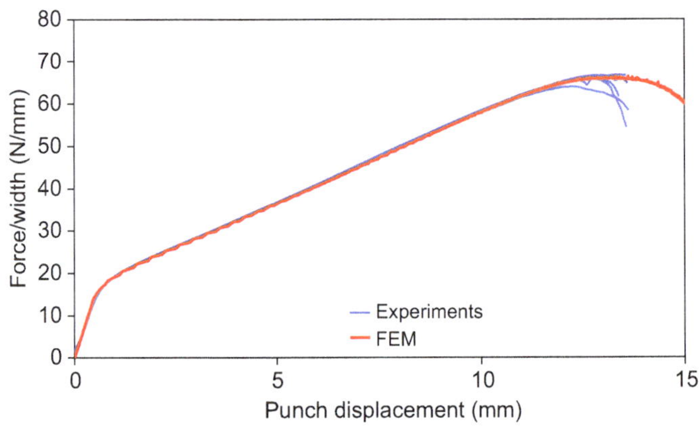

Figure 3.34 Force-displacement diagram of VDA 238 bend test [4].

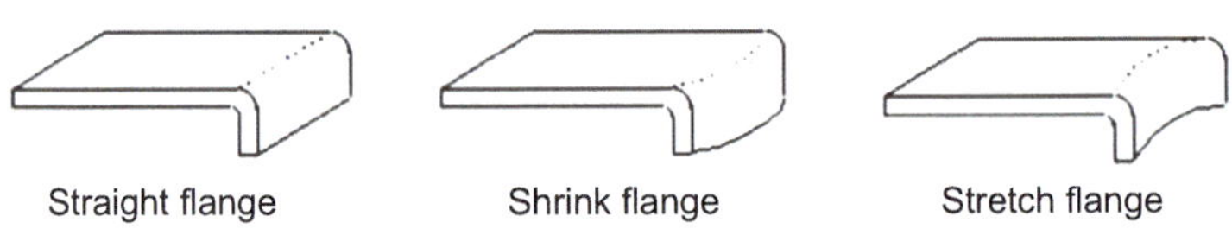

Figure 3.35 Different flanging types.
Source: Picture taken from VM113, Vormgeven van dunne plaat—Buigen; FME; In Dutch.

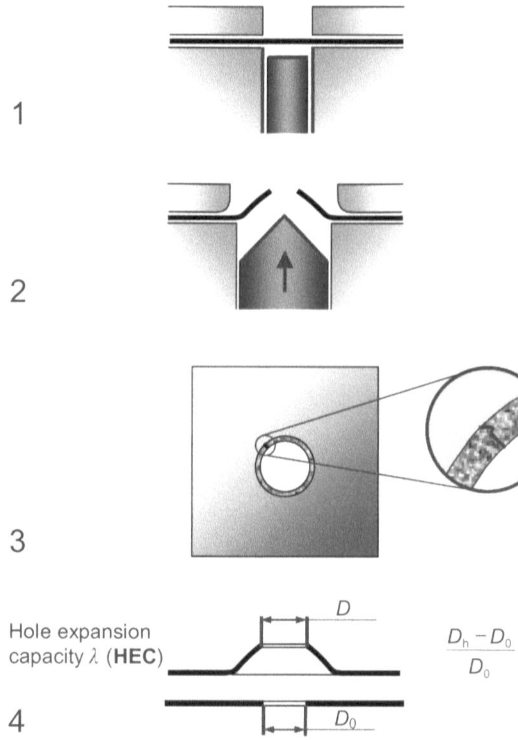

Figure 3.36 HEC test after ISO 16630.
Source: From M. Braun, Determination of the edge cracking sensitivity-estimation of the reduced formability of sheet steel edges produced by shearing using different test methods, in: Forming in Car Body Engineering, 25/26.09.2013, Bad Nauheim, 2013.

As with most aspects this is not an issue for low strength, but problems increase with increasing strength. In this section we will look at stretch flanges only.

A straight flange is just a bending process, and a shrink flange can present wrinkling problems if the material is too thin or the flange too high.

For stretch flangeability most often the Hole Expansion Capacity is measured according to ISO 16630. The test is schematically shown in Fig. 3.36 taken from Ref. [3]. The initial hole diameter is standardized at 10 mm. The standard allows for material thickness up to 6 mm and there the hole diameter to thickness ratio becomes awkward. For thinner gauges, however, say up to 2.5 mm or so, the results are useful.

Typical examples of material performance in HEC can be seen in Fig. 3.37. Although there is continuous ongoing research at all large steel makers to improve the performance, this ranking remains fairly typical.

The hole expansion coefficient obviously depends on the quality of the punched hole. A drilled and reamed hole, or spark eroded (EDM) would show much

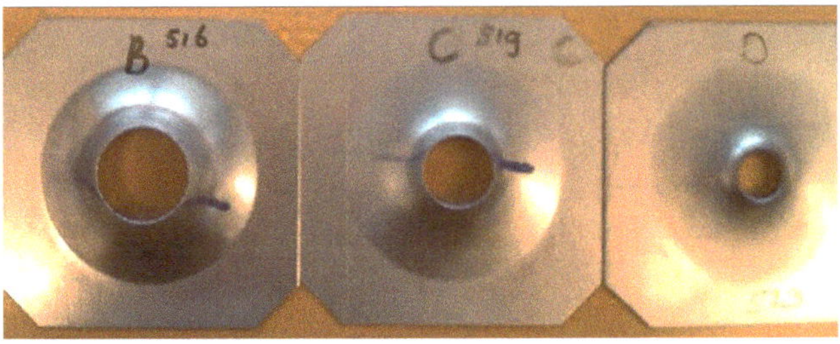

Figure 3.37 HEC examples, left mild steel, middle HSLA steel, right DP steel [4].

higher HEC. These methods however are prohibitively expensive and automotive OEMs would rather change material than invest in such processes. The key issue is to have a cost effective separation method (which at the time of writing means cutting) that in combination with the resistance of the material to fracture produces good HEC. The steel manufacturer SSAB have recently examined the effect of double cutting operations where at the first stage a 9 mm hole is punched and the remaining 1 mm is cut in a second stage (so-called "shaving") [47]. This results in an improvement in HEC for all the materials tested, but most interestingly, material with low HEC also gained the least by this punching method.

The HEC is not well understood theoretically. It can be stated that the edge of the hole is in a uniaxial stress state: no stress perpendicular to the edge exists; there is only circumferential stress. If an edge does not fail in a fracture mode, but develops a necking instability, this could be theoretically calculated, like in a tensile test. Accounting for normal anisotropy (r-value) we arrive at:

$$\mathrm{HEC} = 100\% \cdot e^{n(1+r)} \tag{3.20}$$

Knowing that for AHSS the r-value is close to 1.0, the maximum HEC would correlate with $2n$, see Eqs. (3.13 and 3.16), if fracture occurs this would be expected to take place before necking, or in the neck the maximum HEC will not show a large difference in the two cases. In practice however, CP steels exhibit lower n-values than DP steels (and consequently suffer in stretchability) but have HEC values much higher than theoretically possible. The theory must be flawed, or at least incomplete, and a first attempt to explain comes from Krassovsky and Larour [48]. No doubt in the coming years more light will be shed on this. For this section it must be concluded there is currently no aspect from the basic concepts section that can be applied to HEC.

The HEC measures the circumference of the hole after fracture. Somewhere around the circumference a fracture (either by itself or in a necked zone) will form and even when locally the strains are very high the HEC will not change much, because it is a global measure. There are tests in the literature that attempt to measure the local strain just before fracture, such as KRE and SETi as proposed by the present author [49]. Although it is unlikely that a material doing extremely well

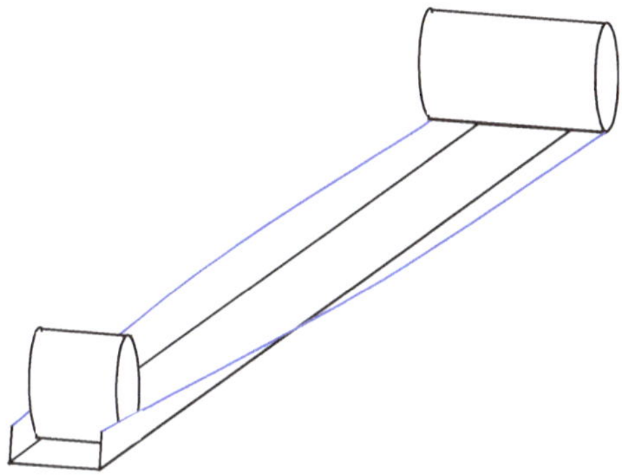

Figure 3.38 Length difference middle (black) to edge (blue) in roll forming.

in one of these tests would show poor HEC, the correlation is not extremely strong and therefore these alternative tests serve a purpose. For the purpose of this chapter the tests will not be detailed; the reader instead is referred to [49].

Besides the tests mentioned above, more tests are available, probably indicative of the struggle researchers have with the subject. In Ref. [3] an incomplete survey of the literature was carried out, and many more tests are described in the literature. There is no need to review them all or even try and understand them, but here we conclude that there is so much ambiguity about the influence of test conditions that researchers feel the need to develop new tests.

3.6.6 Roll forming

For long prismatic profiles a process putting little demand on the stretchability of the material is used, and it is known as roll forming. The process does require bendability, especially when sharp radii are required. However, the final shape is obtained by so many subsequent forming steps that plastic instability does not play a role (Fig. 3.38).

There are other limits to the process, however. The change of shape from one roll set to the next often requires the edges to travel a greater length than the middle. If a steel shows significant hardening, the stresses at the edge will be significantly higher than in the middle. This causes wavy edges. In general terms it can be said for shape tolerances in roll forming the hardening needs to be low, which is quite contrary to the requirement of stretching operations.

Basically only a small stress difference can be allowed and this is an interaction between material hardening and how much deformation is induced in each roll set. In Fig. 3.39 on the left the deformation was applied in 3 roll sets, which admittedly is a very crude process and consequently wavy edges are evident. These can be cured when more steps are introduced as is clear from Fig. 3.39 (right) where the same material was used and roll formed in 6 steps [50].

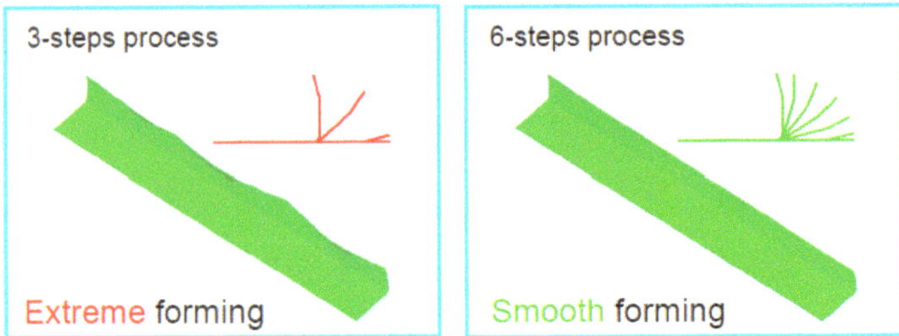

Figure 3.39 Roll forming; influence of number of forming steps [50].

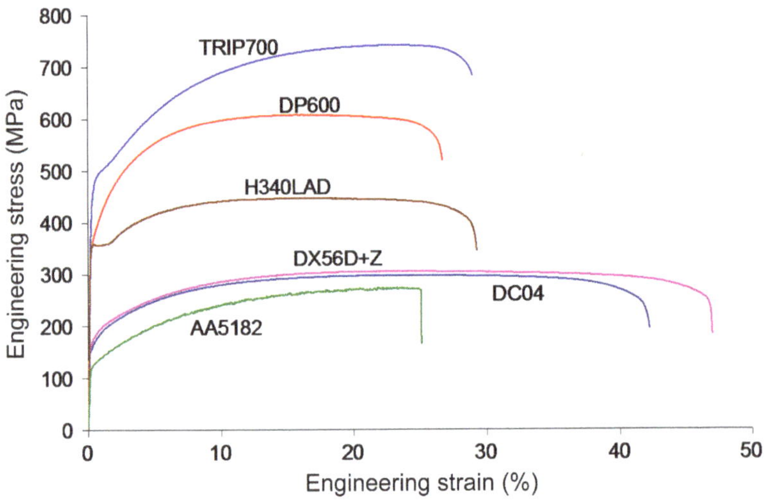

Figure 3.40 Engineering stress-strain curves for a range of materials [13].

3.7 Formability aspects of different steels

As stated earlier, the hardening curve of a material is of paramount importance. From Fig. 3.40 it will be clear that the HSS example does not show much hardening, and thus can be expected to perform less well in stretching. At still higher strengths the DP and especially the TRIP are better in this respect.

Another thing that can be seen is the influence of SRS. The example of aluminum shown here on Fig. 3.40 does have a fair hardening behavior, but absence of strain rate sensitivity causes it to have little post uniform elongation. This is probably not a totally representative example, but still this general observation holds.

As stated before, mild steels, HSS, and even AHSS, predominantly fail in necking mode. This means the FLC is the next aspect that is examined by forming analysts when judging a material on general formability. In Fig. 3.41 an overview

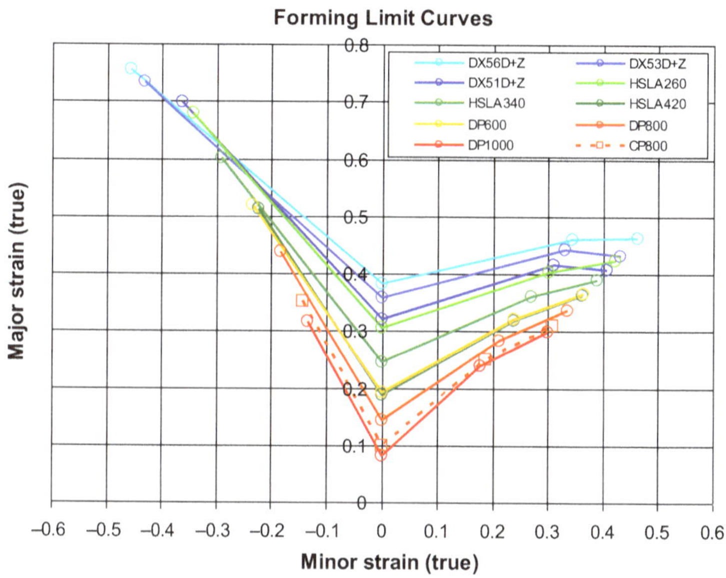

Figure 3.41 Overview of forming limit curves for different steel families.

is seen of the typical FLCs for the different families treated in this chapter. The mild steels selected for this graph (coded DX5_ after norm EN 10142, colored blue) are placed in order of ascending strength (like the others) but the EN coding is done to indicate increasing formability.

All HSS examples (green) shown in Fig. 3.41 are HSLA and it can clearly be seen that with increasing strength the FLC decreases. The same trend is seen within the DP family, but there is a shift relative to HSLA as the DP600 is on the same level as HSLA420 (guaranteed tensile strength 590 vs 470 MPa). Finally, the CP800 is roughly equal to the DP1000, indicating that at the same strength level stretchability is lower for CP when compared to DP.

3.7.1 Mild steels and IF

Despite the ever-increasing use of HSS and AHSS in the car body, a significant amount of mild steels are still being used. This is because the outer panels determine the appearance of the car. So these need to be quite formable to enable the shapes that the car designer intended. Sharp feature lines, for instance, give the car the character that designers and marketers are looking for. And to enable these lines as well as to make them as distinct as possible, very formable steel is needed. For the outer panels these are either low carbon steel grades or (more often) interstitial free steels [1].

These formable steels are very ductile and when too much strain is induced they will show necking first and fracture thereafter. Necking has been explained

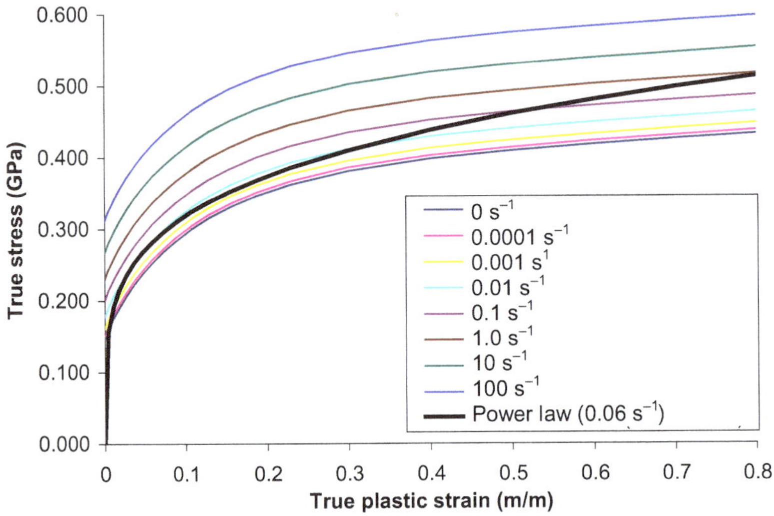

Figure 3.42 Hardening curves at different strain rates for interstitial free steel.

in the Basic Concepts section. The typical hardening behavior of IF steels is shown in Fig. 3.42. It has been shown for different strain rates to illustrate how this effect is relatively large in these steels and this contributes to their general formability.

The IF grades also show strong anisotropy enabling large drawing ratios to be obtained. However, as can be seen in Fig. 3.43, the Hill model (in red) would overestimate the drawability. The green points are measured points for the blue curve (Vegter model) and one can see the Vegter model fits well. So as deep drawable as these materials are, for modeling one should not rely on too simple models as they will still overestimate drawability. The magenta curve in Fig. 3.43 represents the Vegter model for another coil of the same material.

Despite the shortcomings, a constant (i.e., nonevolving) yield surface is capable of simulating sheet metal forming processes to a great accuracy, as shown by the work of Vegter, Sigvant, etc. This, however, is only possible when the yield locus used is sufficiently flexible. Some of the Barlat family of models [51], the BBC2005 [24] and certainly the Vegter [19] criterion fulfill this condition. While the flexibility of the Barlat models may well be sufficient, total freedom of fitting only exists in the Vegter criterion and hence is preferred by the present author.

In the local necking analysis it was noted that the local necking takes place in plane strain mode. It was also established that the local necking phenomenon determines the formability of sheet metal; it can thus be concluded that the plane strain point on the yield locus is ultimately the most important point. The Vegter criterion is the only one that directly incorporates the plane strain point and thus achieves ultimate accuracy as to the limits of forming.

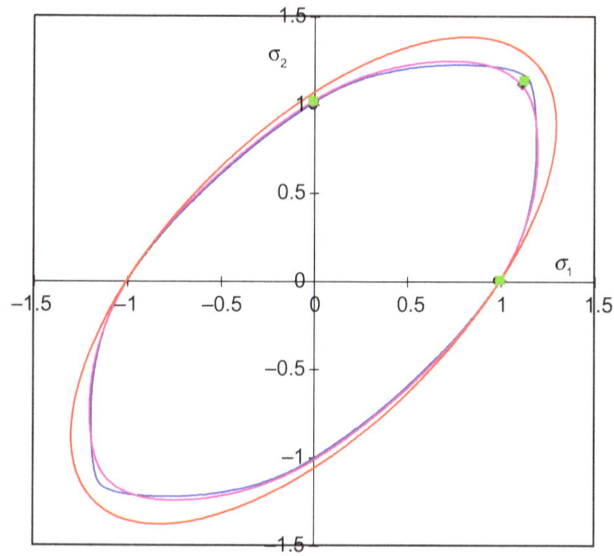

Figure 3.43 Yield loci for two interstitial free steels (magenta and red + blue), two models (Hill red, Vegter blue + Magenta).

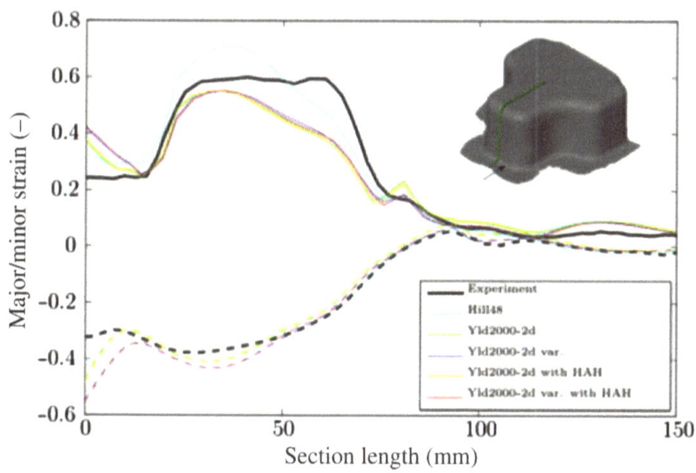

Figure 3.44 Experimental versus simulation results FTF2015 benchmark [30].

The experimental results from a recent benchmark exercise [30] presented in Fig. 3.44 show constant strain (black lines in Fig. 3.2) over the short wall section of the Cross-Die product (from 25 to 70 mm section length) which none of the examined models can reproduce. More detail is provided in [52], where the basic assumption of associated flow, employed in all commonly used models, is questioned. Indeed it is shown that for two different Nakajima tests that otherwise could only be fitted with different material parameters for each test; the proposed approach (i.e., nonassociated) has an advantage. However, it remains to be seen if this is the way forward.

A conclusion that can be drawn, however, is that the evolution of the yield locus with increasing strain (and depending on the strain ratio as well) is the next step to be taken for further improvement in the accuracy of the models. This step may be taken via texture evolution modeling (KUL) or phenomenological. Much research is being carried out at the moment with no final answer yet [7,9].

Flanging and bending is not an issue with these materials. If they fail they do so in a ductile manner, which is entirely predictable from the plasticity modeling and the FLC [13].

3.7.2 High strength steels

Conventional (i.e., single phase) high strength steels mostly consist of HSLA (also known as microalloyed steels) grades. At lower strength levels some other options exist but for the purpose of formability these do not differ much in character. Fig. 3.45 shows a clear Yield Point Elongation (YPE) as well as moderate hardening behavior after that.

For the forming steels we have seen the importance of strain hardening exponent (or n-value) in understanding the necking behavior. It was also stated there that the power law often provides a good fit. Although this is true, it is nevertheless not perfect and hence the determination of the n-value from tensile test results depends on the strain range over which a fit is examined. In the established tensile test norm it has been decided to fit the stress-strain curve from 10% to A_g or 20% (whichever is the smallest) and over at least 2% of strain. This means for any material with $A_g < 12\%$ no fit can be made according to this

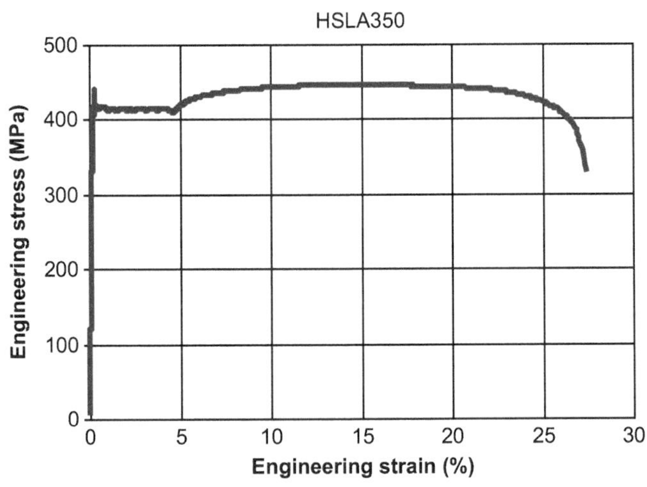

Figure 3.45 Engineering stress-strain curve for HSLA steel [13].

standard and consequently no strain hardening exponent can be reported. This is obviously not an issue for the forming steels, but it can happen for the higher strength HSLAs, for instance.

To nevertheless have an estimate for strain hardening exponent, we take the observation that diffuse necking starts at $\varepsilon = n$ (true strain) and by definition at A_g (engineering strain) an estimate of n-value can be derived from A_g:

$$n = \ln\left(1 + \frac{A_g}{100\%}\right) \quad (3.21)$$

Due to subtleties in determining the n-value as well as in determining A_g in the standards this equation does not hold exactly, but it provides a decent estimate of the n-value.

The lack of hardening (this material can be compared to a DP600 as described in the next section; they have similar yield strength) causes the stretchability to be limited. The overall trend observed within the HSLA family indicates that an increase in strength means a decrease in hardening and HEC. When HSLA350 would have a HEC of 100%; HSLA460 would be around 60%.

The drawability of HSLAs in general is slightly less than would be expected on r-value alone, as was the case for IF, see Fig. 3.46. The normal anisotropy for these steels is around 1.0, with some planar anisotropy evident.

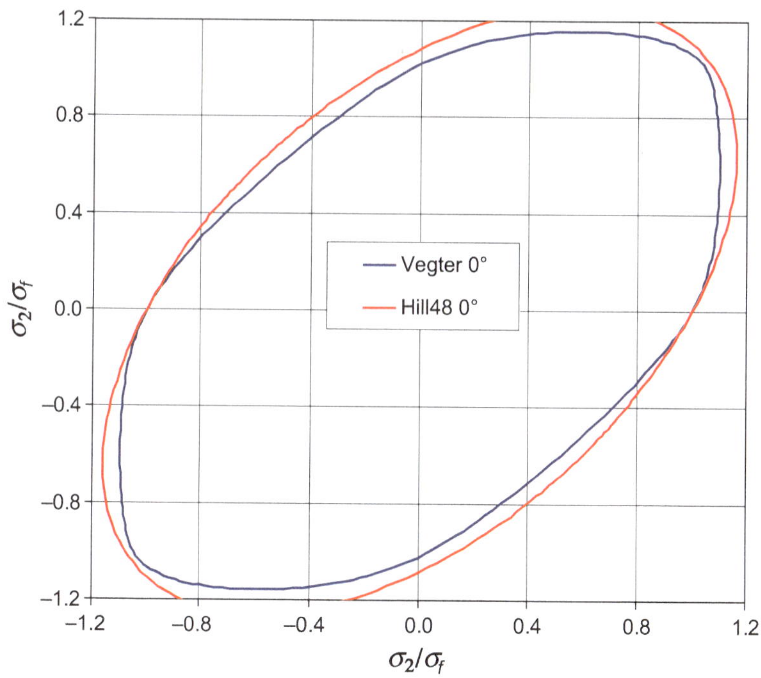

Figure 3.46 Typical yield locus for HSLA.

In terms of bendability, HSS generally pose no problems: they can be folded flat without cracking issues. However, springback does manifests itself in bending. The higher strength is used to enable thinner gauge steels in a car body structure. But higher strength in itself is a cause for higher springback after forming, and so is the decrease in thickness.

3.7.3 Advanced high strength steels

The modern steels commonly referred to as AHSS use a multiphase microstructure to obtain better properties: higher strength and still reasonable formability. Formability, however, has many aspects and when one is improved often others suffer. Speaking in general terms about formability can thus be confusing.

Looking at the tensile curves of DP steels, a good work hardening behavior is evident, Fig. 3.47. From previous sections it is clear that stretch forming of DP steels is excellent, at least at this strength level. For DP steels this comes at the expense of stretch flangeability and bendability. In examining the hardening of CP type of steels of similar strength, Fig. 3.48, it becomes clear that these will perform less well in stretching. The same issue as with HSS arises for CP: the hardening can be so low that the tensile test standard will not give data.

Although the *r*-value is typically low for these AHSS steels the drawability is a little bit better than expected on the *r*-value alone. This is seen in Fig. 3.49 where the Hill model clearly underestimates the drawing potential. So, when judging AHSS for drawability, look beyond *r*-value alone.

Apart from some nitpicking on the test results it is becoming increasingly clear that OEMs will start to specify a minimum bend angle (usually according to VDA238) for material they will want to homologate for their cars. The flangeability is also getting more attention with the chance that in the near future minimum HEC levels will be added to the homologation requirements.

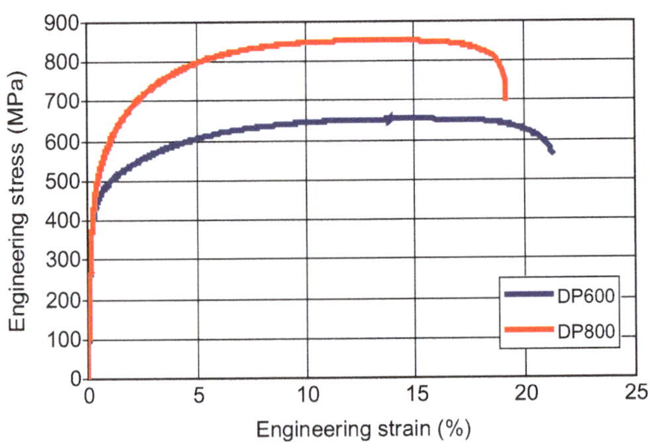

Figure 3.47 Good work hardening behavior of dual phase steels.

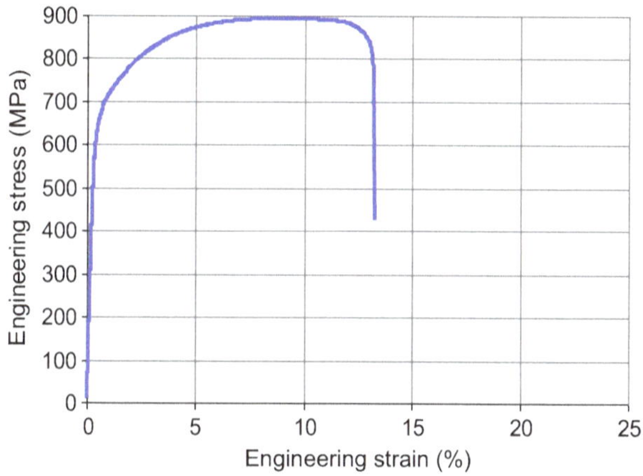

Figure 3.48 Poor work hardening for complex phase steels.

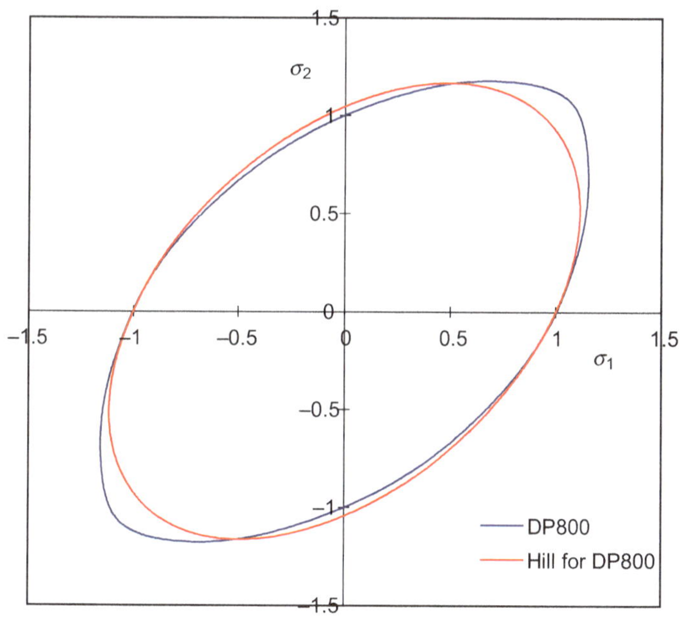

Figure 3.49 Yield locus for DP800.

Higher strength is used to enable thinner gauge steels in a car body structure. Where higher strength in itself is a cause for higher springback after forming, so is the decrease in thickness. Where accurate description of springback is merely desirable for HSS, it is crucial for the process design by simulation of AHSS forming processes. Among many hardening models available for describing the measured effects, the Yoshida-Uemori model can be highlighted. The model is quite versatile and able to capture the Bauschinger effect which is much stronger in DP steels than it is in single phase steels [53].

3.8 Conclusions

It has been shown that many well-known principles of forming still apply to modern steels such as the HSS and even the AHSS. For mild steels more advanced models can mimic the forming behavior ever better. Advanced yield loci, strain rate and temperature sensitivity, and ultimately, developing yield loci, will provide more accuracy.

For HSS, other aspects start to play a role, such as flangeability measured by HEC. For AHSS, limited bendabilty is also a concern. Finally, a brief overview of forming processes has been treated with the link made to the principles introduced in the earlier sections.

Acknowledgments

The author of this chapter is indebted to many colleagues at Tata Steel R&D but particularly those in the Stamping Technology knowledge group. Much of his knowledge was obtained in discussions with the now retired Dr. Henk Vegter.

References

[1] EuroCarBody, Yearly conference organised by Automotive Circle International, in Bad Nauheim, Germany; October. http://www.automotive-circle.com/Conferences/EuroCarBody-2016.

[2] Aachener Karosserietage, Yearly event organised by RWTH Aachen IKA; Aachen, Germany; September; http://www.ika.rwth-aachen.de/de/institut/veranstaltungen/201-upcoming-events/2003-aachener-karosserietage-2016.html.

[3] M. Braun, Determination of the edge cracking sensitivity-estimation of the reduced formability of sheet steel edges produced by shearing using different test methods, in: Forming in Car Body Engineering, 25/26.09.2013, Bad Nauheim, 2013.

[4] E.H. Atzema, Different aspects of formability in AHSS, in: Forming in Car Body Engineering, 26–27 September 2012, Bad Nauheim, 2012.

[5] A. Krabiell, W. Dahl, Zum Einfluss von Temperatur und Dehngeschwindigkeit auf die Streckgrenze von Baustahlen unterschiedlicher Festigkeit', Arch. Eisenhuttenwesen 52 (1981) 429–436.

[6] Advanced High-Strength Steels Application Guidelines Version 5.0, in: S. Keeler, M. Kimchi (Eds.); 2015 WorldAutoSteel. http://www.ika.rwth-aachen.de/de/institut/veranstaltungen/201-upcoming-events/2003-aachener-karosserietage-2016.html.

[7] P. Peters, N. Manopulo, P. Hora, Combination of the strain dependent Yld2000 Model With an Extended HAH Model, in: Proceedings of the 9th Numisheet Conference, January 6–10, 2014, Melbourne, Australia, 2014.
[8] J. Mulder, H. Vegter, Evolving yield loci based on anisotropic hardening curves, in: The 13th International ESAFORM Conference on Material Forming – ESAFORM; University of Brescia, Brescia, Italy, April 7–9, 2010, 2010, pp. 191–194.
[9] J. Mulder, P. Eyckens, A.H. van den Boogaard, Differential hardening in IF steel experimental results and a crystal plasticity based model, in: Forming Technology Forum; June 29 & 30, 2015, Zurich, Switzerland, 2015.
[10] B. Eberle, W. Volk, P. Hora, Automatic approach in the evaluation of the experimental FLC with a full 2D approach based on a time depending method, in: P. Hora (Ed.): Proc. of the NUMISHEET 2008 Conf. (Part B. Bechmark study), Interlaken, Switzerland, 2008, pp. 279–284.
[11] M. Abspoel, M. Scholting, J. Droog, A new method for predicting Forming Limit Curves from mechanical properties, J. Mater. Process. Technol. (2013) 759–769.
[12] M.P. Sklad, E.H. Atzema, F.J. Schouten, M. de Bruine, A. Emrich, Experimental study of forming limits in multistage deformation processes, in: International Deep Drawing Research Group IDDRG 2008 International Conference; 16–18 June 2008; Olofström, Sweden, 2008.
[13] Performance-Product-Process; Internal Tata Steel R&D course.
[14] R. Pearce, Sheet Metal Forming, Adam Hilger, Bristol, Philadelphia and New York, 1991, ISBN 0-7503-0101-5.
[15] Z. Marciniak, J.L. Duncan, S.J. Hu, Mechanics of Sheet Metal Forming, Butterworth-Heinemann, Oxford, UK & Woburn, MA, USA, 2002, ISBN 0 7506 5300 0.
[16] Metallic Materials – Sheet and Strip – Determination of Tensile Strain Hardening Exponent; Standard ISO 10275.
[17] S.P. Keeler, Determination of Forming Limits in Automotive Stampings, Society of Automotive Engineers Technical Paper 650535, 1965, doi:10.4271/650535.
[18] G.M. Goodwin, Application of Strain Analysis to Sheet Metal Forming Problems in the Press Shop, Society of Automotive Engineers Technical Paper 680093, 1968, http://dx.doi:10.4271/680093.
[19] H. Vegter, A.H. van den Boogaard, A plane stress yield function for anisotropic sheet material by interpolation of biaxial stress states, Int. J. Plasticity 22 (2006) 557–580.
[20] H. Vegter, Y.G. An, H. Pijlman, J. Huétink, Advanced mechanical testing on aluminium alloys and low carbon steels for sheet forming, in: J.C. Gelin, P. Picart (Eds.), Proceedings NUMISHEET 99, Besançon, France, 1999, pp. 3–8.
[21] H. Vegter, C.H.L.J. ten Horn, Y.G. An, Modelling of the plastic material behaviour in advanced sheet metal forming simulations, in: R.K. Verma, D. Bhattacharjee (Eds.), Automotive Sheet Metal Forming, 17–18 December 2007, Jamshedpur, India, 2007, ISBN 978-0-07-025218-9.
[22] R. Hill, A theory of the yielding and plastic flow of anisotropic metals, Proc. Royal Soc. London A 193 (1948) 281–297.
[23] M. Sigvant, K. Mattiasson, M. Larsson, Determination of the formability of DP-steels by a combination of experimental methods and FE-simulations, Proceedings of the IDDRG International Conference (2010) 195–204.
[24] D. Banabic, H. Aretz, D.S. Comsa, L. Paraianu, An improved analytical description of orthotropy in metallic sheets, Int. J. Plasticity 21 (2005) 493–512.
[25] L. Kessler, J. Gerlach, The impact of material testing strategies on the determination and calibration of different fem material models, in: Proceedings of IDDRG

International Conference; 2006 "Drawing the Things to Come", Porto, Portogal, 19–21 June 2006, pp. 113–120.

[26] Y. Bergström, A dislocation model for the stress–strain behaviour of polycrystalline Fe with special emphasis on the variation of the densities of mobile and immobile dislocations, Mater. Sci. Eng. 5 (1969) 179–192.

[27] P. Van Liempt, Work hardening and substructural geometry of metals', J. Mater. Process. Technol. 45 (1994) 459–464.

[28] H. Vegter, On the Plastic Behaviour of Steel During Sheet Forming, Ph.D thesis, University of Twente, 1991.

[29] H. Vegter, J. Mulder, P. Peter van Liempt, J. Heijne, Tailored work hardening descriptions in simulation of sheet metal forming, in: J.W. Yoon, T.B. Stoughton, B. Rolfe, J.H. Beynon, P. Hodgson (Eds.), The 9th International Conference and Workshop on Numerical Simulation of 3D Sheet Metal Forming Processes (NUMISHEET 2014), Melbourne, Australia, 6–10 January 2014, pp. 474–481.

[30] P. Hora, P. Peters, N. Manopulo, M. Gorji, Challenges in the accurate modelling of sheet metal forming processes, in: 8th Forming Technology Forum 2015, Zurich, Switzerland, 29 & 30 June 2015, p. 5.

[31] J. Gerlach, L. Kessler, M. Linnepe, A study regarding material aspects for numerical modelling of mild steels, in: IDDRG 2009 International Conference; Golden, Colorado, USA, 1–3 June 2009.

[32] P. Hora, J. Heingärtner, N. Manopulo, L. Tong, D. Hortig, A. Neumann, et al., On the way from an ideal virtual process to the modelling of the real stochastic, in: Forming Technology Forum, Zurich, Switzerland, 17–18 May 2011, 2011.

[33] H. Vegter, C.H.L.J. ten Horn, A.H. van den Boogaard, Influence of the plastic material behaviour on the prediction of forming limits, Proceedings Numiform (2007) 87–92.

[34] H. Vegter, C.H.L.J. ten Horn, M. Abspoel, Modelling of the forming limit curve by MK analysis and FE simulations, In: P. Hora, (Ed.), Proc. of the NUMISHEET 2008 Conf. (Part B. Bechmark study), Interlaken, Switzerland, 2008, pp. 187–192.

[35] C.Chr. Veerman, P.F. Neve, Some aspects of the determination of the forming limit diagram (FLD) – onset of localised necking, Sheet Metal Industries, 1972, p. 421.

[36] T.B. Stoughton, J.W. Yoon, Paradigm change: alternate approaches to constitutive and necking models for sheet metal forming, The 8th International Conference and Workshop on Numerical Simulation of 3D Sheet Metal Forming Processes AIP Conf. Proc 1383 (2011) 15–34. Available from: http://dx.doi.org/10.1063/1.3623589.

[37] D. Jocham, C. Gaber, O. Böttcher, W. Volk, Prediction of formability for multi-linear strain paths, in: 8th Forming Technology Forum, 29–30 June 2015 Zurich, Switserland, 2015, pp. 59–64.

[38] K. Drotleff, M. Liewald, J.-T. Mosebach, Improved failure criterion for nonlinear strain paths, in: 8th Forming Technology Forum, 29–30 June 2015 Zurich, Switserland, 2015, pp. 71–76.

[39] P.T. Vreede, A Finite Element Method for Simulations of 3 Dimensional Simulations of Sheet Metal Forming, PhD thesis, University of Twente, Enschede, the Netherlands, 1992, ISBN 90-9004754-9.

[40] F. Yoshida, T. Uemori, K. Fujiwara, Elastic-plastic behaviour of steel sheets under in-plane cyclic tension-compression at large strain, Int. J. Plasticity 18 (2002) (2002) 633–659.

[41] Spanjer, W., Orthogonal Shearing Model, Internal Tata Steel R&D Report No. 156789; IJTC, 2013.

[42] M. Shih, An innovative shearing process for AHSS edge stretchability improvements, in: Great Designs in Steel. <http://www.autosteel.org/~/media/Files/Autosteel/Great%20Designs%20in%20Steel/GDIS%202012/An%20Innovative%20Shearing%20Process%20for%20AHSS%20Edge%20Stretchability%20Improvements.pdf>, 2012.

[43] C. McEwan, R. Underhill, N. Langerak, G. Botman, M. de Bruine, A new approach to predicting edge splits – the combined FLC/HEC diagram, in: IDDRG 2009 International Conference, Golden, Colorado, USA; 1–3 June 2009, 2009.

[44] Y. An, R. Boterman, E.H. Atzema, M. Abspoel, M. Scholting, Failure orientation in stretch forming and its correlation with a polycrystal plasticity-based material model for a collection of highly formable sheet steels, Metallurg. Mater. Trans. A 47A (2016) 5, ISSN 1073-5623; <http://dx.doi.org/10.1007/s11661-016-3504-9>.

[45] Accounting For Material Scatter In Sheet Metal Forming Processes: A Demonstrator Project; R.J. van der Velde, MSc Thesis; University of Twente, Faculty of Engineering technology, Chair of Applied mechanics; CTW-013/TM-5710; 13 May 2013.

[46] M. Roelofsen, E.H. Atzema, A. Schouws, G. Botman, Ranking formability of materials by means of the Cross Die – value', Internal Tata Steel R&D Report No. 155759, 2012.

[47] T. Müller, Improving edge cracking sensitivity of AHSS with tensile strengths up to 1500 MPa, in: Forming in Car Body Engineering, Bad Nauheim; 29–30 September 2015, 2015.

[48] C. Krempaszky, P. Larour, J. Freudenthaler, E. Werner, Towards more efficient hole expansion testing, in: International Deep Drawing Research Group IDDRG 2014 International Conference, 1–4 June, Paris, France, 2014, pp. 204–209.

[49] E.H. Atzema, P. Seda, Sheared edge tensile test improved: SETi', in: 8th Forming Technology Forum, 29–30 June 2015, Zurich, Switzerland, 2015, pp. 97–102.

[50] Corus RF FEA brochure. <http://www.rollformdefects.com/>, 2007.

[51] F. Barlat, J.W. Yoon, O. Cazacu, On linear transformations of stress tensors for the description of plastic anisotropy, Int. J. Plasticity 23 (2007) 876–896.

[52] N. Manopulo, J. List, M. Gorji, P. Hora, A non-associated flow rule based YLD2000-2D model, in: 8th Forming Technology Forum 2015, Zurich, Switzerland, 29 & 30 June 2015, 2015, pp. 29–33.

[53] H. Ma, Z. Wang, The Influence of the Bauschinger Effect on Springback Prediction for Dual Phase Steel; SAE Technical Paper 2006-01-0145, 2006, http://dx.doi:10.4271/2006-01-0145.

Physical metallurgy of steels: an overview*

4

G. Krauss
Colorado School of Mines, Golden, CO, United States

4.1 Introduction

This chapter presents an overview of the microstructures that comprise an important element of the physical metallurgy of carbon steels: the interrelationships of steel chemistry, processing, microstructure, and mechanical properties. The microstructural components are now combined by alloying and processing into many types of automotive steels, all with unique ranges of mechanical behavior, as described subsequently in this text. Steels are alloys of iron and other elements, with carbon effectively the second most important element, especially in the low-carbon steels emphasized in this chapter. Generally, for any given microstructure the higher the carbon content of the steel, the higher the strength. Low-carbon steels contain up to 0.25 wt.% carbon, have traditionally, as described below, had microstructures largely of ferrite and therefore good combinations of relatively low strength, high formability, and good weldability. Great variations of the combinations of these characteristics are now produced in the new types of automotive steels. Medium-carbon steels contain between 0.25 and 0.50 wt.% carbon and are most frequently heat treated to produce structural components with high strength and hardness by quenching to form martensite and subsequent tempering. High-carbon steels contain more than 0.5 wt.% carbon, are also quenched to martensite and tempered, and by virtue of their high carbon produce microstructures of the highest hardness and strength for applications such as springs and bearings. Fracture resistance however is low, and care in heat treatment and residual compressive stresses introduced by carburizing or induction hardening make high-carbon steels work.

In commercial steels, in addition to carbon, there may be ten [1] or more elements present, some because they are added for beneficial purposes, others residual from steelmaking, all integrated into the microstructure of solid steel parts and sections. Silicon, manganese, and aluminum are used to deoxidize liquid steel during steel making, and may also be used to provide strength to solid microstructures, provide hardenability, or grain size control. For example, aluminum combines with nitrogen to make fine aluminum nitride crystals that prevent austenitic grain growth at certain temperatures. Manganese combines with residual sulfur to make manganese sulfide inclusions. Chromium, molybdenum, and manganese dissolve in austenite and slow the kinetics of the solid state

* Every effort has been made to trace copyright holders and to obtain their permission for the use of copyright material. The publisher apologizes for any errors or omissions in the acknowledgements printed in this book and would be grateful if notified of any corrections that should be incorporated in future reprints or editions.

Automotive Steels. DOI: http://dx.doi.org/10.1016/B978-0-08-100638-2.00004-3

transformations that occur when austenite cools, and are carbide-forming elements that may provide hardening. Nickel does not form carbides, stabilizes austenite, slows phase transformations, and provides toughness. Vanadium, niobium, and titanium are considered to be microalloying elements and very small additions provide strength or control austenitic grain size. Phosphorous residual from steel making is an element responsible for various types of embrittlement and is therefore controlled to low levels. Processing includes operations such as casting from the liquid state, heating and cooling in many defined heat treatments, hot work, cold rolling, forming, and welding. The chemistry and processing together create the microstructures and substructures that produce the mechanical properties and performance of finished steels.

Steels are manufactured, bought, and sold according to chemistry. Heat analyses show the various chemical elements as uniformly dissolved in liquid steel. However during solidification, elements segregate between growing solid dendrites and the surrounding liquid steel, producing a nonuniform microscopic distribution of the elements that might survive hot work processing and might produce variations in microstructure, referred to as banding, in finished steel [2]. Diffusion-controlled solid state phase transformations, often by design, produce localized variations in chemistry and microstructure of steels [3]. These variations in chemistry and microstructure are important aspects in the design and performance of automotive steels.

The great versatility of steel is based on the fact that two quite different phases or crystal structures of iron exist at quite different temperature ranges: austenite at high temperatures, ferrite at low and room temperatures. These two phases are described in detail below and throughout this chapter and book. However at this point it is important to note that austenite has great ductility and steels with austenitic microstructures can be hot worked from very large to small sections or hot forged to complicated shapes. The austenite must transform to other microstructures on cooling, and depending on cooling rate any given steel may be produced to a variety of structures, some desirable during intermediate processing, some desirable in finished products. It is the great variety of austenite transformation product microstructures, as modified by steel chemistry and processing, that produce the great many varieties of steels. The purpose of this chapter is to provide some background for the microstructural components and physical metallurgy involved in development of the automotive steels described in this text.

4.2 The iron-carbon phase diagram

The framework for understanding carbon steels is the iron-carbon diagram, Fig. 4.1, a map of the phases present under equilibrium conditions as a function of carbon content and temperature. Equilibrium assumes there is more than adequate time for attaining phase chemistries and amounts of phases shown in the various fields of the diagram, and in the Fe–C system it is assumed that carbon is present as the metastable phase iron carbide, Fe_3C, also identified by the term cementite, or as more stable graphite. Cementite phase fields are defined by solid lines, graphite phase fields by dashed lines.

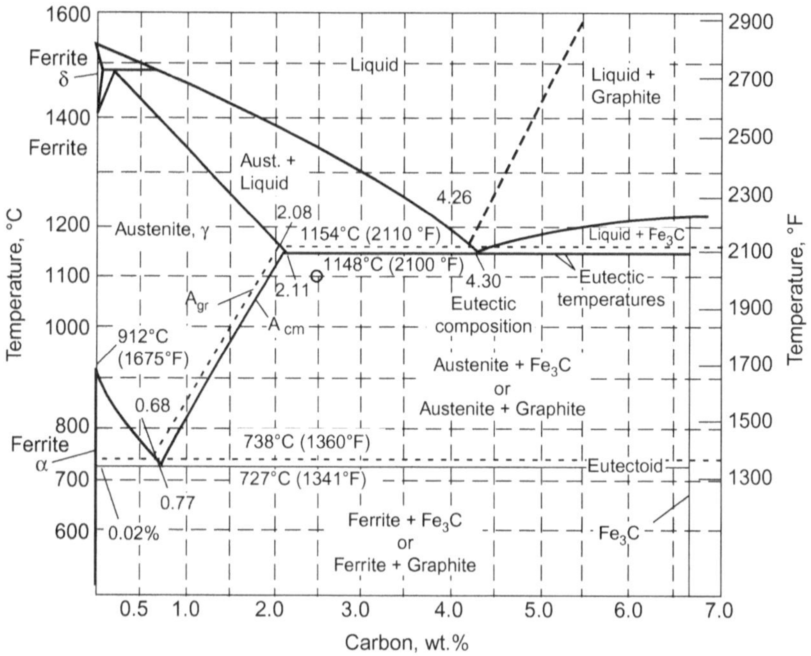

Figure 4.1 The iron-carbon phase diagram [4]. Dashed lines show graphite equilibria, solid lines show cementite equilibria. Steels invariably contain less than 2.00 wt.% carbon and carbon is present as cementite or in solid solution in austenite or ferrite. Iron-carbon alloys with more than 2.00 wt.% carbon are generally known as cast irons and carbon is often present as graphite. R. Elliott, Cast Iron Technology, Butterworths, London, 1988. *Source*: ASM International: Vol 8, Metals Handbook, Eighth Edition, 1973.

Graphite is important when alloy carbon content is high, as in cast irons, but in steels, typically identified as alloys containing less than 2.00 wt.% carbon, graphite rarely forms. Thus in steels carbon is present as cementite, or in solid solution in austenite and ferrite, and the iron-cementite diagram is used for background.

The iron-carbon phase diagram defines single and two-phase fields as a function of carbon content and temperature. Three-phase fields are identified as straight lines, and according to the phase rule, at constant pressure in a two component system, represent invariant equilibria between three phases, i.e., the compositions of the coexisting phases and the temperature at which the phases coexist, are fixed and invariant. The conditions for three coexisting phases are relaxed when other components are added to a two-component alloy system. There are three three-phase equilibria in the iron-carbon diagram shown:

At 1495°C: δ-ferrite + Liquid = γ-iron (austenite), a *peritectic* reaction, defined on cooling as a solid phase combining with liquid phase to form a single solid phase.

At 1148°C: Liquid = Fe_3C + γ-iron, a *eutectic* reaction, defined on cooling as a liquid phase forming two solid phases.

At 727°C: γ-iron = α ferrite + Fe_3C, a *eutectoid* reaction, defined on cooling as a solid phase forming two other solid phases.

The peritectic reaction is important for the solidification of steel, the eutectic reaction is important for cast iron solidification, and for steels, the eutectoid reaction is of major importance.

Although the iron-carbon diagram is an extremely important framework for understanding steels, its value is primarily in predicting the phases that will begin to and ultimately form at a given temperature, given sufficient time at that temperature. Limitations of the iron-carbon diagram for commercial steels include: the arrangements of the phases in microstructures is not given, chemical gradients resulting from solidification or phase transformations are not shown, boundaries of the phase fields are shifted by the addition of other elements, there may not be sufficient time for atom diffusion to produce equilibrium chemistries of the phases, and other phases such as inclusions and alloy or microalloying element carbides and nitrides may be present. Nevertheless these considerations are important in understanding commercial steels and must be evaluated by other theoretical and experimental approaches.

4.3 Austenite

Fig. 4.2 shows the face-centered cubic (fcc) unit cell of austenite. Iron atom positions on the corners and in the centers of the faces of the unit cell are shown by the black metal circles. Two views are shown, one identifying an interstitial octahedral site between iron atoms, the other identifying an interstitial tetrahedral site in the unit cell. High solubility of carbon atoms in octahedral interstitial sites of austenite is a very important characteristic, allowing carbon present in carbides stable at low-temperatures to be dissolved into austenite formed on heating. That carbon is then reconfigured into many other microstructures on cooling of the

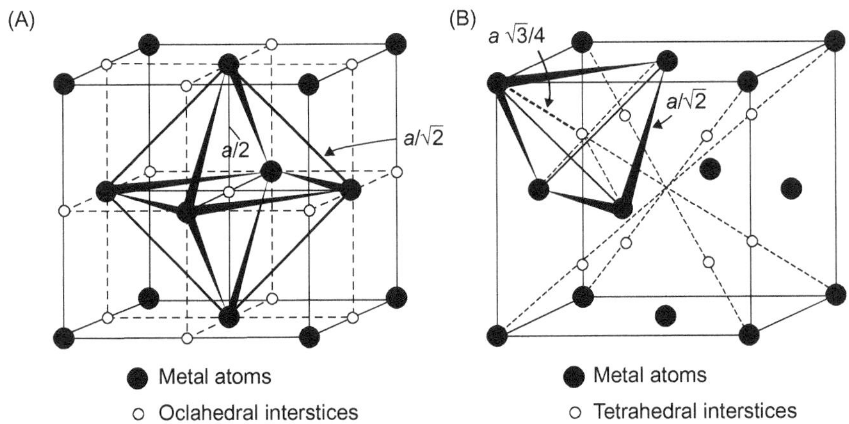

Figure 4.2 Unit cell crystallographic structure of austenite. Octahedral (A) and tetrahedral (B) interstitial sites are shown from G. Krauss, Steels: Processing, Structure, and Performance, second ed., ASM International, 2015.
Source: ASM International: G. Krauss STEELS, Second Edition, 2015. C.S. Barrett, T.B. Massalski, Structure of Metals, third ed., McGraw-Hill, New York, 1966, with permission from Prof. Thaddeus B. Massalski.

austenite. The high solubility of carbon in austenite also produces a single phase microstructure, which in steels austenitized or soaked for hot rolling, is highly deformable and can be readily hot worked.

The fcc unit cell defines the repeating iron atom arrangement in crystals of austenite. Many, many atoms make up a crystal, and another view of austenite is that the crystal structure is made up of close-packed planes of iron atoms stacked in a sequence that repeats every three layers. The orientation of each close-packed plane is defined by three face diagonals in the unit cell. The close-packed atom planes, with Miller indices {111}, are the slip planes on which dislocations move, or constitute the boundaries of annealing twins. The interstitial atom positions shown in Fig. 4.2 are located between the iron atoms in the close-packed planes. Some processing conditions, as described during quenching of austenite to produce martensite in this chapter, and some of the special automotive steels described in other chapters, result in austenite being retained at room temperature. That austenite continues to be ductile and does not undergo the ductile to brittle transition to which body-centered cubic ferritic microstructures are susceptible.

4.4 Ferrite and cementite

The phases ferrite and cementite are coupled in this section because the very low solubility of carbon in ferrite often, but not always, causes carbon to concentrate and form cementite together with ferrite on cooling from austenite. Fig. 4.3 shows the body-centered cubic unit cell of ferrite. Iron atom positions are shown at cell corners

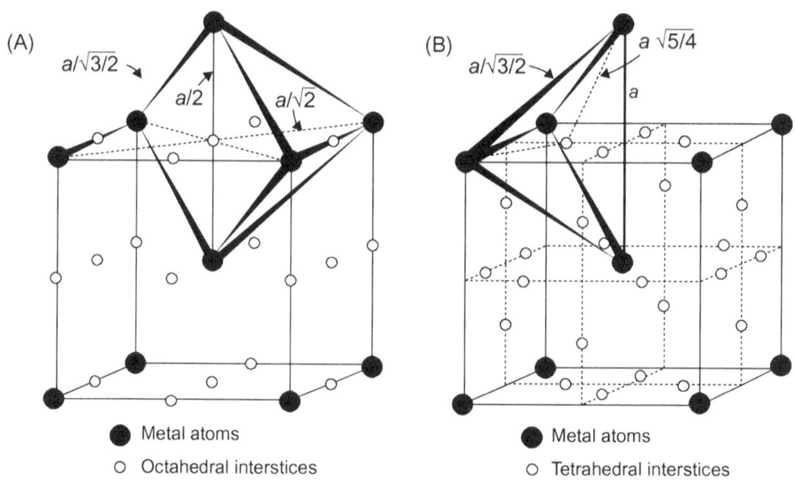

Figure 4.3 Unit cell crystallographic structure of ferrite. Octahedral (A) and tetrahedral (B) interstitial sites are shown from G. Krauss, Steels: Processing, Structure, and Performance, second ed., ASM International, 2015.
Source: ASM International: G. Krauss STEELS, Second Edition, 2015. C.S. Barrett, T.B. Massalski, Structure of Metals, third ed., McGraw-Hill, New York, 1966, with permission from Prof.Thaddeus B. Massalski

and the center-cell position, and again octahedral and tetrahedral interstitial sites are shown. Cementite, a carbide phase of high hardness, has a more complicated orthorhombic crystal unit cell, with a ratio of three iron atoms to one carbon atom [4].

Although ferrite has a more open atom structure than close-packed austenite, the interstitial sites are smaller [3], and this geometric factor contributes to the low solubility of carbon in ferrite. As ferrite forms from austenite on cooling, in many situations carbon is rejected from ferrite, forming cementite, or sometimes enriching austenite, but in some cases carbon atoms are trapped in the ferrite, most probably in octahedral sites. Because of the resulting lattice strains, the carbon atoms find their way to dislocations, pin dislocations by Cottrell locking, and cause discontinuous yielding or Lüders straining of low-carbon sheet steel ferritic microstructures [7]. Strain aging is caused by similar phenomena. Another important consequence of the difference between the unit cells and the atomic density of austenite and ferrite is the expansion of the ferritic structure as it forms, creating strains and possibly distortion of finished parts. Also, under most situations ferrite is a very ductile phase, but in some microstructures subjected to impact loading and low temperatures, ferritic microstructures may undergo a transition to brittle behavior. One explanation of the brittle behavior has been related to the inability of screw dislocations to cross slip because of dissociation of dislocation core structures at low temperatures [8].

4.5 Steel microstructure: general considerations

Major characteristics of the austenite and ferrite phases have been described to this point, but the arrangements of these phases into actual microstructures, together in some cases with cementite, have not been described. All of the phases are crystalline as noted, and it is their crystallinity and metallic bonding that make possible dislocation movement and plastic deformation. Strengthening is created by building obstacles to dislocation motion into steel microstructures. As will be described in subsequent chapters, each high-strength automotive steel is alloyed and processed uniquely to provide a microstructure with balanced strength and ductility dependent on the degree of dislocation mobility.

The arrangements of the phases in steels, e.g., crystal shape and size, the amounts and distribution of various crystals, make up the microstructure of a steel in a given processed condition. Invariably the sizes of the crystals in steels are finer than can be resolved by the unaided eye, 0.1 mm, and thus microscopes must be used for their resolution, hence the term microstructure to describe the arrangements. The light microscope can resolve features on the micron (micrometer) scale, as fine as 0.5 μm (0.0005 mm) in size, and many of the features of steel microstructures, e.g., austenitic grain sizes, are in this size range. However, many microstructural features essential to the performance of steel microstructures, e.g., dislocations and very fine precipitate crystals, are finer than 0.5 μm, on the order of atomic dimensions or nanometers, and these features must be characterized by electron microscopy, scanning electron microscopy (SEM) if etching develops sufficient surface contrast, transmission electron microscopy (TEM) if volume features must be resolved by examination of thin foils transparent to the electron beam.

4.6 Steel microstructures produced by diffusion: ferrite, pearlite, and bainite

As noted in the above sections, austenite must transform to ferrite, cementite, and perhaps other distributions of phases depending on composition. This section describes some of the now rather well-known steel microstructures and their formation in carbon steels, namely ferrite, pearlite, and bainite. Even in these microstructures considerable variations can be produced by modifications in processing conditions and alloying, as is the case for many of the steels described in subsequent chapters.

Fig. 4.4 shows a light micrograph of the ferritic microstructure of an ultralow carbon or interstitial-free steel containing only 0.004 wt.% carbon. The microstructure consists entirely of equiaxed ferrite crystals or grains that have nucleated and grown in austenite cooled after hot work. There is no cementite resolvable in the microstructure because of the very low-carbon content of the alloy. Titanium has been added to tie up small amounts of carbon and nitrogen not removed by vacuum degassing during steelmaking, and very fine Ti precipitates, resolved by electron microscopy, are present in the ferrite grains [9]. The absence of carbon and cementite dispersions in the fully ferritic microstructure makes interstitial-free steels highly

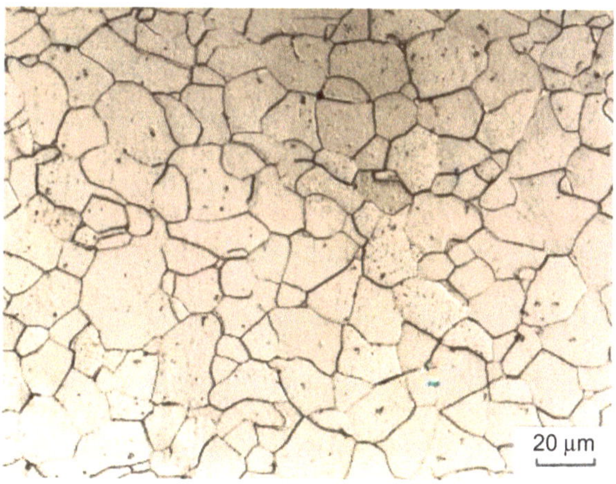

Figure 4.4 Equiaxed ferrite grains in a titanium stabilized interstitial-free steel. Nital etch. Light micrograph.
Source: Courtesy of Diane Oksana Wilshynsky-Dresler while at the Colorado School of Mines.

ductile, especially after cold rolling and annealing that produce formable recrystallization textures [10].

Most commercial steels, those containing more than about 0.02 wt.% carbon, on slow cooling from austenitizing temperatures or after hot work, contain sufficient carbon to produce microstructures containing cementite. As low- and medium-carbon steels containing up to 0.77 wt.% carbon cool, ferrite first forms on austenitic grain boundaries. This ferrite is termed proeutectoid ferrite since it forms above the eutectoid temperature. In view of the low solubility of carbon in ferrite, carbon is rejected from the ferrite and concentrates in the adjacent parent austenite. With decreasing temperature, more and more carbon concentrates into the remaining austenite until the carbon concentration ideally reaches 0.77 wt.% carbon. At that time the eutectoid solid state transformation defined by the horizontal line at 727°C is activated. As noted relative to Fig. 4.1, at that temperature austenite transforms by a eutectoid reaction directly to a mixture of ferrite and cementite. The ferrite and cementite crystal structures, termed lamellae, form parallel to each other in a microstructure termed pearlite. With increasing steel carbon content less proeutectoid ferrite forms and more and more austenite transforms to pearlite. The iron-carbon diagram shows that the austenite of high carbon steels containing about 0.77 wt.% carbon transforms entirely to the ferrite and cementite. These pearlitic microstructures, which when processed to very fine interlamellar spacings of ferrite and cementite produce very high hardness and strength for applications such as rail and high strength wire [3]. Fig. 4.5 shows a light micrograph of ferrite and pearlite that have formed in a 0.4 wt.% carbon steel.

Pearlite, depending on cooling conditions, may form over a range of temperatures below 727°C. More rapid cooling rates allow less time for the diffusion necessary to

Figure 4.5 Proeutectoid ferrite (white) and pearlite formed in austenite of a 0.4 wt.% steel. Nital etch, 500 × magnification, light micrograph.
Source: Courtesy of A. R. Marder and A. Benscoter while at the Bethlehem Steel Corporation.

partition carbon and other elements between the ferrite and cementite lamellae. Therefore with decreasing transformation temperatures, in view of increasing thermodynamic driving force for transformation of the austenite, decreasing diffusivity, and shorter effective diffusion distances, pearlite interlamellar spacing becomes finer. At temperatures below about 550°C, carbon diffusion can no longer sustain the growth of continuous parallel ferrite and cementite lamellae. Mixtures of ferrite and cementite still form, but the cementite crystals are dispersed within the ferrite in a microstructure known as bainite [1].

There are two forms of bainite. Upper bainite forms at temperatures just below those at which pearlite has ceased to form, and lower bainite which forms at lower temperatures close to those at which martensite forms, as described in a later section. Upper bainite forms as parallel lath-like crystals of ferrite with cementite formed between the ferrite crystals. Although discontinuous, the cementite is relatively coarse consistent with still relatively high carbon diffusivity in upper bainite. In lower bainite the cementite precipitates as very fine particles, consistent with reduced diffusivity, within nonparallel acicular crystals of ferrite crystals. Figs. 4.6 and 4.7 show respectively light micrographs of upper and lower bainite. The two quite different ferrite morphologies are shown in a 4150 steel. The carbide dispersions are too fine to be resolved in light micrographs, but have been well documented by transmission electron microscopy [3].

Although the two-phase ferrite-cementite microstructures of upper and lower bainite form readily in medium-carbon steels, in low-carbon steels under similar cooling conditions, bainitic-like microstructures without cementite may form [3,11]. Carbon is rejected from ferritic laths, but steel carbon content is too low to cause cementite formation. Nevertheless austenite adjacent to the ferrite is enriched

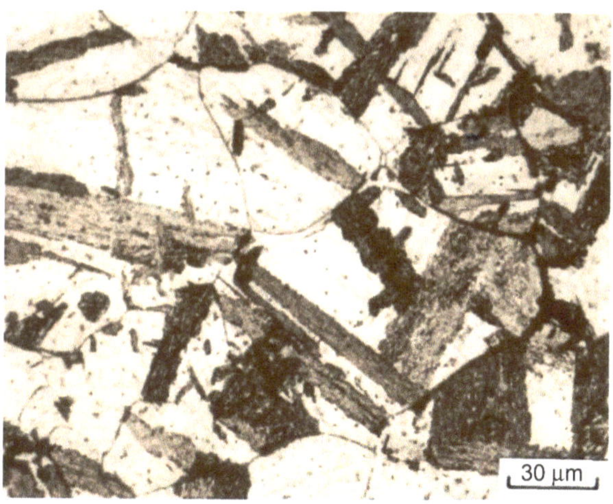

Figure 4.6 Packets of upper bainite laths in austenite grains of 4150 steel. Nital etch, light micrograph.
Source: Courtesy of Florence A. Jacobs while at the Colorado School of Mines.

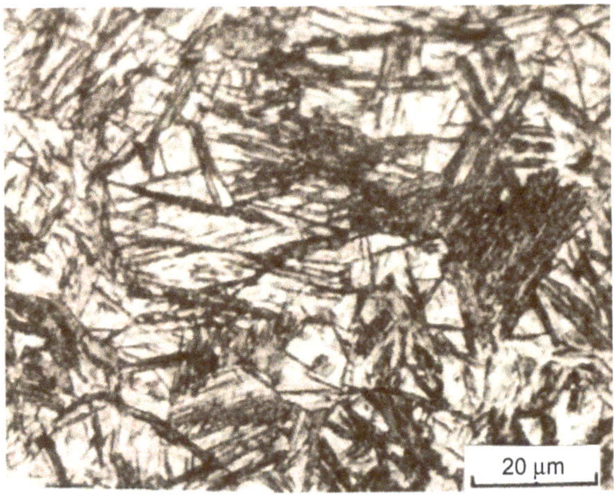

Figure 4.7 Needle- or plate-shaped crystals of lower bainite in 4150 steel. Nital etch, light micrograph.
Source: Courtesy of Florence A. Jacobs while at the Colorado School of Mines.

and stabilized with carbon. Sometimes the interlath austenite transforms partially to martensite, producing a mixture of martensite and retained austenite in a microstructural component termed M/A constituent.

4.7 Diffusionless transformation of austenite: martensite

If cooling rates are fast enough and alloying is sufficient to prevent diffusion-controlled transformation to ferrite, pearlite or bainite, the thermodynamic driving force for the transformation of austenite causes the formation of body-centered crystals by a diffusionless, shear type solid state phase transformation. In honor of an early metallographer, Adolf Martens, this transformation is named the martensitic transformation and the product is termed martensite. Trapped carbon atoms create a body-centered tetragonal unit cell with one axis, designated the c axis, longer than the other two axes, designated the a axes, and the higher the carbon content the greater the tetragonality. Fig. 4.8 shows the body-centered tetragonal unit cell of martensite and octahedral interstitial sites occupied by carbon atoms.

The shear transformation of austenite to martensite produces crystals with planar geometries, lath or board-shaped in low and medium-carbon steels, plate-shaped in high-carbon steels. Fig. 4.9 shows a number of important characteristics of martensite formation. Most important is the effect of shear which effectively acts to produce plain-strain deformation parallel to unique crystallographic planes of the parent austenite termed habit planes. The deformation produces tilting of a

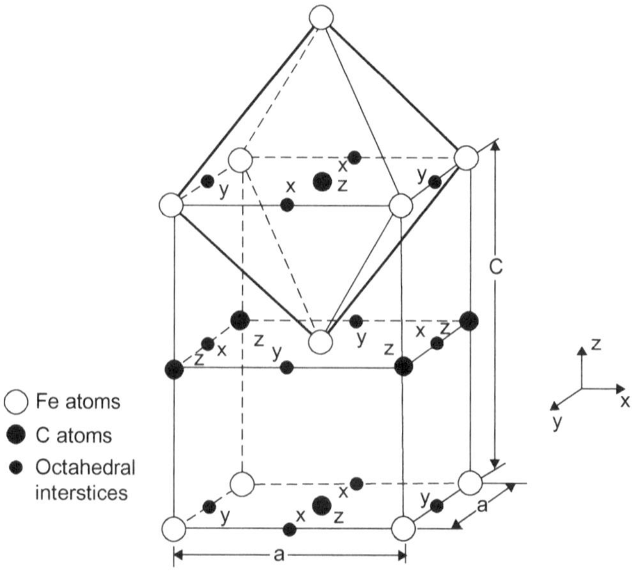

Figure 4.8 Body-centered tetragonal unit cell and octahedral interstitial sites of carbon of martensite in iron-carbon alloys and carbon steels. G. Krauss, Steels: Processing, Structure, and Performance, second ed., ASM International, 2015.

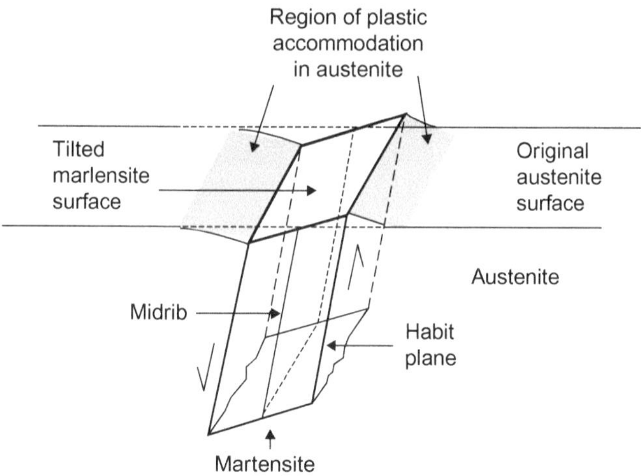

Figure 4.9 Schematic diagram of the effects of the shear formation of a martensite crystal: plane strain deformation, surface tilting, habit plane in parent austenite, and deformation of adjacent parent austenite are illustrated. G. Krauss, Steels: Processing, Structure, and Performance, second ed., ASM International, 2015.
Source: ASM International: G. Krauss STEELS, Second Edition, 2015.

martensitic crystal on a free surface and as shown requires deformation and strain of the adjacent austenite. This strain hardening of the austenite eventually limits growth and results in the thin lath or plate-shaped crystals. Thus the formation of a martensitic microstructure requires repeated nucleation of many thin crystals. Martensite starts to form at a temperature termed M_S below the temperature at which diffusion is no longer effective in forming bainite. M_S temperatures decrease with increasing steel carbon and most alloying elements [3], and since the transformation is diffusionless, time is not important. Increases in volume fraction of martensite are driven only by decreasing temperature below M_S.

Fig. 4.9 illustrates schematically the general morphological features of a martensite crystal. However, also unique to martensite is a fine structure within the crystals. Not only is the crystal structure changed from face-centered cubic austenite to that of the body-centered martensite, a lattice deformation, but in order to maintain the habit plane and shear deformation, the martensite is subjected to a lattice invariant deformation, produced by dislocations or twins, as first identified by the crystallographic theory of the martensitic transformation [13]. Thus a high density of crystal imperfections is an important characteristic of martensite, and in the lath martensites of low-carbon steels dislocation densities as high as 10^{15} m^{-2} have been measured [14].

Not only is the unique shear formation of individual crystals of martensite important, but the transformation of austenite to many martensite crystals in an austenite grain produces unique microstructures. In low-carbon steels lath martensite crystals form on $\{557\}_A$ habit planes and form parallel to each other in packets or blocks [15,16]. Blocks are groups of martensite crystals that have identical or very close crystallographic orientations, and are now readily documented by electron backscattered diffraction [17]. Most of the crystals are too fine to be resolved in the light microscope, and constitute a remarkably fine grain size with a crystal boundary area (in cm^2) per unit volume (in cm^3) of 65,000 cm^{-1} in the lath martensite of an iron-0.2 wt.% carbon alloy [18]. As noted above each crystal may contain a very high density of dislocations. Small amounts of austenite retained between parallel lath crystals even in low-carbon steels are another feature of martensitic microstructures.

Fig. 4.10 shows a light micrograph of lath martensite formed in an iron-0.2 wt.% carbon alloy. The parallel alignment of martensite in packets or blocks in the prior austenite grains is clearly shown, but only the largest crystals are resolved. The light and dark areas may contain many martensite crystals of the same orientation which etch identically. All of the martensite crystals in roughly a single grain of austenite are shown in the TEM micrograph of Fig. 4.11. Several parallel sets of crystals are shown, and the crystals range in size from fractions of a micron to over a micron in width.

In view of the very fine crystal size, and especially because of the high very density of dislocations, martensitic microstructures have the highest strain hardening capability, ultimate tensile strengths, and hardness of any microstructure that can be formed in a given steel. The very high strength and associated low ductility of as-quenched martensite in medium and high carbon steels have traditionally required tempering of the martensite. Depending on property requirements, tempering is performed at temperatures that reduce dislocation densities, cause precipitation of transition carbides or cementite stabilized by carbide-forming alloying elements, and

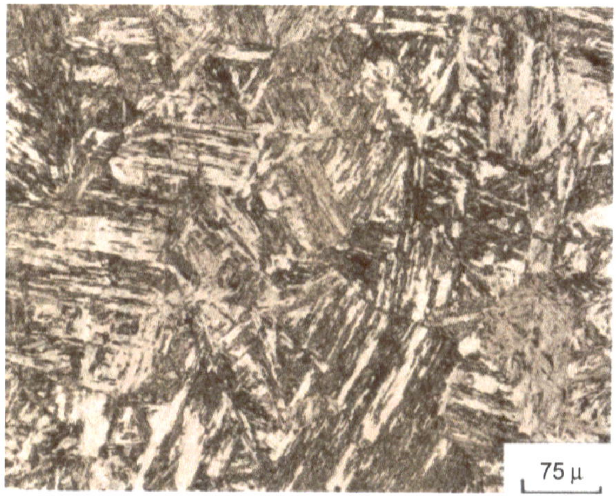

Figure 4.10 Lath martensite in an iron-0.2 wt.% carbon alloy. Nital etch, light micrograph.
Source: Courtesy of Thomas Swarr while at Lehigh University.

Figure 4.11 Martensite crystal distribution in roughly a single grain of parent austenite in an iron-0.2 wt.% carbon alloy. Transmission electron micrograph.
Source: Courtesy of Thomas Swarr while at Lehigh University.

general coarsening of the martensitic microstructure [3,19]. Medium and high-carbon steels were historically and still are extensively hardened by quench and tempering to martensite for machine parts, automotive transmission and drive components, and tools, but as subsequent chapters will show, martensite is now an important microstructural component of advanced low-carbon automotive steels.

4.8 Transformation diagrams and Jominy End Quench Curves

The above sections have briefly described the phases, the phase transformations, and selected examples of ferritic, pearlitic, bainitic, and martensitic microstructures in steels. Every steel can transform to a variation of any one or combination of these microstructures depending on alloy content and thermal processing, namely heating to intercritical temperatures where ferrite and austenite are stable, or heating to full austenitizing temperatures, and subsequently dropping to temperatures where the diffusion-formed microstructures are stable and can nucleate and grow, or cooling through temperature ranges where variations and combinations of the microstructures may form. The thermal processing may be accomplished by continuous or batch annealing, may result from cooling after hot rolling and coiling, or may be accomplished by heating and cooling of separate parts.

All of the time-temperature variations in microstructure of a steel of given composition can be summarized in Continuous Transformation Diagrams, determined by continuous cooling of a number of specimens at various rates, or Isothermal Transformation Diagrams determined by holding a number of specimens as a function of time at a constant temperature. Dilatometry, metallography, and hardness measurements can all be used to experimentally establish the diagrams, the diagrams can be calculated, and atlases with many diagrams for many steels have been published. The same information obtained by continuous cooling of many specimens can be obtained from a single specimen by the Jominy End Quench Test, as described below. Fig. 4.12 compares the results of all three of these approaches to following microstructural changes produced by the transformation of austenite in a medium carbon alloy steel.

Key to the determination of the microstructures produced by the transformation of austenite are the conditions available for diffusion. Faster cooling rates provide less time for diffusion, and therefore less time for ferrite and pearlite formation. Not only is carbon diffusion important, but also the diffusion of alloying elements substitutional to iron atoms must be considered. The diffusion of such elements as manganese, nickel, and chromium is sluggish relative to that of carbon, and if these elements must partition between austenite, ferrite or cementite, because they are austenite, ferrite, or carbide-forming elements, the transformation to diffusion-controlled microstructures is significantly retarded. In fact this retardation is one of the major reasons alloying elements are added to steels: they make it easier to avoid by quenching the formation of microstructures with reduced hardness and instead facilitate the formation of hard martensite. The effect of carbon and alloy element content, relative to cooling conditions, on the formation of martensite is the well-developed subject of hardenability [3]. Fig. 4.12 shows the strong effect of chromium and molybdenum additions on retarding the formation of ferrite and pearlite, thereby increasing hardenability of the steel. Bainite is less affected because of the finer bainitic microstructures and the possible combination of shear as well as diffusion involved as mechanisms for bainite formation [1].

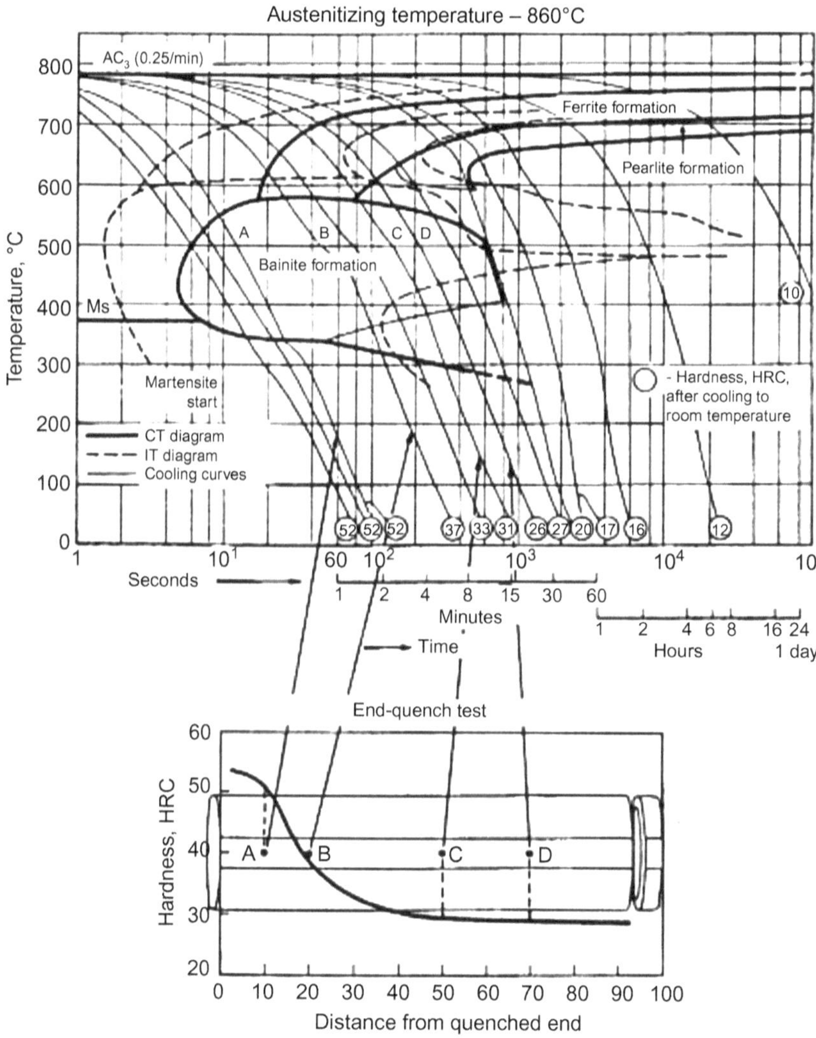

Figure 4.12 Isothermal transformation (IT), continuous transformation (CT), and Jominy End Quench curves for a steel with the German Designation 42CrMo4 containing 0.38 wt.% C, 0.99 wt.% Cr, and 0.16 wt.% Mo. Atlas zur Warmebehandlung der Stahle, Vol. 1–4, Max Plank Institut fur Eisenforshung, in cooperation with the Verein Deutscher Eidrnhuttenleute, Verlag Stahleisen, MBH, Dusseldorf, 1954–1976. Copyright 1954 Verlag Stahleisen GmbH, Düsseldorf, Germany.

The bottom part of Fig. 4.12, labeled End-quench test, shows schematically a Jominy end-quench specimen of the 42MoCr4 steel and the hardness traverse produced along the bar by rapidly water quenching only one end of the one inch diameter round specimen. Cooling rate of the steel is highest at the quenched end, producing high-hardness martensite, and decreases with increasing distance from the end, resulting in decreasing hardness as more and more softer non-martensitic microstructures form at the increasingly slower cooling rates further from the quenched end. This test was developed by Walter Jominy in 1938 [21] and is a very important and effective test to evaluate and compare hardenability of steels. The cooling rates as a function of distance from the quenched end are well known, and as shown in Fig. 4.12, can be related to the microstructural changes that are developed at identical cooling rates in continuous cooling diagrams.

4.9 Summary

The above sections have presented a brief overview of the basis of microstructural formation in plain carbon and low-alloy carbon steels. Much more information on each topic is available in some of the references cited in the vast long-term literature of ferrous metallurgy. The phases austenite, ferrite, cementite, and martensite and the phase transformations that create microstructures incorporating these phases by classical approaches have been described. All of the phases and microstructures of steels must incorporate many other chemical elements beyond the base iron and carbon, and such additions together with modifications of mechanical and thermal processing conditions create the great variety of microstructures and substructures possible in steels. Described in subsequent chapters are the newer processing and alloying approaches that make up excellent combinations of microstructures and properties in the new automotive sheet steels.

References

[1] H.K.D.H. Bhadeshia, *Bainite in Steels*, Book No. 504, The Institute of Materials, London, 1992.

[2] G. Krauss, Solidification, segregation, and banding in carbon and alloy steels, Metall. Trans. B 34B (2003) 781–792.

[3] G. Krauss, Steels: Processing, Structure, and Performance, second ed., ASM International, 2015.

[4] *Metallography, Structures and Phase Diagrams*, Vol. 8, Metals Handbook, eighth ed., American Society for Metals, 1973.

[5] R. Elliott, Cast Iron Technology, Butterworths, London, 1988.

[6] C.S. Barrett, T.B. Massalski, Structure of Metals, third ed., McGraw-Hill, New York, 1966.

[7] W.C. Leslie, The Physical Metallurgy of Steels, McGraw-Hill Book Company, 1981.

[8] J. Hirth, Factors contributing to brittle fracture in metals, in: M. Meshii (Ed.), Mechanical Properties in BCC Metals, TMS-AIME, Warrendale, PA, 1982, pp. 181–187.

[9] D.O. Wilshynsky-Dresler, *Recrystallization Behavior of Interstitial-Free Steels*, PhD Thesis, Colorado School of Mines, Golden, Colorado, 1992.

[10] W.B. Hutchinson, Development and control of annealing textures in low-carbon steels, International Metallurgical Reviews 29 (No. 1) (1984) 25–42.

[11] Special Issue on New Aspects of Microstructure in Modern Low Carbon High Strength Steels, *ISIJ International*, 35 (1995) 937–1041.
[12] M. Cohen, The strengthening of steel, TMS-AIME 224 (1962) p638–656.
[13] M.S. Wechsler, D.S. Lieberman, T.A. Read, On the theory of the formation of martensite, Trans. AIME 197 (1953) 1503–1515.
[14] S. Morito, J. Nishikawa, T. Maki, Dislocation density within lath martensite in Fe–C and Fe–Ni alloys, ISIJ Int. 43 (2003) 1475–1477.
[15] A.R. Marder, G. Krauss, The morphology of martensite in iron-carbon alloys, Trans. ASM 60 (1967) 651–660.
[16] A.R. Marder, G. Krauss, The formation of low-carbon martensite in Fe–C Alloys, Trans. ASM 62 (1969) 957–964.
[17] H. Kitahara, R. Yoshida, N. Tsuji, Y. Minamino, Crystallographic features of lath martensite in low carbon steel, Acta Mater. 54 (2006) 1279–1288.
[18] C.A. Apple, R.N. Caron, G. Krauss, Packet microstructure in an Fe-0.2% C martensite, Metall. Trans. 5 (1975) 593–599.
[19] R.A. Grange, C.R. Hribal, L.F. Porter, Hardness of tempered martensite in carbon and low-alloy steels, Metall. Trans. A 8A (1977) 1775–1785.
[20] *Atlas zur Warmebehandlung der Stahle*, Vol. 1–4, Max Plank Institut fur Eisenforshung, in cooperation with the Verein Deutscher Eidrnhuttenleute, Verlag Stahleisen, MBH, Dusseldorf, 1954–1976.
[21] W.E. Jominy, A.L. Boegehold, A hardenability test for carburizing steel, Trans. ASM 26 (1938) 574–606.

Deep drawable steels*

5

P. Ghosh[1] *and R.K. Ray*[2]
[1]Tata Steel, Jamshedpur, India, [2]Indian Institute of Engineering Science and Technology, Shibpur, West Bengal, India

5.1 Introduction

Steels are widely used in the automobile industry for making auto body parts. Steel sheets are subjected to high to severe deformation for the purpose of producing complex and intricate shapes of different automobile body parts, such as the hood, fender, door inner, door outer, etc. The ability to be deep drawn, or "deep drawability," is one of the most important properties for this kind of application, and the steels used for this purpose are called *Deep Drawable Steels.*

Deep drawing, by definition, is a metalworking process used for shaping flat sheets into cup or box shaped articles. This is done by placing a blank of appropriate size over a shaped die, followed by pressing the blank into the die with a punch [1]. Deep drawing operation is schematically shown in Fig. 5.1 after Dieter [1].

5.1.1 Measure of deep drawing

The Lankford parameter (also known as plastic strain ratio and denoted by the letter *r*) is a measure of the plastic anisotropy and is used extensively as an indicator of the drawability of sheet metals [2]. The Lankford parameter (*r*) is defined as the ratio of true in-plane strain (i.e., strain in width direction) to true through thickness strain. Therefore:

$$r = \frac{\text{true in}-\text{plane strain}}{\text{true through thickness strain}} = \frac{\varepsilon_w}{\varepsilon_t} = \frac{\ln \frac{w}{w_0}}{\ln \frac{t}{t_0}} \tag{5.1}$$

where ε_w *and* ε_t are the true strain in width and in thickness direction, *w* and *t* are the initial width and thickness while w_0 and t_0 are the final width and thickness, respectively.

For sheet metals, the *r* values are normally determined for three different directions of loading in-plane (0 degrees, 45 degrees, and 90 degrees to the rolling direction and

* Every effort has been made to trace copyright holders and to obtain their permission for the use of copyright material. The publisher apologizes for any errors or omissions in the acknowledgements printed in this book and would be grateful if notified of any corrections that should be incorporated in future reprints or editions.

Automotive Steels. DOI: http://dx.doi.org/10.1016/B978-0-08-100638-2.00005-5

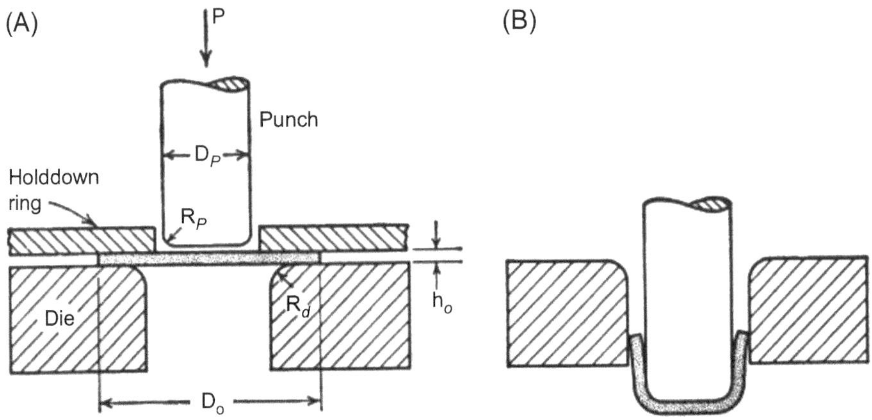

Figure 5.1 Deep drawing of a cylindrical cup. (A) Before drawing; (B) After drawing [1].

denoted by r_0, r_{45}, and r_{90}, respectively) and the normal or average r-value (r_m) or Normal Anisotropy is measured as follows:

$$r_m = \frac{1}{4}(r_0 + 2r_{45} + r_{90}) \tag{5.2}$$

The other important parameter involving r is the planar r value or Planar Anisotropy (r_p) which is a measure of the variation of r with rolling direction, and is expressed as follows:

$$r_p = \frac{1}{2}(r_0 - 2r_{45} + r_{90}) \tag{5.3}$$

In deep-drawing applications, both normal as well as planar anisotropy are important design parameters. The higher the Normal Anisotropy and the lower the Planar Anisotropy, the better is the sheet for forming operation.

It has been found by several researchers that the deep drawability of a low carbon steel strip is significantly influenced by the crystallographic texture of the steel. It is clear from this plot that the higher the volume fraction of {111} planes in the rolling plane (i.e., the γ fiber), the higher the deep drawability [3].

Besides the Lankford parameter, the other important factor for sheet metal formability is its ability to strain harden during forming operation. However, the strain hardening ability of a material plays very little role in the process of deep drawing, and thus stretch formability has to be measured. This is measured by the strain hardening exponent (n) which is defined by the slope of the true stress–true strain curve in a tensile test and expressed as:

$$\sigma = k\varepsilon^n, \text{ and } n = \frac{\triangle \log\sigma}{\triangle \log\varepsilon} \tag{5.4}$$

where σ is the true stress, ε is the true strain, k is the strength coefficient, and n is the strain hardening exponent. Higher the n, better is the homogeneous distribution

Table 5.1 **BS EN 10130:1991 cold—rolled carbon steel flat products for cold forming [6]**

Grade	R_m, N/mm^2	A_{80}, % min	r_{90}, min	n_{90}, min
Fe P01	270/410	28		
Fe P03	270/370	34	1.3	
Fe P04	270/350	38	1.6	0.18
Fe P05	270/330	40	1.9	0.20
			r_m, min	n_m, min
Fe P06	270/350	38	1.8	0.22

of strain across the sheet metal and in turn higher is the resistance to localized thinning, leading to superior formability.

5.1.2 Classification of steels based on drawing capacity

Low and ultra-low carbon sheet steels in cold rolled and annealed conditions are generally used where forming operation is needed. Depending on their forming ability, low and extra low carbon steels are classified as follows [4]:

- CQ (commercial quality) or CS (commercial steel)
- DQ (drawing quality) or DS (drawing steel)
- DDQ (deep drawing quality) or DDS (deep drawing steel)
- EDDQ (extra deep drawing quality) or EDDS (extra deep drawing steel)
- Superformable sheet steel

The drawability and elongation of formable low carbon steels increase from CQ through superformable sheet steel grades [5].

The cold rolled and annealed low carbon steel flat products supplied for cold forming operation in Europe are covered by European standard EN 10130:1991 and designated as Fe P01 to Fe P06 but with Fe P02 missing. The highest grade, Fe P06 has the highest formability while Fe P01 has the lowest. Therefore, Fe P06 needs the stringent requirements for *r*-value and *n*-value. Apart from formability, the difference between the grades lies in their acceptable strain aging limit. As expected, grade Fe P06 is considered as being completely nonaging, whereas a small degree of aging is acceptable for Fe P03 to Fe P05, provided the steel is suitable for use within up to 6 months of storing. Fe P01 has even higher acceptable degree of aging [6]. The mechanical properties of these grades are tabulated in Table 5.1.

5.1.3 Grades of deep drawing steels

The formable low carbon steels are generally subdivided into two categories, on the basis of composition:

- Al-killed (AK) steels: These steels are more of a traditional kind of formable steels in which carbon is relatively free and may form iron carbides. Al is intentionally added in these steels to control the final properties, using aluminum nitride (AlN) precipitation. Apart from this Al also acts as deoxidizer

- Interstitial Free (IF) Steels: These steels are a kind of special steels and contain very a low amount of carbon (and nitrogen). On top of it, this carbon (and nitrogen) is tied up with carbide (and nitride) forming elements, such as Ti and/or Nb to make the matrix free of interstitial carbon (and nitrogen)

R_m, A_{80}, r_{90}, n_{90}, r_m, n_m stand for tensile strength (minimum and maximum value), elongation at fracture with 80 mm initial length, Lankford Parameter measured 90 degrees to rolling direction, Strain Hardening Exponent measured along transverse direction, average Lankford parameter, average Strain Hardening Exponent; respectively.

IF steels typically contain very low carbon, below 0.005 wt.%, whereas AK steels usually have a carbon content above 0.015 wt.%. IF steels are nonaging type but AK steels can be nonaging or partially aging type, depending on their compositions and the processing variables. The IF steels are the only type of steel that can satisfy the formability requirement of the Fe P06 specification. AK steels are generally used to satisfy the requirements of the Fe P01 and Fe P03 grades, and sometimes can also be applied to satisfy the demands of Fe P05 grade as well. However, IF steels usually always satisfy the requirements of the Fe P05 grade.

Although both come under the gambit of low carbon steels, the underlying methods of manufacturing AK steels and IF steels are quite different. Both AK and IF steels are produced via batch annealing (BA) and continuous annealing (CA) routes. However, depending on the annealing route chosen, the final property can be distinctly different for steels, even with similar composition and prior processing history. For example, better property of AK steels can be achieved in the BA route whereas it is much easier to obtain higher formability by the CA route in the case of IF steels. Details of these *Deep Drawable Steels* are discussed further in the following sections.

5.2 Aluminum killed (AK) steels

The main elements which influence the properties of an AK steel are its aluminum, nitrogen, carbon, and manganese content. A typical process route for a low carbon AK steel is shown in Fig. 5.2. At slab reheating temperature (SRT), typically between 1200°C and 1250°C, AlN is in solid solution and generally remains so till hot rolling finishes. During the coiling stage of hot rolled strips, temperature plays a very critical role. If coiling is performed at relatively high temperature, i.e., around 710°C, the large, tightly wound coil with high thermal mass will cool very slowly and afford the opportunity for precipitation of AlN. On the contrary, if the hot rolled material is cooled quickly to a temperature of about 560°C before coiling, the precipitation of AlN is suppressed and both aluminum and nitrogen will be retained in solid solution after cooling to room temperature [7].

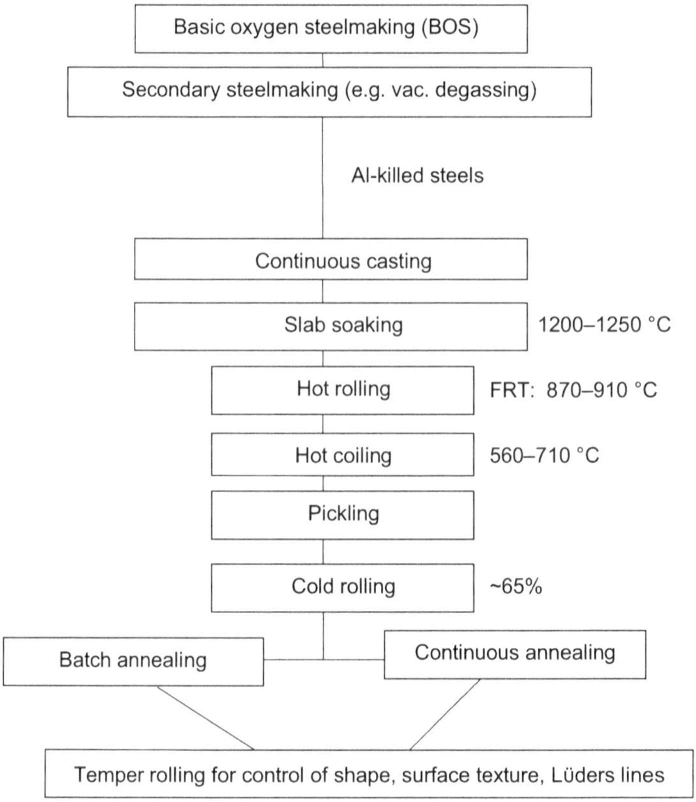

Figure 5.2 Schematic representation of a typical process route for low carbon-aluminum killed steel [7].

Cold rolled hot bands of AK steel can be subjected to either BA or CA. However, the underlying mechanisms are slightly different as far as the AlN precipitation is concerned. These aspects are discussed in detail in the following sections.

5.2.1 Batch annealed AK steels

The batch annealed aluminum killed steels are characterized by:

1. A strong {111} texture.
2. A large ferrite grain size.
3. Low solute carbon and nitrogen content.

As already stated above, good deep drawability in low carbon steels can be achieved by controlling crystallographic texture. Development of strong {111} cube-on-corner and reduction of {100} cube-on-face texture components result in

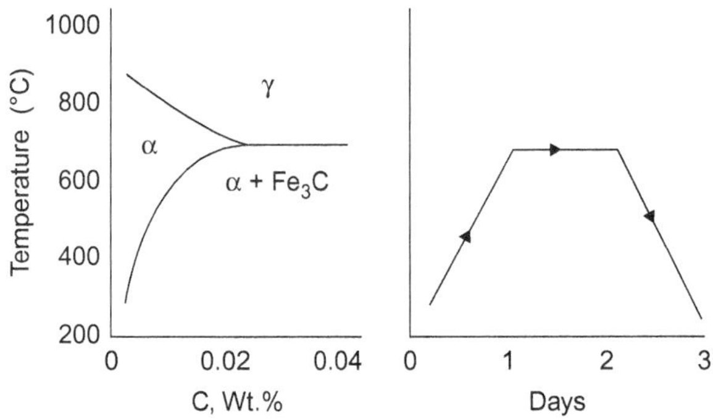

Figure 5.3 A typical batch annealing cycle for Al-killed steel [7].

good r_m value. In AK steels, texture control is carried out by controlling the precipitation of fine AlN precipitates. This is obtained by keeping Al in solid solution prior to the BA treatment. Thus, AK BA steels are cooled quickly from hot rolling temperature to coiling stage in order to avoid the precipitation of AlN during cooling. Further, precipitation of AlN during coiling is suppressed by keeping coiling temperature (CT) low (~560°C).

During BA, due to their large mass, the cold rolled strips experience very slow cooling and heating rates. Thus BA operation can be continued for several days depending on the level of formability required. The strips are heated to a temperature of 700°C, close to Ac_1 temperature, soaked at that temperature for several days and then cooled to ambient temperature slowly. The BA process of AK steel is shown in Fig. 5.3. In general, the recrystallization of the deformed structure starts in the temperature range of 500–550°C. However, during the slow heating period to annealing temperature, AlN precipitation takes place and retards the recrystallization process substantially. AlN preferentially forms at sub-grain boundaries and inhibits recovery and sub grain growth, thereby increasing the recrystallization temperature of the strips. This helps the beneficial {111} component to be the strongest component in the recrystallized texture. However, the best texture, measured by the highest r_m value, is obtained by an optimum inhibition of recrystallization. This is caused by a drag on grain boundary movement due to AlN particles, which decays slowly as a result of particle coarsening [8]. This slow coarsening of AlN particles leads to preferential development of {111} texture and a coarser grain size due to restricted nucleation of other components during soaking period. Batch annealed AK steel develops a relatively coarser grain size of ASTM 5-6. The higher limit of soaking temperature is restricted to ~730°C in order to avoid both the formation of coarse carbide and the risk of sticking of adjacent laps in the coil. The highest solubility of carbon in ferrite in AK steels is found at around a temperature of 700°C. A slow rate of cooling from 700°C to room temperature results in the precipitation of most of the carbon from the solid solution, in the form of iron carbide which aid in development of strong {111} texture in the recrystallized annealed steel [6,7].

5.2.1.1 Importance of chemical composition

The most important factor in BA of AK steels is the control of precipitation of AlN, and composition is the key. It has been found that the optimum nitrogen and soluble aluminum contents are 0.005–0.01 wt.% and 0.025–0.04 wt.%, respectively. A favorable combination of nitrogen and aluminum not only inhibits the recrystallization process by the formation of AlN precipitates, but also results in coarsening of the particles in a way that will allow the drag on grain boundary movement to decay with the optimum speed, as the recrystallization process progresses. The end result is the development of favorable {111} texture in the final annealed steels and a high r_m value. Table 5.2 lists a typical composition of Batch Annealed AK steels [4].

Apart from nitrogen and aluminum, manganese has an effect, mainly detrimental, on r_m value. However, its effect is greatly influenced by the carbon content as can be understood from Fig. 5.4. This can be attributed to the tendency of Mn forming dipoles with carbon atoms and subsequent impairment of {111} texture formation [10].

Table 5.2 Typical composition range of important alloying elements in BA AK steel

Elements (in ppm)	C	N	Al	Mn
	200–500	<100 (typically 40)	200–500	1500–2000

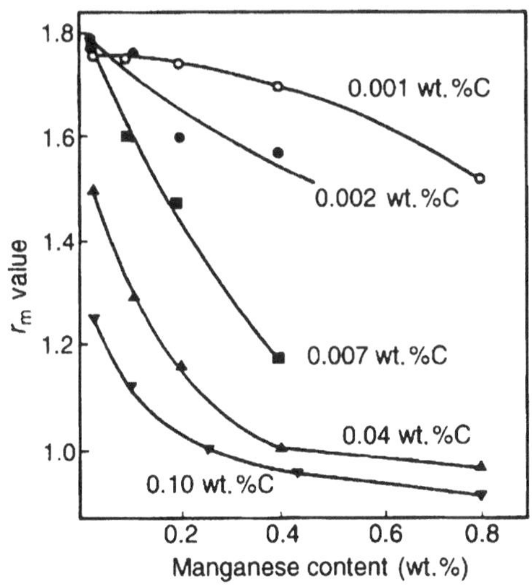

Figure 5.4 Effect of manganese content on r_m values of cold rolled and annealed steels with various carbon contents, heating rate 100°C/h, held at 700°C for 1 h [10].

5.2.1.2 Importance of processing parameters

In conjunction with right amount of alloying elements, processing parameters play a very important role in deciding the final outcome of these steels. As expected, every step during the processing is critical to the final properties of these steels.

Slab reheating temperature (SRT)

In order to take advantage of the AlN precipitation, it must be dissolved at SRT and the same must be held there during and after coiling and cold rolling, i.e., before the BA process.

The solubility of AlN in austenite is given by [11]:

$$\log[\mathrm{Al}][\mathrm{N}] = -6770/T + 1.033 \quad (5.5)$$

where T is the temperature in Kelvin and [X] represents the weight percentage of element X in solid solution. As clear from the above equation, relatively high SRT, commonly 1250°C, is required to take most or all of aluminum and nitrogen in solid solution.

Finish rolling temperature (FRT)

FRT during hot rolling should be in the single phase austenite region to avoid similar fall in r_m value, as shown in Fig. 5.5 [12]. When FRT is lower than the austenite to ferrite transformation temperature (i.e., Ar_3), abnormal coarse grain structure is developed which adversely affects the {111} recrystallization structure of low carbon steels after cold rolling and annealing process, resulting in deterioration of deep drawability. The ideal FRT should be kept below no recrystallization (T_{nr}) temperature of austenite in order to produce pancaked austenite grains. Ferrite grains transformed from pancaked austenite structure are finer, and this leads to better drawability [12].

Coiling temperature (CT)

Low CTs below 600°C are used to avoid substantial precipitation of AlN in the hot band, otherwise a deterioration of formability (i.e., r_m value) is noticed,

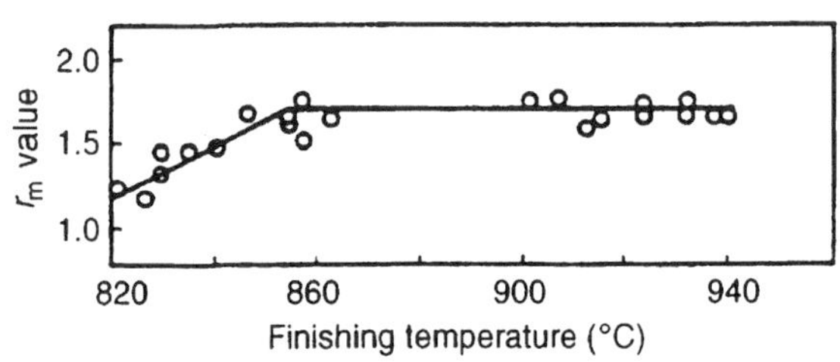

Figure 5.5 Variation of r_m value with finishing temperature for batch annealed, aluminum killed steel coiled at 525–565°C. T.R. Parayil, I. Gupta, in: L.G. Kuhn, F.A. Hultgren (Eds.), Mechanical Working and Steel Processing XXVII, AIME, Warrendale, 1989, p. 314. (13), with permisssion from AIST.

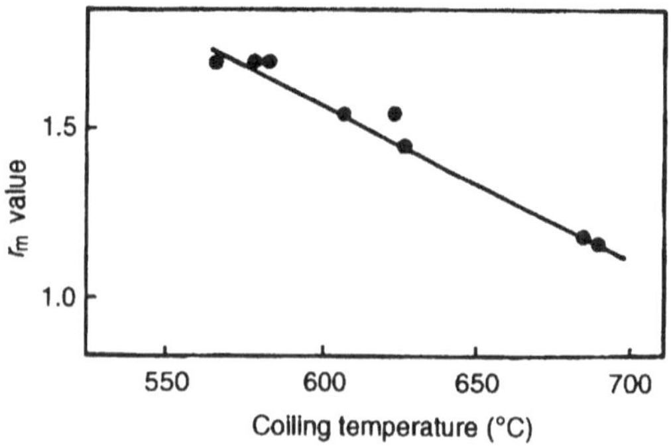

Figure 5.6 Effect of coiling temperature on the r_m value of a batch annealed aluminum killed steel with Al content of 0.036 wt.% [13].

as shown in Fig. 5.6 [13]. Although the solubility of AlN is very low in ferrite, its nucleation and growth kinetics are governed by the diffusivity of Al which is very limited below 600°C.

Batch annealing heating rate

The other important processing variable is the heating rate to BA temperature. The general tendency for r_m value is to decrease with increasing heating rate. The optimum heating rate, which results in larger grain size and a decrease in yield strength (YS), is found to be dependent largely on Al content as well as on the nitrogen, manganese and the amount of cold reduction. An idea of optimum heating rate can be obtained from the following empirical equation [14]:

$$\log(\text{optimum heating rate}) = 18.3 + 2.7\log[\text{N}][\text{Mn}]/\text{CR} \tag{5.6}$$

where optimum heating rate is expressed in °C/hour and [X] represents the weight percentage of element X in solid solution.

5.2.2 Continuously annealed AK steels

The basic difference between BA and CA of AK steels is the very short time available in the latter process with respect to the former. This lack of time affects the following to a great deal:

- Production of large grain size
- Formation of strong {111} texture
- Total or nearly total precipitation of N and C to make the matrix relatively free of the interstitials

However, the above three are precisely the necessary conditions to be met in order to match the properties of the batch annealed AK steels. A change in composition and processing parameters must be undertaken so that CA may result in properties comparable to those of BA steels. In BA, precipitation of AlN takes

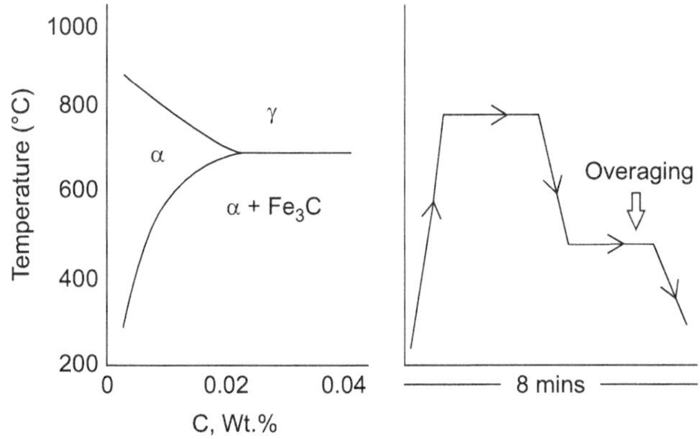

Figure 5.7 A typical annealing cycle for continuously annealed aluminum killed steels [7].

place during the slow heating step. However, due to the fast kinetics of CA, precipitation of AlN can't take place during heating. In order to reduce the level of nitrogen in solid solution, the hot rolled strip, scheduled to undergo CA, is coiled at higher temperature [7]. The grain size problem is taken care of by employing a number of scavenging reactions involving aluminum, nitrogen, carbon, manganese, and sulfur to purify the iron matrix and by using a higher annealing temperature, normally above Ac_1 [15]. Because of the short time and continuous nature of the strips going through the furnace in CA process, the problem of sticking and appreciable thermal deformation do not take place in the strips in spite of ~100–150°C higher temperature than that of BA process. In the same way suitable texture can also be obtained in CA-AK steels. An extra step, so called over-aging, is incorporated in the CA process to take care of the carbon in solution where the strip is kept at 350–450°C for a few minutes followed by a slow cooling from that temperature.

A typical industrial CA process for AK steels is depicted in Fig. 5.7 [7]. Typically, the stages in the cycle include a fast heating to 700–850°C in less than a minute, holding there for 40 seconds, cooling to an over-aging temperature range of 350–450°C, holding there for around 3 minutes, followed by cooling to ambient temperature.

5.2.2.1 Importance of chemical composition

As already stated, the short annealing time during CA process imposes close control over alloying elements. Stricter composition control in CA-AK steels helps steel manufacturers to obtain the right combination of grain size, texture, and microstructure for deep drawing property.

Amongst alloying elements, Carbon is very crucial. Free carbon in solution directly affects {111} texture formation adversely. It has been observed that in spite of the over-aging treatment, some amount of carbon can still remain in solid solution. Ono et al. [16] have shown that n-value progressively increases with reduction of carbon in solution but r_m value initially increases and reaches a plateau at

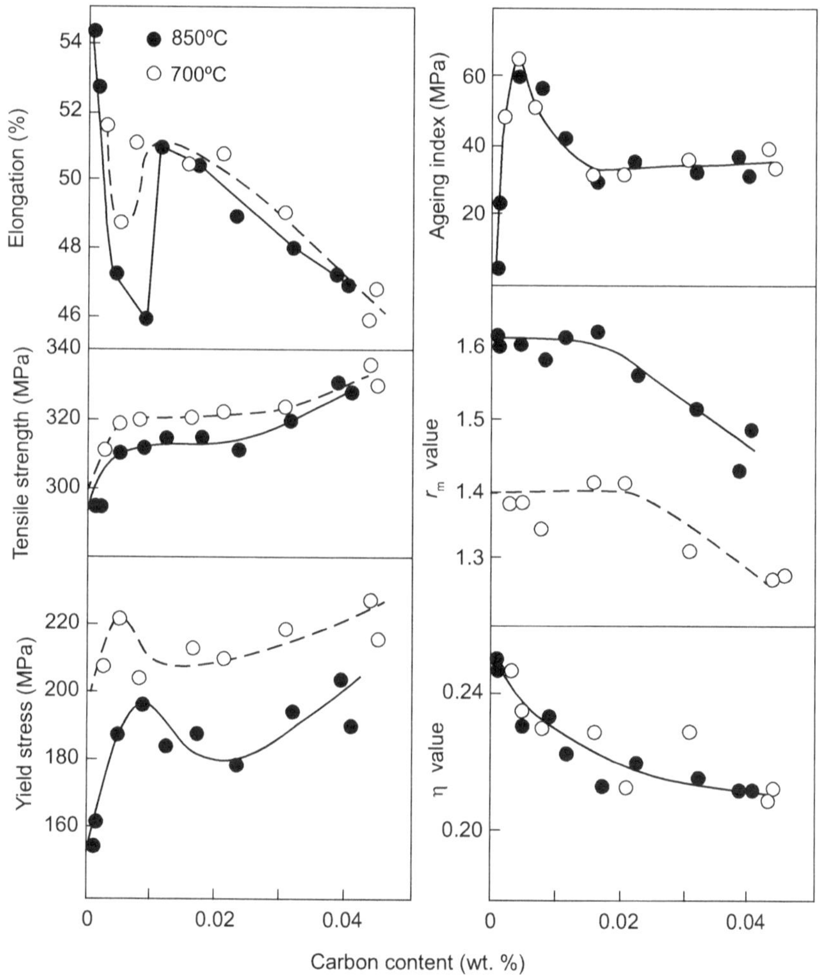

Figure 5.8 Variation of the mechanical properties of aluminum killed steel with carbon content, continuously annealed at 700°C and 850°C. S. Ono, O. Nozoe, T. Shimomura, K. Matsudo, B.L. Bramfitt, P.L. Manganon (Eds.), Metallurgy of Continuously Annealed Sheet Steel, AIME, Dallas, 1982, p. 99.
Source: Reprinted with permission of The Minerals, Metals & Materials Society.

C ~0.02 wt.% and below. Moreover, reduction in carbon below this level causes an increase in the aging index, and yield and tensile strength and a reduction in the elongation value. These are attributed to the low driving force of carbide precipitation during over-aging process of strips containing a very low solute carbon. Taking the above facts into consideration, the carbon content in CA steels is normally kept to a level of around 0.02–0.03 wt.%, which is 0.01–0.02 wt.% less than that of BA AK steels. The effects of carbon content on the mechanical properties of CA-aluminum killed steels are illustrated in Fig. 5.8.

In low carbon steel, smaller grain size is found to be detrimental to r_m value. Presence of excess solutes makes grain boundary movement difficult by solute drag effect. In CA, due to the very short annealing time, this problem is more and thus CA steels generally yield smaller grain sizes. Apart from increasing the annealing temperature, the other effective way to compensate for this effect is the removal of solid solution elements from the matrix. Both of these practices promote grain growth.

In this context, it is beneficial to keep manganese content to a level just adequate to fix the amount of sulfur which is present as impurity elements in steels. Toda et al. [17] defined a parameter, called K, which is related to Mn and S (both in wt.%) content as follows:

$$K = [\mathrm{Mn}] - 55/32[\mathrm{S}] - 55/16[\mathrm{O}] \tag{5.7}$$

So, when $K = 0$, manganese is at stoichiometric level to fix all the sulfur and oxygen present in solution. When K is negative, some sulfur and oxygen will still be there in the matrix, whereas when K is positive, excess manganese will be in solid solution. Although the above equation was developed for rimming steels, it was found to be equally applicable for aluminum killed steels as well, provided the oxygen term is set at 0.

5.2.2.2 Importance of processing parameters

Apart from composition control, processing parameters have to be designed very carefully to obtain good formability in aluminum killed steels. Very often it is found that the requirement of CA steels is just the opposite of the BA aluminum killed steels. Effects of processing parameters in CA-AK steels are briefly discussed below.

Slab reheating temperature (SRT)

The requirements of SRT in BA and CA AK steels are the complete opposite. As already discussed, SRT should be high in BA-aluminum killed steel in order to take AlN in solid solution By contrast, in case of CA-aluminum killed steels SRT is kept somewhat low so that some of the AlN particles remain in the structure to act as nucleation sites for fresh AlN during coiling after hot rolling. Thus, in CA-aluminum killed steels a progressive decrease in yield stress is noticed with decreasing SRT, as shown in Fig. 5.9 [18].

Finish rolling temperature (FRT)

A FRT in fully austenitic region is preferred in CA-aluminum killed steels as well.

Coiling temperature (CT)

Unlike BA-AK steels, higher CT is used in CA process to fix nitrogen with aluminum during the coiling stage itself and thus solute nitrogen is no longer

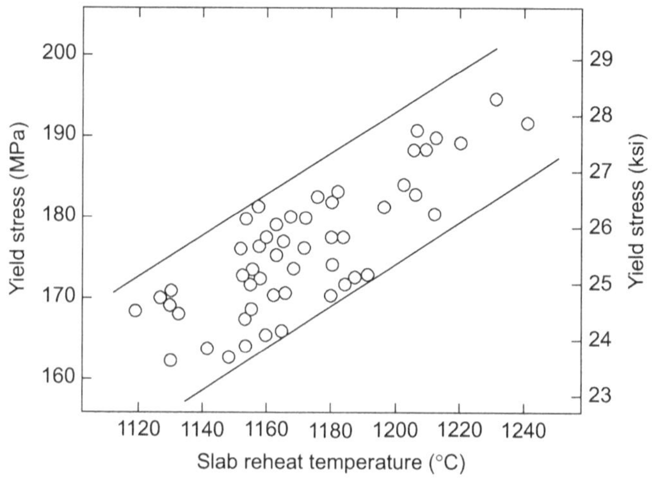

Figure 5.9 Variation of yield strength with reheating temperature for continuously annealed aluminum killed extra low carbon steel. N. Prum, U. Meers, H. Mathy, P. Messien, V. Leroy, in: R. Pradhan, G. Ludkovsky (Eds.), Proceedings of Symposium on Hot and Cold Rolled Sheet Steels, 3, The Metallurgical Society, Cincinnati, 1988.
Source: Reprinted with permission of The Minerals, Metals & Materials Society.

available to affect the grain growth process during recrystallization annealing. Fig. 5.10 shows these effects schematically. As can be clearly seen the effect is greater in high carbon steel than in the low carbon variety [16].

Cold rolling and annealing

It has been found that a very high cold reduction level, as high as 90%, may be required to achieve highest formability in continuous annealed aluminum killed steels.

However, it is the annealing cycle which is primarily responsible for attaining a very high r_m value in CA-aluminum killed steels. Thus it is needless to say that each part of annealing cycle has to be designed very carefully to attain the required properties. Suitable texture and high grain size are attained by scavenging the solute elements in combination with adopting a high annealing temperature. A high annealing temperature around 850°C is generally employed to yield better r_m value. The effect of annealing temperature on the final properties of CA-aluminum killed steels is given in Fig. 5.11 [16]. However it is not only the temperature during annealing but also the cooling strategy after annealing which is equally important. The aim of this part of the annealing cycle is to re-precipitate carbon, which is inevitably taken into solution during annealing, down to a sufficiently low level to give a very small strain aging effect at room temperature. This is normally done by rapid cooling to over-aging temperature from 700°C. Rapid cooling from 700°C ensures higher carbon in solid solution just before the over-aging step, which results

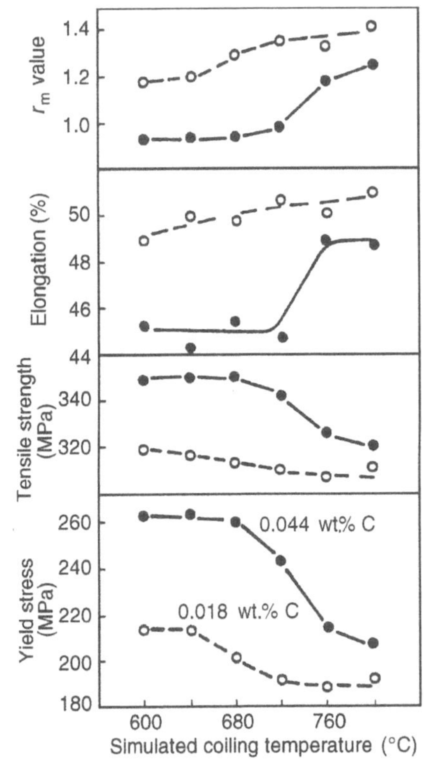

Figure 5.10 Effect of coiling temperature on the mechanical properties of continuously annealed aluminum killed steels with carbon content of 0.018 and 0.044 wt.%. S. Ono, O. Nozoe, T. Shimomura, K. Matsudo, B.L. Bramfitt, P.L. Manganon (Eds.), Metallurgy of Continuously Annealed Sheet Steel, AIME, Dallas, 1982, p. 99.
Source: Reprinted with permission of The Minerals, Metals & Materials Society.

in re-precipitation of freshly produced smaller carbide particles on grain boundaries as well as inside the grains of ferrite. Otherwise, carbide precipitation will occur during slow cooling from the annealing temperature to the over-aging temperature, resulting in less amount of carbon in solution just prior to over-aging stage. This leads to precipitation of carbide only on the grain boundaries of ferrite, which is a time consuming process due to involvement of long-range diffusion. This is the opposite of rapid cooling process where short-range diffusion of carbon takes place due to higher amount of the same in the solution. In short, without the rapid cooling step, carbon may be left in iron matrix after the annealing treatment, which will lead to strain aging at room temperature. Over-aging temperature should also be chosen carefully. The re-precipitation proceeds quickly at high over-aging temperature of ~450°C due to high diffusion coefficient but is very sluggish at around 250°C due to the low rate of diffusivity of carbon. However, higher over-aging temperature results in relatively bigger sized carbides precipitates spaced far apart. This is

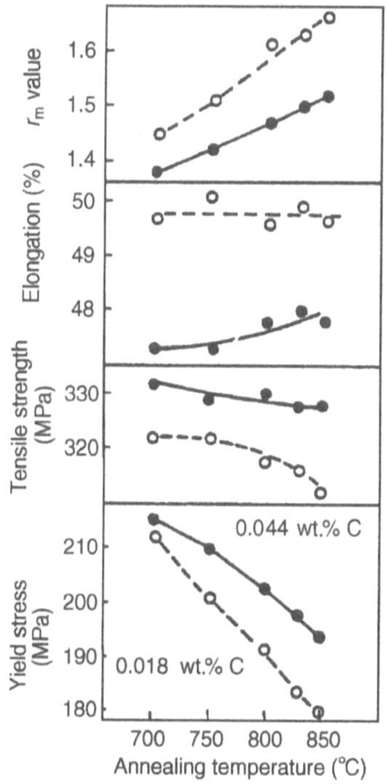

Figure 5.11 Variation of mechanical properties of an aluminum killed steel annealed for 1.5 min at different annealing temperatures. S. Ono, O. Nozoe, T. Shimomura, K. Matsudo, B.L. Bramfitt, P.L. Manganon (Eds.), Metallurgy of Continuously Annealed Sheet Steel, AIME, Dallas, 1982, p. 99.
Source: Reprinted with permission of The Minerals, Metals & Materials Society.

disadvantageous as it produces steels with lower YS. In order to take advantage of both high and low over-aging temperatures, strips are fast cooled to below 300°C to allow precipitation of closely spaced carbide nuclei and then reheated slightly above 350°C, to take advantage of the high diffusion rate of carbon [19].

5.3 Interstitial free (IF) and interstitial free high strength (IFHS) steels

It has been known for quite some time that reducing the carbon content of steel to a very low level is in itself sufficient to improve the r_m value. The development and application of vacuum degassing technology in the steel industry and the increasing demand for deep-drawable coated steels by the auto manufacturers worldwide led to the development of a new class of steels, known as the IF steels. Vacuum degassing technology ensures the requirement of the interstitial elements, carbon and nitrogen, to be present in ultra-low amounts down to ~30–50 ppm to be met in these steels. Further addition of carbide and nitride forming elements such as Ti and/or Nb makes the matrix nearly completely free of the interstitials. The absence of carbon in these IF steels imparts their characteristic nonaging property. As carbon is already tied up with Ti and/or Nb during the annealing process, overaging treatment needs to be given during the CA stage. This makes the processing by CA more efficient for IF steels than for aluminum killed steels.

IF steels generally show low YS in the range of 140–150 MPa, which is attributed to the lack of interstitial carbon in the matrix. Their ultimate tensile strength (UTS) normally hovers around 280–290 MPa. The higher strength variety of IF steels, known as high strength IF (IFHS) steel has UTS around 390–440 MPa. Typically their YS is in between 220 and 320 MPa. As will be discussed later, the higher strength is achieved by the addition of solid solution strengthening elements such as Mn, P, and Si to normal IF steels. These additions increase strength but at the cost of formability and deep drawability.

IF steels are produced by the continuous casting—hot rolling—cold rolling—annealing route. As stated earlier, vacuum degassing technique is used to reduce the level of interstitial elements to a very low level, preferably below 30 ppm. Alloying elements, such as Ti and/or Nb, P, Mn, Si, are added depending on the final property requirement. Continuously cast slabs are reheated to the desired temperature, hot rolled and coiled. The hot rolled strips are further cold rolled and subjected to annealing treatment. Both BA and CA can be employed, although CA generally gives better r_m values.

Each and every step in the production process has its influence on the final microstructure and properties of IF steels. The most important factor in achieving the final properties in these steels is the precipitation process. Grain size, texture, mechanical properties, etc. are directly dependent on the precipitates formed in these steels. Thus a proper control of total precipitation behavior is expected to yield the desired properties. Keeping these facts in mind, the formation and role of precipitates in IF steels have been discussed in detail in a separate section.

5.3.1 Importance of chemical composition

Increasing the solute carbon level in IF steels leads to a decrement in the intensity ratio of {111}/{100} texture and thus reduces the r_m value drastically. It has also been observed that the above ratio steadily decreases with increasing carbon/sulfur ratio [4]. Fig. 5.12 shows the influence of C on the r_m value. Presence of interstitial elements degrades the beneficial {111} recrystallization texture in favor of the deleterious {110} and {100} type texture components [20]. Thus, in practice, the solute carbon level is kept below 0.003 wt.% in IF steels. Nitrogen has similar effect like carbon and therefore should also be kept below 0.003 wt.% in IF steels [21].

A minimum value of Ti and/or Nb is required to stabilize the total carbon and nitrogen in IF steels. Normally these elements are added in quantities higher than the stoichiometric amount required. The excess Ti and/or Nb, over the stoichiometric amounts, is considered to be the most important compositional parameter that

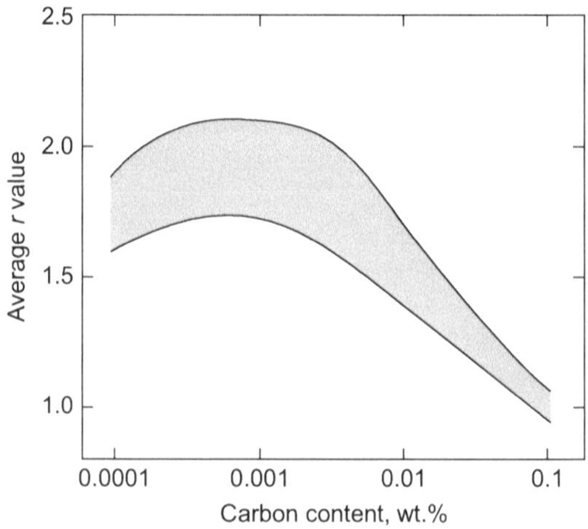

Figure 5.12 Effect of carbon content on the r_m value of low carbon steel. W.B. Huchinson, K.I. Nilson, J. Hirsch, Metallurgy of Vacuum Degassed Steel Products, TMS, Warrendale, PA, 1990, 109–126.
Source: Reprinted with permission of The Minerals, Metals & Materials Society.

determines the behavior of IF steels. A little excess of Ti and/or Nb seems to be beneficial in attaining high r_m value, however, too much of an excessive amount may deteriorate properties, especially in phosphorous added IF steels [22].

While the primary role of Ti is to tie up with carbon and nitrogen, making the iron matrix free of interstitial elements, Nb seems to play other roles too. Although Nb forms carbide, its effectiveness in tying up with carbon is much less than that of Ti. However, Nb retards the austenite recrystallization temperature, which has a direct influence on the r_m value. The {100} component of the transformation texture of ferrite results from the texture of recrystallized parent austenite. However, presence of solute Nb suppresses the austenite recrystallization process. The resulting hot band texture will then have a lower {100} component and thus a higher ratio of {111}/{100} in the product ferrite phase, giving rise to higher r_m and better deep drawability [23–24].

A small amount of sulfur has often been found to be somewhat beneficial in IF steels. Sulfur reacts with Ti and C to form carbosulfide precipitates at relatively high temperature. Formation of these precipitates over other carbon containing precipitates (i.e., TiC) is always more beneficial, as these are bigger in size and therefore do not adversely influence recrystallization and the subsequent grain growth process [25].

Manganese and phosphorus are the two most common alloying elements added to IF steel to enhance its mechanical strength, and the resulting grade is called the high strength version of the IF steel or IFHS steel. The addition of 0.1 wt.% P leads to an increment of more than 100 MPa in strength. Sometimes Si, Cr, and Cu are also used to increase strength but their effect is limited to 4–8 MPa per 0.1 wt.% addition. Although, the effect of Mn addition on the strength is quite similar to

Si/Cr/Cu but Mn is preferred over them because of its inexpensiveness as well as its less deleterious effect on overall properties of steels [21,26].

In general, normal IF steel contains very low amounts of Mn and P. Mn forms MnS and may act as nucleation sites for TiC precipitates upon cooling. Apart from that, Mn has a tendency to form dipoles with carbon atoms, which may reduce carbon availability for carbide/carbosulfide precipitation. This could possibly mean that a substantial amount of interstitial carbon atoms will remain in Fe-matrix due to high Mn addition. This will degrade r_m value by adversely affecting the {111} texture formation [26] in high Mn IF steels.

P does not have much effect due to its low concentration in IF steels. However, when their amounts are increased to higher levels, as in IFHS steels, they exert a huge effect on the precipitation behavior and texture formation and thereby on the r_m value. There is a higher risk of cold work embrittlement with increasing P content. This problem is taken care of to a great extent by the addition of a small amount of boron. Boron segregates to grain boundaries and forces P to remain in solid solution. However, in high P IFHS steels, the major influence of P lies on the precipitation behavior. Phosphorus forms a ternary precipitate FeTiP in abundance in IFHS steels. Formation of this precipitate takes out a lot of Ti from the matrix and leaves only little or none to tie up with carbon. As a result, there can be a lot of free carbon in the iron matrix which negatively influences the {111} texture formation. Very often it is found that these precipitates are fine enough to hinder the recrystallization and/or grain growth process. In both the ways higher P can degrade the r_m value, and thus the key to success in achieving high drawability in IFHS steels is to control FeTiP precipitation very effectively.

5.3.2 Importance of processing parameters

The performances of IF and IFHS steels are greatly influenced by the processing history of the steels, starting from slab casting through hot rolling, cold rolling, and annealing stage. During the steel-making process, it is important to ensure the stringent compositional requirements of these grades. Carbon and nitrogen contents are reduced to very low levels by vacuum degassing technique. On top of it, Ti and Nb addition is controlled in such a way that their amounts just reach the stoichiometric levels with respect to C and N or are just a little above it. Therefore, very careful control of alloying elements is absolutely necessary for obtaining satisfactory texture in the end product.

5.3.2.1 Steel making

Control of cleanliness is particularly important for the steel strips used for external parts of car bodies, where premium surface quality as well as excellent formability are required [27]. Ladle holding time and the time gap between finish of RH treatment to the start of casting is also important. During the holding period, oxide particles originating from deoxidation of the steel will have another chance to float up in the slag layer. To promote their removal, ladle holding time may be extended to longer than the 30 minutes period. Care has also to be taken to reduce reoxidation by air and entrapment of slag droplets in the initial stages of casting. Tundish sealing, argon gas purging and control of the casting start temperature are also important for this purpose. Moreover, a basic tundish flux has to be applied to increase the removal rate of alumina inclusions [28].

5.3.2.2 Slab reheating temperature (SRT)

Decreasing SRT is beneficial for improving the formability of IF steels. Low SRT results in smaller hot band grain size and reduces the number of fine precipitates; both of these aid in sharpening the {111} texture in the final product [29–31].

5.3.2.3 Finish rolling temperature (FRT)

Deep drawability gets improved when the FRT within the austenitic region decreases. This is generally attributed to the refinement and uniform distribution of the grain structure [32]. However, there are some ambiguities on this effect when FRT is either in the two-phase region or in the ferrite. Nevertheless, generally it is observed that finish rolling in ferritic region, at least, does not deteriorate the formability [26].

5.3.2.4 Coiling temperature (CT)

Higher CT is beneficial for improving the formability of IF steels. A higher CT produces bigger sized precipitates that do not interfere in the recrystallization and growth of {111} grains during annealing. Fig. 5.13 shows the effect of CT on the r_m value of an industrially produced IF steel [33].

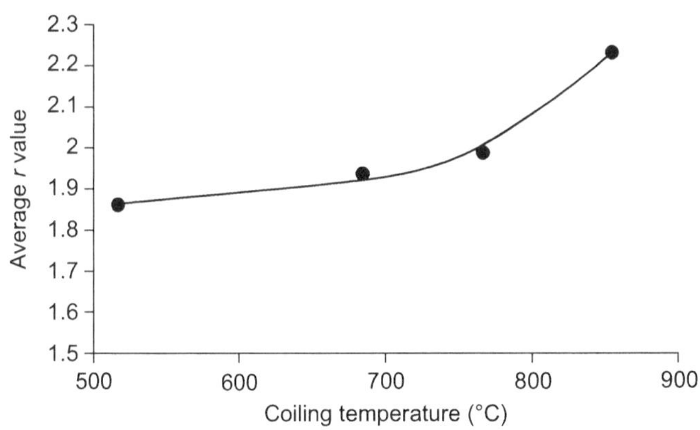

Figure 5.13 Effect of coiling temperature on the r_m value of an industrially produced IF steel [33].

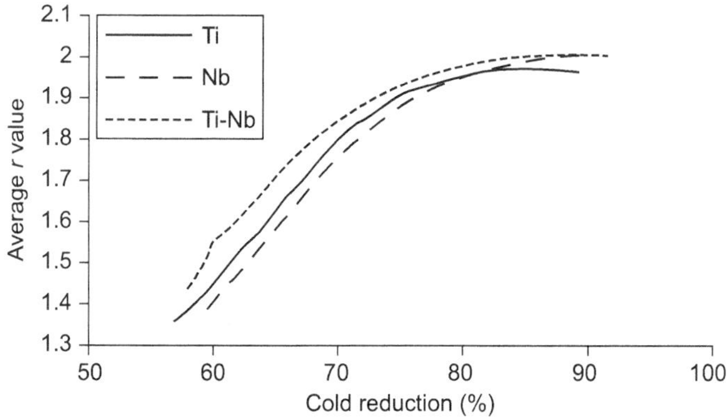

Figure 5.14 Effect of cold rolling reduction on the r_m values of Ti, Ti + Nb, and Nb IF steels. Y. Tokunaga, M. Tamada, US patent 4,504,326.
Source: United States Patent and Trademark Office, www.uspto.gov.

5.3.2.5 Cold rolling and annealing

Cold rolling reduction is an important factor that affects drawability of IF steel. It has been observed that increasing cold rolling reduction increases the r_m value, at least up to 90% [34]. This effect has been shown in Fig. 5.14. Perhaps the most important parameters in the production of a strong {111} recrystallization texture are the annealing temperature and the time of holding [35]. These two together, along with the amount of cold deformation, will determine the final texture. IF and IFHS steels with larger recrystallized grain sizes will favor attainment of higher r_m values, irrespective of the type of annealing treatment, i.e., batch or continuous. In case of continuously annealed IF and IFHS steels, the r_m usually increases with increasing annealing temperature until the transformation temperature is reached.

5.3.2.6 Role of precipitates in interstitial free steels

As stated earlier, the precipitation behavior in these steels plays a very crucial role in the formation of texture, which has a strong influence on the final properties. Precipitates do not only affect individually and directly, but synergistically and indirectly as well. Thus it is of utmost importance to understand the behavior of each type of precipitate, as well as of the combined effect of different types of precipitate particles. Again, any and every processing step that has a direct or indirect effect on precipitation, is worthy of consideration, in order to understand how to achieve the best properties in these steels.

In IF and IFHS steels, some alloying elements are deliberately added to form precipitates, especially in combination with the interstitials, carbon and nitrogen. Titanium and niobium are the most important of these precipitate forming elements.

There are other elements, such as sulfur, manganese, and phosphorus, which are inevitably present in steel and which can form precipitates during the course of processing. Normal levels of Ti and Nb in these steels range between 300–800 ppm and 0–300 ppm, respectively. Sulfur, Al, Mn, and P should be present within the range 20–100 ppm, 500 ppm, 700–1500 ppm, and 70–800 ppm, respectively. The precipitates that are generally observed in these steels are TiN, TiC, NbC, Nb(CN),Ti(CN), $Ti_4C_2S_2$, TiS, AlN, FeTiP, etc., depending upon the alloying elements and the processing parameters.

Although a similarity between the precipitation behaviors of IF and IFHS steels is quite expected, in reality there are substantial differences, which are discussed below.

Precipitation in IF steels

The normal IF steels are quite "clean," primarily due to the addition of the carbide forming element Ti. Titanium stabilizes carbon in two ways, by formation of $Ti_4C_2S_2$ and TiC. The amount of Ti required to stabilize the steels fully is given by the following relationships [34,36]:

$$Ti_{(STAB)} = 4C + 3.42N + 1.5S \tag{5.8}$$

assuming that the C is tied up as TiC, N as TiN, and S as TiS.

When stabilization occurs by the formation of $Ti_4C_2S_2$ and TiN, the amount of Ti required is given by:

$$Ti_{(STAB)} = 3.42N + 2(1.5S) \tag{5.9}$$

The amounts of the chemical elements mentioned above are all in wt.%.

However, when Ti content is even less than the amount required for stabilization through $Ti_4C_2S_2$, a completely different situation may develop. A subcritical amount of Ti will result in the presence of free carbon, which would then be available for the formation of a new and additional array of particles, formed as strain induced precipitates in the austenite during rolling or as precipitates in ferrite during coiling [37]. The dissolution of TiS and formation of $Ti_4C_2S_2$ are interdependent and very often the coexistence of sandwich like particles of TiS and $Ti_4C_2S_2$ are observed. Nb is generally added along with Ti, and preferably not as a single carbide forming element. Its main role is to refine the grain size by increasing the austenite recrystallization temperature, and to improve the ductile-brittle transition behavior by preferentially segregating at the grain boundaries, at the expense of P. It has been observed that as much as 67% of Nb added to steel can be in solid solution in a 0.06 wt.% Ti-IF steel [38]. In reality, a rather insignificant amount of Nb is actually present as precipitates [37].

The TiN particles form first, mainly during reheating and subsequent cooling to hot rolling temperature, followed by TiS, which very often forms epitaxially over TiN particles. With further decrease in temperature, carbon starts to diffuse

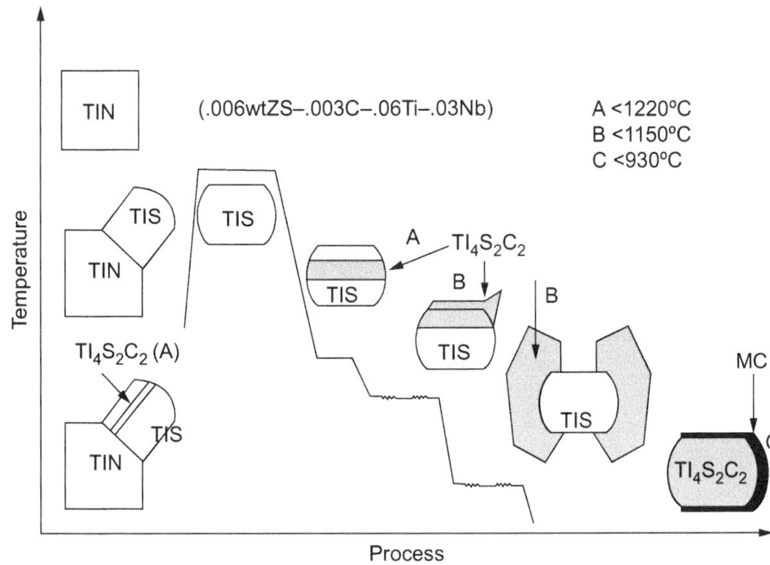

Figure 5.15 Schematic representation of precipitation sequence in Nb + Ti IF steel [37].

in TiS and forms $Ti_4C_2S_2$. On further lowering of the temperature to around 900°C, carbides form mainly on the existing particles of other precipitates. Apart from that, free-standing carbides may also form as strain induced precipitates during hot rolling of austenite or during the coiling operation in ferrite. A schematic representation of the precipitation sequence in IF steels is depicted in Fig. 5.15 [37].

Stabilization of carbon by $Ti_4C_2S_2$ over TiC formation is always beneficial for achieving higher r_m value in IF steels. Carbo-sulfide precipitates are generally bigger in size as they form at relatively high temperature as compared to TiC particles. Due to their bigger size, they pose little difficulty during recrystallization annealing and do not hinder the growth of beneficial {111} grains, thereby improving r_m value.

Precipitation in IFHS steels

Unlike the normal IF steels, IFHS steels can have numerous precipitates, depending on the P and Ti contents, and therefore these are certainly not as "clean" as the normal IF steels. The precipitation behavior in these steels is also a bit complicated. The primary difference in precipitation in these two grades of IF steels arises from the presence of FeTiP-type particles in the higher strength version. In IFHS steels, with higher amounts of P and Ti, the possibility of formation of these precipitates becomes pretty high, leading to a decrement in the r_m value. Fig. 5.16 shows this effect in two Ti added IFHS steels, with the help of calculated solubility diagram of different precipitates [39]. Increasing Ti content clearly leads to an increase in the volume fraction of FeTiP precipitates. Not only that, with increasing Ti content,

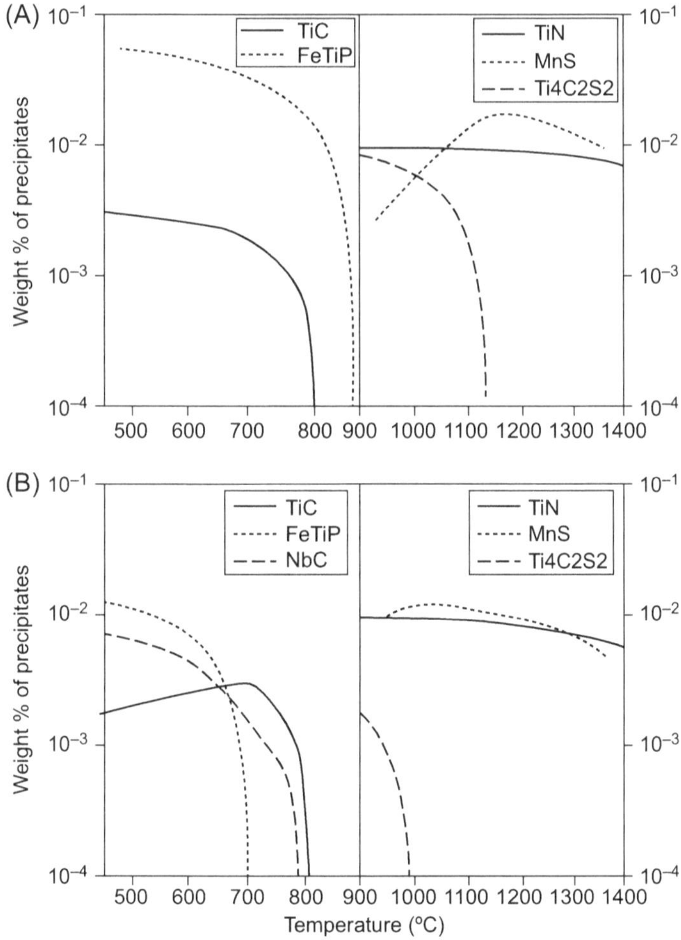

Figure 5.16 Calculated weight fraction of different precipitates formed at different temperatures (A) High Ti and (B) Low Ti content [39].

FeTiP tends to form at higher temperature and before the formation of TiC. Ghosh et al. [40] have shown that increasing P content has also a similar effect on the formation of FeTiP precipitates. An IFHS steel with high P content yields copious FeTiP-type precipitates in the final microstructure and this will significantly lessen the chance of TiC formation. As a result of this, carbon may be left free in the iron matrix and this will degrade the {111} texture. An abundance of small (normally 20–200 nm in size) FeTiP particles may pose a hindrance to the {111} grain growth process. Fig. 5.17 clearly depicts one such instance in a micrograph taken with a transmission electron microscope [41]. These two factors, namely, presence of free carbon in the matrix and pinning of grain boundaries around {111} grains during growth, together result in the formation of a poor gamma fiber,

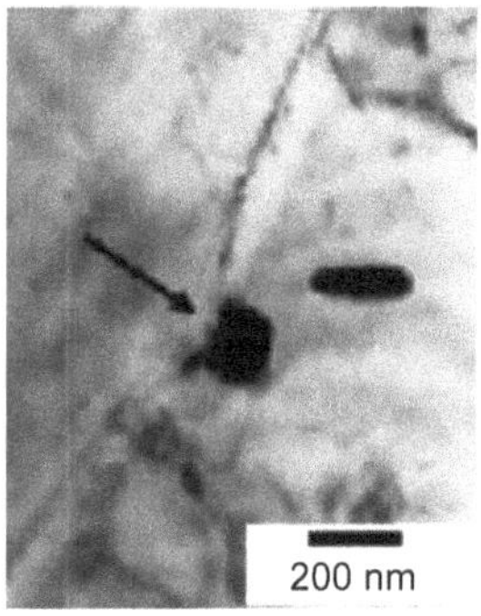

Figure 5.17 A TEM micrograph of a Ti-IFHS steel showing pinning of grain boundary by a FeTiP-type of precipitate [41].

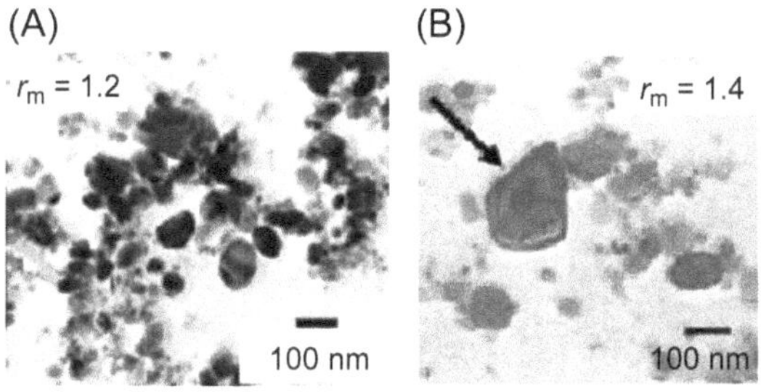

Figure 5.18 Images (TEM) showing perceptibly higher amount of FeTiP-type precipitates (A) in steel containing higher amount of P and Ti as compared to (B) steel containing lesser amount of P and Ti [40].

thus reducing the r_m value substantially. Fig. 5.18 illustrates the precipitation of FeTiP-type particles in two IFHS steels with varying P and Ti content. Presence of lower P and Ti clearly leads to lesser volume fraction of FeTiP-type of particles in the microstructure [40]. This is directly reflected in the increased r_m value, signifying superior drawability.

The solubility product versus temperature plots of FeTiP and TiC are found to be very close as shown in Fig. 5.19 [42–43]. Thus the sequence of formation of these two precipitates becomes very sensitive to composition and processing variables. Slight change in these variables may result in the formation of FeTiP precipitates before TiC has a chance to form. Formation of FeTiP particles will mean that most of the Ti will be exhausted, and it is quite possible that insufficient Ti may be left in the matrix to tie up with carbon. This will lead to the presence of free carbon in the iron matrix and therefore poor {111} texture development, thus impairing the deep drawability.

In another study by Ghosh et al. [41], presence of FeTiP was observed only after BA and not after CA treatment. BA was carried out at ~700°C for 24–25 hours

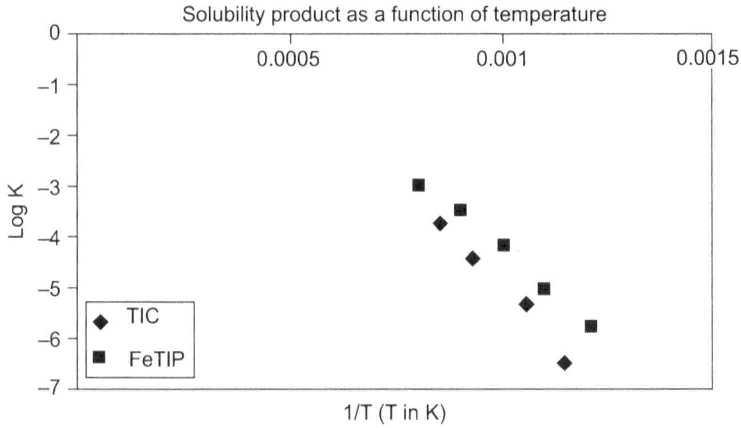

Figure 5.19 Solubility product of TiC and FeTiP as a function of temperature [42].
T. Gladman, Physical Metallurgy of Microalloyed Steels, Springer, Berlin, 1977.
Source: Reproduced by permission of Taylor and Francis Group, LLC, a division of Informa plc.

whereas CA was undertaken at the higher temperature of ~830°C and only for 1.5–2 minutes. Presence of FeTiP particles was not observed in the CA condition, instead small Fe-Ti type agglomerates were found occasionally. Due to their rather small size, the crystal structure of these agglomerates could not be identified. It is quite possible that at high temperature, there is an instantaneous formation of complexes of Fe-Ti, which then get enriched with P slowly diffusing into Fe-Ti from the matrix. This mechanism will lead to the formation of FeTiP when sufficient time is available, as in BA. The situation is quite similar to the formation of TiS-$Ti_4C_2S_2$ sandwich particles in normal IF steels. However, in CA condition P does not get the time to diffuse in Fe-Ti agglomerates, which usually remain as particles of Fe-Ti with some P in them. Thus BA-IFHS steels reveal numerous FeTiP-type particles in the microstructure, whereas CA-IFHS steels do not. However, Wang et al. [44] reported the presence of FeTiP particles in their CA-IFHS samples as well. They attributed the lower r_m value associated with the BA-IFHS steel with respect to their CA counterpart to the size of the individual particles, which is smaller in case of the BA steel.

The presence of Mn and the heating rate also play some important roles in IFHS steels. Higher heating rate, comparable to CA, is found to increase r_m value and strength simply by suppressing the formation of FeTiP. The presence of FeTiP not only degrades r_m but also reduces the strength of IFHS steels by removing the solid solution strengthening element P out of the matrix. Mn appears to suppress the formation of TiS and carbo-sulfides, forms Mn-C dipoles and increases solubility of carbon in Fe-matrix. All these adversely affect carbon tie up and indirectly promote FeTiP formation. A schematic diagram depicting the formation of different precipitates in a Ti-IFHS steel is shown in Fig. 5.20 [45]. TiN forms during steel making

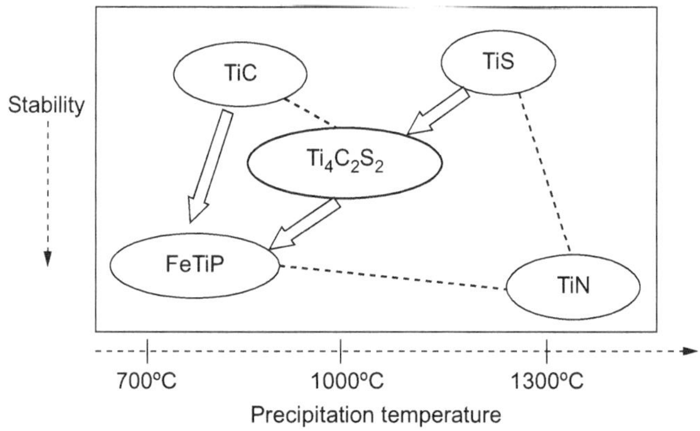

Figure 5.20 Schematic illustration of stability versus precipitation temperature of various Ti compounds in IFHS steels. A. Okamoto, N. Mizui, in: R. Pradhan (Ed.), Metallurgy of Vacuum Degassed Steel Products, TMS, Warrendale, PA, 1990, pp. 161–180.
Source: Copyright 1990 by The Minerals, Metals & Materials Society. Reprinted with permission.

and reheating and does not dissolve thereafter. During subsequent slab reheating, TiS may form depending on the sulfur and Mn content. During coiling MnS formation normally takes place. The most important precipitation though takes place during annealing. Annealing at 700°C results in the precipitation of FeTiP and TiC particles, while annealing at 800°C may reveal a complete absence of FeTiP and presence of numerous $Ti_4C_2S_2$ type precipitates, depending on composition and processing history. A complete absence of FeTiP and presence of $Ti_4C_2S_2$ also yields better r_m values [46].

5.4 Bake hardening (BH) steels

Bake hardening (BH) steel is a special grade of low carbon steels (and some other steels as well) in which the static strain aging property of free carbon is exploited in a controlled way. In these steels a small amount of free carbon is intentionally kept in the iron matrix after the final annealing. The amount of this free carbon is insignificant enough not to influence the formability of the steels. The steel strips are then formed into critical shaped parts, followed by painting. The deep drawing process generates a lot of dislocations that can be locked by the free carbon atoms present in the steel, during the accelerated aging associated with the paint baking treatment of the parts. This process can increase the YS of the steel by 30–90 MPa [4].

BH steels are exclusively used in the production of automotive outer panels, due to their higher dent resistance, caused by BH. An increase in yield stress of about 40 MPa would compensate for a reduction in gage of about 0.1 mm to give the same dent resistance [47]. However, a strength increment of more than 60 MPa,

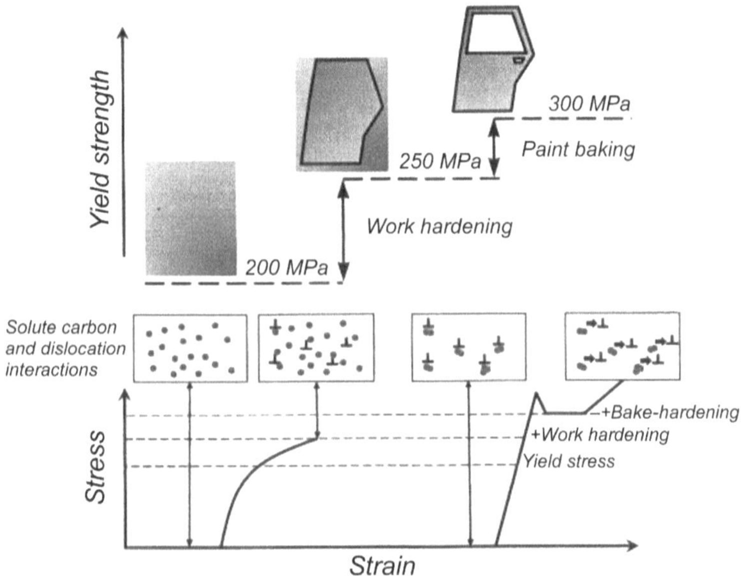

Figure 5.21 Schematic representation of the bake hardening (BH) process and its effect on yield strength of steel [4].

caused by excessive room temperature strain aging, may lead to the generation of stretcher strains during forming process [6]. Thus, BH steels, with BH indices more than this, are only suitable for just-in-time delivery before forming.

The BH process is illustrated in Fig. 5.21 [4]. Two mechanisms may be responsible for the BH effect:

1. Dislocation pinning: Dislocations generated during the press forming process will lead to the formation of Cottrell atmosphere due to the presence of free carbon atoms at the vicinity of dislocation cores. This is a fast process and is controlled by the amount as well as the temperature dependent diffusivity of the interstitial carbon atoms.
2. Precipitation hardening: At a later stage of the aging process some carbide may also form to contribute to the YS.

BH property can be achieved in both batch and the continuous annealed steels. Both low carbon AK and IF steels possess the capability of achieving the BH effect, provided the steels' compositions are designed properly.

5.4.1 Bake hardenable aluminum killed steel

Most of the industrial batch annealed bake-hardening steels are the AK ones. Carbon content above 300 ppm in such steels results in complete re-precipitation of carbon during cooling from the BA temperature, and therefore a non-aging non-bake hardenable property is achieved. However, when the carbon content is around 150 ppm and below, re-precipitation process is difficult and a small amount of carbon remains in solution giving rise to the BH effect. In a similar way, in case of continuous annealed steels, the precise control of carbon is done through over-aging and the cooling strategy. A carbon content <300 ppm is suitable for CA BH low carbon grade.

5.4.2 Bake hardenable interstitial free steel

As already stated, IF steels can be suitably designed to give the BH effect. It is generally done by adding carbide forming elements, in slightly lower amounts than the stoichiometric requirement, to tie up with the total carbon. This makes way for a small amount of free carbon to remain in solution to provide the BH effect. This procedure is normally adopted for batch annealed steels. However, for continuous annealed steels, even compositions with Ti addition higher than the stoichiometric requirement to fix the total carbon can be manipulated to give rise to the BH effect. In this case, a higher annealing temperature of around 850°C is used to take carbon back into solid solution and cooled quickly to minimize the reformation of precipitates on cooling. A cooling rate around 30°C/s is generally suitable to obtain sufficient BH property in continuously annealed IF steels.

5.5 Summary and conclusions

Deep drawable steels are essentially low to ultra-low carbon steels that are characterized by their excellent r_m value. In general, lowering the carbon content itself always improves {111} texture formation and deep drawability. Thus AK steels always possesses lower r_m value than IF steels due to the 10 times higher carbon content in the former.

The primary use of these grades is in the manufacture of automobile body panels. These body panels have very intricate shapes, and therefore the steels used require a great deal of deep drawability, which can be provided by only the low carbon grades of steel.

IF steels require very low levels of carbon and sulfur. Unlike AK steels, a stringent compositional control is very much essential for IF steel production. This is only achieved by the technological development of vacuum degassing technique.

Although both the above grades can be manufactured by batch and/or CA routes, the AK steels are more suitable by the BA route, and the IF steels by the CA route. In general the highest r_m value obtained in AK steels is ~1.8 whereas for IF steels it is ~2.2. However, IF steels possess lower strength than the AK steels due to lesser amounts of carbon and other elements present in the former. The high strength version of IF steels, the IFHS steels containing P, have been developed where the advantage of higher drawability of IF steels and solid solution strengthening ability of P are utilized.

The YS of these low carbon steels, in service condition, can be further improved by taking advantage of the high temperature strain aging of free carbon, which is available in the BH grades. In these grades a very small amount of carbon

(10–20 ppm) is deliberately kept free in the Fe matrix after the final annealing treatment. Since this carbon in solution is very little, it does not compromise the formability of steel strips to any significant extent and therefore these can be formed into intricate shapes. However, it is during the baking treatment after painting, normally carried out at around 170°C for 20 minutes, when accelerated aging of carbon takes place. Carbon atoms diffuse into the dislocations generated during the forming process and lock them. This way the in-service YS of the steel increases and dent-resistance property improves.

In all low and extra low carbon grades, precipitation plays a very important role. In aluminum killed steel AlN is the most important precipitate which actually controls the recrystallization process and final texture. The precipitation strategy has to be designed, depending on the processing route to be followed. For example, in batch annealed AK steel, coiling, after hot rolling, is done at a lower temperature in order to keep all the Al and N in solid solution. These Al and N atoms are then allowed to precipitate as AlN during the slow heating to the BA temperature. This way the final texture can be controlled. By contrast, in the CA process of AK steel, the aluminum and nitrogen contents are reduced to a lower level and the AlN precipitation is allowed to take place during high temperature coiling, unlike in the BA process. The other important precipitates are of course the carbides, carbonitrides, and carbosulfides, which can also be suitably controlled in order to obtain the desired properties.

In IF and IFHS steels, the precipitation process is more complex than in aluminum killed steels. Precipitates such as TiN, TiS, MnS, $Ti_4C_2S_2$, TiC, and FeTiP are the important ones that are observed in these grades, and precise control over these precipitates is very much needed to achieve the desired properties in the steels. In this context the roles of the compositional and processing parameters, which have definite effects on precipitate formation, must be considered, allowing some role of synergistic effects also. Very often it has been observed that the probabilities of formation of two or more precipitates are very close and, as a result, contradictory reports on the sequence of formation of the precipitates are available in the literature. For example, Ghosh et al. [41] reported the absence of the harmful FeTiP precipitates in continuous annealed IFHS steels, whereas Wang et al. [44] reported the presence of the same in CA IFHS steels as well. This is probably because of these precipitates having very close solid solubility in steels. Slight change in composition and/or processing parameters may therefore give rise to totally different results. Thus the composition and the processing variables should be very carefully designed, keeping an eye on the targeted properties.

References

[1] G.E. Dieter, Mechanical Metallurgy, McGraw Hill Book Co, London UK, 1988.

[2] W.T. Lankford, S.C. Snyder, J.A. Bausher, New criteria for predicting the press performance of deep drawing sheets trans ASM 1950;42:1197–1205.

[3] F.J. Held, Mechanical working and steel processing conference ISS 1965:3–38.

[4] B.C.D. Cooman, J.G. Speer, Fundamentals of steel product physical metallurgy: AIST, Association for Iron and Steel Technology.
[5] O. Akisue, International Symposium on Ultra High Purity Base Metals, UHPM-94 1994. p. 238–247.
[6] D.T. Llewellyn, R.C. Hudd, Steels Metallurgy and Applications, third ed., Butterworth-Heinemann, Oxford, UK, 1998.
[7] D.T. Llewellyn, Steels Metallurgy and Applications, second ed., Butterworth-Heinemann Ltd, Oxford, UK, 1992.
[8] Y. Meyzaud, B. Michaut, P. Parniere, Texture and the Properties of Materials. 4th International Texture Conference, Cambridge: The metals society, London, 1975, p. 255.
[9] D.T. Gawne, Steel Research, British Steel Corporation 1976. (11).
[10] K. Osawa, K. Matsudo, K. Kurihara, T. Suzuki, Tetsu-to-Hagane, &0, S552 (1984) (12).
[11] W.C. Leslie, R.L. Rickett, C.L. Doston, C.S. Walton, Trans ASM 46 (9) (1954).
[12] T.R. Parayil, I. Gupta, in: L.G. Kuhn, F.A. Hultgren (Eds.), Mechanical Working and Steel Processing XXVII, AIME, Warrendale, 1989, p. 314. (13).
[13] W.L. Whitely, D.E. Wise, Flat Rolled Products III, Interscience, New York, 1962, p. 10.
[14] M. Takahashi, A. Okamoto, Sumitomo Metals 27 (1974) 40.
[15] M. Takahashi, M. Abe, O. Akisue, H. Katoh, Metallurgy of Continuously Annealed Sheet Steel, AIME, Dallas, 1982, p. 51.
[16] S. Ono, O. Nozoe, T. Shimomura, K. Matsudo, B.L. Bramfitt, P.L. Manganon (Eds.), Metallurgy of Continuously Annealed Sheet Steel, AIME, Dallas, 1982.
[17] K. Toda, H. Gondoh, H. Takechi, M. Abe, N. Uehara, K. Komita, Trans Iron and Steel Institute. Japan, 15 (1975) 305.
[18] N. Prum, U. Meers, H. Mathy, P. Messien, V. Leroy, in: R. Pradhan, G. Ludkovsky (Eds.), Proceedings of Symposium on Hot and Cold Rolled Sheet Steels, 3, The Metallurgical Society, Cincinnati, 1988.
[19] H. Katoh, K. Koyama, K. Kawasaki, in: R. Pradhan (Ed.), Proceedings of Symposium on Metallurgy of Continuously Annealed Cold Rolled Sheet Steel, AIME, Detroit, 1984, p. 79.
[20] W.B. Huchinson, K.I. Nilson, J. Hirsch, Metallurgy of Vacuum Degassed Steel Products, TMS, Warrendale, PA, 1990.
[21] L. Meyer, W. Bleck, W. Mueschenborn, Proceedings of International Forum for Physical Metallurgy of IF Steels, ISIJ, Tokyo, Japan, 1994, pp. 203–222.
[22] B. Huchinson, Mater. Sci. Forum (1994) 157–162, 1917–1928.
[23] L.J. Ruiz-Aparicio, C.I. Garcia, A.J. DeArdo, Metall. Mater. Trans. A 32A (2001) 417.
[24] L.J. Ruiz-Aparicio, C.I. Garcia, A.J. DeArdo, Proceedings of IF Steels, Warrandale, PA, USA, 2000, ISS, 85.
[25] N. Mizui, Proceedings of International Symposium on 'Modern LC and ULC sheet steels for cold forming: processing and properties', (W. Bleck Ed.), 169–178; 1998, Aachen.
[26] R.K. Ray, P. Ghosh, D. Bhattacharjee, Materials Science and Technology 25 (9) (2009) 1154–1167.
[27] O. Kwon, K.Y. Lee, M.H. Hong, Proceedings of International Forum for Properties and Application of IF Steels, ISIJ, Tokyo, Japan, 2003, pp. 39–46.
[28] W. Cho, P. Fan, K. Lee, W. Huh, Steelmaking Conference of Proceedings, 2002, 453–461.
[29] L.F. Verdeja, J.P. Sancho, J.I. Verdeja, Rev. Minas, 200, 19, 33–40.
[30] M. Abe, Y. Kokabu, Y. Hayashi, S. Hayami, Trans. Jpn Inst. Met. 23 (1982) 718.
[31] S. Sanagi, CAMP-ISIJ, 3 (1990) 1768.

[32] O. Kwon, K.Z. Min, Proceedings of International Forum for Physical Metallurgy of IF Steels, ISIJ, Tokyo, Japan, 1994, pp. 9–12.
[33] O. Hashimoto, Proceedings of International Conference on 'Advances in Physical Metall and Applications of Steel', The metals society of Great Britain, London, 1982, Book 284.
[34] Y. Tokunaga, M. Tamada, US patent 4,504,326.
[35] I. Gupta, D. Bhattacharya, in: R. Pradhan (Ed.), Metallurgy of Vacuum Degassed Steel Products, TMS, The Minerals, Metals and Materials Society, 1990, pp. 43–72.
[36] M. Hua, C.I. Garcia, A.J. DeArdo, Scr. Metall. Mater. 28 (1993) 973–978.
[37] G. Tither, C.I. Garcia, M. Hua, A.J. DeArdo, Proceedings of International Forum for Physical Metallurgy of IF Steels, ISIJ, Tokyo, Japan, 1994, pp. 293–322.
[38] A.J. DeArdo, IF Steels Proceedings (2000) 125–136.
[39] K. Lee, S.-H. Han, Y.-S. Jin, International Forum for the Properties and Application of IF Steels (2003) 286–289.
[40] P. Ghosh, R.K. Ray, B. Bhattacharya, S. Bhargava, Precipitation and texture formation in two cold rolled and batch annealed interstitial-free high strength steels, Scr. Mater. 55 (3) (2006) 271–274.
[41] P. Ghosh, B. Bhattacharya, R.K. Ray, Comparative study of precipitation behavior and texture formation in cold rolled-batch annealed and cold rolled-continuous annealed interstitial free high strength steels, Scr. Mater. 56 (2007) 657–660.
[42] O. Hamart, T. Lung, S. Lanteri, Proceedings of 40thMWSP Conference, 1998, 189.
[43] T. Gladman, Physical Metallurgy of Microalloyed Steels, Springer, Berlin, 1977.
[44] L. Wang, I. Jin, J. Liu, Proceedings of International Forum for Properties and Application of IF Steels, ISIJ, Tokyo, Japan, 2003, pp. 298–301.
[45] A. Okamoto, N. Mizui, in: R. Pradhan (Ed.), Metallurgy of Vacuum Degassed Steel Products, TMS, Warrendale, PA, 1990, pp. 161–180.
[46] P. Ghosh, C. Ghosh, R.K. Ray, D. Bhattacharjee, Precipitation behavior and texture formation at different stages of processing in an interstitial free high strength steel, Scr. Meter. 59 (2008) 276–278.
[47] H. Takechi, in: R. Pradhan, G. Ludkovsky (Eds.), Proceedings of Symposium Hot and cold rolled sheet steel, The Metallurgical Society, Cincinnati, 1988, p. 117.

High strength low alloyed (HSLA) steels*

6

C.I. Garcia
University of Pittsburgh, Pittsburgh, PA, United States

6.1 History and definition

Two excellent historical overviews clearly document the genesis of high strength low alloyed (HSLA) or Microalloyed (MA) steels [1,2]. According to these reviews, in 1934, a paper describing the properties of V-bearing steel in the as-hot rolled condition was published [3]. About a quarter of a century later, in 1958, the first commercial series of HSLA steels called GLX-W with Nb-additions was produced by the National Steel Corporation of the United States. These examples represent, perhaps, the discovery stage and the commercial introduction of a family of steels that has subsequently come to be known as HSLA steels. Since their commercial inception this family of steels has created a great deal of interest around the world. The alloy design philosophy and thermomechanical processing necessary to optimize the microstructure-property relation of HSLA steels can be considered one of the most disruptive technologies in the steel industry. The interest of using HSLA steels in the automotive industry became a necessity due to the oil crisis in 1973. The industry was faced with the challenge of reducing the weight of vehicles to decrease fuel consumption [4,5]. During the period of 1972–75 the weight in a standard-size car increased by about 159 kg (350 lbs) [5] to satisfy consumer life style and expectations.

Since the first oil crisis to date, the types of HSLA steels used in vehicles have changed considerably. While the original objective of weight reduction still remains the major driving force, other requirements such as environmental protection, i.e., reduction of CO_2 emissions, increased crash resistance, better formability performance, weldability, and paintability among others, have been added to the steels used in modern vehicles. Fig. 6.1 shows the potential reduction of CO_2 emissions with vehicle weight savings. A few examples of HSLA steels used in the auto industry, its critical functional properties, main applications and expected surface quality are shown in Table 6.1 [6].

The worldwide use of HSLA steels in other industries such as oil and gas extraction, transportation, construction, heavy equipment, and defense has required the steel industry to develop a large selection of steels and processes for the development of high strength hot-rolled and cold-rolled sheet steels.

* Every effort has been made to trace copyright holders and to obtain their permission for the use of copyright material. The publisher apologizes for any errors or omissions in the acknowledgements printed in this book and would be grateful if notified of any corrections that should be incorporated in future reprints or editions.

Automotive Steels. DOI: http://dx.doi.org/10.1016/B978-0-08-100638-2.00006-7

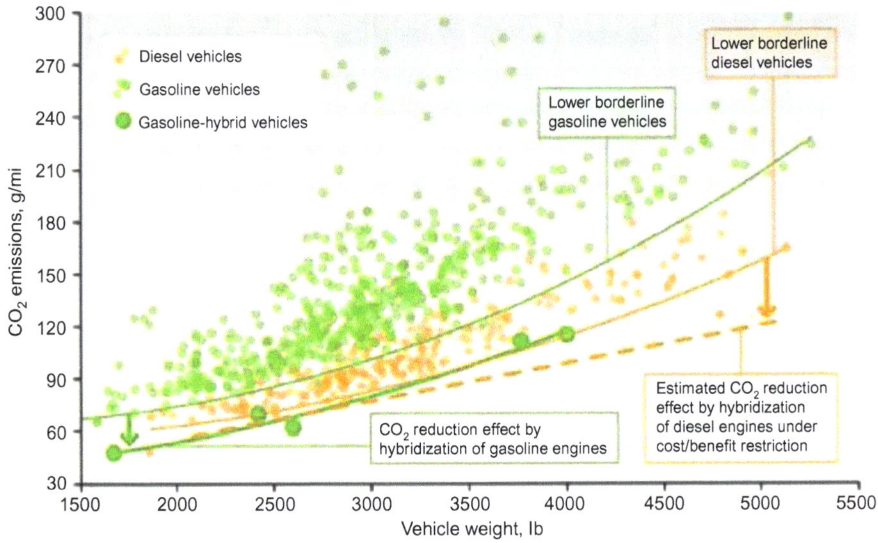

CO_2 emissions levels for gasoline and diesel vehicles showing the effect of hybridization.

Figure 6.1 shows the reduction of CO_2 emissions per vehicle weight savings [7].

Table 6.1 Examples of HSLA sheet steels and their applications in the auto industry [6]

Grade	Product type	Functional properties	Main applications	Surface quality
HSLA 260-420 HSLA 550 HSLA 750	Hot rolled Hot rolled Hot rolled	Cold forming Cold forming Cold forming	Chassis, wheels, rim and disc, and suspension parts	Non exposed and Semi-exposed (wheels)
HSLA 320 HSLA 420 HSLA 590	Cold rolling for hot dip galvanizing	Cold forming	Automotive parts that require certain levels of stretch-forming and deep drawing characteristics	Non exposed Non exposed

Regarding automotive applications, the entire range of HSLA steels are suitable for structural components of cars, trucks and trailers such as suspension systems, chassis, and reinforcement parts. HSLA steels offer good fatigue, torsional rigidity, and impact strengths. Given these characteristics, this class of steels offers weight reduction for reinforcement and structural components [8–10].

The HSLA range of steel products is available in hot and cold rolled grades. The specific types of grades are classified by their yield strength. Table 6.2 shows the

Table 6.2 Required mechanical properties[1] of HSLA hot-rolled and cold rolled uncoated and coated sheet steels[2]

SAE J2340 grade designation and type	Yield strength[3] (MPa) minimum	Yield strength[3] (MPa) maximum	Tensile strength MPa minimum	$\%E_{total}$ minimum cold-reduced	$\%E_{total}$ minimum hot-rolled[4]
300 S	300	400	390	24	26
300 X	300	400	370	24	28
300 Y	300	400	400	21	25
340 S	340	440	440	22	24
340 X	340	440	410	22	25
340 Y	340	440	440	20	24
380 X	380	480	450	20	23
380 Y	380	480	480	18	22
420 X	420	520	490	18	22
420 Y	420	520	520	18	19
490 X	490	590	560	14	20
490 Y	490	590	590	12	19
550 X	550	680	620	12	18
550 Y	550	680	650	12	18

[1]The mechanical property requirements shall be determined in the longitudinal direction unless otherwise specified.
[2]User and producer should agree regarding the selection of specific steel grade and welding process optimization.
[3]Yield strength is 0.2% offset or, in the presence of yield point elongation, lower yield point.
[4]For thickness less than 2.5 mm, minimum percent elongation is permitted to be 2% less than the value shown.

typical mechanical properties for HSLA steels according to SAE J2340 for hot and cold rolled steels [11].

Hot-rolled HSLA sheet steels are used extensively for applications such as chassis parts, front-side rail, engine mount, wheels, rims and disc, and suspension components. Cold-rolled HSLA steel parts are often used in applications that do not require high levels of formability. These types of steels have been available to the automotive industry for over 40 years. The required yield strength is in the range of 260–520 MPa. The typical mechanical properties according to the SAE J2340 are shown in Table 6.2.

The application of HSLA steels in the automotive industry requires an in-depth understanding of the part and its functionality. For example, the performance of each part depends on a set of design (shape), property (strength-loading, denting, fatigue, corrosion, spring-back, and weldability) parameters [5].

HSLA steels are a special category of plain carbon steels with microalloying elements, such as vanadium (V), niobium (Nb), or titanium (Ti), and possess superior mechanical properties. The yield strength of some HSLA steels can reach as high as 690 MPa, which is more than two times higher than the strength (in the range 170–250 MPa) of typical plain carbon steels. The high strength of HSLA steels is believed to stem from their microstructural factors such as grain refinement [12,13], precipitation hardening, and inclusion shape control [14,15]. As a specific example, V-microalloyed HSLA (denoted as HSLA-V) steels with ultrafine grains have been found to exhibit an excellent balance of strength and ductility [16,17], and consequently, have been employed in a wide range of applications (e.g., bridges, buildings, vehicles, and pipelines) where materials reliability, environmental issues, and fabrication costs are main factors of design considerations [18–20,21]. Similarly modern vehicle bodies make extensive use of HSLA steel grades to meet the economic and technical demands of weight reduction and superior mechanical performance. To achieve these goals the great majority of HSLA steels contain Nb-additions. HSLA-Nb bearing steels exhibit higher strength primarily by grain refinement. In interstitial free high strength steels the presence of Nb acts as stabilizing element.

Microalloyed HSLA steels have characteristically a high yield strength to ultimate tensile strength ratio and thus a low work hardening coefficient. This characteristic makes the local yield strength rather insensitive to the level of strain during forming operations. Other characteristics of HSLA steels are the quasi-isotropic behavior (planar anisotropy parameter or Δr-value $\sim$ 0) and a good fatigue resistance [21,22].

6.2 Structure–property relationships: effect of microstructure on the mechanical properties of HSLA steels

HSLA steels require a range of properties in order to be technologically and economically successful. The principal properties of interest in HSLA steels are strength, formability, weldability, and fatigue resistance. Many steel end users often require steels with superior combinations of strength and formability, to achieve these combinations it is essential to have a good understanding of the composition-processing-microstructure relationship. For example, if the yield strength of HSLA steels needs to be increased, this can be relatively straight forward to accomplish. Dislocation motion which is essential for plastic deformation must be inhibited. The inhibition of dislocation motion can be achieved by the presence of obstacles to their motion on their slip plane. Similarly the presence of high angle grain boundaries with high energy and high misorientation are more effective to inhibit dislocation motion from one slip plane to another, hence increasing the yield strength. High angle grain boundaries are somewhat similar to incoherent interfaces. In summary, the strength of steels is derived from the interaction of several structural factors to dislocation motion. There are several linear equations based on the theory of polycrystalline strengthening and expressed as an expanded version of the Hall–Petch equation. [23,24] One of the most accepted and used versions can be expressed as follows:

$$\sigma = \sigma_0 + \Delta\sigma_s + \Delta\sigma_T + \Delta\sigma_P + \Delta\sigma_D + k_y D\alpha^{-1/2} \tag{6.1}$$

where σ_0 is the intrinsic strength of ferrite due to the lattice friction stress that the dislocations experience when moving on their glide plane. $\Delta\sigma_s$ is the solid solution strengthening due to substitutional and interstitial elements. $\Delta\sigma_T$ and $\Delta\sigma_P$ are strengthening increments caused by texture and precipitation. The increment due to texture is mainly due to the redistribution of strength. The precipitates might be coherent or incoherent with the lattice. $\Delta\sigma_D$ is the strengthening that results when dislocation motion becomes inhibited as a result of increased dislocation density.

The last term $k_y\ D\alpha^{-1/2}$ is the contribution to strength by the grain size in the polycrystalline steel. The strengthening results in the direct barrier effect of grain boundaries to dislocation pile-ups.

Table 6.3 gives examples of the influence of chemical composition on the properties of steels and strengthening mechanisms [25,26].

Table 6.3 Examples of steel strengthening mechanisms in steels

Strengthening mechanism		Composition dependence	Example	Effect
Solid solution	Substitutional atoms: immobile	Strong to moderate strongly related to the solute atom diameter.	Lattice distortion by substitutional atom.	Moderate
	Interstitial atoms: mobile	Strong to moderate depends on carbon in solid solution and segregation at grain boundaries	Grain boundary hardening and strain aging	Moderate
Precipitation	Coherent precipitates	Strong dependence on the volume fraction, size, and inter-particle spacing.	Small coherent Cu-rich precipitates	Moderate
	Incoherent precipitates		Microalloying precipitates: NbC, V(C,N), TiC	High
Dislocations	Work hardening	Indirect	All steel grades	High

(Continued)

Table 6.3 (Continued)

Strengthening mechanism		Composition dependence	Example	Effect
Grain refinement	Grain size and high angle grain boundaries	Low for plain carbon steels high for HSLA or micro-alloyed steels.	All steel grades	Moderate strengthening increases as the grain size decreases

6.3 Fundamental metallurgical principles of thermomechanical processing

The continuous HSLA steel developments around the world for the automotive and energy sectors provide both an opportunity and challenge to develop steels with superior properties and performance. To fulfill these requirements a fundamental understanding of the composition-processing-microstructure-property relationship is needed. The logical path to achieving the required objectives starts with the alloy design (steel composition). The chemical composition and the content of microalloying elements influences the well-known three critical temperatures of austenite; (1) the grain coarsening temperature (T_{GC}); (2) the temperature on non-recrystallization (T_{nr}); and (3) the γ to α transformation temperature (T_{Ar3}).

The grain coarsening temperature (T_{GC}) is defined to be that temperature above which grain coarsening by secondary recrystallization commences. This temperature is not a fixed temperature; it is in fact a range of temperatures. The lower end marks the start of the T_{GC} and indicates to the temperature at which partial dissolution of the microalloying precipitates starts and the upper end is the temperature above which the undissolved precipitates can no longer suppress grain growth, i.e., $F_{\text{grain growth}} > F_{\text{pinning}}$, where F is either the grain growth or the pinning force. The influence of various microalloying elements on grain growth during reheating is shown in Fig. 6.2 [27]. The most common microalloying element used to control the T_{GC} of steels is Ti. This element is most effective when used in sub-stoichiometric additions with N, i.e., Ti/N $\sim$2.1–2.5.

The extent to which alloying elements can be maintained in solid solution in austenite is governed by the phase solubility as a function of temperature and can be presented by the following equation: [28]

$$\log[\mathrm{M}]\cdot[\mathrm{X}] = B - A/T \tag{6.2}$$

where: [M] and [X] are the mass fraction of microalloying additions and C and/or N dissolved in austenite at T(K) temperature, respectively. A and B are constants related to the free enthalpy of MX carbide or nitride formation. The values of the

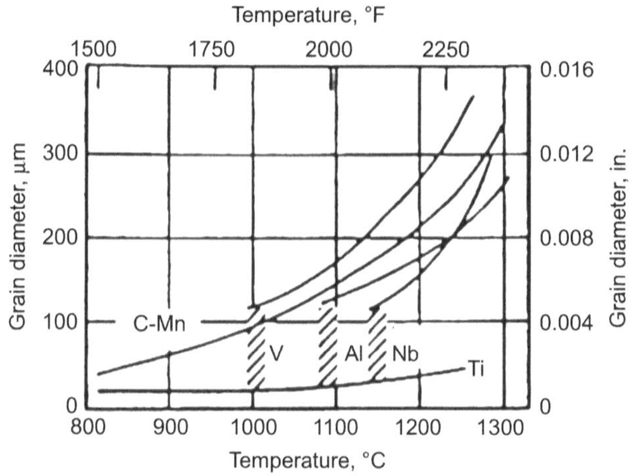

Figure 6.2 Austenite grain coarsening behavior in steels containing various microalloying additions. G.R. Speich, et al., in: A. Marder, J. Goldstein (Eds.), Phase Transformations in Ferrous Alloys, The Metallurgical Society of AIME, 1984, p. 341. Reprinted with permission of The Minerals, Metals & Materials Society.

Table 6.4 Values of constants *A* and *B* for selected carbides and nitrides [28]

MX compound	A	B
AlN	7184	1.79
VC	9500	6.72
VN	7840	3.02
TiC	10,745	5.33
TiN	14,400	5.00
NbC	7290	3.04
NbN	8500	2.80

constants *A* and *B* for selected carbides and nitrides are shown in Table 6.4 [28]. It is important to mention that the values of the constants *A* and *B* depend on the method of determination [28,29].

Because the solubility of microalloying elements varies in austenite, each of these elements and compounds has a particular role during the thermomechanical processing (TMP) of HSLA steels. TiN is used to control the grain coarsening of austenite during reheating, NbC and/or Nb(C,N) are used to control the temperature of non-recrystallization (T_{nr}) and the transformation temperature T_{Ar3}. VC and VN are mainly used for precipitation strengthening during the austenite to ferrite transformation under controlled cooling conditions.

The temperature of non-recrystallization T_{nr} is perhaps one of the most important tools during thermomechanical processing of HSLA steels. The control and understanding of this temperature permit the microstructural control of the austenite in

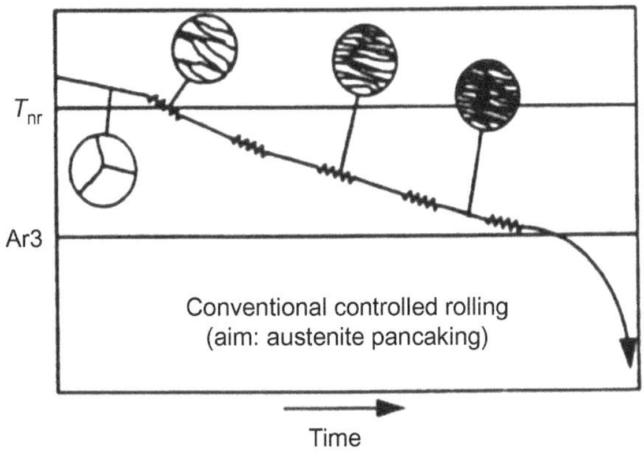

Figure 6.3 Schematic representation of the effect of deformation between the T_{nr} and the T_{Ar3} on the austenite microstructure prior to γ to α transformation [30]. Siciliano, F. & Jonas, J.J. Metall and Mat Trans A (2000) 31: 511. http://dx.doi.org/10.1007/s11661-000-0287-8 [30a].

terms of: (1) microstructural uniformity in the transfer bar, prior to entering the finishing mill; (2) substantial deformation between the T_{nr} and the T_{Ar3} to produce a large number of intercrystalline defects (deformed grain boundaries, deformation bands, and annealing twin boundaries) in austenite which act as nucleation sites for ferrite. Fig. 6.3 [30] shows a schematic diagram of a conventional controlled rolling processing of HSLA steels. Deformations above the T_{nr} will produce a fully recrystallized austenite with uniform grain size, while deformations between the T_{nr} and the T_{Ar3} will result in a deformed austenite microstructure.

There are several equations published in the literature to calculate the T_{nr} and T_{Ar3} of steels, one of the most popular equations is the one proposed by Borato et al. [30] and shown below:

$$T_{nr} = 887 + 464[C] + \left(6445[Nb] - 644\sqrt{Nb}\right) + \left(732[V] - 230\sqrt{V}\right) + 890[Ti] + 363[Al] - 357[Si] \tag{6.3}$$

$$T_{Ar3}(^{\circ}C) = 910 - 310C - 80Mn - 20Cu - 55Ni - 15Cr - 80Mo + 35[(Cu) - 8] \tag{6.4}$$

where [X] is in wt.% and the T in °C.

The final stage to the microstructural refinement of HSLA steels is the interplay between the finishing rolling temperature and the coiling temperature (T_{coil}). The proper selection of these two temperatures will dictate the final as-hot rolled microstructure and hence the mechanical properties. The effect of the finishing rolling temperature and T_{coil} on the uniformity of the ferrite is presented in Fig. 6.4 [6]. The information presented in this figure clearly reveals that finer ferrite microstructures in HSLA steels will occur when the finishing rolling temperature is coupled with lower coiling temperatures.

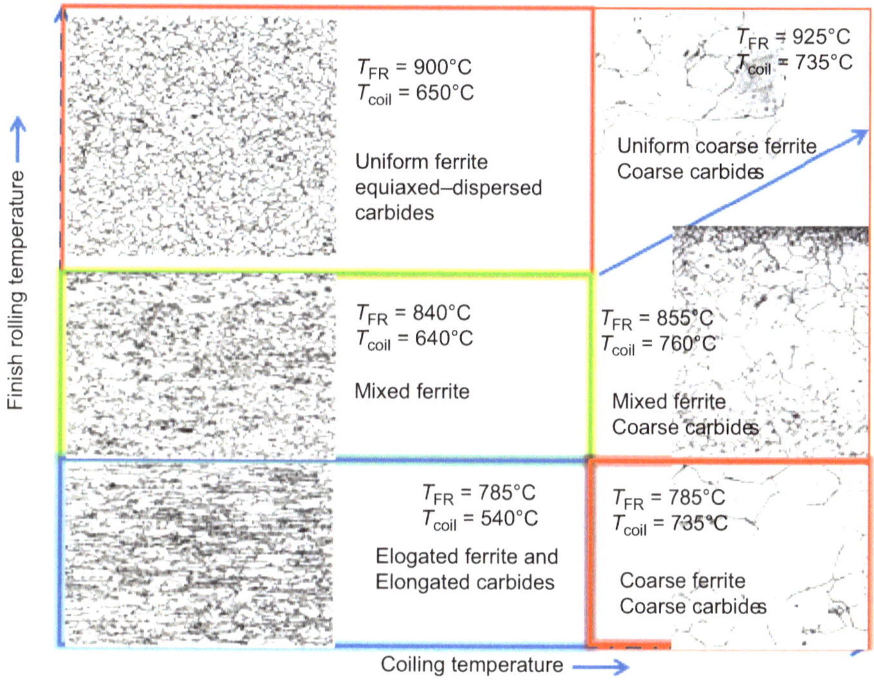

Figure 6.4 Effect of finishing rolling temperature and coiling temperature on the uniformity of the ferrite grain size in HSLA steels [6].

6.4 Examples of hot and cold rolled HSLA steels used in the transportation industry

The mechanical properties of hot rolled HSLA steels and their excellent cold forming performance and low-temperature brittle fracture resistance and weldability provide the technical opportunities to redesign components with lower weight, thickness, and size. Typical applications of hot rolled HSLA steels include chassis frames, wheels, slide rails, and cross-members. The major required properties are strength, fatigue, flexural resistance, and moment of inertia. The typical required mechanical properties are shown below: [31]

6.4.1 Mechanical properties

Yield limit 500–600 Nmm^{-2}
Tensile strength Max. 730 Nmm^{-2}
Extension Min 21%
Hardness 200 Hv
Impact strength Min. 18 J @ RT. (Fig. 6.5).

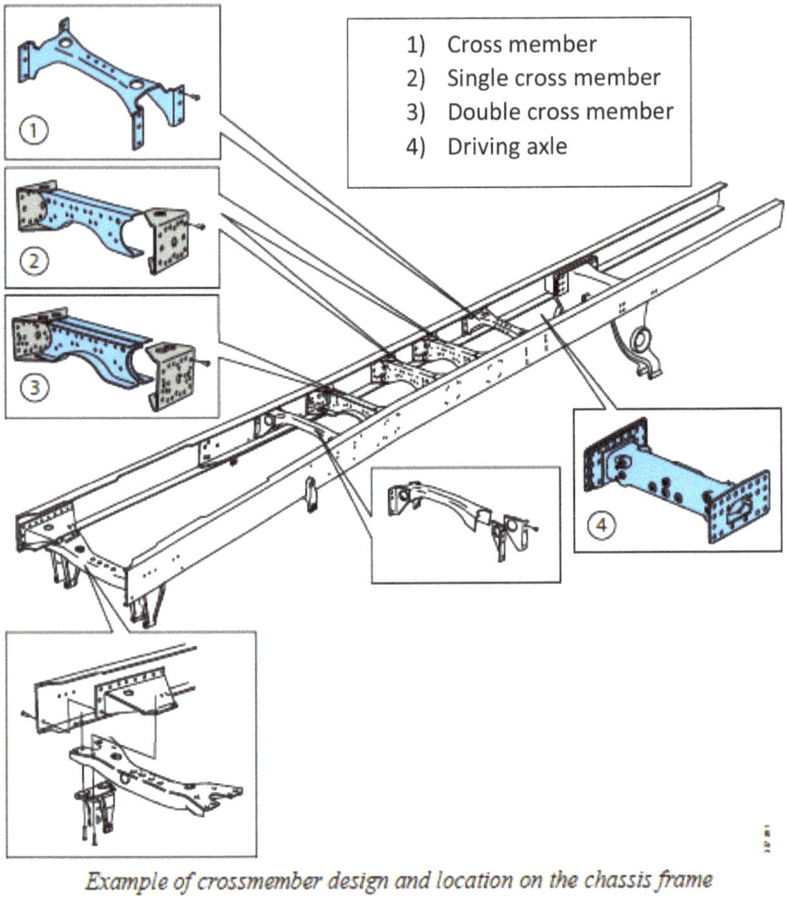

Figure 6.5 An example of the design of a chassis frame [31].

Typical examples of cold rolled HSLA steels are a reinforcement Center Body Pillar and a Pillar Body Lock Inner are shown in Fig. 6.6. The HSLA steel grade selected for these applications is a Ti-Nb bearing HSLA Gr350 with a required thickness of 1.50 and 1.30 mm, respectively. The final properties are achieved by cold rolling 50–70% the hot band and subsequent batch annealing (BA) at 650°C. The BA process is required to eliminate the large amount of dislocations that are generated in the matrix during cold rolling. These dislocations would pile up along their preferred slip planes and cause undesired work-hardening in cold-rolled steel sheets. The annealing process removes residual stresses, reduces anisotropy of mechanical properties, and develops a fully recrystallized ferrite grain size of the cold rolled HSLA steel.

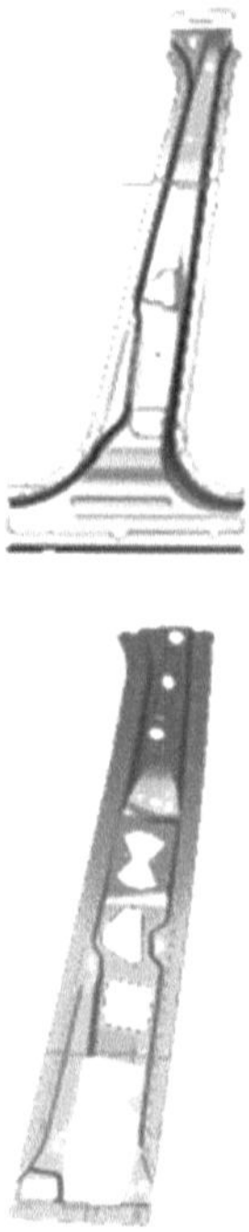

Reinforcement center body pillar (1.50 mm—HSLA Gr350)

Pillar body lock inner (1.30 mm—HSLA Gr350)

Figure 6.6 Example of application of HSLA cold rolled and annealed steels [6].

6.5 Transformation behavior

Optimization of mechanical properties, such as strength and ductility, are pivotal in the development of HSLA and other steel systems. These mechanical properties are governed by the microstructural components of the steel such as grain size, phase volume fraction, composition, precipitation, and other well-known structural factors. Hence, understanding and controlling the microstructure-property relationship remain the overall goal of the steel companies. In order to control these properties, the steel is often subject to many different rolling processes to induce deformation at high temperatures, which subsequently upon transformation to low temperature changes the microstructure and mechanical properties of the steel. In many cases, dependent variables during hot deformation such as reheating temperature, rolling schedule, i.e., deformation temperature, amount of strain per pass, number of passes, strain rate, accumulated strain, T_{nr} and T_{Ar3}, thermal path (cooling scheme in the run-out-table start) and T_{coil} will have a great influence on the ferrite transformation and precipitation kinetics. These transformations cannot be directly measured, in which case the steel microstructure's overall resistance to deformation can be directly correlated back to the high temperature microstructure, i.e., austenite. Predictive modeling can be used in order to calculate the effects of thermomechanical processing conditions and cooling rate on the final microstructure of the steel. Based upon these predictions, the mill can then be adjusted to reproduce these processing conditions to obtain the desired microstructure.

The logical structural path for developing a predictive model for microstructural changes during thermomechanical processing involves various components that contain austenite grain coarsening during reheating, deformation of austenite, recrystallization, recovery, and transformation. Each component plays an important role in the determination of the final microstructure of the steel that controls the mechanical properties of the final as-hot rolled product. The process can be briefly outlined with the following steps:

1. *Slab Reheating*: Initial austenite grain size as well as subsequent grain growth is determined.
2. *Hot Rolling*: Once the slab is reheated at the desired temperature and conditions, the hot rolling begins, in which the steel is deformed to its desired hot rolling shape. During hot rolling austenite might undergo the following structural events, deformation, recovery, recrystallization, precipitation, and non-recrystallization. After deformation and transformation, recovery, recrystallization, and additional precipitation might occur in the coil.
3. *Transformation Modeling*: Calculation of the cooling behavior using the transformation kinetics of the material and using the grain size and hot rolling conditions as the input in order to find the final microstructure.
4. *Mechanical Prediction*: The mechanical properties can be predicted by using the calculated microstructure.

6.5.1 Slab reheating

The quality of slab reheating plays a large role in the grain size and overall grain uniformity throughout the material. The final ferrite grain size that is observed after hot rolling is highly dependent upon the initial austenite grain size established during the reheating stage. The grain growth occurs as the slab is held in the reheating furnace and is dependent upon reheating temperature as well as the holding time.

Slab reheating also plays a very important role in controlling the dissolution of microalloying elements, which also can affect the grain size control during the recrystallization and precipitation hardening, depending on whether or not there is a presence of precipitates. The first case which assumes that there is no precipitation, models the austenite grain size as the following [32]:

$$(\overline{D})^n - (\overline{D}_o)^n = k_1 \times t \tag{6.5}$$

where:

$\overline{D}$ = average austenite grain diameter
$\overline{D}_o$ = inital austenite grain diameter
t = time of particle growth
n, k = constants

The constant n can either be two or three, depending on whether the steel under consideration is single or dispersion in structure, respectively. In the case where there is precipitation, the dissolution of the elements must be taken into account due to their effect on the grain growth. Such phenomena include Zener pinning or solute drag. Zener pinning describes a fine dispersion of particles hindering the movement of grain boundaries by exerting a pressure opposite of the driving force for growth. Solute drag results when the solutes migrate toward the grain boundaries in order to lower the activation energy lag behind in the movement of the grains causing a dragging pressure that impedes boundary movement [33]. Both pinning and solute drag have a significant effect on the recovery, recrystallization, and grain growth so to take them into account the following relation is implemented [34]:

$$\overline{R} = \frac{4\overline{r}}{9f} \tag{6.6}$$

where:

$\overline{R}$ = average grain radius
$\overline{r}$ = average radius of precipitated particles
f = volume fraction of the dispersed particles

While taking into account the phenomenon of Ostwald ripening, and along Eq. (6.6), the following relation is found which includes precipitation [34]:

$$(\overline{R})^3 - (\overline{R}_o)^3 = \left(\frac{4\overline{r}}{9f}\right)k_o \times t \tag{6.7}$$

where:

$\overline{R}$ = average grain radius
$\overline{r}$ = average radius of precipitated particles
f = volume fraction of the dispersed particles
k_o = solubility constant

6.5.2 Hot deformation model

The hot deformation model is the plastic deformation of the reheated slab while the material has less resistance to deformation due to being held at a high temperature, and occurs during the roughing stage of the slab deformation process. The slab coming out of the reheating stage has a high temperature equal to or higher than 1250°C, making the slab temperature uniform. As plastic deformation occurs and the slab begins to cool, strong microstructural changes take place such as deformation, recovery, recrystallization, and possible grain growth.

6.5.2.1 Recovery

Prior to recrystallization, there is a period of time where gradual changes in the microstructure take place which results in the restoration of material properties. The recovery stage is due to the changes in the dislocation structure of the material during high temperature deformation. Recovery is not exclusive to materials that undergo plastic deformation and this can occur in any crystal that contains a high concentration of point or line defects, such is the case after quenching or irradiation. The recovery process is not a single event, but instead it is described as a series of micromechanisms in which all or some may occur during or after deformation, known as dynamic and static recovery respectively [35]. The extent of recovery is often very hard to measure directly due to the very small microstructural changes that occur, but recovery still plays a role in the overall kinetics of recrystallization. The stored energy from the deformed state provides the driving force for both recovery and recrystallization, this causes a competing effect between the two phenomena. As the recovery occurs prior to the recrystallization phase, it will subsequently lower the driving force for recrystallization.

6.5.2.2 *Recrystallization*

During hot rolling, nucleation and nucleus growth have the largest effect on the grain refinement and is described collectively as the process of recrystallization. Similarly to recovery, recrystallization can occur both statically and dynamically depending upon the amount of strain and strain rate during deformation. Dynamic recrystallization begins to occur when a critical strain value is reached during deformation. It is important to introduce the Zener–Holloman parameter, also known as the temperature compensated strain rate, which is defined as [34]:

$$Z = \dot{\varepsilon} \times \exp\left(\frac{Q_{\mathrm{Def}}}{RT_{\mathrm{def}}}\right) \tag{6.8}$$

where:

$\dot{\varepsilon}$ = strain rate
Q_{Def} = activation energy for deformation
R = gas constant
T_{def} = absolute temperature of deformation

The peak and critical strain values are directly related to both the grain size as well as the Zener–Holloman parameter, and show a similar trend for all steels as shown by Sellars [36]:

$$\varepsilon_{\mathrm{p}} = AZ^m d_{\mathrm{o}}^n \tag{6.9}$$

$$\varepsilon_{\mathrm{c}} = a\varepsilon_{\mathrm{p}} \tag{6.10}$$

where:

ε_{p} = peak strain value
ε_{c} = critical strain value
Z = Zener − Holloman Parameter
d_{o} = Grain size
A, m, n = constants

In which case, using Eqs. (6.9–6.10) [36], the kinetics of dynamic recrystallization can be expressed in Avrami type form:

$$X_{\text{dyn}} = 1 - \exp\left[-k\left(\frac{\varepsilon - \varepsilon_c}{\varepsilon_p}\right)^m \right] \tag{6.11}$$

where:

X_{dyn} = volume fraction of dynamic recrystallization
ε_p = peak strain value
ε_c = critical strain value
ε = strain
k, a, m = material dependent constants

The static recrystallization occurs between subsequent rolling and between deformation processes and it is in this time period that the majority of the grain refinement takes place. Following the deformation, there is still sufficient amount of stored energy in the dislocations for recrystallization to take place, and it is the timing between subsequent rolling that is essential in the accurate modeling of the static recrystallization kinetics. Similar to dynamic recrystallization, static recrystallization kinetics can be modeled in the form of the familiar Avrami equation [37,38] according to Beynon and Sellars [39]:

$$X_{\text{st}} = 1 - \exp\left[-.693\left(\frac{t - t_s}{t_{0.5}}\right)^m \right] \tag{6.12}$$

where:

X_{st} = volume fraction of static recrystallization
t_s = static recrystallization time
$t_{0.5}$ = time to reach 50% static recrystallization
m = constant

The deformation parameters have little effect on the constant m, typically will range between one and two, and are mostly dependent upon the steel composition, however the 50% recrystallization time is greatly affected and can change by an order of magnitude dependent upon the conditions that occur during the hot rolling period.

Work by Sellars and Whiteman [40] shows that the recrystallized grain size for both static and dynamic recrystallization is unaffected by the temperature of deformation and can be expressed in terms of the Zener–Holloman parameter:

$$d = 25\left(\frac{1}{6.7 \times 10^{-2}} \ln \frac{Z}{8.5 \times 10^{9}}\right)^{-\frac{2}{3}} \varepsilon^{-1} d_o^{\frac{1}{3}} (\varepsilon \le \varepsilon_c) \tag{6.13}$$

$$d = 3.6 \times 10^3 C Z^{-0.15} (\varepsilon \ge \varepsilon^*) \tag{6.14}$$

where:

d = grain diameter
d_o = initial grain size
Z = Zener − Holloman parameter
ε = strain
ε^* = critical strain for metadynamic recrystallization
ε_c = crititcal strain for dynamic recrystallization

The grain growth after the deformation, recovery and recrystallization is quite different from the grain growth described previously during the slab reheating process. The growth occurs much faster immediately after recrystallization due to the increase in the difference in dislocation density, which is also accentuated at higher temperatures. The grain growth is a function of temperature and time and can be represented as [40]:

$$d^{10} = d_{\text{rex}}^{10} + A \times \exp\left(\frac{-Q_{\text{gg}}}{RT}\right) \quad (6.15)$$

where:

A = constant
Q_{gg} = activation energy for grain growth
R = gas constant
T = temperature
d_{rex} = starting grain size

Eq. (6.15) [40] is applicable mostly for short time periods at high temperatures but after the final stage of rolling, the growth rate is far too high to be measured accurately due to a very small starting grain size and low temperature.

6.5.3 Transformation model

Upon the completion of the hot deformation, the transformation behavior of austenite to a lower transformation product will go through the process of cooling on the run out table, where the refined austenite structure will begin to transform into the sub phases of ferrite, pearlite, bainite, and martensite, depending upon the cooling rate. Transformation models predict the changes within the microstructure as the material cools so that the final microstructure can be found at the end of its cooling cycle, which in turn determines the mechanical properties of the material. Most transformation models are based on the early work done by Avrami on the transformation kinetics, described by the equation below [37,38,41]:

$$X = 1 - \exp(-kt^n) \quad (6.16)$$

where:

X = volume of transformed phase
k, n = material parameters

When using the Avrami equation, it is important to accurately portray the composition as well as the cooling conditions as the parameter k is highly dependent upon composition and temperature, while the parameter n is mostly dependent upon the transformation mechanism. The parameters can be found directly from the appropriate isothermal transformation diagram for the material. Variations of this general model such as the Johnson-Mehl, Avrami, Kolmogorov model (JMAK) as well as the Cahn model incorporate different nucleation and growth theories. In the case of modeling transformation kinetics after deformation, it is most appropriate to use the Cahn model or its extensions because it incorporates grain boundary nucleation theory, whereas the JMAK model assumes homogeneous nucleation, which is not always an appropriate assumption [42]. The Cahn model accounts for the nucleation site and growth mechanism, in which the most generalized forms of these equations will simplify to either the JMAK or Avrami equations. The following equations are based upon the assumption that the nucleation site is at the surface of the grain boundaries and the transformation rates are either based upon the nucleation and growth of a new phase, or that the growth of the new phase only occurs when the nucleation sites are completely saturated by the new phase. The first case is shown by the following from Suehiro et al. [43]:

$$X = 1 - \exp\left(\frac{-\pi}{3ISG^3t^4}\right) \tag{6.17}$$

$$\frac{dX}{dt} = 4\left(\frac{\pi}{3}\right)^{\frac{1}{4}}(IS)^{\frac{1}{4}}G^{\frac{3}{4}}\left(\ln\frac{1}{1-X}\right)^{\frac{3}{4}}(1-X) \tag{6.18}$$

where:

X = fraction transformed
I = nucleation rate per unit volume
S = grain boundary area of nucleation site
G = growth rate

Similarly the transformed fraction and transformation rate of the second case where the new phase only occurs when the nucleation sites are completely saturated by the new phase is given by:

$$X = 1 - (-2SGt) \tag{6.19}$$

$$\frac{dX}{dt} = 2SG(1 - X) \tag{6.20}$$

The importance of using the time differential of the transformed fraction during thermomechanical processing is because the transformation rate can then be obtained using the continuous-cooling-transformation diagram (CCT) as opposed to the time-temperature-transformation diagram (TTT), which will provide more accurate measurements for the various cooling rates associated with hot rolling.

Eqs. (6.17–6.20) are applicable for the transformation of austenite into ferrite, pearlite, and bainite, while the transformation into martensite is typically more common during high rate cooling conditions such as quenching. In the case of quenching, the fraction of austenite transformed to martensite can be added with the following relation:

$$X = 1 - \exp(-.0110(Ms - T)) \tag{6.21}$$

where:

Ms = martensite start temperature
T = temperature below martensite temperature

The start times for the transformations of ferrite, pearlite, and bainite are quite different depending upon the chemical composition of the material. The ferrite transformation begins when the temperature of the steel drops below the A_{e3} temperature, while the start of the pearlite transformation is dependent upon the carbon concentration. Carbon enrichment occurs during the ferritic transformation and once the carbon content of the steel reaches the A_{cm} line in the iron-carbon diagram, the transformation to pearlite begins. The starting temperature and necessary carbon concentration for the ferrite and pearlite transformations can be calculated from thermodynamics, however, the bainite start temperature (B_S) is most accurately found using experimental data.

6.5.4 Prediction of mechanical properties

The structure–property relationship of steel is used for the predictive modeling of the mechanical properties. Early work in this area involved extending the Hall–Petch equation in order to take into account the effects of alloying elements with regard to yield strength of the material. Extensive work was performed by Gladman and Pickering [44] as well as Irvine and Pickering [45] in which the following relations between the alloying elements and yield strength, tensile strength, and impact transition temperature as is seen in the following equations for ferrite-pearlite steels:

$$\mathrm{YS}(\mathrm{Nmm}^{-2}) = 53.9 + 32.3\%\mathrm{Mn} + 83.2\%\mathrm{Si} + 354\%N_f + 17.4d^{-1/2} \tag{6.22}$$

$$\mathrm{TS}(\mathrm{Nmm}^{-2}) = 294 + 27.7\%\mathrm{Mn} + 83.2\%\mathrm{Si} + 3.85\%X_p + 7.7d^{-1/2} \tag{6.23}$$

$$\mathrm{ITT}(^\circ C) = -19 + 44\%\mathrm{Si} + 700\sqrt{(\%N_f)} + 2.2\%X_p - 11.5d^{-1/2} \tag{6.24}$$

where:

d = mean ferrite grain zize
X_p = volume fraction of pearlite
N_f = free(soluble)Nitrogen

These relations were formulated using large amounts of empirical data in order to develop regression equations for the mechanical properties prediction. The tensile strength prediction in Eq. (6.23) is highly dependent upon the transformation temperature, as well as the change in the transformation temperature which is caused by the processing parameters in the mill such as cooling rate. Trzaska [46] and Suehiro [43] have shown that the hardness of the microstructural constituents can be calculated in a similar manner, and depending upon the chemical composition, the cooling rate may or may not have a large effect on the microstructural hardness values. In Trzaska's work, the regressions presented were derived from over 100 tests on 40 industrial steel grades to create a robust set of equations. In Suehro's work, however, he found that the cooling rate did not have a significant effect on the microstructural hardness of each phase and that it had a nearly linear relationship with the average transformation temperature. The relationship results in the following simple equations for the microstructural hardness:

$$H_F = 361 - .357T_F + 50\%\mathrm{Si} \tag{6.25}$$

$$H_F = 175 \tag{6.26}$$

$$H_B = 508 - .588T_B + 50\%\mathrm{Si} \tag{6.27}$$

where:

H_F, H_P, H_B = Hardness of ferrite, pearlite, and bainite
T_F, T_B = average transformation temperature of corresponding phase

Then in order to find the tensile strength, the law of mixtures for hardness is implemented, for example:

$$\mathrm{TS} = aX_F\left(H_F + bd_\alpha^{-\frac{1}{2}}\right) + X_PH_P + X_BH_B \tag{6.28}$$

where:

a, b = constants
X_F, X_P, X_B = fraction of ferrite, pearlite, bainite
d_α = the ferrite grain size

The above relations are not applicable for steels that contain elements that form precipitates, and in this case the precipitation hardening should be taken into account. Tamura et al. [34], shows the following relations for accelerated cooling that incorporates the effect of precipitate hardening on the tensile and yield strength:

$$\Delta\mathrm{YS} = \Delta\left(K_yd^{-\frac{1}{2}}\right) + \Delta\sigma_{\mathrm{ppt}} + \alpha \tag{6.29}$$

$$\Delta\mathrm{TS} = \Delta(Kd^{-\frac{1}{2}}) + \Delta\sigma'_{\mathrm{ppt}} + K_BX_B + \beta \tag{6.30}$$

where:

K, K_y = coefficient for grain size dependence of strength
$\Delta\sigma_{ppt}$, $\Delta\sigma'_{ppt}$ = increments of precipitation hardening due to accelerated cooling
$\propto, \beta$ = correction terms
K_B = coefficient of strengthening due to bainite
X_B = volume fraction of bainite

6.5.5 Current state of predictive material modeling

More difficult problems are involved when considering the predictive modeling for steels that have a very low carbon content such as in interstitial free steel. The transformation mechanism for steel that contains very low carbon content is controlled by the interface movement across the ferrite-austenite phase boundary, in which the diffusion type transformation processes are not appropriate. Problems of this matter are being dealt with by producing large amounts of empirical data in order to develop proper theoretical models. Some of the most interesting advances in material predictions have occurred during the rapid development of more complex and accurate mesoscopic models for microstructural evolution. This type of modeling provides insight into the microstructural evolution on the grain size level, allowing for a deeper understanding to the phenomena of grain growth, solidification, recrystallization and solid-state phase transformations. Common examples of computational mesoscopic models include phase field, cellular automation and Monte Carlo models, which provide an abundance of information that can be coupled to macroscopic thermos-mechanical models. The mesoscopic models offer the advantage of simulating multiple microstructure evolutions at the same time, which is of utmost importance in thermomechanical processing where some of the phenomena can overlap. The further development of the mesoscopic models will aid in the improvement of material prediction accuracy to enhance the mechanical properties of continuously cast steel products.

6.5.6 Application of predictive models

The application of the aforementioned constitutive equations and models was first shown by Sellars and Whiteman [36,40] with their modeling of the microstructural change that occurs in a steel slab during the hot rolling process. The computational model follows the process of slab reheating all the way through to ferrite transformation, including the subsequent rolling and deformation that is supplementary to recrystallization and grain growth in the slab. Tamura et al. [34] outlined the process of predicting the austenite grain size during multiple pass rolling in several steps: first, the strain, strain rate and Zener–Holloman parameter must be calculated from the initial conditions of the material in order to determine whether the criterion for dynamic recrystallization is met. In the event that these conditions are met, the calculation of dynamically recrystallized grains size ensues. However, if these conditions are not met, then the statically recrystallized grain sizes are calculated. The next step involves the case where the recrystallization is not completed during the intermediate time between rolling. In this case separate calculations are performed for the grains in which full recrystallization occurred and those in which it did not. In the latter case, the retained strain during each pass is used as an input into the strain for future passes. It is possible that areas of fully recrystallized and non-recrystallized grains coincide, making separate calculations of this area difficult. To treat this type of phenomenon, Sellars and Whiteman calculated the mean grain size of the recrystallized fraction, and used the assumption that the non-recrystallized grain size is the same as the recrystallized grain size, just prior to the preceding pass. Finally, an expression for ferrite nucleation sites is formulated by using the information regarding the deformation and grain boundaries. The formation of ferrite in recrystallized austenite occurs strictly at the grain boundaries, however, the case of deformed austenite, the ferrite formation occurs at the grain boundaries and also at the deformation bands. This methodology for the calculation of recrystallized grain size sets forth a blue print on how these simulations can be carried out. Priadi et al. [47] followed a similar path to calculate austenite grain growth during hot rolling, while making slight modifications to Eq. (6.15), in order to match closer with empirical data associated with the material under consideration, 0.028 wt.% steel. In any case, the methodology proposed provides reasonably accurate results with empirical results using constitutive modeling.

Orend et al. [48] provided a comprehensive review of the different models for microstructural evolution during hot rolling, stemming from the constitutive type models mentioned above, to the more intricate models such as cellular automata and Monte Carlo methods. Constitutive models provide a much shorter computational time as opposed to the more complex mesoscale models, but the results from the latter provide more detailed data of the phenomena. Orend et al. propose a method in which the gap between the two types of models can be filled by using an ensemble model. The proposed method uses Lagrangian formulation to describe grain growth using the energy changes in the system. The model calculates the flow stress as well as the average grain diameter under transient conditions with an abrupt change in the strain rate from 5 to 1 s^{-1}. They found that during the first rolling pass, there is a refinement in the average grain size occurring at critical strain which is much lower that the peak strain value. Once this begins, the data shows oscillatory behavior in terms of flow stress and grain diameter, a phenomenon also observed in experiment [49].

6.6 Summary

This chapter discusses the importance of HSLA or Microalloyed steels in many industrial applications. The major emphasis is the use of HSLA steels in the automotive industry. The major drivers for their development and usage are discussed. It is discussed that HSLA steels are designed and processed to meet specific mechanical properties for a given automotive application. The chapter discusses the alloy design, thermomechanical processing and transformation behavior to obtain the required structure–property relationship. It also includes a special section on transformation behavior and modeling.

References

[1] W.B. Morrison Overview of microalloying in steel. The proceedings of the vanitec symposium, Guilin, China 2000, the Vanadium International Technical Committee, Vanitec Limited, Westerham Kent, England, pp. 25–35.
[2] W.B. Morrison, Microalloy steels—the beginning, Mater. Sci. Technol. 25 (9) (2009) 1066–1073.
[3] E.F. Cone, Steel 1934, September, 41.
[4] J.R. Pierce, Scientific American, 232 (1975) 35–44.
[5] S. Dinda, Using microalloyed steels to reduce weight of automotive parts.
[6] R. Bruna, Product Technologist, Ternium Corporation—Private Communication.
[7] April 2007 aei-online.org.
[8] http://automotive.arcelormittal.com/europe/products/HYTSS/HSLA/EN Automotive Worldwide.
[9] Todd M. Link, U.S. Steel, Private Communication.
[10] https://bodybuilderhomepage.scania.com/index.aspx.
[11] SAE J2340 October 1999.
[12] M.Y. Chen, M. Gouné, M. Verdier, Y. Bréchet, J.R. Yang, Mater. Sci. Eng. A 464 (2007) 186–191.
[13] A. Galibois, M.R. Krishnadev, A. Dubé, Control of grain size and sub-structure in plain carbon and high strength low alloy (HSLA) steels—the problem and the prospect, Metall. Trans. A 10 (1979) 985–995.
[14] M.Y. Chen, M. Gouné, M. Verdier, Y. Bréchet, J.R. Yang, Interphase precipitation in vanadium-alloyed steels: strengthening contribution and morphological variability with austenite to ferrite transformation, Acta Mater. 64 (2014) 78–92.
[15] M.A. Altuna, A. Iza-Mendia, I. Gutiérrez, Precipitation of Nb in ferrite after austenite conditioning. Part II: strengthening contribution in high-strength low-alloy (HSLA) steels, Metall. Trans. A 43 (2012) 4571–4586.
[16] L. Luyckx, J.R. Bell, A. McLean, M. Korchynsky, Metall. Trans. 1 (1970) 3341–3350.
[17] S.K. Das, S. Chatterjee, S. Tarafder, Effect of microstructures on deformation behaviour of high-strength low—alloy steel, J. Mater. Sci. 44 (2009) 1094–1100.
[18] T.N. Baker, Processes, microstructure and properties of vanadium microalloyed steels, Mater. Sci. Technol. 25 (2009) 1083–1107.
[19] B.K. Show, R. Veerababu, R. Balamuralikrishnan, G. Malakondaiah, Effect of vanadium and titanium modification on the microstructure and mechanical properties of a microalloyed HSLA steel, Mater. Sci. Eng. A 527 (2010) 1595–1604.
[20] R. Lagneborg, T. Siwecki, S. Zajac, B. Hutchinson, The role of vanadium in microalloyed steels, Sc. J. Metall. 28 (1999) 186–241.
[21] Applications of Microalloyed HSLA Steels. http://www.totalmateria.com/.
[22] J.A. Davidson, A Review of the Fatigue Properties of Spot-Welded Sheet Steels, SAE Technical Paper Series 830033.
[23] E.O. Hall, Proceeding of the Physical Society B64 (1951) 747–753.
[24] N.J. Petch, The cleavage strength of polycrystals, J. Iron Steel Inst. 174 (1953) 25–28.

[25] S.K. Mandal, Steel Metallurgy: Properties, Specifications, and Applications, McGraw-Hill Education (India) Private Limited, New Delhi, 2014.

[26] B.C. De Cooman, J.G. Speer, Fundamentals of Steel Product Physical Metallurgy, Published by AIST, Association for Iron & Steel Technology, 2008.

[27] G.R. Speich, et al., in: A. Marder, J. Goldstein (Eds.), Phase Transformations in Ferrous Alloys, The Metallurgical Society of AIME, 1984, p. 341.

[28] H. Adrian, Thermodynamic Calculations of Carbonitride Precipitations as a Guide for Alloy Design of Microalloyed Steels. Proceedings of the International Conference Microalloyed'95, Iron and Steel Soc., Pittsburgh, PA, 1995, pp. 285–305.

[29] E.J. Palmiere, Precipitation Phenomena In Microalloyed Steels. Proceedings of the International Conference Microalloyed'95, Iron and Steel Soc., Pittsburgh, PA, 1995, pp. 307–320.

[30] F. Borato, R. Barbosa, S. Yue, J.J. Jonas, in: I. Tamura (Ed.), Proceedings Thermec'88, Iron and Steel Inst., Tokyo, 1988, p. 383.

[30a] F. Siciliano & J.J. Jonas, Metall and Mat Trans A (2000) 31:511. Available from: http://dx.doi.org/10.1007/s11661-000-0287-8.

[31] C.V. Scania AB 2014. Sweden.

[32] J. Burke, Grain Control in Industrial Metallurgy, ASM, 1949.

[33] Y. Huang, F. Humphreys, The effect of solutes on grain boundary mobility during recrystallization and grain growth in some single-phase aluminium alloys, Mater. Chem. Phys. (2012) 166–174.

[34] I. Tamura, C. Ouchi, R. Tanaka, H. Sekine, Thermomechanical Processing of High Strength Low Alloy Steels, Borough Green, Butterworths, 1988.

[35] F. Humphreys, M. Hatherly, Recrystallization and Related Annealing Phenomena, Pergamon, New York, 1996.

[36] C. Sellars, Hot-working and Forming Processes. The Metals Society (1980) p. 3.

[37] M. Avrami, Kinetics of phase change I, J. Chem. Phys. 7 (1939) 1103–1112.

[38] M. Avrami, Kinetics of phase change II: transformation-time relations for random distribution of nuclei, J. Chem. Phys. 8 (1940) 212–224.

[39] J. Beynon, M. Sellars, Modelling microstructure and its effects during multipass hot rolling, ISIJ Int. (1992) 359–367.

[40] C. Sellars, J. Whiteman, Recrystallization and grain growth in rolling, Metal Sci. (1979) 187–194.

[41] M. Avrami, Granulation, phase change, and microstructure: kinetics of phase change III, J. Chem. Phys. 9 (1941) 177–184.

[42] E. Jagle, E. Mittemeijer, The kinetics of grain-boundary nucleated phase transformations: simulations and modeling, Acta Mater. (2011) 5775–5786.

[43] M. Suehiro, K. Sato, Y. Tsukano, H. Yada, T. Senuma, Y. Matsumura, computer modeling of microstructural change and strength of low carbon steel in hot strip rolling, Trans. ISIJ 32 (1992) 439–445.

[44] F. Pickering, T. Gladman, ISI Special Report 81 (1961).

[45] K. Irvine, F. Pickering, Low-carbon bainitic steels, JISI 187 (1957) 292–309.

[46] J. Trzaska, Calculation of the steel hardness after continuous cooling, Int. Sci. J. (2013) 87–92.

[47] D. Priadi, R.A. Napitupulu, E.S. Siradj, Austenite grain growth calculation during hot rolling of 028% Nb steel, J. Mater. Sci. Eng. (2011) 678–683.

[48] J. Orend, F. Hagemann, F. Klose, B. Maas, H. Palkowski, A new unified approach for modeling recrystallization during hot rolling of steel, Mater. Sci. Eng. A 647 (2015) 191–200.

[49] T. Sakai, J. Jonas, Overview no. 35 dynamic recrystallization: mechanical and microstructural considerations., Acta Metall. (1984) 189–209.

Dual-phase steels*

7

N. Fonstein
ArcelorMittal Global R&D, East Chicago Labs, United States

7.1 Introduction

Dual-phase (DP) steels were the first family of high strength automotive sheet steels which manufacturing involved specific phase transformations. Featuring a high balance of strength and ductility (formability), they range in tensile strength (TS) from 450 to 1200 MPa. Their important advantages are very high strain hardening and additional high gain of the yield strength (YS) during paint baking of the finished parts (baking hardenability, BH).

These steels are and will remain the most consumed steels in existing and future cars (Fig. 7.1). It is necessary to emphasize that intercritical annealing used as the basic of DP steel manufacturing is applied as well for production of transformation induced plasticity (TRIP) steels, medium Mn steels and partially for steels with carbide free bainite (CFB or TBF steels), as well as steels processed using a quenching and partitioning cycle (Q&P steels) [2].

The rise of DP steels in 1975 and their development for the initial 25 years did not go above tensile strength of 600–700 MPa as was summarized, in particular, in the book "Low-alloyed dual-phase steels," published in 1986 (Ref. [3], in Russian), whereas current requirements from leading carmakers assume DP steels of up to 1200 MPa tensile strength with ductility approaching the achieved level for DP steels of TS ~ 600–800 MPa. That requires revising all sources of steel strengthening keeping their high formability, reconsidering all possible effects of structure constituents and alloying/microalloying.

Consequently the current chapter will describe the structure features of DP steels in their relationship with various mechanical characteristics, phase transformations allowing to control the ratio of structure constituents, effects of alloying on both phase transformation and properties of individual-phases, as well as (briefly) the application of DP steels.

* Every effort has been made to trace copyright holders and to obtain their permission for the use of copyright material. The publisher apologizes for any errors or omissions in the acknowledgements printed in this book and would be grateful if notified of any corrections that should be incorporated in future reprints or editions.

Automotive Steels. DOI: http://dx.doi.org/10.1016/B978-0-08-100638-2.00007-9

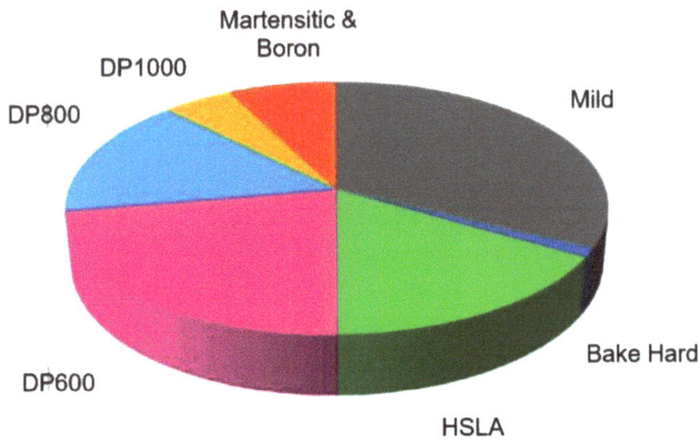

Figure 7.1 Shares of different materials in 2015 Ford Edge car [1].

7.2 Effect of structure on mechanical properties of dual-phase steels

As mentioned above, DP steels are characterized by high combination of strength and formability that includes high strain hardening, high uniform elongation (UE), and low YS/TS ratio. As will be shown below, these features as well as high BH are controlled by different parameters of main phases, ferrite and martensite.

7.2.1 Existing models of strength of heterogeneous materials

The presence of high strength "hard" martensite islands in ductile ferrite matrix makes a heterogeneous microstructure similar to the natural composite, the properties of which are determined by the properties of the individual-phases and their stereological parameters.

The rule of mixtures used for two-phase materials assumes:

$$\sigma_c = \sigma_1 V_1 + \sigma_2 V_2 \tag{7.1}$$

where σ_c, σ_1, and σ_2 are the strengths of the composite and of each of the two phases, respectively, with the phase volume fractions $V_1 + V_2 = 1$.

In fact, two extreme cases of phase behavior were considered [4], one case being when the *strains* in phases I and II are equal, as in fibrous materials under loading, parallel to the fibers, and in the other case, when the *stresses* are equal, as in lamellar composites at applied stress perpendicular to the composite layers.

For a random distribution of hard particles in a soft matrix, as in the case of DP steels, an intermediate alternative mechanical behavior would take place, where neither strains nor stresses are identical in phases, but are distributed between the two phases in some way. In this case the softer phase is more deformed than the harder one. Consequently the stresses in the harder phase are lower than the average stress of the mixture.

Eventually, in spite of numerous modifications, it appeared that the rule of mixture presented, in particular, by Mileiko [5] could not be used for an adequate description of the properties of DP steel, because of the unequal strains in the ferrite and martensite, which has been revealed experimentally (Refs. [6–8]). As a result, the stresses in the ferrite and the martensite in Eq. (7.1) should refer to different strains:

$$\varepsilon_c = \varepsilon_F V_1 + \varepsilon_M V_2 \tag{7.2}$$

where ε_c, ε_F, and ε_M are strains in composite, ferrite and martensite respectively. Martensite deformation is very small, but martensite experiences relatively high stresses and so provides the major portion of the load carrying capacity of the DP steel.

Tamura and Tomota [9] as well as Araki [10] have found that the strain ratio $m = \varepsilon_M/\varepsilon_F$ varies from $m = 0$ for equal stresses to 1 for equal strains, depending on the ratio of hardness of phases.

In an investigation by Speich and Miller [11], who studied the DP steels group, a satisfactory agreement with the Tomota–Tamura model was found using $m \approx 0.01$. In this case near TS of DP steel (at UE = 21%) the strain of martensite was approximately 0.2%.

As mentioned, both models of Mileiko and Araki assume perfect cohesion between phases up to the onset of plastic instability. The deviations from this assumption are of paramount importance for fracture behavior. The experimentally observed origination of voids at the ferrite-martensite interface in DP steels before martensite fracture suggests that in that moment the ultimate tensile strength level had not been achieved for the hard phase.

Bortsov and Fonstein had confirmed the strain distribution between the phases of DP steel containing 0.072C–1.38Mn – 0.49Si – 0.71Cr–0.003B[1] [12]. Heat treated samples were cold rolled from 2.5 mm to thicknesses of 1.5, 0.9, 0.6, and 0.35 mm, which corresponded (with corrections for widening) to the true strains in the rolling direction of 0.42, 0.97, 1.31, and 1.81, respectively. The effect of martensite hardness on the strain distribution in the phases was evaluated by a comparative study of quenched samples with those tempered at 400°C for 1 hour prior to cold rolling. The martensite hardness values were 7680 MPa for the quenched and 5950 MPa for the quenched and tempered specimens.

Fig. 7.2 depicts microstructures of DP steels with different martensite hardness after rolling up to a true strain of 1.81. The relation between the strain of the hard phase and its hardness is quite obvious.

Quantitative measurements of the strains in the martensite were made using image analyzer and software developed based on the method proposed by Gurland [8]. Mean values of l/w ratio were calculated, assuming the shape of martensite particles close to ellipse where *l* is the big axis of ellipse, and *w* is its small axis.

Fig. 7.3 presents the calculated strain in the martensite as a function of specimen deformation. The dashed line corresponds to the condition of equal strains in the ferrite and the martensite when $m = \varepsilon_M/\varepsilon_F = 1$.

[1] Here and after in weight percent

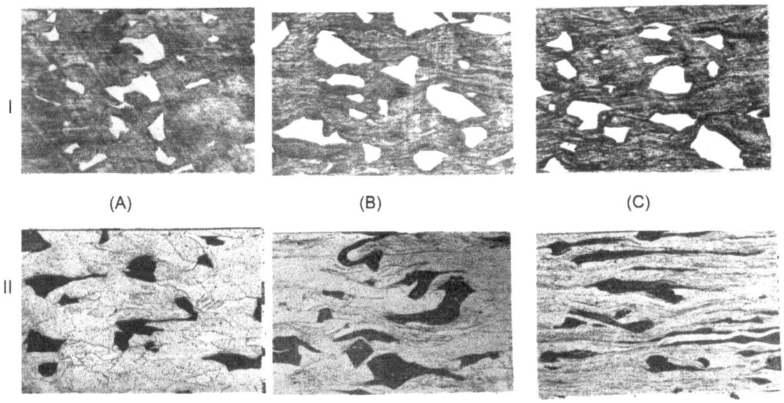

Figure 7.2 Microstructure of dual-phase steels after true deformation $\varepsilon = 0.4$ (A), 1.0 (B), and 1.81 (C):I- nontempered martensite (using Le Pera etching); II- martensite tempered at 400°C (etched by Nital) OM, $\times 200$ [12].

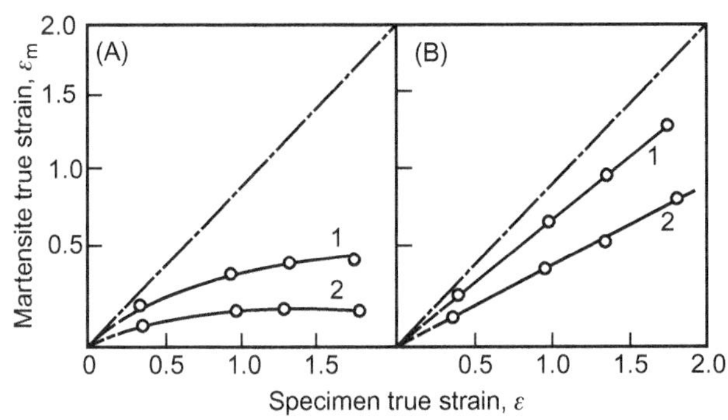

Figure 7.3 Relationship of deformation of nontempered (A) and tempered (B) martensite and the total deformation; 1: calculations without deduction of rotation; 2: calculations taking account the contribution of martensite grain rotation [12].

As shown, for the DP steel, tempered at 400°C, the martensite strain ε_M increases progressively with the total strain of specimens ε, when ε_M being approximately 0.5–0.7 of ε. For untempered martensite at $\varepsilon < 1$ the martensite strain is a maximum 0.2 of the total strain, meaning that the stronger martensite the lower *m*.

More discussions regarding existing models describing mechanical behavior of DP steels can be found elsewhere [2,6,7,13–15].

7.2.2 Factors affecting strength characteristics of DP steels

Strength of DP steels is determined by contributions from both phases, ferrite and martensite.

Many authors noted a linear relationship between TS of DP steel and the volume fraction of martensite. Serious debates continue as to whether this relationship should depend on the properties of martensite or ferrite matrix.

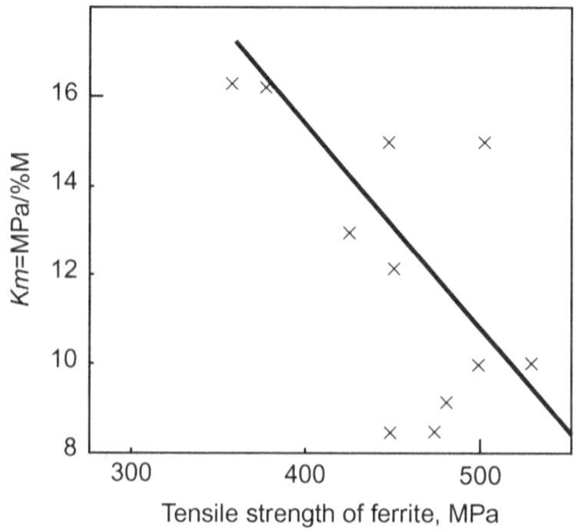

Figure 7.4 Relationship between extrapolated tensile strength of ferrite in DP steel and martensite hardening rate, K_m.

Re-arranging the basic Eq. (7.1) gives

$$TS_{DP} = TS_F V_F + TS_M V_M = TS_F(1 - V_M) + TS_M V_M = TS_F + (TS_M - TS_F)V_M \tag{7.3}$$

Consequently, the strengthening of ferrite by grain refinement or solid solution strengthening by Si or P should decrease strengthening rate of DP steels per 1%[2] martensite, which depends on the difference $(TS_M - TS_F)$.

Based on known regression equations for DP steels of different compositions [16–23], Fig. 7.4 shows the relationship between the martensite hardening rate, k_M, and the ferrite tensile strength. The last one was obtained by cited researchers using the extrapolation of the tensile strength of DP steel to zero martensite content. As shown, there is some trend for k_M to decrease with increasing ferrite tensile strength, but there is no direct correlation between the two.

Fig. 7.5 presents the experimental data of measured TS (solid line) of quenched CrMnSiB steels with different carbon content versus volume fraction of martensite [24]. Dashed lines correspond to TS calculated using a modified equation for the law of mixture for corresponding carbon content in steel.

As shown, the rate of experimentally observed strengthening by martensite was found to be three times higher than the value calculated from the rule of mixture (11.5 MPa/% and 3.5 MPa/%, respectively).

As illustrated in Fig. 7.5, the tensile strength of DP steels is independent of the intrinsic strength of martensite. Similar trends were also observed by others [18,20,25]. In particular, no effect of C_M (carbon content in martensite, i.e., its strength) on the rate of strengthening of DP steel by of martensite was noted by Davies [17] in the study of steels containing 0.06–0.47% C quenched from different temperatures in the intercritical range.

[2] Phase contents here and afterwards are in volume percent.

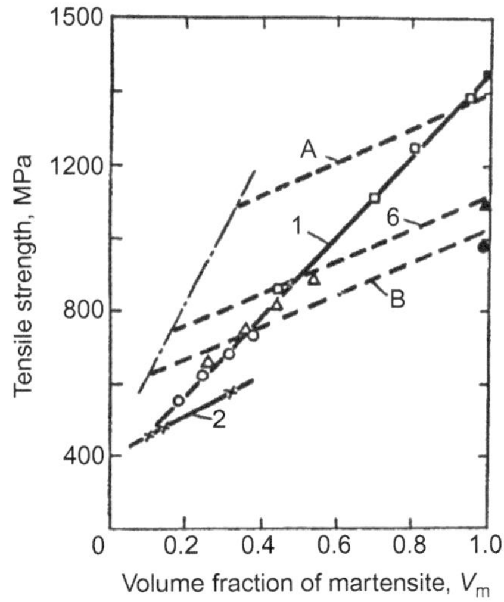

Figure 7.5 Effect of volume fraction of martensite on tensile strength of CrMnSiB DP steels with various content of carbon: ○- steel with 0.07% C; Δ-0.10% C; □—0.18% C; solid symbols indicate fully martensitic microstructure; dashed lines : calculations using rule of mixtures [12].

Very few data are available regarding the effect of bainite when ferrite-bainite mixture is formed or martensite in DP steels is partially replaced by bainite. Bhadeshia and Edmonds stated that the effects of martensite and bainite on the strength of DP steels are described by a common curve [14]. Some data suggest that hardening by bainite is substantially weaker than that by martensite [26]. Similar results were obtained when a part of martensite was replaced by bainite showing that strengthening by bainite was 4 MPa/% lower than by martensite [27].

The absence of physical discontinuous yield strength (yield point elongation, YPE) and low YS/TS ratio, which can be as low as 0.4–0.5, are intrinsic features of DP steels that are a result of the high density of mobile dislocations created due to local deformation of ferrite under effects of expansion of adjacent grains due to austenite-martensite transformation.

Contributions of internal tensile residual stresses due to ferrite contraction during γ-α' transformation act in the same direction [28–32].

Depending on steel composition, the minimum amount of martensite necessary to suppress YPE should be a function of the M_S (martensite start) temperature and thus of the level of residual stresses and can range from 3 to up to 10–15% [25,33–35].

YPE can return after tempering of DP steels due to relaxation of residual stresses and decrease in density of mobile dislocations. As a result of tempering the YS/TS ratio of DP steels increases from 0.45–0.5 to 0.65–0.7. Takada et al. [36] showed that low YS/TS ratio and the absence of YPE require the hardness of martensite H_M to be much higher than that of ferrite, H_F, $H_M/H_F \geq 2.2$.

It was found that the effect of V_M on YS (after achieving the minimum due to suppression of YPE) can be described by a single curve independent on strength of martensite (determined by its carbon content). The insensitivity of YS of untempered DP steels to strength of martensite is explained by the fact that at small offset strains (0.2%) the martensite islands deform elastically so that the actual stresses in martensite are essentially independent of its strength (C_M).

At martensite volume above 0.45–0.60 (at TS of DP steel about 1000–1200 MPa with martensite skeleton), the properties of martensite began to affect the dependence of yield strength and tensile strength of DP steels on martensite strength more closely to the predictions of the law of mixture, because the actual ability of steel to bear the loads became more dependent on martensite strength.

Numerous publications concerning the effects of ferrite grain size on flow stress of DP steels can be found. Hall–Petch type equation:

$$\sigma = \sigma_0 + K_y d^{-1/2} \tag{7.4}$$

was derived based on inverse proportionality of YS to dislocation free path in ferrite of DP steels $\lambda_{<\alpha>}$ instead of ferrite grain size d. (σ = strength, σ_0 = frictional stress).

The physical meaning of the Hall–Petch coefficient K_y for tempered DP steels differs from that for quenched DP steels because the tempering is accompanied by pinning of dislocations in ferrite and by relaxation of residual stresses. These conditions agree with the usual meaning K_y as the intensity of stress necessary to generate new dislocations in the adjacent ferrite grain.

n investigation of untempered DP 0.08C–1.45Mn – 0.2Si steel [37] it was found that the Hall–Petch coefficient K_y strongly depends also on true strain, increasing from about 0 close to yield strength with fast growth at strain increase (Fig. 7.6). At low strain (0.2%) there was no effect of grain size due to high volume of mobile dislocations facilitating the initial deformation regardless of grain size. The shown dependence of yield strength (at 0.5% offset strain) on ferrite grain size, in author's opinion is explained by rapid movement and interaction of unpinned dislocations, i.e., by significant contribution of strain hardening starting from very small strains.

Qualitatively, similar results were obtained later when studying 0.1C–0.15Si – 1.5Mn – 0.8 (Cr + Mo) steel [38]. The authors also analyzed the strength difference in the DP steel with different grain size of ferrite at different levels of plastic strain. Although some increase in YS with grain refinement was found, $K_y = 9.88\ \mathrm{Nmm}^{-3/2}$ appeared to be remarkably smaller than the value of $17.4\ \mathrm{Nmm}^{-3/2}$ reported for ferritic steel. The authors explained their data by lower stresses required to move dislocations, due to existence of unlocked dislocations sources in quenched DP steels. Similar conclusions were made by Bleck et al. who found $K_y = 4\ \mathrm{Nmm}^{-3/2}$ for YS of DP steels compared to $12\ \mathrm{Nmm}^{-3/2}$ for YS of Al-killed mild steels after temper rolling [39].

7.2.3 Strain hardening of dual-phase steels

An extremely important advantage of ferrite—martensite steels is their high strain hardening rate compared to other low-alloy steels. Low yield strength in the initial condition facilitates cold forming but high strain hardening rate determines the possibility of a significant increase in steel strength after deformation. Understanding

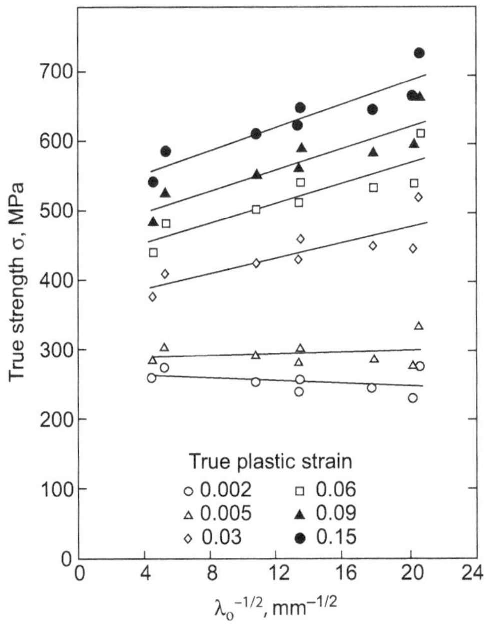

Figure 7.6 Hall–Petch analysis of flow stress data, steel 0.08C-1.45Mn-0.2Si oil quenched after holding at 760°C for 345 s; [37].

the factors controlling strain hardening behavior is the key to optimization of the microstructure of DP steels.

There are more than twenty different models to approximate a stress—strain curve including those proposed by Ludwig [40], Swift [41], Voce [42], as well as the modified Jaoul–Crussard model [43]. The most common parameter of strain hardening is the exponent n in the parabolic approximation of stress–strain σ-ε curve introduced by Hollomon [44]:

$$\sigma = K\,\varepsilon^n \tag{7.5}$$

where σ and ε are the true stress and true strain, respectively. The coefficients k and n can be obtained by plotting the true stress-true strain curve on a logarithmic scale

$$\log \sigma = \log k + n \log \varepsilon \tag{7.6}$$

then

$$n = \Delta \log \sigma / \Delta \log \varepsilon \tag{7.7}$$

When a strain hardening curve is plotted on a logarithmic scale, several hardening stages can often be identified [45]. Sometimes the strain ranges for evaluation of n are prescribed, e.g., from 6 to 12% elongation, from 10% up to ultimate uniform

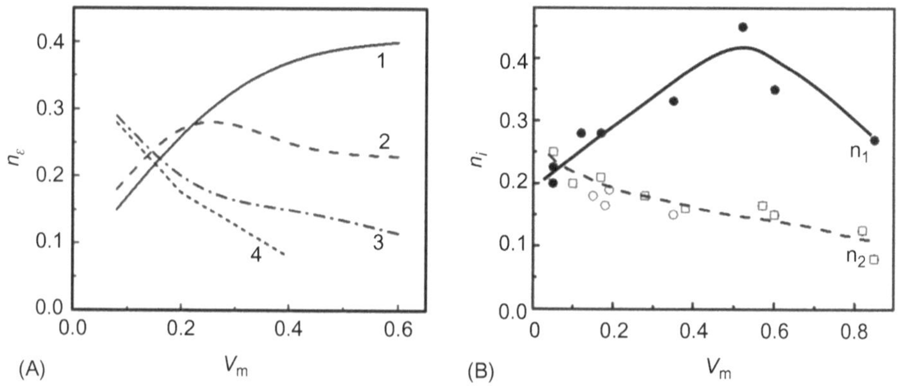

Figure 7.7 Effect of (A) True strain on n_i-values of 0.06C-CrMnSiB steel: 1−0.01, 2−0.02, 3−0.05, 4−0.10 and (B) Volume fraction martensite in DP steels on strain hardening exponent of 10Mn2V steels.
Source: N.M. Fonstein, Factors affecting the strain hardening of low-alloy steels of ferritemartensite structure, Steel USSR, Metally 3 (1991) 118−123. With permission from American Iron and Steel Institute [47].

elongation (i.e. total uniform elongation), etc. The complete plot of instantaneous n (i.e. n_i) values can be obtained using:

$$n_i = \frac{d\sigma}{d\varepsilon_{\varepsilon-\varepsilon_i}} \times \frac{\varepsilon_i}{\sigma_i} \tag{7.8}$$

In fact, the physical significance of the point of ultimate tensile strength TS is the attainment of the condition when the true tensile stress at the current decreased cross-section of the specimen (geometrical softening) becomes equal to strain hardening rate $d\sigma/d\varepsilon$, and necking starts. In this moment the instantaneous n_i value becomes equal to the true ultimate uniform elongation. This equality was confirmed experimentally by Davies [46].

The strain hardening rate of DP steels is the highest in strain range of $\varepsilon = 0.01-0.05$ (Fig. 7.7A). This is interpreted as a result of initial high density of free dislocations and hence higher rates of dislocation accumulation and multiplication within this strain range. The retained austenite, if present, also transforms to martensite typically at this stage [19,26,48]. It is believed that local transformation of retained austenite increases the dislocation density, creates additional nonuniformity of deformation and hence the rate of strain hardening.

The decrease in strain hardening rate with increasing strain was attributed to the formation of dislocation forest and further to more ordered dislocation (cell) substructure.

Some researchers interpreted the strain hardening behavior of DP steels in terms of hardening by nondeformable dispersed particles that cannot be cut by dislocations. As shown in Ashby's theory [49], which is often employed in this approach [6,50−53], displacement of grain boundaries and phase interfaces under external loading involves the formation of dislocation pile-ups necessary to sustain the continuity (accommodation) between nondeformable particles and matrix.

According to Ashby, the strain hardening is proportional to the ratio of the volume fraction to the size of particles, or to the dislocation free path in the matrix, i.e., to the-particle spacing (martensite particle spacing in DP steels):

$$\sigma - \sigma_0 = kG\sqrt{\frac{bV_\beta \varepsilon_{\mathrm{pl}}}{0.41\overline{L}_\beta}} \tag{7.9}$$

where σ_0 is a constant, G is the shear modulus of the matrix ($\approx$ 82,400 MPa for ferrite), b is the Burgers vector of the matrix dislocations (about 0.247 nm), V_β is the volume fraction of hard phase, $\overline{L}_\beta$ is the mean particle size in mm, $\varepsilon_{\mathrm{pl}}$ is the plastic strain and k is a constant close to 1.

Differentiation of Eq. (7.8) gives:

$$\frac{d_\sigma}{d_\varepsilon} = 0.78\, k\frac{Gb^{1/2}}{\varepsilon^{1/2}}\sqrt{\frac{V_\beta}{\overline{L}_\beta}} \tag{7.10}$$

Substitution of constants gives, e.g., for $\varepsilon = 0.2$ (about 20% engineering strain)

$$\frac{d_\sigma}{d_{\varepsilon 0.2}} = 2200\sqrt{\frac{V_\beta}{\overline{L}_\beta}} \tag{7.11}$$

Numerous studies considered effects of volume fraction, sizes and morphology of martensite phase on strain hardening of DP steels.

As shown in Fig. 7.7B, n_i-values measured in the range of $\varepsilon = 0.01-0.05$ (n_1) increase with volume fraction of martensite, V_{M}, until the formation of martensite skeleton (at about 50–55% of martensite) is formed. In contrast, the n values at final stage of uniform deformation n_2 (at strains above 0.05 and up to ultimate uniform elongation) decrease with V_{M} within the entire range of martensite volume fractions.

In accordance with Ashby theory strain hardening of DP steels should be higher at higher difference in hardness of microstructure constituents when an increase in the strength of martensite (C_{M}) should enhance the work hardening by increasing plastic incompatibility between the phases, especially at initial strain [54]. This is experimentally confirmed by decrease in strain hardening observed after tempering of DP steels when the difference in hardness between phases becomes smaller.

It is reasonable to assume that substitution of martensite by the same volume fraction of less hard bainite would reduce strain hardening rate of steel. This assumption is confirmed, in particular, by higher YS/TS ratio of DP steels with bainite [55].

From the Ashby theory (Eq. 7.9) it follows that strain hardening rate of DP steels should increase with higher V_{M} and lower $\overline{L}_{\mathrm{M}}$. Although this model, which is based on the assumption of nondeformable equiaxed particles, cannot quantitatively describe strain hardening of DP steels, qualitatively it aligns well with the experimental data for DP steels with various volume fractions and sizes of martensite phase. An important point is that refinement of martensite particles should be

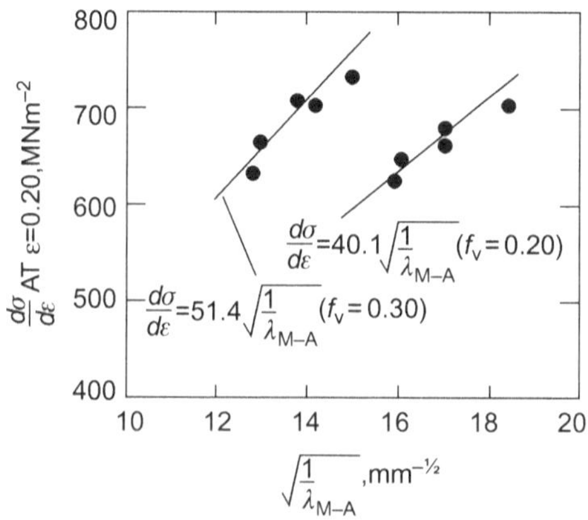

Figure 7.8 Relationship between work hardening rate at $\varepsilon = 0.20$ and size of martensite particles [53]. f_v and λ_{M-A} stand for volume fraction and interparticle spacing of martensite-austenite (M–A) constituents, respectively.

beneficial for strain hardening and therefore for attainment of a higher uniform elongation of DP steels.

In particular, it agrees with data presented in Fig. 7.7B, showing that initial strain hardening rate increases with V_M (in this case, up to the formation of martensite skeleton). Maxima in $d\sigma/d\varepsilon - V_M$ plots were reported also by Adamczyk and Graijcar [4] and De Cooman et al. [56].

It was shown theoretically [53] that the initial strain hardening rate should be higher in case of small, well dispersed martensite particles than in the case of coarse ones. Some of their calculations are presented in Fig. 7.8 [53].

As was observed, the strengthening of ferrite by additions of elements promoting higher strain hardening of ferrite, as silicon, phosphorus, or their combination, increases the strain hardening rate and hence the uniform elongation of DP steels [57].

A decrease in ferrite grain size is almost always accompanied by some refinement of martensite particles that should contribute to enhanced strain hardening. Increase in the initial strain hardening rate was noted in studying of 0.17C–1.49Mn – 0.22Si steel, where grain refinement of ferrite from 12.4 to 1.2 μm at constant martensite content was produced [58]. The authors attributed the observed effect to early onset of dislocation interactions, higher numbers of dislocation sources and back stresses from martensite islands.

It is reasonable to point out the findings about positive effects of fine and ultrafine ferrite grains [59] on strain hardening and consequently plasticity of DP steels. The presence of the second phase makes that effect quite different in comparison with the known negative influence of grain refinement on single-phase ferrite of mild steels [60].

In conclusion, it should be noted that in spite of the imperfect microstructure models for the strain hardening of DP steels a few facts seem to be beyond doubt. Namely, the strain hardening rate of the two-phase structures during initial deformation (at strains below 4–5%) is increased by refinement of martensite islands, by an increase in the volume fraction of martensite, provided the martensite particles do not coarsen, and to a lesser extent by an increase in martensite strength. The same effect is induced by factors favoring increase in the strain hardening rate of ferrite, as well as increase in the amount of retained austenite.

7.2.4 The ductility parameters of dual-phase steels

High ductility at given strength level is one of the greatest advantages of DP steels.

As already noted, uniform elongation, UE, is governed by strain hardening rate. Higher uniform elongation UE is facilitated by all factors that increase *n*-value, or $d\sigma/d\varepsilon$, or simply the difference TS−YS.

However, as shown in Fig. 7.9, an increase in volume fraction of martensite, V_M, progressively lowers UE. It agrees with data presented in Fig. 7.7 pointing out that increase in V_M at strains above 0.1 (approaching uniform elongation) decreases strain hardening.

There are some discrepancies concerning the effect of the martensite strength. In Fig. 7.9, all the UE-V_M data fit, within experimental scatter, a common curve irrespectively of the C_M values [24]. Similar results were obtained by Davies [61] and Marder [48].

A decreasing ductility of DP steels with increasing V_M at higher relative strength of martensite (at higher YS_M/YS_F ratio) was reported in other studies [62] as a result of the premature formation of voids in steels with significantly harder martensite due to incompatibility of deformation of phases. Authors derived an

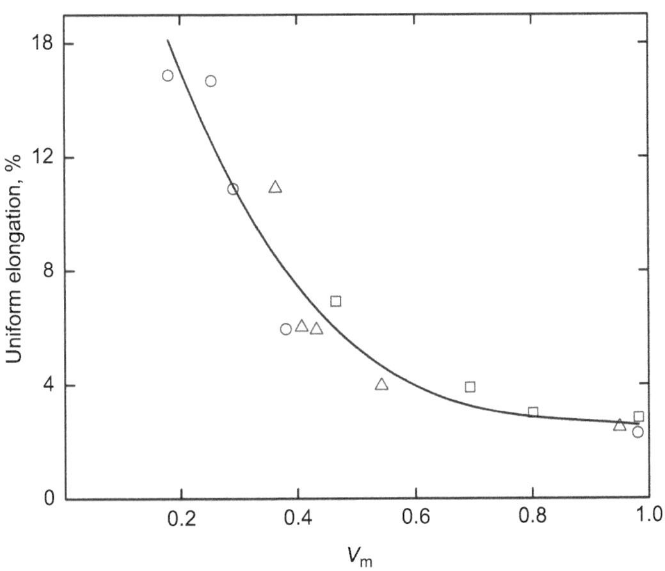

Figure 7.9 Effect of volume fraction of martensite on uniform elongation of Cr-Mn-Si-B steel with various amounts of carbon (○-steel with 0.07% C; Δ-0.10% C; □-0.18% C).

empirical equation correlating the true uniform strain of DP steel, ε_u^{DP}, and true uniform strain of ferrite, ε_u^F, on the one hand, with the volume fraction of martensite, V_M, and its carbon content, C_M, on the other hand:

$$\frac{\varepsilon_u^{DP}}{\varepsilon_u^F} = 1 - 2.2\, C_M \sqrt{V_M} \tag{7.12}$$

This dependence suggests, in particular, that uniform elongation can be increased by reducing C_M without increasing V_M, i.e.,, by decreasing carbon content in the steel. Further analyzing Eq. (7.12), the authors concluded that increasing ε_u^F is the most efficient way to achieve higher UE in DP steels. This can, in particular, be accomplished by adding silicon or phosphorus [61], which are believed to increase $d\sigma/d\varepsilon$ of DP steels.

Total elongation, TE, decreases with increasing V_M of DP steels. This can be interpreted in terms of both decrease in UE and smaller post-necking strain (TE−UE).

The duration of the latter stage of localized deformation is determined by the rate of nucleation and amount of microvoids, whereas the moment of their coalescence to produce a main crack is defined by the distance between martensite islands and properties of ductile ferrite matrix.

Irrespectively of the void nucleation mechanism, by decohesion along grain boundaries or due to fracture of martensite particles, an increase in V_M reduces the void spacing and hence the plastic strain necessary for their coalescence. It was shown that the difference TE−UE, i.e., the span of localized deformation, decreases with increasing C_M (and hardness of martensite) as a result of easier decohesion or martensite fracture.

At a constant volume of martensite the ductility of a DP steel as a composite material is seemly governed by the properties (ductility) of ferrite matrix.

The differences in TE observed in DP steels at roughly constant TS (and the same V_M) point at the important influence of martensite particle sizes, geometry, distribution, and continuity on the initial size of microvoids and the length of the stage of localized (post-necking) deformation before coalescence of microvoids. DP steels with finely dispersed martensite particles exhibit the highest total ductility because of both higher uniform elongation and smaller voids and consequently their delayed coalescence.

The data on the reduction of area (RA) for DP steels are sparse, primarily because these steels are mostly intended for sheet applications. At the same time, since a correlation was found between RA and such an important parameter as hole expansion [63], the interest in this characteristic of localized deformation of AHSS (advanced high strength steels) is currently growing.

The increase in RA with decreasing strength of martensite was noted in particular by Speich and Miller [62]. With this respect, higher tempering temperature markedly improves RA.

According to Sugimoto [64], ferrite (structure) refinement resulted in favorable higher RA due to relative suppression of size related critical stresses and therefore due to later void formation around martensite islands.

7.2.5 Quench and strain aging

DP ferrite-martensite steels, as materials for cold forming applications, should not possess any propensity for static aging during long storage at ambient temperatures. On the other hand, the service (in-use) characteristics of the final parts manufactured from DP steels, such as dent resistance, can be improved by hardening during deliberate strain aging at slightly elevated temperatures that makes such hardening highly desirable.

In DP steels, ferrite that undergoes accelerated cooling from the $\alpha + \gamma$ region is supersaturated with interstitials (see Section 7.3). Nevertheless, as shown by many researchers, DP steels do not exhibit any aging effects such as the development of YPE or a yield strength increase during holding at room temperature for as long as more than one year.

In reheating of nondeformed DP steels, several stages must be considered: (1) formation of Cottrell atmospheres in ferrite; (2) carbon clustering in ferrite; (3) precipitations in ferrite, and (4) tempering of martensite—volume contraction, changes in martensite strength and additional carbon clustering or precipitations near ferrite/martensite interfaces where the dislocation density is high.

Pre-strain accelerates all of the above processes, but does not change the nature of the phenomena.

The data in Fig. 7.10 depict YS increase after various temperatures of reheating vs. the extent of pre-strain and over-aging time.

As noted above, DP steels are characterized by nonhomogeneous dislocation substructures, which implies that the dislocation density in ferrite, while being high in the vicinity of martensite/ferrite interface, decreases sharply toward the interior of ferrite grains [29]. Therefore a long-range diffusion of interstitials from internal areas of ferrite grains to their periphery is required to produce sensible dislocation pinning. As a result, at relatively low temperatures of reheating many dislocations remain unpinned, preventing aging, which could induce the YPE or substantial strengthening.

It is generally accepted that this nonhomogeneous dislocation substructure is preserved during the temper rolling of DP steels, as opposed to tensile deformation that leads to a more uniform dislocation distribution. In the latter case many active

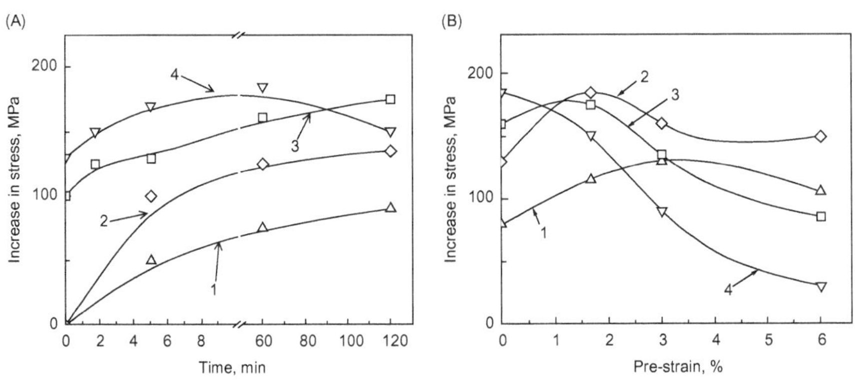

Figure 7.10 Effect of aging time (A) and prestrain (B) on YS increase of dual-phase 0.08C-1.5Mn-0.5Si steel heated at: 1–100, 2–200, 3–300, 4–400°C [64].

sources emerge within the grains, with the short-range diffusion being sufficient for dislocation pinning.

As a result, DP steels typically exhibit no yield-point elongation after aging at temperatures of up to 260°C if the annealed steel has been temper rolled with reductions of up to 10%. In 0.1C–1.4Mn steel deformed in tension, strengthening achieved maximum 80 MPa at strain of 5% and the YPE developed at 260°C after 10–15 seconds of holding [66].

As shown in Fig. 7.10, the increase in the aging temperature to a level required for baking corrosion-resistant paint coatings, i.e., to 170–220°C, results in substantial strengthening of DP steels [31,67,68].

The necessity of raising the temperatures above 100°C to reach the required aging indicates the key role of carbon atoms, since the diffusion rates of nitrogen atoms are fairly high even at room temperature and are believed to be responsible for aging at temperatures below 100°C observed in mild steels [69].

Considering the contribution from residual stresses and dislocation density in ferrite, the strengthening produced by aging should depend on the volume fraction of martensite V_M. As V_M increased from 5 to 25%, the strengthening effect increased from 35 to 100 MPa in a DP steel quenched from 785°C and deformed in tension to 2%, with subsequent tempering at 260°C. At pre-strain of 4% in steel of 0.17C–1.4Mn – 0.4Si with 14 and 22% martensite obtained after annealing at 735°C and 780°C, the maximum strengthening effect after aging at 200°C was 65 and 85 MPa, respectively [70].

Grain refinement reduces the effect of nonhomogeneity of dislocation distribution responsible for delaying natural aging. Together with shorter diffusion paths, grain refinements should accelerate aging and increase the resultant strengthening [71].

7.2.6 Tempering of dual-phase steels

Tensile strength of DP steels decreases upon tempering being mostly controlled by changes in the strength of martensite phase.

As shown in Fig. 7.11, the conventional short-term tempering of DP steels in continuous annealing lines (CALs) at temperatures of ~200°C mostly results in slight

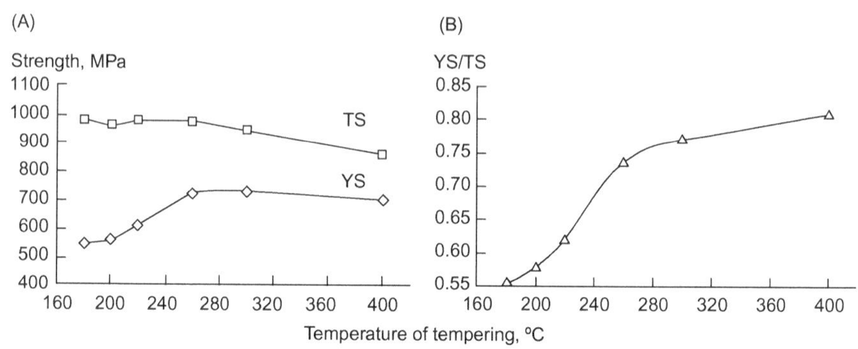

Figure 7.11 Effect of tempering temperature on strength characteristics (YS, TS, and YS/TS ratio) of 0.08C-2.1Mn-0.7Si-Mo steel heated to 800°C and water quenched from 760°C [72].

yield strength increase without changes in TS. Further heating leads to the decrease in TS of DP steels mainly due to softening of martensite. Both factors result in the increase in YS/TS ratio. However, as shown in Section 7.4, the actual level and the rate of softening depend on alloying composition that affects the behavior of martensite during tempering (precipitation and coarsening of carbides) [73].

Sometimes, an improvement of ductility of DP steels is observed after tempering at temperatures above 300°C. This is believed to be primarily due to larger post-uniform deformation [74] as a result of better strain compatibility between the phases and the suppression of microvoid nucleation and growth.

The increased ductility along with lower strain hardening and smaller differences in hardness of ferrite and martensite with increase in temperature of tempering result in substantial improvement of stretch flangeability evaluated by hole expansion. This improvement of behavior of DP steel during localized deformation (cut edge, hole expansion) is becoming more evident at higher tempering temperatures (up to 450–500°C).

7.2.7 Fracture behavior of dual-phase steels

Applications of high strength materials are often restricted by their fracture behavior.

Many parts stamped from DP steels are used in service under alternating loads. This, in addition, calls for the importance of the fatigue fracture resistance characterized, e.g., by the fatigue limit and low-cycle fatigue life.

A general approach to predicting the effects of microstructure parameters on the fracture behavior of heterogeneous materials has not been fully formulated and needs experimental verification.

Non-homogeneity of plastic deformation has the strongest impact on crack initiation as it accelerates void nucleation. At the same time, however, plastic nonhomogeneity often increases the resistance to crack propagation. Under certain conditions the nonhomogeneous microstructure can either accelerate or delay crack growth, e.g., through crack branching. Another possibility is the local strain-induced martensitic transformation of retained austenite at the crack front in DP steels containing some amount of austenite. The resulting volume change generates internal compressive stresses in martensite thus reducing the level of the applied tensile stresses.

7.2.7.1 Toughness

In general, the toughness of DP steels depends on three characteristics of martensite: its volume fraction, morphology (shape, size, distribution), and toughness, which is determined by carbon content in martensite, along with the properties of ferrite. The effect of the martensite particle dispersion on toughness of DP steels was discussed by Tomota and Tamura [75] who showed that coarse DP microstructures exhibit lower toughness levels evaluated by K_{1c}.

Increase in strength at improved ductility due to microstructure refinement is attributable to the combined effect of strengthened ferrite and enhanced toughness of martensite. This leads to less severe stress/strain partitioning and better interface cohesion [58] so that crack propagation becomes more sluggish.

7.2.7.2 Fatigue

The yield strengths of DP steel with 0.05C–1.1Mn – 0.3Cr and HSLA steel of 0.08C–0.9Mn – 0.4Si–Ti produced using batch annealing were 285 and 315 MPa and the tensile strengths were 545 and 515 MPa, respectively. The maximum tensile stress amplitudes after 10^6 cycle fatigue test at 20–60 Hz with $R = 0.1$ differed only by 10 MPa in as-heat treated condition, but by 40 MPa after 5% deformation and tempering at 200°C for 1 hour. At maximum stress amplitude of 420 MPa the endurance limit of these steels after that work hardening and tempering differed by more than two orders of magnitude. Similar data result in a wide commercial application of as rolled DP steels for car wheel disks.

7.2.7.3 Hydrogen embrittlement

High strength materials are often susceptible to hydrogen embrittlement. In particular, Davies [76] studied dual-phase steel with TS = 690 MPa and found that hydrogen pickup resulted in a loss of ductility of steel and in a reduction in its long-term strength, σ_T, during testing for 50–100 hours, with the fracture mechanism changing from dimple-type to transcrystalline cleavage. It was suggested that hydrogen embrittlement is brought about by the presence of 15–20% of high carbon martensite (with $C_{\mathrm{M}} \approx 0.6\%$). However, the extent of this hydrogen embrittlement in DP steels was not very significant. After 50–100 hours of holding in hydrogen, delayed fracture was observed at stresses higher than the yield strength with no intergranular fracture.

Interesting results were obtained when studying of the effects of cold working usually employed for low-carbon steels with ferrite-martensite structures. Additional dislocations introduced by plastic deformation increased the load carrying capacity of steel in a hydrogen environment by 10–20%.

Morphology of microstructure of DP steels is also important. The best resistance against hydrogen embrittlement was demonstrated by steel with ferrite-martensite microstructure subjected to pre-quenching, repeated heating and subsequent controlled cooling at 30–40°C s^{-1} from the γ-region. The resultant microstructure had fine ferrite grain size and maximum amount of low-carbon uniformly distributed needle-like martensite (up to 45%). Coarser ferrite grain size, lower amount and therefore higher carbon content in martensite impaired the resistance to hydrogen embrittlement of DP steels.

Similarly in an earlier study [76], low temperature tempering at 180–300°C substantially improved the resistance to hydrogen embrittlement. The beneficial effect of low temperature tempering is generally attributed to deeper relaxation of internal stresses that usually facilitate hydrogen embrittlement.

Usually the susceptibility to hydrogen embrittlement increases with tensile strength of steel. Results of the study of DP steel with TS > 1180 MPa [77] showed that hydrogen embrittlement was facilitated by higher stress, higher strain and higher hydrogen content during testing, as well as by using the U-bending mode of strain as opposed to cup drawing. On the other hand, it was shown that the content of initially charged diffusible hydrogen (about 1 ppm) decreased fast at elevated temperatures becoming equal 0.1 ppm after 3 minutes of holding at 170°C, which is typically applied after painting car parts.

In general, for accurate evaluation of delayed fracture of automotive parts it is important to clarify the type of deformation they experienced, the sequence of heat treatment and potential hydrogen pick-up, as well as to consider the relevance of hydrogen content obtained by forced charging to the possible hydrogen accumulation during manufacturing and service.

The concentration of hydrogen absorbed, e.g., during pretreatment in the painting cycle is much lower than the critical concentration under all of the analyzed conditions [78]. Moreover, so far no hydrogen-induced delayed cracks have been detected during service of parts manufactured from DP steels.

7.3 Obtaining dual-phase steels by transformations of austenite using controlled cooling from the intercritical region

Heat treatment from two-phase $\alpha + \gamma$ temperature range, more often referred to as the intercritical region, is the main feature of DP steels production, but currently it is the main manufacturing tool for most of cold rolled AHSS including some of the third generation. The process implies controlled cooling from two-phase range between the critical A_{c1} and A_{c3} temperatures, i.e., between the onset of austenitization and its completion, respectively.

This kind of heat treatment allows obtaining a combination of microstructure constituents, which is often impossible to reach for steel of the same composition cooled after full austenitization. Annealing in the intercritical region and phase transformations during subsequent cooling are featured by three main factors. First of all, a substantial enrichment of austenite phase with carbon, as shown schematically in Fig. 7.12, accompanied by a certain partitioning of austenite forming alloying elements. Second, the transformation upon cooling occurs under the condition of coexisting austenite and ferrite phases and thus at the presence of austenite-ferrite interfaces that significantly facilitate the formation of ferrite at austenite decomposition. Last but not least, due to the short holding time at the annealing temperature, typical for continuous annealing/coating lines, austenite in the two-phase mixture is characterized by a considerable heterogeneity of its chemical composition.

As shown above, mechanical characteristics of the microstructure consisting of two and more phases depend on the ratio of their fractions, morphology and properties of individual microstructure constituents obtained under given annealing conditions. To produce dual- or multi-phase steel with a desired combination of phases

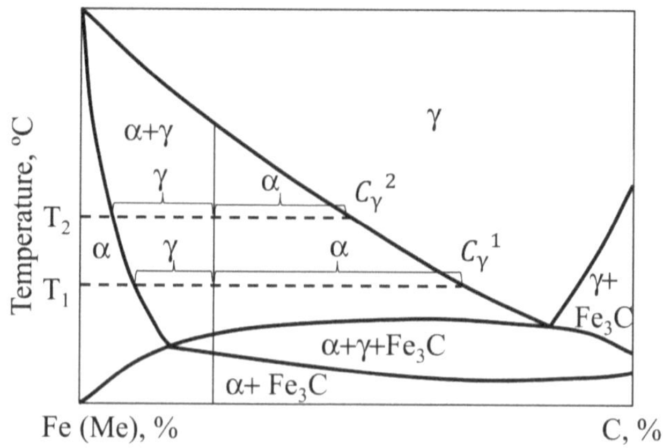

Figure 7.12 Pseudo-binary Fe(Me)-C phase diagram (Me—a second alloying element in Fe).

in their desired state, it is necessary to keep in mind the effects of various heat treatment parameters such as heating rate, annealing temperature, cooling profile, tempering conditions, etc. Special attention should be paid to the holding time of intercritical annealing which influences the isothermal growth of austenite and its chemical homogeneity. It is also important to consider the initial microstructure of steel prior to annealing that can affect the austenitization kinetics and the morphology of the resultant ferrite-austenite mixture before cooling.

The pseudo-binary phase diagram in Fig. 7.12 applies to the equilibrium conditions so that it can be used only for tentative estimation of the carbon content and the amount of austenite that could be approached in the end of long holding time at selected intercritical temperatures. Apart from insufficient time to achieve equilibrium, the temperature that determines the desired phase ratio can be shifted under the influences of heating rate, initial microstructure, and concentrations of alloying elements and their distribution. Features related to specific design of various continuous lines were presented in a few papers of the proceedings of "Metallurgy of Continuous-Annealed Sheet Steels," 1982, and summarized in [2].

Various aspects of the austenitization including heating and holding within the "intercritical" A_{c1}-A_{c3} temperature range have been addressed and described in such fundamental publications as Refs. [3,11,79–82].

When steel is cooled from the intercritical temperature region, the products of austenite transformation emerge within the parent γ-phase. Thus, the final DP microstructure is governed by the initial preferential sites of austenite nucleation at heating and by the final geometry of the formed austenite grains.

Kinetics of austenite growth at various carbon content in steel (0.06, 0.12, and 0.20%), temperatures and holding times in the intercritical temperature range was studied in details by Speich et al. [11]. The isothermal austenitization curves for these steels are summarized in Fig. 7.13. As can be seen, approaching the "equilibrium amount of austenite" was observed only at low temperatures of intercritical region and only for steels with 0.06 and 0.12% C, and required more time than 15 minutes.

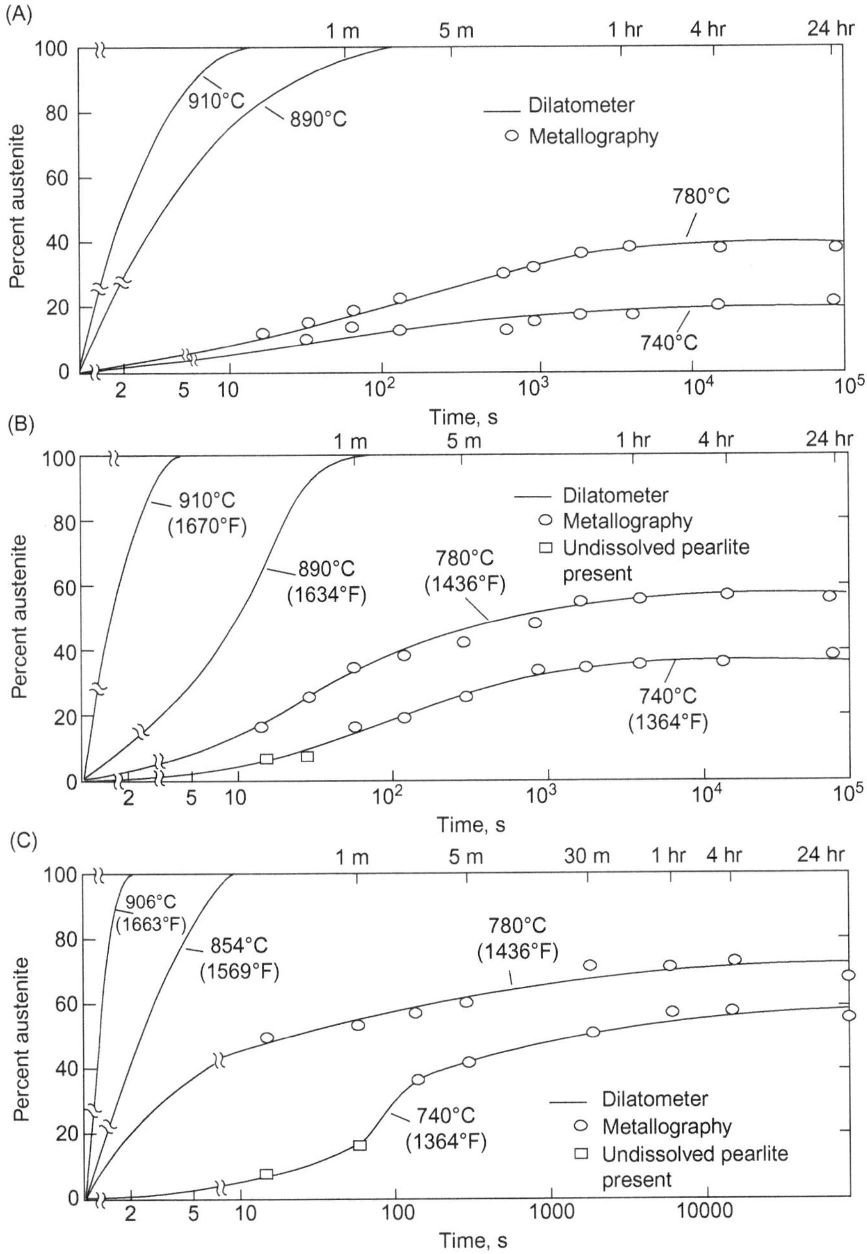

Figure 7.13 Kinetics of isothermal austenitization at different temperatures: (A) Steel 0.06C-1.5Mn (B) Steel 0.12C-1.5Mn (C) Steel 0.20C-1.5Mn [11].

Maximum austenitization rate is always observed in the beginning of isothermal holding (IH) and then it progressively decreases. In accordance with the data obtained using the 0.15C−1.42Mn − 0.37Si − 0.03 Nb steel, the austenitization rate reaches maximum near the finishing temperature of pearlite "dissolution" [83].

The effect of steel composition on the process of austenitization encompasses a number of factors: the changes in activation energy of $\alpha \rightarrow \gamma$ transformation, carbon diffusivity, thermodynamic activity of carbon in austenite, carbon concentration gradients between α and γ phases as well as the values of A_{c1} and A_{c3} critical temperatures. These will be discussed in Section 7.4 of this chapter.

7.3.1 The effect of annealing temperature in the two-phase region

The heating temperature is the key factor controlling the volume fraction of austenite and therefore its average carbon content. That in turn has a crucial effect on the stability of the austenite and the final volume and structure of martensite formed during subsequent cooling.

As shown, with initial ferrite-pearlite microstructure, the "dissolution" rate of pearlite increases by three orders of magnitude when annealing temperature is raised from 730°C to 780°C [11].

In general, both the rate of austenitization and the amount of austenite increase with intercritical heating temperature. For example, in 0.06C−1.2Mn − 0.5Cr − 0.003B steel, a 5 minutes holding produced 20% of austenite at 750°C, 35% at 780°C, 50% at 800°C, and 80% at 830°C. The formation of equal amounts of austenite (e.g., 20%) at 750°C and 800°C required 300 and 24 seconds, respectively. [84].

7.3.1.1 Effect of initial microstructure

The initial microstructure controls the austenitization kinetics, nucleation sites and the morphology of austenite-ferrite mixture, which is later inherited by the final microstructure of steel after cooling.

In terms of austenitization, the effect of the original microstructure manifests itself as a shift in the critical temperatures of phase transformations and in the rate of austenite growth at increasing holding temperature or time. In order of increasing rate of austenitization, the initial microstructures are categorized as follows: ferrite with spheroidized carbides, ferrite with lamellar pearlite, as-quenched and cold-deformed microstructures [81].

Higher austenitization rate was found during reheating of initially martensitic microstructure compared to that for bainitic microstructure reheated under identical conditions. Microstructure containing spheroidized carbides demonstrated the biggest shift in A_{c1} and A_{c3} temperatures to higher values, as well as the slowest rate of austenitization. Since austenite usually nucleates at carbide/ferrite interfaces, the amount of γ-phase and its growth rate increase with increasing specific area of ferrite-carbide interfaces, i.e., with refinement of carbide particle and/or ferrite grains [85].

The existing data regarding effects of the initial microstructure on *austenitization after cold rolling* are controversial. For example, Shirasawa and Thompson employed special heat treatment of hot bands of homogenized 0.07C–1.5Mn – 1Si steel to obtain speroidized carbides, ferrite-pearlite, bainite and fully martensite microstructures prior to cold rolling with 50% reduction [86]. After annealing of cold rolled specimens at 800°C for 1 minute to get ~40% austenite, cooling at 10°C s^{-1} to 750°C followed by water quenching and tempering at 200°C the authors found negligible differences in tensile properties regardless of initial microstructure.

These results differ from the data obtained by the author using steel with various initial microstructures that simulated the microstructure after hot rolling under different conditions. It was shown that the difference in amount of austenite became really negligible at long holding times at temperatures close to A_{c3}. However, when holding times less than 5 minutes were used, the initial (pre-deformation) microstructure appeared to have an important effect on the austenitization rate. With short holding times (30–60 seconds) the maximum amount of obtained austenite corresponded to the initial microstructure containing coarse-lamellar pearlite that simulated high coiling temperatures and slow cooling of a hot-rolled strip. The minimum amount of austenite was produced from the microstructure containing martensite tempered at high temperature that simulated low coiling temperature involving fast early cooling of strip (quenching) and subsequent long tempering during coil cooling. These results are in good agreement with data obtained in a later work [87] where effect of coiling temperature on austenitization kinetics was studied.

Banded ferrite-pearlite microstructures after high temperature coiling are usually featured by manganese segregation in pearlite bands. Therefore, the local A_{c1} in high manganese regions (pearlite) can be lower than in surrounding volumes of ferrite that should additionally accelerate pearlite–austenite transformation. The following growth of the forming austenite into ferrite should be slower in a coarsely banded microstructure than in a refined one due to shorter carbon diffusion paths through ferrite matrix in the latter. Eventually the banded structure of hot rolled steel will be inherited by the final DP structure.

7.3.1.2 Recrystallization of ferrite

Nucleation and growth of austenite proceed differently depending on the completion of recrystallization in ferrite before $\alpha \rightarrow \gamma$ transformation [88].

During heating of a cold rolled low carbon unalloyed steel, ferrite has enough time to recrystallize well before the intercritical temperature is reached. The presence of Mn, Cr, Mo, Nb, as well as higher heating rates or insufficient soaking time, raises the temperature, at which recrystallization is complete, to 730–800°C. Therefore the heating to lower temperatures results only in recovery of ferrite [89,90].

Studies on cold rolled 0.08C–1.45Mn – 0.21Si steel heated into the intercritical region showed that, regardless of the initial microstructure, the recrystallization of ferrite was complete only after 30 seconds at 760°C. This was accompanied, in particular, by the change in aspect ratio of ferrite grains from about 10:1 to (1.5–2):1 [91].

The difference in the initial ferrite grain sizes prior to cold rolling remained after recrystallization: the finer the ferrite grains in the hot rolled steel the finer the ferrite grains in the DP steel after cooling of cold rolled steel from the intercritical region.

As the holding time or heating temperature in the intercritical range is increased, the effective ferrite grain size seems to shrink as a result of ferrite "consumption" by the growing γ-phase.

7.3.2 Effect of cooling rate

The austenite transformations during cooling from the intercritical temperature range control all major characteristics of a DP steel such as volume fractions of martensite and new ferrite, final carbon content in martensite and therefore the expansion of martensite during transformation and hence the residual stresses and dislocation density in ferrite. The cooling rate also determines the content of interstitials in ferrite, the presence of retained austenite and possible amount of bainite.

The enrichment of austenite with carbon during heating into two phase region substantially increases the austenite stability compared with austenite of the same steel cooled from a temperature above A_{c3}. This is an extremely important aspect of the intercritical heat treatment, which allows for austenite of low-carbon steel to undergo transformations that are mostly typical for medium and high carbon structural steel. The important peculiarity of transformations during cooling of the ferrite-austenite mixture is also brought about by facilitated formation of ferrite or by its epitaxial growth at the pre-existing ferrite interface without the nucleation stage [92]. This promotes further enrichment of the remaining portion of austenite with carbon. As a result, cooling of austenite phase from the $\alpha + \gamma$ temperature range can substantially delay the pearlite transformation compared to cooling of the fully austenitized steel of the same composition.

Fig. 7.14 presents superposition of two CCT (continuous cooling transformation) diagrams for 0.14C–1.2Mn – 0.5Si – 0.5Cr – 0.003B steel, obtained at cooling from γ region and from the intercritical temperature range.

Evidently, the enrichment of austenite with carbon during intercritical heating produces a significant shift of pearlite transformation toward lower temperatures and slower cooling rates, suppression of bainite region and consequently the decrease in M_S temperature. In contrast, the starting temperature of ferrite formation increases.

Another comparison of CCT diagrams for 0.06C–1.8Mn steel after full and incomplete austenitization is depicted in Fig. 7.15. As can be seen, steel

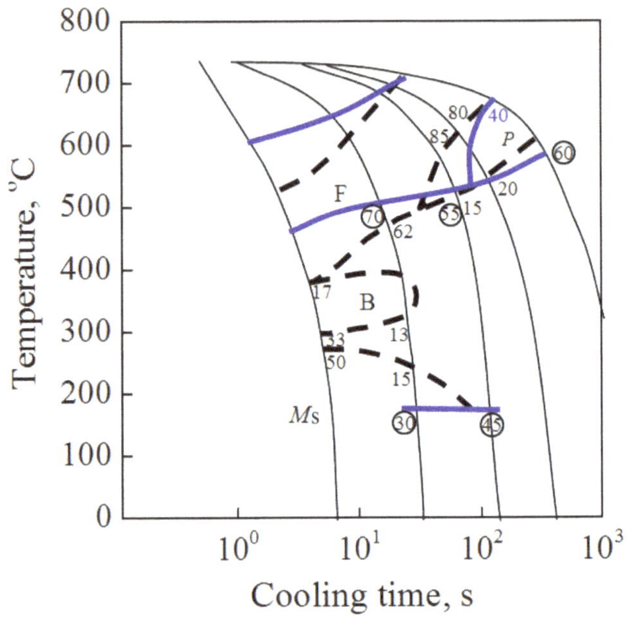

Figure 7.14 CCT diagrams of steel with 0.14C-1.0Mn-0.5Cr-0.003B cooled from $\alpha + \gamma$ region (solid lines) and after full austenitization (dashed lines).

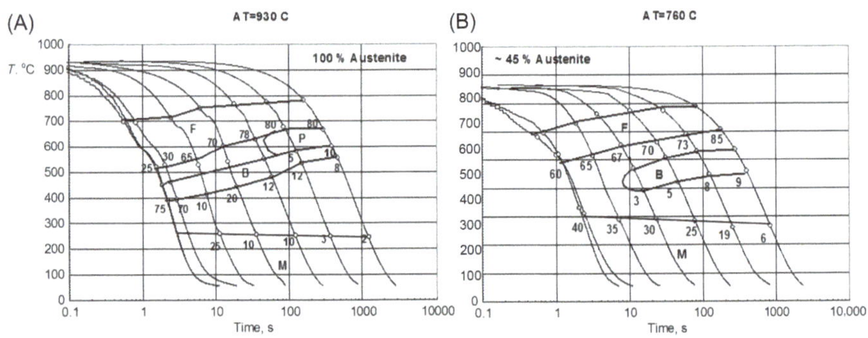

Figure 7.15 CCT diagrams of 0.06C-1.8Mn steel cooled from fully austenitic area (A) and from intercritical region (B) [132].

composition can change the rate of new ferrite formation but enrichment of austenite with carbon shifts bainite transformation to lower temperatures and slower cooling rates and results in slightly decreasing M_S temperature.

As shown above, at cooling from the $\alpha + \gamma$ region, the stability of γ-phase in the two-phase mixture is greater than at cooling 100% austenite of the same steel. At the same time this stability appeared to be substantially lower in terms of resistance

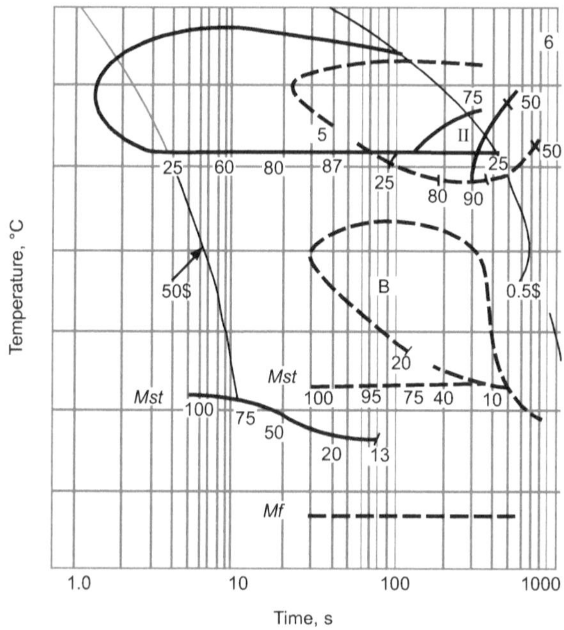

Figure 7.16 Comparison of CCT diagrams of 20% austenite of 0.06C-Cr-Mn-Si steel cooled from 780°C (solid lines) and of fully austenitized Mn-Cr-Si steels with 0.30% C.

to ferrite formation than that of austenite with the same carbon content in fully austenitized steel (Fig. 7.16). The dominant role should be attributed to pre-existing austenite-ferrite interfaces that facilitate ferrite formation dramatically.

As shown in Fig. 7.17, the initial increase in annealing temperature results in higher M_S value due to decrease in carbon content in larger volume of austenite. However, at further increase in austenite fraction and decrease in its carbon content the hardenability of γ-phase can become insufficient for being fully transformed to martensite at given cooling rate. Its decomposition with formation of the "new" ferrite results in a gradual increase in carbon content in the remaining portion of austenite, and consequent decrease in M_S temperature.

The effects of heating temperature in the intercritical range and those of the critical cooling rate are interrelated in processing of DP steels. For a given steel composition, the lower the heating temperature in the $(\alpha+\gamma)$ region the higher the $C\gamma$ (carbon content in austenite) and the lower the cooling rate required for martensitic transformation of austenite phase.

With high cooling rate, DP steels can be produced using lower carbon and/or at substantially reduced alloy content [93]. However in rapid quenching the amount of the formed austenite and therefore of the corresponding martensite, as well as the properties of DP steels, are very sensitive to heating temperature. In contrast, at relatively slow cooling rates (2–30°C s^{-1}) the formation of "new" ferrite can take place. The new ferrite plays a critical "adjusting" role: its amount

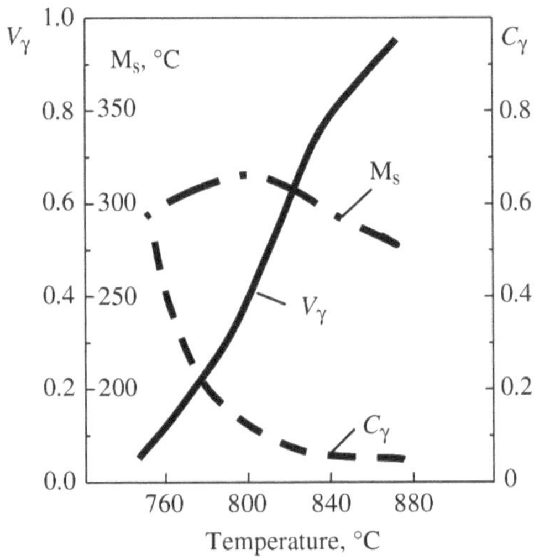

Figure 7.17 Effect of annealing temperature in two-phase region on the austenite fraction, its carbon content and M_S temperature at cooling rate of 30°C s^{-1}, 0.06C-Mn-Cr-B steel, holding for 10 min at each temperature.

grows with increasing volume of austenite, which therefore has lower carbon content and hardenability, and vice-versa [94]. This self-control of the final DP microstructure at relatively slow cooling was observed by several researchers [18,46,95]. Consequently, at a given slow cooling rate, DP steels can be produced with roughly constant amount of martensite (and hence with the same strength) over a reasonably wide range of heating temperatures of 40–80°C. As shown in Fig. 7.18, this adjusting effect of "new ferrite" depends on chemical composition of the steel and applied cooling rate.

A decrease in cooling rate reduces concentrations of dissolved carbon C_f and nitrogen N_f in ferrite due to partitioning of interstitials from ferrite to the remaining austenite and due to precipitation of fine cementite and nitride particles. Lower concentration of interstitials in ferrite enhances its ductility.

When slow cooling is employed, the final portion of remaining austenite is enriched with carbon due to its partitioning from ferrite to the γ-phase. This induces a decrease in the martensite transformation temperature (M_S). Some amount of retained austenite can stem from the increased stability of the final portion of γ-phase at slower cooling [26,96].

7.3.3 Austempering cycle

Some annealing lines are designed so that isothermal or pseudo IH is unavoidable in processing of DP steels. This applies to annealing line equipped with

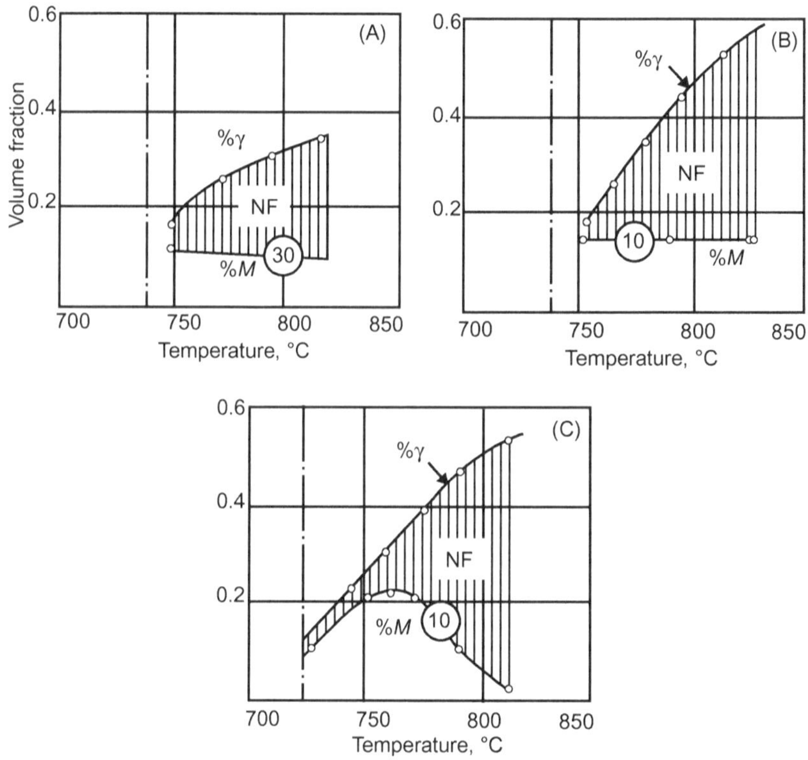

Figure 7.18 Effect of annealing temperature on amount of formed γ-phase and the martensite fraction depending on cooling rate: (A) 0.08C-1.35Mn-0.35Si-0.08 V steel, cooling 30°C s^{-1} rate; (B) 0.065C-1.21Mn-0.6Si-0.003B steel, cooling rate 10°C s^{-1}; (C) 0.065C-1.2Mn-0.2Cr-0.1Mo steel, cooling rate 10°C s^{-1}.

direct transfer of strip cooled by mist or cold rolls to the so-called over-aging furnace [97] (Fig. 7.19).

When cooled from intercritical temperatures, the initial austenite is enriched with carbon so that, depending on the initial cooling rate(s), a substantial portion of austenite can stay untransformed down to the temperature of IH. The temperature of IH then dictates whether either upper or lower bainite or martensite can be generated during the holding or final cooling.

In CALs the temperature of IH is typically below 400°C and can be controlled in the desired range. Final microstructures can vary from ferrite-bainite to ferrite-martensite or to ferrite plus mixture of bainite and martensite, which can be fully controlled by the relationship between processing parameters and austenite stability determined by its composition.

Various combinations of annealing parameters with respect to critical regions of phase transformation are illustrated in Fig. 7.19. Initial cooling rate can control the

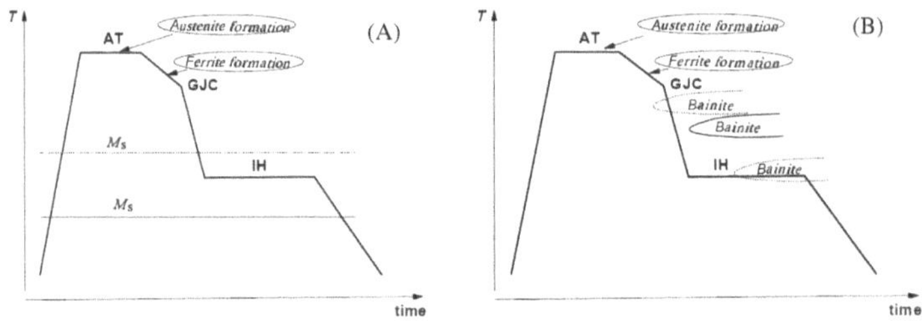

Figure 7.19 Phase transformation during annealing using austempering cycle: (A) Formation of ferrite-martensite microstructure; (B) Formation of ferrite-bainite-martensite microstructure. Dashed line designates conditions for steels with insufficient hardenability [132].

generation of "new ferrite" and therefore the additional enrichment of austenite with carbon. The latter, in turn, lowers the M_S temperature and slows down the bainite reaction, similar to the effects of certain additional alloying. For example, at cooling of a "low C" – Mn steel of 45–50% of austenite plus ferrite the M_S temperature was detected to be 270–310°C [132].

To obtain ferrite-martensite structure, using this type of thermal cycle, the temperature of IH should be above M_S as in Fig. 7.19B and the austenite to martensite transformation should take place after IH to avoid tempering of martensite as depicted in Fig. 7.19A when an IH below M_S is used.

To avoid the formation of bainite, the time of bainite reaction start in comparison with IH duration should be delayed by appropriate alloying, and the cooling rate to IH temperature should be higher than critical rate of bainite region for given steel composition (Fig. 7.19B).

The majority of CGL (continuous galvanizing line) have over-aging sections of various lengths (holding time) where the steel strip is kept at melted Zn temperature. As was found for typical DP steel with C-Mn-Cr(Mo) composition, the nose of the bainite C-curve can be near 450°C, i.e., very close to a typical over-aging temperature of CGL [98] that should facilitate bainite reaction.

To slow down bainite reaction the lower annealing temperature (higher carbon content in the initial austenite) [99] or higher alloying is used.

The elevated strip temperature, higher than Zn bath before its submersion, is also possible, promoting "quenching" austenite fraction, but requires special means of Zn bath cooling such as, e.g., the ZinQuench system [100,101].

A few of the most modern lines have the capability of initial strip cooling below Zn pot temperature. That allows the complete avoiding of undesired bainite and, depending on the temperature of the end of cooling with respect to M_S temperature, the final microstructure can contain tempered/untempered martensite, bainite and/or retained austenite [102]. The corresponding process is used as a basis for so called Q&P (quenching and partitioning) technology and will be discussed in Chapter 10.

7.4 Obtaining as-rolled dual-phase microstructure by cooling of deformed austenite

DP microstructure can be obtained by thermomechanical processing that involves deformation in the two-phase region (intercritical rolling) and subsequent rapid cooling [103]. This, however, results in strain hardened ferrite that impairs ductility of ferrite matrix of the DP microstructure. In addition, fast cooling or quenching of deformed ferrite-austenite microstructures produces a highly anisotropic banded or fibrous microstructure. Therefore, a better way to produce DP steels is deformation in the γ region under the conditions that allow obtaining a desired amount of ferrite during subsequent cooling to transform the remaining austenite into martensite.

For example, in manufacturing a hot rolled DP steel with tensile strength of $\sim$600 MPa (so far, the most used commercial DP steel grade) the key is to ensure a large amount of ferrite (80–85%) to emerge within a short period of time while the steel strip is still moving down the run-out table.

It is well known that a faster generation of ferrite during cooling is promoted by low carbon content, heavy deformation of the austenite close to A_{r3} temperature and by refinement of austenite grain size, i.e., by all factors reducing the hardenability of austenite and/or facilitating the nucleation and growth of ferrite [104].

This approach implies that stability of the remaining 15–20% of austenite, enriched by carbon by 5–6 times, should be sufficiently high to ensure the transformation to martensite during the following cooling.

Modern hot strip mills utilize the capability for interrupted cooling at temperatures close to "ferrite nose," to ensure the fastest formation of the necessary portion of ferrite. This is followed by fast cooling and coiling in bainite region or below M_S temperature aiming at final ferrite-bainite or ferrite-martensite microstructures, respectively (Fig. 7.20). This approach allows using lean steel compositions.

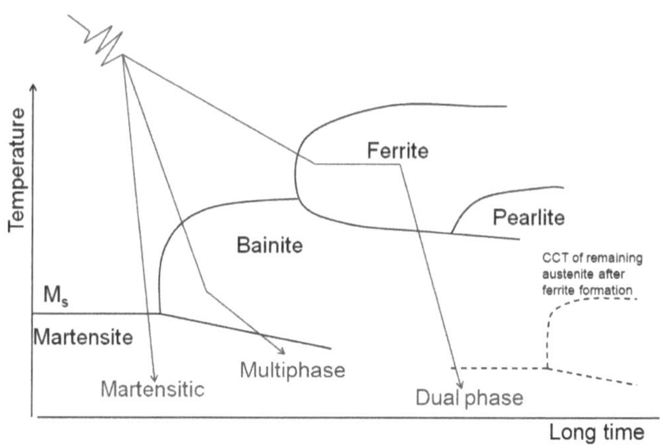

Figure 7.20 Scheme of hot rolling aiming at production of dual-phase steel.

7.5 Effects of chemical composition on dual-phase steels

The wide ranges of capabilities offered by heat treatment process and various annealing equipment led to development of DP steels with similar final properties using different chemical compositions.

The present section considers the role of steel composition in producing ferrite-martensite microstructure under specified heat treatment conditions, the effects of steel composition on the properties of individual phases, and mechanical behavior of two-phase microstructure.

7.5.1 Effect of steel composition on phase transformations during annealing in two-phase region

Effects of chemical composition in processing of DP steels are related to their influence on steel's responses during both heating in the $\alpha + \gamma$ temperature range and decomposition of the formed austenite in cooling.

The most significant role of alloying elements at heating is their influence on $\alpha + \gamma$ temperature range ($A_{C1}-A_{C3}$) and the corresponding sensitivity of the amount of γ-phase to the heating temperature. For example, significant increase in volume fraction of austenite at the same temperature of annealing caused by increasing manganese content is induced by gradual lowering of the A_{c1} temperature. This is opposite to the effects of Si and Al additions that raise A_{c1} (Fig. 7.21).

Carbon has the greatest effect on the $A_{c1}-A_{c3}$ range. In fact, a decrease in carbon content in Cr-Mn-Si-B steel from 0.18% to 0.10% and then to 0.07% reduced the rate of variation in the amount of γ-phase with temperature from 6%/°C to 3.3%/°C and then to 2.3%/°C, respectively within the temperature range of 760–850°C.

Among the alloying elements used in DP steels, silicon and especially aluminum elevate the intercritical temperature range dramatically and therefore reduce the rate of variation of austenite volume fraction with temperature [105].

The variation of austenite fraction with heating temperature tends to decrease in the upper half of the intercritical range regardless of steel composition.

Another factor affecting the amount of austenite is the reduction of carbon content of eutectoid point. As shown, practically all alloying elements reduce eutectoid carbon content [106]. Then, the increase in particular in Mn content, will result in additional increase in austenite volume fraction due to that factor.

In cooling of ferrite-austenite mixture, the main factors of austenite composition effects are hardenability of the austenite and martensite transformation temperature.

The typical criterion of austenite stability is the cooling rate from the $\alpha + \gamma$ region, which allows producing a DP steel with ferrite-martensite microstructure without pearlite constituent (and, as very often desired, without bainite as well).

Hashiguchi and Nashida [107] while calculating theoretical CCT diagrams for steels containing 0.5% C (simulating ~20% austenite in 0.1% C steel) with grain

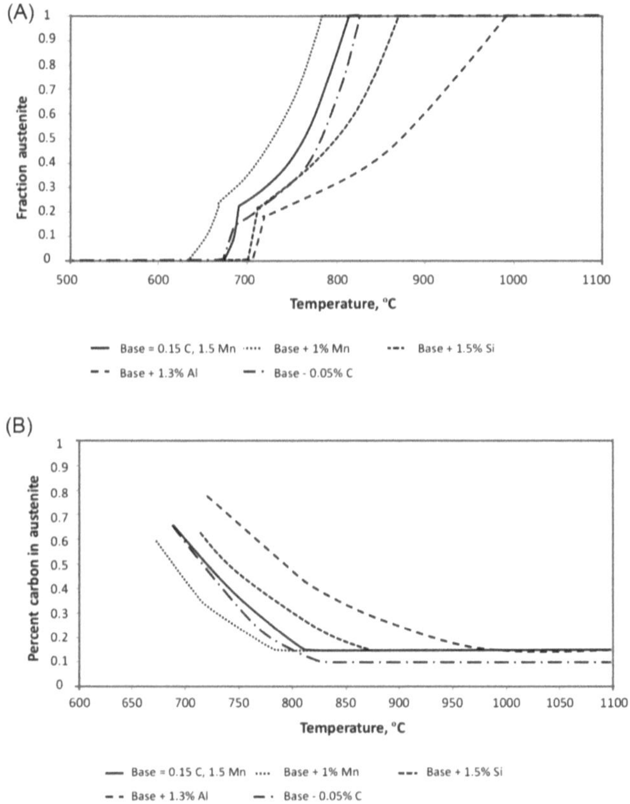

Figure 7.21 Effect of changes in steel composition (C, Mn, Si, and Al) on: (A) The amount of volume fraction of austenite formed during heating in the intercritical range; (B) Carbon content in austenite (Thermocalc, orthoequilibrium calculations).

size of 9 μm had found that the strongest positive effect on the austenite stability, that is the strongest decrease in the critical cooling rate, is caused by molybdenum and then the effect weakens for, in the sequence, chromium, manganese, copper, and nickel. The results at constant manganese content of 1.2% were summarized in the regression equation:

$$\log V_{cr}(^{\circ}\mathrm{Cs}^{-1}) = -1.73\mathrm{Mn}_{eq}(\%) + 3.95 \quad (7.13)$$

where V_{cr} is the critical cooling rate capable to produce a ferrite-martensite microstructure; the combined contribution of alloying elements was expressed in terms of Mn equivalent:

$$\begin{aligned}\mathrm{Mn}_{eq} = {} & (\%)\mathrm{Mn} + 3.28(\%\mathrm{Mo}) + 1.29(\%\mathrm{Cr}) + 0.46(\%\mathrm{Cu}) \\ & + 0.37(\%\mathrm{Ni}) + 0.07(\%\mathrm{Si})\end{aligned} \quad (7.14)$$

Regression analysis based on microstructure of DP steels cooled from various intercritical temperatures targeting at least 5% of martensite indicated a significant contribution of silicon to austenite hardenability [108]:

$$\log V_{5\%M}, (^{\circ}\mathrm{Cs}^{-1}) = 4.93 - 1.70(\%\mathrm{Mn}) - 1.34(\%\mathrm{Si}) - 5.68C\gamma \tag{7.15}$$

These equations imply constant effective carbon content in austenite (0.5%), which in reality can change with carbon content in steel and heating temperature. When modern DP steels with 0.08–0.12% C are heated to obtain the 30–50% austenite or more, almost double decrease in $C\gamma$ versus 0.5%) can alter the quantitative effect of a particular alloying element on hardenability of austenite.

Considering detailed effects of individual elements, carbon content in austenite is the major factor governing phase stability during cooling from the $\alpha+\gamma$ region. The effective carbon content in the γ- phase, $C\gamma$, should be considered in two different aspects: (1) carbon content in the initial austenite phase determined by the heating parameters and the presence of carbide forming elements [109], and (2) carbon content in the final portion of γ-phase before the martensitic transformation; that can be affected by the amount of newly transformed ferrite.

Manganese is known as a strong austenite stabilizer promoting higher hardenability of austenite. Increase in manganese moves the range of new ferrite formation to lower temperatures and slower cooling rates.

It was observed that additions of manganese are accompanied with austenite grain refinement. This effect of Mn is induced by: (1) decrease in A_{c1} and hence in the required annealing temperatures, (2) broadening of the $\alpha+\gamma+$ cementite triple-phase region where grain growth is inhibited, (3) refinement of cementite in the initial microstructure, which causes a stronger pinning effect, (4) decreasing grain boundary mobility due to solute drag effect [110].

Manganese and silicon are the most typical alloying elements of DP steels and often their combined effects are observed when the influence of silicon can depend on the Mn content. According to Nouri et al. [111] in case of steels with 0.14C–1.6Mn and (0.34–2.26)Si the enhanced hardenability of austenite at increased Si content could be explained by significant enhancement of Mn partitioning between ferrite and austenite.

As presented above, the formation of new ferrite during austenite decomposition in cooling at moderate rates is an important factor for stabilization of the remaining portion of γ-phase due to carbon partitioning and the resultant increase in effective carbon content $C\gamma$. As shown the formation of new ferrite is accelerated by an increase in aluminum content. As a result, the higher carbon content in the remaining austenite suppresses bainite and with 1.1% of Al the ferrite-martensite microstructure can be obtained at a cooling rate as low as $5^{\circ}\mathrm{C\ s}^{-1}$ [105].

Molybdenum has the strongest effect on austenite stability facilitating austenite-to-martensite transformation at the lowest cooling rate. Its effect is 2.6 and 1.3 times stronger than that of Mn and Cr, respectively [112]. Mo shifts the ferritic transformation to lower cooling rates, lowers the bainite start temperatures and/or extends the holding time for the onset of bainitic transformation, thus allowing for avoiding bainite in the final microstructure of coated DP steels.

Chromium effect is similar to that of Mo in terms of tendency but weaker. Additions of Cr enhance the hardenability of austenite, slow down the kinetics of ferrite transformation as well.

It was found that boron additions increase hardenability of austenite fraction facilitating the larger fraction of austenite transformed to martensite on account of decreased volume fraction of new ferrite.

The martensite transformation start temperature M_S controls the substructure of martensite, potential auto-tempering during cooling and the possibility to preserve some amounts of retained austenite in final microstructure. The importance of M_S cannot be overestimated because the most critical mechanical properties of DP steels are governed by the number of mobile dislocations and the level of residual stresses generated in local $\gamma \rightarrow \alpha$ transformation (see Section 7.2). As was shown in numerous equations, carbon, manganese, chromium, and silicon, in that order, have the greatest effects on M_S as shown in the following equations, based on Ref. [113] and Ref. [114] respectively:

$$M_S(^\circ C) = 571 - 474.4C - 33Mn - 17Ni - 17Cr - 21Mo \quad (7.16)$$

$$M_S(^\circ C) = 538 - 317C - 33Mn - 11Si + 30Al \quad (7.17)$$

Carbon content in austenite fraction by the moment of its transformation in cooling from $\alpha + \gamma$ region is the main factor determining the M_S temperature and hence the properties/strength of martensite, the level of residual stresses, dislocation density in ferrite, etc.

Depending on the carbon content in typical DP steels, as well as on the annealing temperature and the amount of newly transformed ferrite, the carbon content in the γ-phase undergoing the martensite transformation, $C\gamma$, can vary from 0.1% to 0.6%. This leads to wide variations in M_S.

Indirect effects of alloying elements on M_S temperature are defined by the permissible cooling rate, the amount of newly transformed ferrite and the final carbon enrichment of γ-phase.

In particular, in accordance with Eq. (7.17) Al should raise the M_S temperature at about 30°C per wt.% Al. It is true for quenching from supercritical region. However, at relatively slow cooling an increase in Al content facilitates the formation of new ferrite thus increasing carbon content in the remaining portion of austenite, and actually decreasing the M_S temperature [105,115].

On the contrary, increase in content of alloying elements that substantially increase hardenability of austenite and prevent the formation of new ferrite can raise M_S temperature compared to the less alloyed steels due to lower carbon enrichment in remaining portion of austenite.

The dominating role of enrichment of remaining portions of austenite by carbon due to formation of new ferrite is confirmed by an observed decrease in the M_S temperature at lower cooling rate, as shown in particular in Fig. 7.16.

7.5.2 Effect of chemical composition on mechanical properties of dual-phase steels

The effect of composition on mechanical properties of DP steels is the result of the variation in their microstructural parameters such as the volume fraction of martensite, as well as the strength of martensite and ferrite.

Considering the potential role of solid solution strengthening of ferrite phase by substitutional elements, it is important to point out the important advantage of the subsequent opportunity to reach required strength of DP steels at lower volume fraction of martensite (at lower carbon content in steels). In this case the potential decrease in ferrite ductility can be substantially compensated by beneficial effects of lower V_M (see Section 7.2).

Regarding growing requirements for improved flangeability of DP steels it is worth referring to data that reducing the strength difference between the two phases (in particular, on account of higher strength of ferrite) improves the steel behavior during localized deformation, particularly in hole expansion [63].

Substitutional elements such as Mn, Si, and P strengthen ferrite (Nakaoka et al., 1979; [61]). According to Pickering [116], an addition of 1% of Mn, Si, and P increases the tensile strength of ferrite by 20–35, 80–120, and 600 MPa, respectively.

The advantage of using silicon is also the increase in strain hardening of ferrite and thus overall strain hardening of DP steels.

If ferrite strengthening is not desirable due to possible decrease in ferrite ductility, addition of Cr for enhanced hardenability is especially advantageous as Cr has the least effect on ferrite strengthening.

The current trends in demand for DP steels with significantly higher strength dramatically change the attitude toward utilization of microalloying elements and corresponding precipitation hardening.

In steels containing carbide or nitride forming elements, such as niobium, titanium, zirconium, aluminum, molybdenum and vanadium, fine particles of special carbides and carbonitrides can precipitate during heating or holding in the intercritical temperature range, as well as during subsequent cooling. These particles tend to precipitate primarily in ferrite due to substantially lower solubility of the carbides and nitrides in α-phase compared to their solubility in austenite.

Fig. 7.22 shows the total amount of Nb in precipitates after hot and cold rolling and annealing at various temperatures. It can be seen that about 50–65% of the total amount of Nb precipitated after coiling at 620°C (Fig. 7.22A), while it was less than 10% after coiling at 500°C (Fig. 7.22B).

Upon heating the cold rolled steels to 790°C, additional precipitation of Nb particles took place irrespective of coiling temperature, but was more pronounced after coiling at 500°C with less preexisting Nb precipitates (Fig. 7.22B). After IH of steel with 0.02% Nb at 790°C for 150 seconds (as shown by the red square symbols), almost all Nb precipitated in ferrite regardless of coiling temperature [117].

The parameters of heat treatment used to produce cold rolled DP steels reduce potential strengthening by dispersed particles since Nb containing pre-existing precipitates coarsen. Nevertheless the strengthening by Nb is significant and together

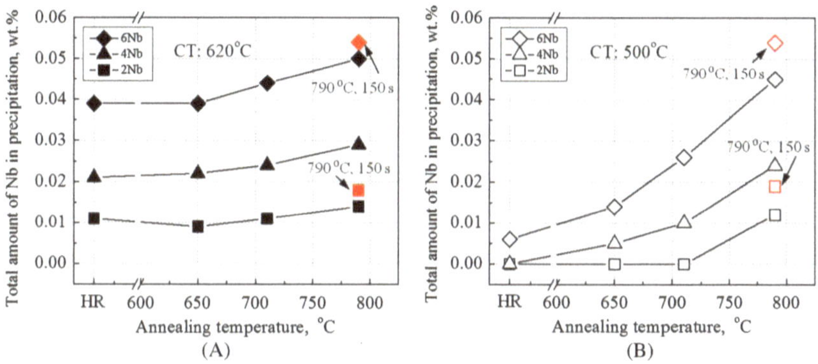

Figure 7.22 Total amount of Nb precipitated during heating and isothermal holding at 790°C (A) Coiling temperature (CT) of 620°C; (B) Coiling temperature of 500°C. Red symbols present Nb amount after isothermal holding at 790°C for 150 s [117].

with observed structure refinement results in achievement of higher strength at the same or higher ductility as shown below.

For proper design of DP steels it is very important to know the potential contributions of different alloying elements to mechanical behavior of dual-phase mixture which are summarized below.

The effect of carbon on tensile properties of ferrite-martensite microstructure is defined by its influence on the achievable volume fraction of martensite, V_M, containing certain amount (concentration) of carbon (C_M). As was discussed in Section 7.2, the majority of published data show that the strength of DP steels is linearly dependent on V_M.

With regard to the strength of martensite, it is worth recalling that the final properties are independent of carbon content in martensite when V_M is below 40–45%. At higher V_M, the microstructure stereology changes so that instead of martensite islands embedded into soft ferrite matrix with possible partitioning of strain and stresses, a skeleton type martensite becomes prevalent.

The effect of manganese itself on the strength of ferrite phase is not very significant and is about 25–35 MPa per wt.% Mn [116]. At the same time, at a given temperature of heating in the $\alpha + \gamma$ region the increase in manganese content results in a higher volume fraction of austenite, V_γ, as Mn lowers the A_{c1} temperature and the eutectoid carbon content. This typically produces larger volume fractions of martensite also due to higher hardenability of austenite, more essential under slower cooling, see Fig. 7.23. Thus, the overall observed influence of Mn on the strength of DP steel can be far stronger than just merely on account of ferrite solute solution strengthening.

Takada et al. [36] noted that the YS/TS ratio decreases with increasing Mn content. They attributed this behavior to higher dispersion of microstructure, emphasizing the important role of the higher ratio of the number of martensite particles to that of ferrite particles [36]. However, probably the dominant role of Mn is the enhancement of the austenite stability which resulted in lower M_S temperature. The transformation of austenite at lower temperatures should increase the density of

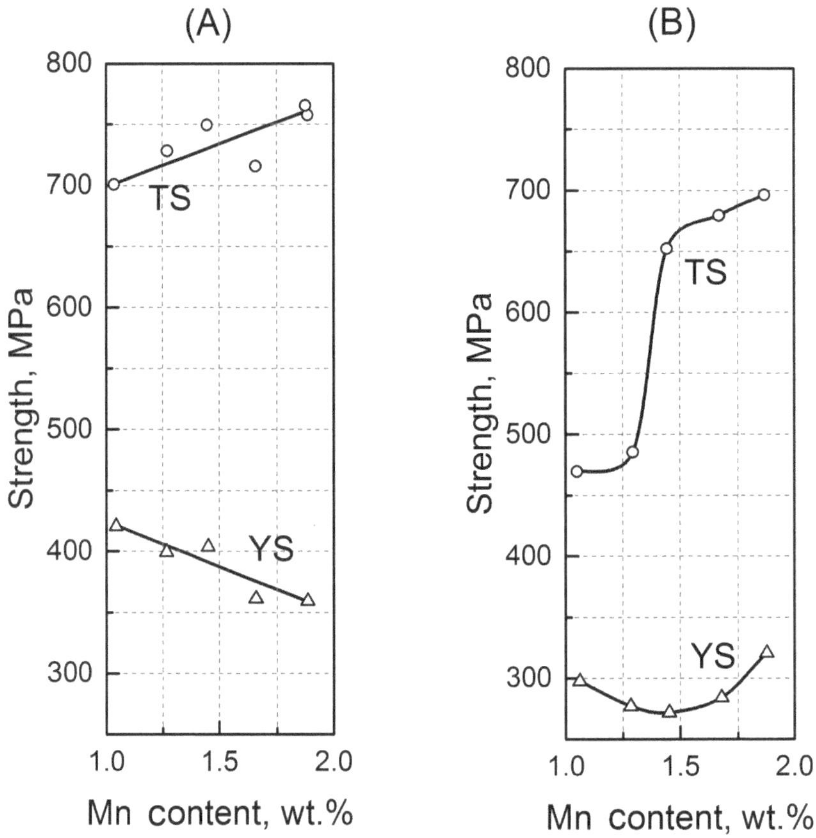

Figure 7.23 Effect of manganese on mechanical properties of 0.08% C-0.5% Si steel cooled from 750°C by water quenching (A) and by forced air at 30°C s^{-1} (B).

mobile dislocations near to the martensite/ferrite interface as well as residual stresses in ferrite, which should contribute to lower YS/TS ratio [72].

Alloying with silicon induces the substantial increase in both yield strength and tensile strengths of DP steels due to solid solution strengthening of ferrite, but simultaneously the increase in Si content is accompanied by an increase in balance of uniform/total elongation and tensile strength of DP steels [61,118,119].

The retention of high ductility in silicon bearing DP steel has been often attributed to enhanced work hardening rate of silicon-alloyed ferrite within the entire strain range. In accordance with Hironaka et al. [120], who used 0.15C–2.2Mn DP steels with Si ranging from 0.01% to 1.6%, in low Si steels the dislocation cell structure was observed already at 8% tensile strain whereas in high Si steels dislocation tangles dominated that resulted in significantly higher rate of increase in dislocation density in ferrite during tensile strain. Higher strain hardening is responsible for higher uniform elongation and better tensile strength-ductility balance of DP steels at higher Si content.

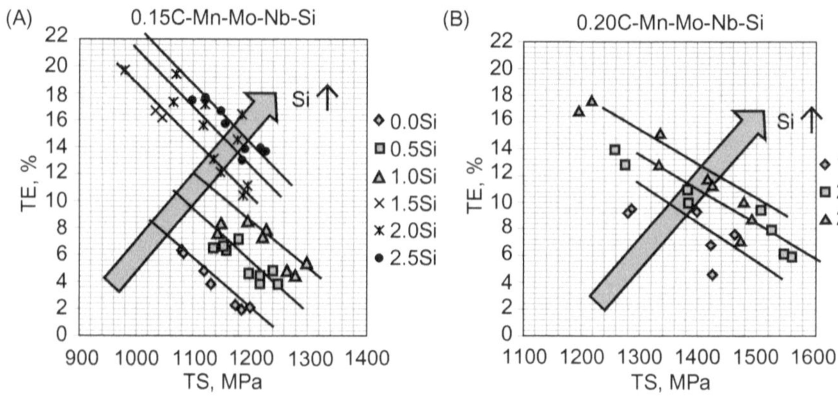

Figure 7.24 Effect of Si on the balance between strength and ductility for DP steels with (A) 0.15 C and (B) 0.20 C [119].

The general trend of better balance between tensile strength and elongation due to increased Si content is illustrated in Fig. 7.24.

Multi-phase steels containing up to 2.5% Si are attracting a growing practical interest for development of steels targeting properties of third generation AHSS.

Direct data about aluminum effect on tensile properties of DP steels are very limited. Since Al significantly raises the A_{c3} temperature, the annealing of steels with increased Al content at given intercritical temperature is accompanied by progressively decreasing volume fraction of the formed austenite and hence of martensite after cooling.

Steels with the base composition of 0.09C–2.0Mn − 0.6(Cr + Mo) − 0.03 (Ti + Nb) containing 0.042, 0.40, 0.70, 0.88, and 1.10Al, were annealed at different temperatures to obtain the same range of tensile strength. As was found, the total elongation increased with higher Al content [105]. Better balance of TS and TE due to higher Al content was explained by refined ferrite structure, by smaller sizes of discontinuous martensite islands due to larger portions of new ferrite formation, and by lower M_S values of remaining austenite.

The effects of chromium on the tensile properties are determined primarily by the type of microstructure formed during cooling of austenite, i.e., by the effect of Cr on austenite hardenability. Another contribution is due to lowering the M_S temperature, which can reduce the YS/TS ratio to as low as 0.40 [107]. Some publications noted the refinement of microstructure by Cr additions but the data are inconsistent and the opposite effect was also reported.

The effect of molybdenum increasing strength of DP steels after relatively slow cooling can be explained by higher hardenability of austenite [104,107] that results in a higher amount of martensite in the final structure.

According to Pottore et al., additions of 0.16–0.30 Mo to 0.09C–1.6Mn – 0.1Si–0.035Ti-B steel substantially improved the balance between tensile strength and total elongation of water-quenched 980DP steel that was explained by microstructure refinement inherited from hot rolling [121].

As pointed out by Shen and Preistner (Fig. 7.25), the key effect of boron on mechanical properties of DP steels is determined by its corresponding influence on suppression of ferrite formation and an effective fraction of martensite formed at austenite transformation at given parameters of annealing (temperature and cooling rate).

Effect of microalloying of DP steels by Nb and/or Ti can be related to the combined influence of structure refinement and precipitation hardening.

There are numerous data indicating that additions of Nb increase yield strength and tensile strength of DP steels [117,123].

After annealing in the lower part of the intercritical temperature region, Nb is responsible for the combination of strengthening effects: the illustrated increase in volume fraction of austenite/martensite with higher Nb content, full or partial inhibition of ferrite recrystallization, general structure refinement and, to lesser extent, precipitation hardening. At higher temperatures closer to A_{c3}, the recrystallization of ferrite is complete and the austenite fraction becomes closer to "equilibrium" so that the

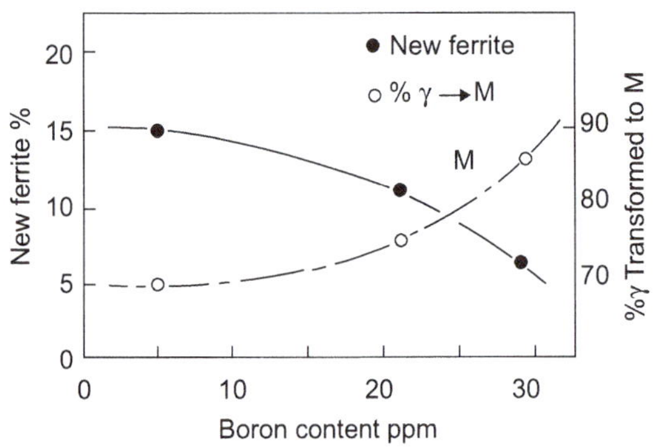

Figure 7.25 Effect of boron on the volume of new ferrite and the fraction of austenite transformed to martensite after oil quenching a 0.1C-1.35Mn-0.5Si-0.005Ti steel from 800°C [122].

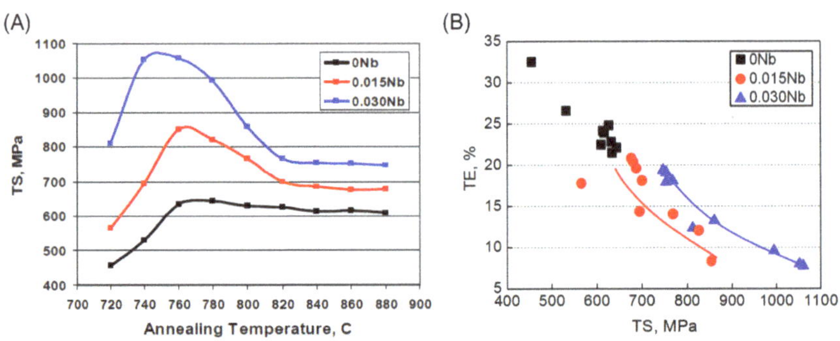

Figure 7.26 Effect of Nb on (A) Tensile strength of 0.085C-2Mn-0.5Cr steel depending on annealing temperature (cooling rate 10°C s^{-1}) and balance of TS and TE (B) [55].

strengthening effect due to Nb becomes less significant. Nevertheless, grain refinement, slightly bigger portions of martensite and higher contributions from precipitation hardening produce together a ~ 140 MPa increase in strength in steel with 0.03% Nb (Fig. 7.26A). As shown in Fig. 7.26B, strengthening by Nb is featured also by a better balance of TS and TE that can be explained by the dominating role of structure refinement resulting in higher strain hardening and hence in higher uniform elongation.

7.5.3 Effect of steel composition on strain aging and tempering behavior of DP steels

Alloying affects interstitials in ferrite in many different aspects, such as a) decrease in interstitial solubility in ferrite, e.g., by manganese; b) variation of the thermodynamic activity of carbon in iron, e.g., by silicon, which can affect carbon partitioning between α- and γ-phases; c) chemical interaction of, e.g., carbonitride-forming elements with interstitial elements that leads to precipitations.

Strain aging (bake hardening, BH) of DP steels and the factors controlling the aging kinetics and resultant hardening have been discussed above in Section 7.2.

The effects of alloying elements on aging of DP steels are determined by their impact on a) equilibrium carbon and nitrogen solubility in ferrite; b) ferrite grain size (diffusion path of interstitials to regions of higher dislocation density); c) amount of martensite; d) the M_S temperature (that, along with the amount of martensite determines the level of residual stresses and density of free dislocations), and last but not least, e) the minimum cooling rate from the $\alpha + \gamma$ region allowing for obtaining ferrite-martensite microstructure and therefore determining the actual amount of interstitials in solution.

Microalloying with Nb and/or Ti and the associated grain refinement should induce higher BH effect along with lower propensity to aging at ambient temperatures due to binding of extra nitrogen. Experimental confirmations of this effect are limited [124].

Effect of alloying elements at tempering of DP microstructure is fully determined by their effect on tempered martensite, i.e., by retardation of martensite softening. Grange et al. compared the effects of various elements on low and medium carbon martensite [125]. It was shown that Mn, Si, Cr, Mo, P, and V not only differed in their ability to retard softening of martensite, but also in the temperature ranges in which the effects of these elements are maximized.

It was found, in particular, that fine cementite in steels alloyed with Cr and/or with higher Mn content results in high resistance to softening of martensite compared to leaner chemistries featured by severe decomposition of martensite and forming coarser cementite [126]. DP steels microalloyed with V + Nb also demonstrate significantly lower softening after tempering at up to 600°C due to strengthening by precipitation of carbonitrides [127].

Chromium as a carbide forming element becomes more efficient after tempering at temperatures above the formation of alloy carbides (>200°C) providing maximum strength near 425°C, which then decreases because of coalescence of carbides.

Molybdenum partitions to carbide phase at elevated temperatures and thus keeps the carbides small and numerous; therefore its strengthening effects are especially strong after tempering closer to 500°C.

Silicon, which increases the hardness of martensite in the wide temperature range due to solid solution strengthening effect, was found to have much stronger effect after tempering at temperatures around 325°C that was explained by its well-known effect to inhibit the conversion of ε-carbide to cementite.

Vanadium and niobium form stronger carbides than those of Cr and/or Mo. Therefore, these elements can be expected to affect the strength of tempered martensite retarding its softening.

7.6 Application of dual-phase steels in modern cars

DP steels offer an outstanding combination of strength and drawability as a result of their high combination of strength and ductility, as well as their high strain hardenability which gives them good strain redistribution capacity [128]. For some applications it is important that the initial yield strength of steels is further increased by work hardening and the paint baking (BH) process.

Steels with ferrite-martensite structure cover tensile strength ranging from 490 to 1180 MPa that allow using them for numerous car parts requiring resistance to denting, fatigue, intrusion, impact, and so on.

High tensile strength results in high fatigue properties [129]. The high strength and strain hardening capacity of these steels combined with a strong BH effect give them excellent potential for reducing the weight of structural and closure parts. As shown, based on their high energy absorption capacity and fatigue strength, cold rolled dual-phase steels are particularly well suited for automotive structural and safety parts such as longitudinal beams, cross members and reinforcements.

A wide application of DP steels in modern cars is well illustrated, in particular, in SAAB publication (Fig. 7.27).

As presented in Fig. 7.28—an example of GM models—the application of DP steels is permanently growing.

7.7 Summary

Historically, DP steels with ferrite-martensite structure were the first family of AHSS, the concept of which was developed in the 1970s. Four decades have passed since their initial development and the strength of DP has extended from 500–700 to 1000–1200 MPa, giving rise to some of their modifications such as, DP steels with improved hole expansion, with elevated YS, with TRIP effect. The importance of DP steels and the basics of their design and processing are defined by their dominance of AHSS applied in modern cars as well as by use of the same approach to alloying and critical features of processing for development and producing candidates of steels for the third generation AHSS.

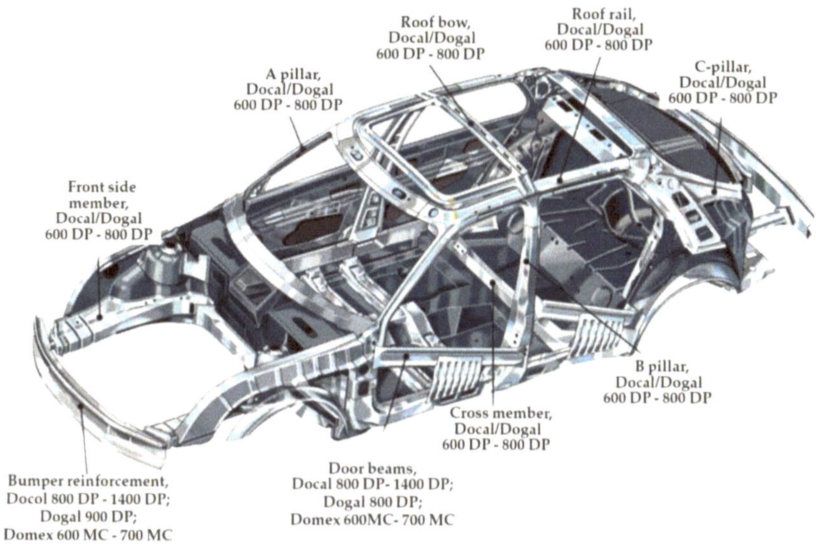

Figure 7.27 Examples of car parts manufactured from various DP grades of SAAB [131].

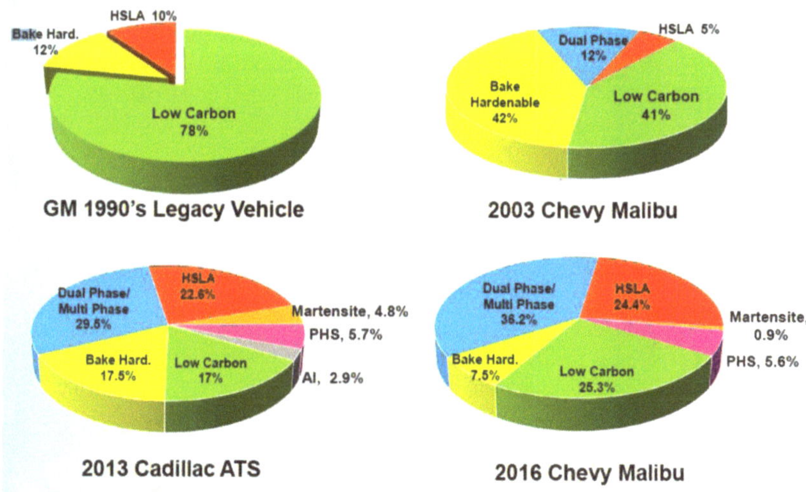

Figure 7.28 Evolution of application of DP steels in cars of Generals Motors.

References

[1] J. Reed, AHSS Technology in the 2015 Ford Edge. In: *Great Design in Steel*. Detroit, MI, 2015. http://www.autosteel.org/~/media/Files/Autosteel/Great%20Designs%20in%20Steel/GDIS%202015/Track%202%20-%20Reed.pdf

[2] N. Fonstein, Advanced High Strength Sheet Steels: Physical Metallurgy, Design, Processing, and Properties, Springer, 2015, 396 p.

[3] S.A. Golovanenko, N.M. Fonstein, Dual-phase low alloyed steels. M. Metall. (1986) 206 p.
[4] J. Adamczyk, A. Graijcar, Heat treatment and mechanical properties of low-carbon steel with dual-phase microstructure, J. Achiev. Mater. Manuf. Eng. 22 (5) (2007) 13–20.
[5] C.T. Mileiko, The tensile strength and ductility of continuous fibre composites, J. Mat. Sci. 4 (11) (1969) 974–977.
[6] N.K. Balliger, T. Gladman, Work hardening of dual-phase steel, Met. Sci. 15 (3) (1981) 95–108.
[7] D.A. Korzakwa, R.D. Lawson, D.K. Matlock, G. Krauss, A consideration of models describing the strength and ductility of dual-phase steels, Scr. Metall. 14 (9) (1980) 1023–1028.
[8] A.F. Szewczyk, I. Gurland, A study of the deformation and fracture of a dual-phase steel, Met Trans 13A (10) (1982) 1821–1826.
[9] I. Tamura, Y. Tomota, On the strength and ductility of true phase iron alloys, Trans. Iron Steel Inst. Jpn. 13 (4) (1973) 283–292.
[10] K. Araki, Y. Takada, K. Nakaoka, Work hardening of continuously annealed dual-phase steels, Trans. ISIJ 17 (12) (1977) 710–717.
[11] G. Speich, V. Demarest, R. Miller, Formation of austenite during intercritical annealing of dual-phase steels, Metall. Mater. Trans. A 12 (8) (1981) 1419–1428.
[12] A.N. Bortsov, N.M. Fonstein, The distribution of strain between the phases of ferrite-martensite steel, Phys. Metall. Metallography (USSR) 61 (2) (1986) 74–81.
[13] Y. Tomota, I. Tamura, Mechanical properties of dual-phase steels, Trans. ISIJ 22 (5) (1982) 677–689.
[14] H.K.D.H. Bhadeshia, D.V. Edmonds, Analysis of mechanical properties and microstructure of high-silicon dual-phase steel, Metall. Sci. 14 (2) (1980) 41–49.
[15] J.Y. Koo, M.I. Young, G. Thomas, On the law of mixtures and dual-pahse steels, Metall. Trans. 11 A (5) (1980) 852–854.
[16] J.H. Bucher, E.G. Hamburg, J.F. Butler, Property characterization of VAN-QN dual-phase steels. In: *Structure and Properties of Dual-Phase Steels*, (1979) 346–59. New Orleans.
[17] R.G. Davies, Influence of martensite composition and content on the properties of dual-phase steels, Metall. Trans. 9A (5) (1978) 671–679.
[18] G.T. Eldis, Influence of microstructure and testing procedure on the measured mechanical properties of heat treated dual-phase steels, Conference of Structure and Properties of Dual-Phase Steels, Symposium at the AIME Annual Meeting, Metallurgical Society of AIME, New Orleans, LA, USA, 1981, pp. 202–220, Warrendale, PA.
[19] W.R. Gribb, I.M. Rigsbee, Work-hardening behavior and its relationship to the microstructure and mechanical properties of dual-phase steels. In *Structure and Properties of Dual-Phase Steels*, (1979) 91–117.
[20] A.R. Marder, B.L. Bramfitt, processing of molybdenum-bearing dual-phase steel. In *Structure and Properties of Dual-Phase Steels*, (1979) 242–59.
[21] R.R. Mould, C.C. Skena, Formable HSLA and dual-phase steels, Metallurgical Society of AIME, Metallurgical Society of AIME, New York, 1977, pp. 181–204.
[22] R.D. Lawson, D.K. Matlock, G. Krauss, The effect of microstructure on the deformation behavior and mechanical properties of dual-phase steel. In *Fundamentals of Dual-Phase Steels*, (1981) 347–81.
[23] A.R. Marder, Deformation characteristics of dual-phase steels, Metall. Trans. 13A (1) (1982) 85–92.

[24] A.N. Bortsov, N.M. Fonstein, Influence of carbon concentration on the mechanical properties of the low-carbon ferrite-martensite steels, Phys. Metall. Metallography (USSR) 57 (4) (1984) 142–148.
[25] S.S. Hansen, R.R. Pradhan, Structure/property relationships and continuous yielding behavior in dual-phase steels, Conference of Fundamentals of Dual-Phase Steels, Symposium at the 110th AIME Annual Meeting, Metallurgical Society of AIME, Chicago, IL, 1981, pp. 113–144, Warrendale, PA, USA.
[26] J.M. Rigsbee, J.K. Abraham, A.T. Davenport, J.E. Franklin, J.W. Pickens, Structure-processing and structure-property relationships in commercially processed dual-phase steels. In *Conference of Structure and Properties of Dual-Phase Steels, Symposium at the AIME Annual Meeting*, (1981) 304–29. Struct and Prop of Dual-Phase Steels, Proc of a Symp at the AIME Annu Meet. New Orleans, LA, USA.
[27] N. Fonstein, H.J. Jun, G. Huang, et al. Effect of bainite on mechanical properties of multi-phase ferrite-bainite-martensite steels." In *MST'11* (2011).
[28] R. Stevenson, D.J. Bailey, G. Thomas, High strength low carbon sheet steel by thermo-mechanical treatment: II. Microstructure, Metall. Trans. A 10 (1) (1979) 57–62.
[29] T.Tanaka, M. Nishida, K. Hachiguchi, T. Kato, Formation and properties of ferrite plus martensite dual-phase structures. In *Structure and Properties of Dual-Phase Steels*, (1979) 221–41.
[30] K. Araki, S.H. Fukunaka, K. Uchida, Development of continuously annealed high strength cold rolled sheet steels, Trans. ISIJ 17 (12) (1977) 701–709.
[31] D.I. Bailey, R. Stevenson, High strength low carbon sheet steel by thermomechanical treatment: I. Strengthening mechanisms, Metall. Trans. 10A (1) (1979) 47–55.
[32] A. Nishimoto, Y. Hosoya, K. Nakaoka, A new type of dual-phase steel sheet for automobile, Trans. ISIJ 21 (11) (1981) 778–782.
[33] A.R. Marder, B.L. Bramfitt, Processing of a molybdenum-bearing dual-phase steel, Conference of Structure and Properties of Dual-Phase Steels, Symposium at the AIME Annual Meeting, Metallurgical Society of AIME, New Orleans, LA, USA, 1981, pp. 242–259, Warrendale, PA.
[34] T. Greday, H. Mathy, P. Messien, About different ways to obtain multi-phase steels. In *Structure and Properties of Dual-Phase Steels*, (1979) 260–80.
[35] T. Sakaki, K. Sugimoto, T. Fuluzato, The role of internal stresses for continuous yielding of dual-phase steels, Acta Metall. 31 (10) (1983) 1737–1746.
[36] Y. Takada, Y. Hosoya, K. Nakaoka, Possibilities of achieving low yield ratio with low manganese dual-phase steels, Metallurgy of Continuously Annealed Steels, Metal/ Society of AIME, 1982, pp. 251–269.
[37] D.A. Budford, D.K. Matlock, G. Krauss, Effect of microstructural refinement on the deformation behavior of dual-phase steels. In *Strength of Metals and Alloys*, 1: (1985) 189–94. Montreal, Canada.
[38] P. Tsipouridis, E. Werner, C. Krempaszky, E. Tragl, Formability of high-strength dual-phase steels, Steel Res. Int. 77 (10) (2006) 654–667.
[39] W. Bleck, K. Phiu-On, Grain refinement and mechanical properties in advanced high strength sheet steels." In *HSLA'* 2005, 50–57. Sanya, China.
[40] P. Ludwig, Element Der Technologichen Mechanick, Springer-Verlag, Berlin, 1909.
[41] H.W. Swift, J. Mech. Phys. Solids 1 (1952) 1–32.
[42] E. Voce, The relationship between stress Abd Strain for homegenious deformation, J. Inst. Metals 74 (1948) 537–549.
[43] B. Jaoul, J. Mech. Phys. Solids 5 (1) (1957) 95–114.
[44] J.H. Hollomon, Tensile deformation, Trans. ASM no. 162 (1945) 268–290.

[45] R. Matsuda, T. Shimomura, Production of high strength cold-rolled steel sheets by the NKK continuous annealing line process (NKK-CAL Process), Nippon Koknn Tech. Rep. 20 (1980) 1–9.
[46] R.G. Davies, The deformation behavior of a vanadium-strengthened dual-phase steel, Metall. Trans. A 9 (1) (1978) 41–52.
[47] N.M. Fonstein, Factors affecting the strain hardening of low-alloy steels of ferrite-martensite structure, Steel USSR, Metally 3 (1991) 118–123.
[48] A.R. Marder, Factors affecting the ductility of 'dual-phase' alloys. In *Formable HSLA and Dual-Phase Steels, Metallurgical Society of AIME*, (1977) 87–98.
[49] M.F. Ashby, Work hardening of dispersion-hardened crystals, Phil. Mag. 14 (132) (1966) 1157–1178.
[50] A. Rizk, D.L. Bourell, Dislocation density contribution to strength of dual phase steels, Scr. Met. 16 (12) (1982) 1321–1324.
[51] A.M. Sarosiek, W.S. Owen, On the importance of extrinsic transformation accommodation hardening in dual-phase steels, Scr. Met. 17 (2) (1983) 227–231.
[52] L.M. Brown, W.M. Stobbs, Work hardening of cupper silica. II The role of plastic relaxation, Phil. Mag. 23 (185) (1971) 1201–1233.
[53] C.A. Lanzilotto, F.B. Pickering, Structure–property relationship in dual-phase steels, Met. Sci. 16 (8) (1982) 371–382.
[54] L.F. Ramos, D.K. Matlock, G. Krauss, On the deformation behavior of dual-phase steels, Metall. Trans. 10A (2) (1979) 259–261.
[55] O. Girina, N. Fonstein, D. Bhattacharya, Effect of Nb on the phase transition and mechanical properties of advanced high-strength dual-phase steels. In *New Development on Metallurgy and Applications of High Strength Steels*, Paper #13. Buenos Aires, Argentina, 2008.
[56] B.C. De Cooman, S. Lee, Y. Estrin, Strain hardening control in AHSS for automotive applications, Proceedings of International Symposium: New Developments in AHSS, Veil, CO, USA, 2013.
[57] R.G. Davies, On the ductility of dual-phase steels. In *Formable HSLA and Dual-Phase Steels*, (1977) 25–39.
[58] M. Calcagnotto, Y. Adachi, D. Ponge, D. Raabe, Deformation and Fracture Mechanism in Fine- and ultrafined-grained ferrite/martensite dual-phase steels and effect of aging, Acta Mater. 59 (2011) 658–670.
[59] R. Song, D. Ponge, D. Raabe, Improvement of the work hardening rate of ultrafine grained steels through second phase particles, Scr. Mater. 52 (2005) 1075–1080.
[60] W.B. Morrison, The effect of grain size on stress-strain relationship in low carbon steels, Trans. ASM 59 (1966) 224–246.
[61] R.G. Davies, Influence of silicon and phosphorus on the mechanical properties of both HSLA and dual-phase steels, Metall. Trans. 10a (1) (1979) 113–118.
[62] G.R. Speich, R.L. Miller, Mechanical properties of ferrite-martensite steels. In *Structure and Properties of Dual-Phase Steels*, (1979) 145–82. New Orleans, LA, USA.
[63] K.-I. Sugimoto, J. Sakaguchi, T. Ida, T. Kashima, Stretch-flangeability of a high-strength TRIP Type bainitic sheet steel, ISIJ Int. 40 (9) (2000), 920–026.
[64] K. Sugimoto, T. Sakaki, T. Fukusato, O. Miyagawa, Influence of martensite morphology on initial yielding and strain hardening in a 0.11C-1.36Mn dual-phase steel, J. ISIJ 71 (8) (1985) 70–77.
[65] L.M. Storozeva, N.M. Fonstein, S.A. Golovanenko, Examination of Quench and strain aging of low-carbon ferritic-martensitic steels, Russ. Metall. (Metally) no. 1 (1985) 89–93.

[66] R.P. Krupitzer, Strain aging behavior in continuously-annealed dual-phase steel. In *Fundamentals of Dual-Phase Steels*, (1981) 315–30.
[67] R.G. Davies, The mechanical properties of zero carbon ferrite plus martensite structure, Metall. Trans. 9A (3) (1978) 451–466.
[68] K. Nakaoka, K. Araki, K. Kurihara, Strength, ductility and aging properties of continuously-annealed dual-phase high strength sheet steels. In *Formable HSLA and Dual-Phase Steels, Metallurgical Society of AIME*, (1977) 126–141.
[69] J.D. Baird, Strain aging in steel—critical review, Iron Steel 36 (7) (1963) 326–334.
[70] S. Gunduz, B. Demir, R. Kacar, Effect of aging temperature and martensite by volume on strain aging behavior of dual-phase steel, Ironmaking Steelmaking 35 (1) (2008) 63–68.
[71] A.H. Yegneswaran, K. Tangri, Strain distribution and load transfer characteristics of a Cu-9.6 percent Al dual-phase structure, Z. Metallkunde 74 (8) (1983) 521–524.
[72] N.M. Fonstein, M. Kapustin, N. Pottore, I. Gupta, O. Yakubovsky, Factors that determine the level of the yield strength and the return of the yield-point elongation in low-alloy ferrite-martensite steels, Phys. Metals Metall. 104 (3) (2007), 3324–323.
[73] G. Krauss, Steels: Processing, Structure and Performance. TMS, 2015.
[74] R. Pradhan, Metallurgical aspects of quenched and tempered dual-phase steels produced via continuous annealing, Technology of Continuously Annealed Cold -Rolled Sheet Steel, Metal/Society of AIME, Detroit, MI, 1985, pp. 297–318.
[75] Y. Tomota, I. Tamura, Mechanical properties of ductile two-phase steels, J. Iron Steel Inst. Jpn. 67 (13) (1981) 439–455.
[76] R.G. Davies, Hydrogen embrittlement of dual-phase steel, Metall. Trans. 12A (9) (1981) 1667–1672.
[77] S. Takagi, Y. Toji, M. Yoshino, K. Hasegawa, Hydrogen embrittlement resistance evaluation of ultra-high strength steel sheets for automobiles, ISIJ Int. 52 (2) (2012) 316–322.
[78] G. Lovicu, M. Bottazi, F. D'Aiuto, M. DeSanctis, Hydrogen embrittlement of automotive advanced high-strength steel, Metall. Mater. Trans. A 43 (11) (2012) 4075–4087.
[79] C.I. Garcia, A.J. DeArdo, Formation of austenite in low-alloy steels, Conference of Structure and Properties of Dual-Phase Steels, Symposium at the AIME Annual Meeting, Metallurgical Society of AIME, New Orleans, LA, USA, 1979, pp. 40–61, Warrendale, PA.
[80] P. Wycliffe, G.R. Purdy, J.D. Embury, Austenite growth in the intercritical annealing of ternary and quaternary dual-phase steels, Conference of Fundamentals of Dual-Phase Steels, Symposium at the 110th AIME Annual Meeting, Metallurgical Society of AIME, Chicago, IL, USA, 1981, pp. 59–83, Warrendale, PA, USA.
[81] S.S. Dyachenko, The austenite formation in Fe-C alloys, M. Metallurgia (1982) 128.
[82] H. Azizi-Alizamini, M. Militzer, W.J. Poole, Austenite formation in plain low-carbon steels, Metall. Mater. Trans. A 42 (6) (2011) 1544–1557.
[83] F.L.G. Oliveira, M.S. Andrade, A.B. Cota, Kinetics of austenite formation during continuous heating in a low carbon steel, Mater. Charact. 58 (3) (2007) 256–261.
[84] N.M. Fonstein, Heat treatment for obtaining a controlled ferritic-martensitic structure in steel, Metal Science and Heat Treatment (English Translation of Metallovedenie I Termicheskaya Obrabotka 27 (7–8) (1985) 610–616.
[85] N.C.A. Law, D.V.B. Edmonds, The formation of austenite in a low-alloy steel, Metall. Mater. Trans. A 11 (1) (1980) 33–46.
[86] H. Shirasawa, J.G. Thomson, Effect of hot band microstructure on strength and ductility of cold rolled dual-phase steel, Trans. Iron Steel Inst. Jpn. 27 (5) (1987) 360–365.

[87] R.R. Mohanty, O.A. Girina, Effect of coiling temperature on kinetics of austenite formation in cold rolled advanced high strength steels, Mater. Sci. Forum (2012) 2112–2117, 706–709.

[88] R.R. Mohanty, O.A. Girina, N.M. Fonstein, Effect of heating rate on the austenite formation in low-carbon high-strength steels annealed in the intercritical region, Metall. Mater. Trans. A 42 (12) (2011) 3680–3690.

[89] R. Pradhan, Continuously annealed cold-rolled microalloyed steels with different microstructures, International Conference on Technology and Applications of HSLA Steels, ASM, Philadelphia, PA, USA, 1984, pp. 193–201, Metals Park, OH, USA.

[90] C.I. Garcia, A.J. Deardo, Formation of austenite in 1.5% Mn steels, Metall. Trans. A 12 A (3) (1981) 521–530.

[91] D.Z. Yang, E.L. Brown, D.K. Matlock, G. Krauss, Ferrite recrystallization and austenite formation in cold-rolled intercritically annealed steel, Metall. Trans. A 16 (8) (1985) 1385–1392.

[92] M.D. Geib, D.K. Matlock, G. Krauss, The effect of intercritical annealing temperature on the structure of niobium microalloyed dualphase steel, Metall. Trans. A 11 (10) (1980) 1683–1689.

[93] K. Nakaoka, Y. Hosoya, M. Ohmura, A. Nishimoto, Reassesment of the water-quenching process as a means of producing formable steel sheets. In *Conference of Structure and Properties of Dual-Phase Steels, Symposium at the AIME Annual Meeting*, (1981) 330–345.

[94] R. Lagneborg, Structure-property relationship in dual-phase steels. In *Dual-Phase and Cold Pressing Vanadium Steels, Seminar of Vanitec*, (1978) 43–52.

[95] D.K. Matlock, G. Krauss, L.F. Romas, G.S. Huppi, A correlation of processing variables with deformation behavior of dual-phase steels, Conference of Structure and Properties of Dual-Phase Steels, Symposium at the AIME Annual Meeting, Metallurgical Society of AIME, New Orleans, LA, USA, 1981, pp. 62–90, Warrendale, PA.

[96] N. Bangaru-Rao, A.K. Sachdev, Influence of cooling rate on the microstructure and retained austenite in an intercritically annealed vanadium containing HSLA steel, Metall. Mater. Trans. A 13 (11) (1982) 1899–1906.

[97] K. Kakuta, N. Takahashi, Technology of nippon steel's continuous annealing and processing line (C.A.P.L.), Symposium of The Metal Society of AIME, Metal/Society of AIME, 1982, pp. 373–377.

[98] A. Pichler, S. Traint, G. Arnoldner. Phase transformation during annealing of a cold-rolled dual-phase steel grade. In *Proceedings of 42nd MWSP Conference* XXXVIII, (2000) 573–93.

[99] D.P. Hoydick, D.M. Haezenbrouck, Effect of annealing parameters on structure and propertsies of intercritically annealed steels intented for hot-dip dual-phase application. In *45th MWSP Conference*, XLI, (2003) 421–35.

[100] S. Radtke, P. Sippola, S. Makimattila, ZINQUECH a new process for coated steels, Iron Steel Eng. 63 (6) (1986) 32–35.

[101] R. Patil, O. Girina, D. Bhattacharya, Development of a dual-phase high strength galvanized steel using Zinquench Technolog®. In *Galvatech '04: 6th International Conference on Zinc and Zinc Alloy Coated Steel Sheet*, edited by Baker M.A, (2004) 439–47. Chicago, IL.

[102] H. Mohrbacher, T. Schulz, V. Flaxa, M. Pohl. Low carbon microalloyed cold rolled hot dip galvanized dual-phase steel for larger cross-sectional areas with improved properties. In *Galvatech'11 6th International Conference on Zinc and Zinc Alloy Coated Steel Sheet—Conference Proceedings*. Genova, (2011).

[103] E. Ahmad, T. Manzoor, N. Hussein, Thermomechanical processing in the intercritical region and tensile properties of dual-phase steel, Mater. Sci. Eng. A 508 (1–2) (2009) 259–265.
[104] A.P. Coldren, G.T. Eldis, Using CCT diagrams to optimize the composition of an as-rolled dual-phase steel, J. Met. 32 (3) (1980) 41–48.
[105] O. Girina, N. Fonstein, Influence of Al additions on austenite decomposition in a continuously annealed dual-phase steels." In *MS&T'05*, (2005) 65–76.
[106] ASM Handbook; Heat Treating. 1991. Vol. 4. ASM International.
[107] K. Hashiguchi, M. Nashida, Effect of alloying element and cooling rate after annealing on mechanical properties of dual-phase steel." *Kawasaki Steel Technology Report*, no. 1 (1980) 70–78.
[108] P. Messien, J.C. Herman, T. Greday, Phase transformation and microstructures of intercritically annealed dual-phase steels, Conference of Fundamentals of Dual-Phase Steels, Symposium at the 110th AIME Annual Meeting, Metallurgical Society of AIME, Chicago, IL, USA, 1981, pp. 161–180, Warrendale, PA, USA.
[109] C.I. Garcia, et al., Influence of critical carbide dissolution temperature during intercritical annealing on hardenability of austenite and mechanical properties of DP-980 steels, ISIJ Int. 51 (6) (2011) 969–974.
[110] M. Calcagnotto, D. Ponge, D. Raabe, On the effect of manganese on grain size stability and hardenability in ultra-fine-grained ferrite-martensite dual-phase steels, Metall. Mater. Trans. A 43A (1) (2012) 37–42.
[111] A. Nouri, H. Saghafian, Sh Kheirandish, Effects of silicon content and intercrtical annealing on manganese partitioning in dual-phase steels, J. Iron Steel Res. Int. 17 (5) (2010) 44–50.
[112] T. Irie, S. Satoh, K. Hashiguchi, I. Takahashi, O. Hashimoto, 1981. Kawasaki Steel *Technology Report*, no. 2: 14.
[113] W. Steven, A.G. Haynes, The temperature of formation of martensite and bainite in low-alloy steels, J. Iron Steel Inst. 183 (8) (1956) 349–359.
[114] N. Imai, N. Komatsubara, K. Kunishige, Effect of alloying element and microstructure on mechanical properties of low-alloy TRIP-steels, CAMP-ISIJ 8 (1995) 572–575.
[115] N. Fonstein, Effect of Al, C, and Mn on structure and mechanical properties of dual-phase and trip steels. In *HSLA* '05. Sanya, China, (2005).
[116] F.B. Pickering, Physical Metallurgy and the Design of Steels. (1978).
[117] R. Song, N. Fonstein, H.J. Jun, D.B. Bhattacharya, S. Jansto, N. Pottore, Effects of Nb on microstructural evolution and mechanical properties of low-carbon cold-rolled dual-phase steels, Metall. Microstruct. Anal. 3 (2014) 174–184.
[118] J.Y. Koo, G. Thomas, Design of duplex Fe/X/0.1C steels for improved mechanical properties, Metall. Trans. A 8 (3) (1977) 525–528.
[119] N. Fonstein, H.J. Jun, O. Yakubovsky, R. Song, N. Pottore. Evolution of advanced high strength steels (AHSS) to meet automotive challenges. In. Vail, CO, 2013.
[120] S. Hironaka, H. Tanaka, T. Matsumoto, Effect of Si on mechanical property of galvannealed dual-phase steel. In *Materials Science Forum*, 638–642 (2010) 3260–65.
[121] N. Pottore, I. Gupta, R. Pradhan, Effect of composition and processing in cold-rolled dual-phase steels for automotive applications. In *MS&T'06*, 653–63. Cincinnati, Ohio, 2006.
[122] X.P. Shen, R. Priestner, Effect of boron on the microstructure and tensile properties of dual-phase steel, Metall. Trans. 21A (9) (1990) 2547–2553.
[123] K. Hausmann, D. Krizan, K. Spiradek-Hahn, A. Pichler, E. Werner, The influence of Nb on transformation behavior and mechanical properties of TRIP-assisted bainitic-ferritic sheet steels, Mater. Sci. Eng. A 588 (2013) 142–150.

[124] S. Hashimoto, Effect of Nb on Zn-coated dual-phase steel sheet. In Materials *Science Forum*, 539-543 (2007) 4411–4416.
[125] R.A. Grange, C.R. Hribal, L.F. Porter, Hardness of tempered martensite in carbon and low alloy steels, Metall. Trans. 8A (11) (1977) 1775–1785.
[126] V.H. Baltazar Hernanadez, S.S. Nayak, Y. Zhou, Tempering of martensite in dual-phase steels and its effects on softening behavior, Metall. Mater. Trans. (2011).
[127] S. Gunduz, Effect of chemical composition. martensite volume and tempering on tensile behavior of dual-phase steels, Mater. Lett. 63 (2009) 2381–2383.
[128] S. Sadagopan, B. Yan, M. Huang, Characterization of press formability of advanced high strength steels using laboratory tests. SAE 2004-01-0506, 2004.
[129] Arcelor Mittal Automotive Product Catalogue. 2016. www.arcelormittal.com.
[130] N. Fonstein, Effect of annealing parameters on austenite decomposition in a continuously annealed dual-phase steel." In *Proceedings of 41st MWSP*, 2003 403–414.
[131] K. Olsson, M. Gladh, J.E. Hedin, J. Larsson, Microalloyed high-strength steels, Adv. High-Strength Steels no. 8 (2006) 44–46.
[132] O.A. Girina, N.M. Fonstein, D. Bhattacharya, Effect of annealing parameters on austenite decomposition in a continuously annealed Dual-phase steel, In proc. of 41st MWSP, 2003, 403–414.

TRIP aided and complex phase steels*

8

K. Sugimoto[1] *and M. Mukherjee*[2]
[1]Shinshu University, Wakasato, Nagano, Japan, [2]Tata Steel Ltd., Jamshedpur, Jharkhand, India

8.1 Introduction

The introduction of stringent vehicular safety and environmental regulations in the past few decades, as well as the increasing competition from other new and lighter materials, has led to the development of a new grade of steels called Advanced High Strength Steels or AHSS. AHSS are essentially "multiphase steels" (at least two different microstructural components) that combine attractive strength and formability properties. Typical microstructures of AHSS are illustrated in Fig. 8.1 and Table 8.1. However, the volume fractions of different microstructural constituents in industrially produced AHSS might slightly differ from these typical values.

Among the various steels belonging to the AHSS family, TRIP aided steels possess probably the best mechanical properties in terms of a high strength/elongation combination and high strain hardening exponent value [2] (Fig. 8.2). This engenders excellent formability for manufacturing and makes TRIP aided steels promising replacements for structural automotive components [4]. In addition, better passenger safety is ensured by the improved crash-worthiness of these steels [5].

The acronym TRIP for Transformation Induced Plasticity was first coined by Zackay et al. [6] to classify all alloys with a unique combination of high strength and enhanced ductility imparted by the deformation-induced transformation of austenite to martensite which results in an increase in the strain hardening rate leading to a delay in the onset of necking (the TRIP effect). The alloys investigated by this group were essentially high alloy metastable austenitic steels, which unfortunately could not become commercially popular because they were not economically viable. Then around 1990, researchers of Nippon Steel Corporation (NSC), Japan, used a novel process to develop a low alloy, Si bearing, medium carbon steel [7], which had ferrite as the matrix and bainite and metastable retained austenite as the other phases in its microstructure. These steels also demonstrated the TRIP effect. Since then, a lot of research has been done to improve the original NSC concept and at present this new variety of low alloy high strength formable steels is called TRIP assisted multiphase steels or simply TRIP aided steels.

* Every effort has been made to trace copyright holders and to obtain their permission for the use of copyright material. The publisher apologizes for any errors or omissions in the acknowledgements printed in this book and would be grateful if notified of any corrections that should be incorporated in future reprints or editions.

Automotive Steels. DOI: http://dx.doi.org/10.1016/B978-0-08-100638-2.00008-0

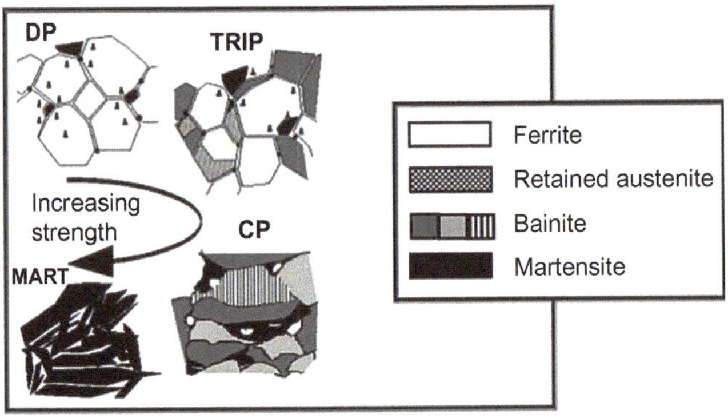

Figure 8.1 Illustration of typical microstructures of various AHSS. (*DP*: dual phase steel; *TRIP*: TRIP aided steel; *CP*: complex phase steel; *MART*: martensitic steel) [1].

Table 8.1 Typical proportions of different phases in AHSS [1a] (phase contents are in volume fraction)

Steel	Ferrite	Martensite	Bainite	Retained austenite
DP	0.80–0.85	0.15–0.20	–	–
TRIP	0.55–0.65	–	0.25–0.35	0.05–0.20
CP	0.05–0.10	0.05–0.10	0.80–0.90	–
MART	–	1.0	–	–

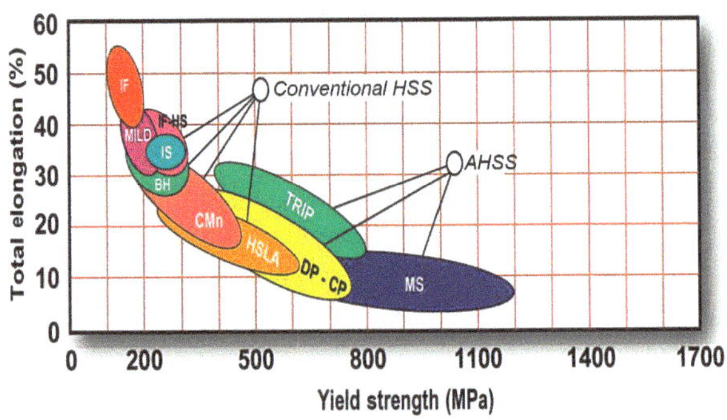

Figure 8.2 Strength/total elongation properties of various automotive steel sheets. (*HSS*: high strength steel; *AHSS*: advanced high strength steel; *IF*: interstitial free steel; *IF-HS*: high strength IF steel; *MILD*: mild steel; *IS*: isotropic steel; *BH*: bake hardenable steel; *CMn*: carbon manganese steel; *HSLA*: high strength low alloy steel; *TRIP*: transformation induced plasticity-aided steel; *DP*: dual-phase steel; *CP*: complex phase steel; *MS*: martensitic steel) [3].

Another variety of AHSS is the complex phase or CP steels which can be regarded as a transition from dual phase (DP) steels to ultrahigh strength steels with a tensile strength (TS) above 800 MPa, good formability, high impact energy absorption and hole-expansion capacity [8]. These steels are strengthened by solid-solution, precipitation, grain refinement, and phase transformation mechanisms [9] and have been used for parts that require high energy-absorption capacity, such as bumpers and B-pillar reinforcements [9]. These steels are also expected to be quite attractive because their TS can be enhanced up to 1500–2000 MPa by replacing the matrix structure to lath bainitic ferrite and lath martensite [10,11], though the chemical composition of these varieties are different from each other.

8.2 Processing route and microstructure

8.2.1 TRIP aided steels

The microstructure of present day TRIP aided steels typically consists of a continuous ferrite matrix with bainite and 0.05–0.20 volume fraction of metastable retained austenite as other phases. The main aim of any process route is therefore to obtain this minimum required amount of retained austenite in the final microstructure at room temperature. This is possible only if the highest temperature at which austenite transforms to martensite on cooling, i.e., the martensite start temperature or M_S, is somehow reduced to a value below room temperature. Considering the significant influence of carbon in reducing M_S and the fact that carbon is the cheapest source of austenite stabilization [12], the process routes are usually designed so as to ensure carbon enrichment and consequent retention of austenite at room temperature.

In general, a two-step heat treatment as shown in Fig. 8.3 is employed. The first stage of heat treatment is carried out in the two-phase ferrite (α) + austenite (γ) region, and is called the Intercritical Annealing (IA) treatment. During this treatment recrystallization of the cold rolled structure, dissolution of pearlite and carbide particles and formation of austenite take place. The soaking temperature and time are adjusted to obtain approximately 0.50 volume fraction ferrite and 0.50 volume

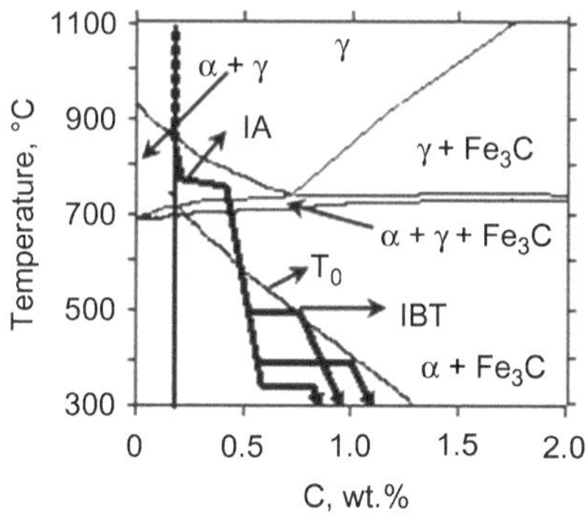

Figure 8.3 Two-step heat treatment for producing TRIP aided steel superimposed on a Thermo-Calc [13] calculated phase diagram of C-1.5Mn-1.5Si (wt.%) steel. (*IA*: Intercritical Annealing; *IBT*: Isothermal Bainitic Transformation) [14].

fraction austenite [15]. Nucleation and growth of austenite is accompanied by its carbon enrichment. However, the maximum carbon enrichment during IA is not sufficient to prevent martensitic transformation on cooling. This necessitates a second heat treatment. Hence, IA is followed by fast cooling (which may cause some austenite to further transform to pro-eutectoid ferrite) to an intermediate temperature at which the final transformation is carried out isothermally in the bainite formation region and this is called the Isothermal Bainitic Transformation (IBT) treatment. During the isothermal holding, supersaturated bainite forms as a first step. This supersaturation is relieved either by partitioning of carbon to austenite or formation of carbides. Some amount of carbon may even remain trapped in the dislocated structure of the newly formed bainite [16]. In TRIP aided steels, however, carbide formation is restricted by adding suitable alloying elements like Si/Al/P. Therefore, it is more likely that most of the carbon partitions to austenite and hence increases its stability [17].

The microstructural development at different stages of heat treatment have been elucidated by Chen et al. [18] as shown in Fig. 8.4. The different phase transformations taking place during the various stages of processing are clearly brought out. TRIP aided steels as described above can be produced either through cold rolling (CR) route or through hot rolling (HR), but most of the modern TRIP aided steels are generally produced by the cold rolling route.

8.2.1.1 Cold rolling route

The slabs are first reheated and soaked at around 1200°C and then hot rolled. After finish rolling at around 950°C the strips are air cooled and coiled. The microstructure at this point consists of ferrite and pearlite. The strips are then cold rolled and subsequently, the cold rolled sheets are reheated in the continuous annealing line (CAL) to the intercritical temperature of around 780–860°C (Fig. 8.3). After IA, the sheets are cooled rapidly to around 450°C and held isothermally at this temperature for the IBT treatment. This can be realized only in a CAL with an overaging section or a galvanizing and/or galvannealing line [15]. The ensuing microstructure consists of a ferrite matrix and ~0.20 volume fraction second phase with 0.06–0.14 volume fraction of retained austenite [19].

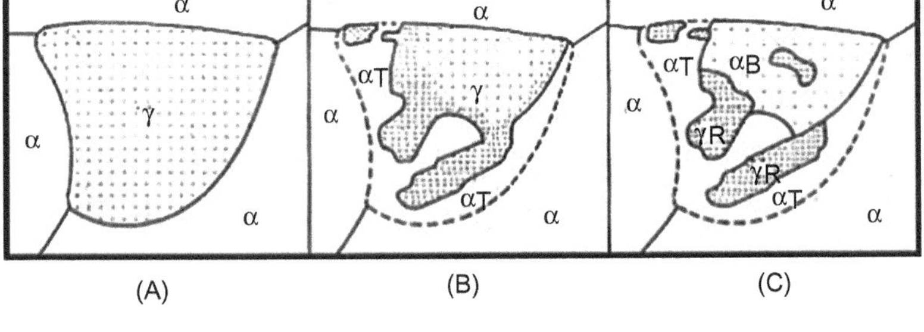

Figure 8.4 Microstructural changes during two-step heat treatment: (A) IA stage, (B) Cooling from IA temperature to IBT temperature, (C) IBT stage. (α: ferrite; α_B: bainite; γ: austenite; α_T: pro-eutectoid ferrite formed on cooling from IA temperature to IBT temperature; γ_R: retained austenite) [18].

8.2.1.2 Hot rolling route

TRIP aided steels can also be produced in the as hot rolled condition. After finish rolling at the intercritical temperature of around 780–860°C [14,20] or even higher than Ac_3 in some cases [21], coiling is carried out at controlled temperatures. The coiling step is equivalent to the IBT step in the CR route. This is followed by air cooling which results in the desired microstructure. Hot rolled TRIP aided steels of up to 780 MPa TS have been commercially produced [22].

8.2.1.3 Alternative microstructures

Though most TRIP aided steels have the ferrite/bainite/retained austenite microstructure described above, a lot of work has also been done to develop alternate microstructures in response to the complex formability requirements of the present day auto-body manufacturers. For example, Sugimoto et al. [23,24] have produced two alternate TRIP aided steels, one with bainitic ferrite as matrix and another with annealed martensite as matrix. Whereas the TRIP-aided steel with bainitic ferrite matrix is produced by complete austenitization as a first step (instead of IA) followed by IBT [23], the TRIP-aided steel with annealed martensite matrix is produced by subjecting the steel to an additional austenitizing and oil quenching step prior to the conventional treatment of IA and IBT [24]. The typical microstructures obtained by these process routes are shown in Fig. 8.5.

On the other hand, Haidemenopoulos [26], Laquerbe et al. [27] and Eberle et al. [28] have developed a novel process route that uses Mn enrichment to stabilize austenite at room temperature. This process involves a three-step heat treatment cycle (Fig. 8.6). As a first step, the hot rolled steel is quenched to produce upper bainite. This is followed by a soft annealing treatment, wherein the upper bainite is tempered at around 640°C for nearly 40 hours in order to produce manganese enriched coarse cementite. Finally, the steel is cold rolled and intercritically annealed for a short duration which ensures transformation of Mn rich cementite to austenite under para-equilibrium conditions such that the austenite inherits the entire Mn content of the cementite and gets chemically stabilized. If sufficient Mn enrichment of

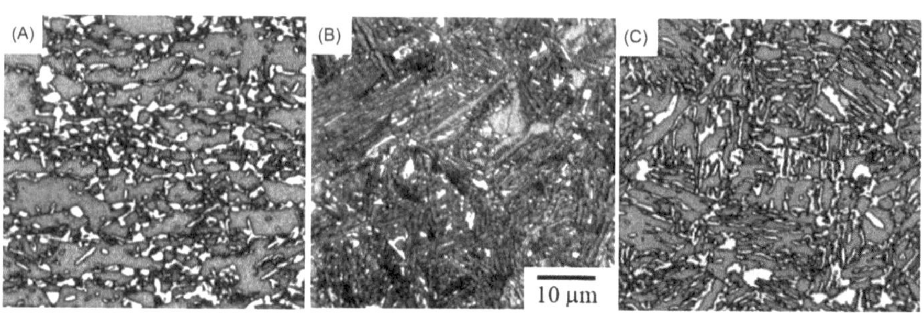

Figure 8.5 Microstructures (Le Pera etched) of TRIP aided steels with (A) Polygonal ferrite, (B) Bainitic ferrite, and (C) Annealed martensite matrices. K. Sugimoto, Advanced high- and ultra high-strength TRIP-aided steels, Trans. Indian Inst. Met. 59 (2006), pp. 725–738, with permission from The Indian Institute of Metals.

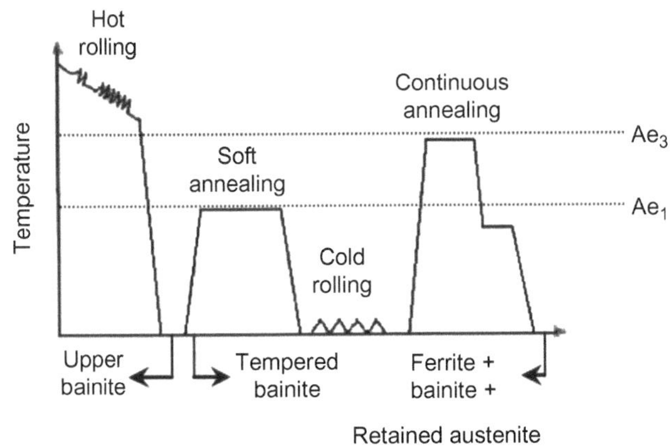

Figure 8.6 Alternative process route utilizing Mn enrichment for stabilizing austenite [28].

austenite is obtained, the isothermal holding step can be completely eliminated. This then permits the use of compositions free of carbide inhibiting elements like Si/Al/P.

Yet another process route called the "Quenching and Partitioning" route has been developed [29] to produce a higher fraction of retained austenite by carbon partitioning from supersaturated martensite. This process involves an initial full/partial austenitization step, followed by "quenching" and then "partitioning" treatments. This process route is further discussed in detail in Chapter 10, Bainitic and Quenching and Partitioning Steels.

Though the conventional TRIP aided steels with polygonal ferrite as matrix as well as the TRIP aided steel with bainitic ferrite matrix can be produced by conventional processing routes like HR and cold rolling as described above, the commercial viability of the other varieties of TRIP aided steels is yet to be established.

8.2.2 CP steels

CP steels possess a microstructure consisting of ferrite, bainite, martensite, and tempered martensite [30]. They contain <0.15 wt.% C and use many of the same alloying elements found in DP and TRIP aided steels but additionally have small quantities of Nb, Ti, and/or V to form fine carbides and carbonitrides to strengthen the matrix. Typically these steels have no retained austenite in their microstructure [31]. It is due to the inherent complexity of the microstructure of these steels, wherein it is extremely difficult to distinguish between matrix and second phase, that these steels are called "complex phase" steels [30]. Typical microstructure of CP steel is shown in Fig. 8.7B. For comparison, microstructure of DP steel is also shown in Fig. 8.7A. A variant of the conventional CP steel is the bainitic CP steel in which the polygonal ferrite is replaced by bainitic ferrite [31].

8.2.2.1 Hot rolling route

The processing route for hot rolled CP steels, as shown in Fig. 8.8, is essentially similar to that for hot rolled TRIP aided steels [9,33]. However, the cooling practice

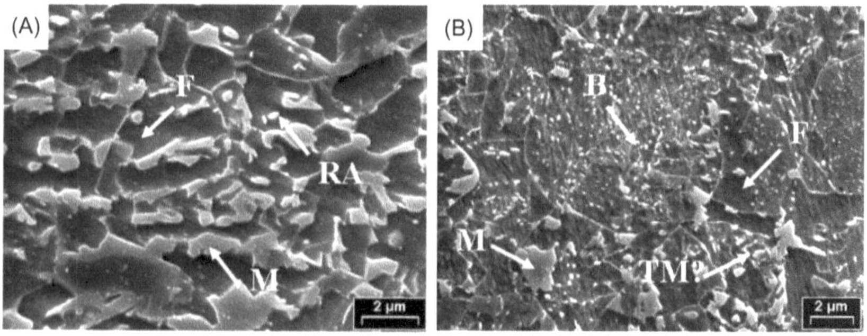

Figure 8.7 SEM images of (A) Dual phase and (B) Complex phase steels with tensile strength of 800 MPa [Nital etched; *F*: ferrite; *B*: bainite; *M*: martensite; *TM*: tempered martensite; *RA*: retained austenite] [32].

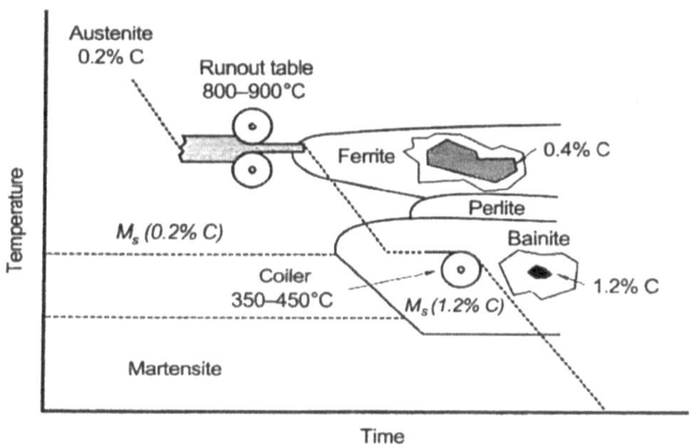

Figure 8.8 Cooling schedule in the production of the CP steel sheet [9].

is less stringent as no retained austenite is required in the final microstructure [31]. Typically for producing hot rolled CP steels, after finish rolling the steel is fast cooled to the ferrite transformation region and then held for some time to allow ferrite formation and then again the steel is fast cooled to the coiling temperature. Selection of the right coiling temperature is very important because on the one hand it should be low enough to allow bainite to form and refine the microstructure, but at the same time it should be high enough to maximize the formation of precipitates to ensure adequate precipitation strengthening [34].

8.2.2.2 Cold rolling route

Cold rolled CP steels, on the other hand, are produced by IA of the cold rolled strips, followed by fast cooling to a temperature above M_S, where it is isothermally held for some time followed by cooling to room temperature. The ferrite volume fraction is decided by the IA temperature and time. During the subsequent cooling and isothermal steps, the intercritical austenite transforms to bainite, and the austenite remaining after the isothermal holding transforms to martensite after the final cooling step. Some untransformed austenite may also be retained after this final cooling step [35].

8.3 Alloy design

Both TRIP aided steels and CP steels are characterized by a very low alloy content with the total amount in typical 800 MPa grades summing up to 3–4 wt.%. The main alloying elements used in TRIP aided steels are C, Mn, and Si. But, alternate concepts with Al or P partially or completely replacing Si have also been designed to address the coatability issues of Si containing TRIP aided steels. The chemical composition of CP steels is similar to that of TRIP aided steels, but they also contain microalloying elements like Nb, Ti, and V for precipitation strengthening effect. Microalloy addition is also incorporated in present day TRIP aided steels especially when the strength requirement is 1 GPa or more.

Selection of appropriate alloying elements is crucial for attaining the desired microstructure and mechanical properties. Therefore, for proper alloy design, the effect of different alloying elements on the transformation behavior of steels (as is illustrated schematically in Fig. 8.9) is analyzed in order to design suitable processing routes. A brief overview of the different alloying elements used in TRIP aided and CP steels is presented in this section.

8.3.1 Carbon

Carbon has a very strong solid solution strengthening effect and an adequate amount of carbon is necessary to ensure that the desired strength levels are reached. In studies

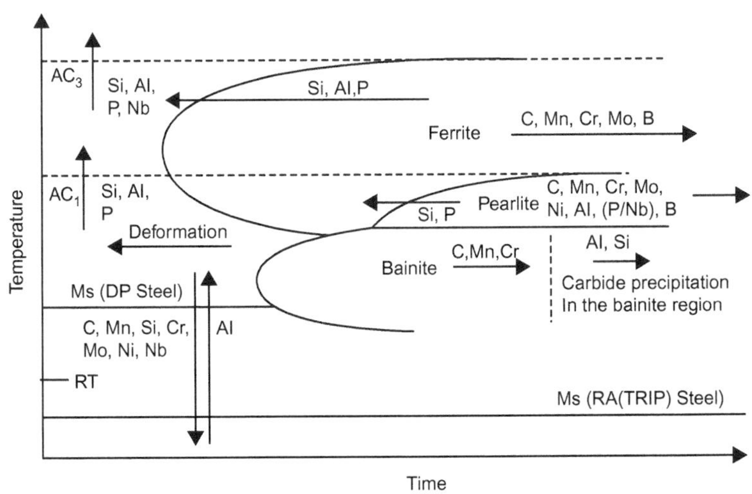

Figure 8.9 Schematic illustration of the effect of alloying elements on the TTT behavior in TRIP aided steel [36].

pertaining to bainitic steels [37] and TRIP aided steels [38], it has been found that every 0.1 wt.% increase in carbon content increases the TS by 80–100 MPa respectively. For TRIP aided steels, the carbon content is even more crucial considering that increasing carbon content results in increasing the volume fraction and stability of retained austenite which contributes to its high strength/ductility balance. However, carbon content cannot be increased beyond 0.2 wt.% as it causes problems during welding.

8.3.2 Manganese

Manganese addition in AHSS varies between 0.4 and 2.0 wt.%. Manganese dissolves in the ferrite matrix resulting in the deformation of the crystal lattice and generation of an elastic strain field [39]. This strain field hinders the movement of dislocations and hence imparts solid solution strengthening to the ferrite. Manganese also increases the hardenability of the steel by reducing the driving force and increasing the critical activation energy for nucleation of ferrite. This leads to suppression of the $\gamma \rightarrow \alpha$ transformation temperature [40], and hence significant refinement of the ferrite grain size. Larger additions in the range of 1.5–2.0 wt.% slow down the pearlite reaction significantly and promote the formation of acicular ferrite or bainitic transformation products which also leads to improvement in the strength of these steels.

In TRIP aided steels Mn plays an important role in increasing the stability of austenite by lowering the M_S temperature. But at the same time Mn lowers the T_0 temperature as well [19] which implies that during the IBT treatment, a higher amount of austenite can be obtained but the maximum carbon content of this austenite is lower. Lower carbon content raises the M_S temperature and the retained austenite becomes less stable. These two opposing effects should be kept in mind while deciding the Mn content of the TRIP aided steel.

High Mn additions (>2.5 wt.%) are avoided as it results in the formation of a banded structure wherein bands of low temperature transformation products are formed due to segregation of Mn. This is not desirable since it renders the structure brittle. High Si:Mn ratio can however restrict the formation of band structure [41]. This can be understood from the opposing effects of Si and Mn on the Ar_3 temperature (which is one of the primary factors responsible for the occurrence of banding in steels) [42], as a result of which a higher Si content reduces the severity of banding caused by Mn segregation.

8.3.3 Silicon

Silicon is a ferrite stabilizer and refines the ferrite grain size too, by increasing the nucleation rate of ferrite from austenite [43]. It also imparts considerable solid solution strengthening to ferrite. Apart from the above factors, the most significant characteristic of Si which results in its extensive use in TRIP aided steels is its ability to restrict carbide precipitation during isothermal holding resulting in the desired carbon partitioning to austenite and its stabilization. This in turn ensures formation of a large amount of stable retained austenite in these steels.

The ability of Si to suppress cementite formation during IBT is explained on the basis of its relative insolubility in cementite. Bainite transformation takes place under para-equilibrium conditions [44] and so, long range diffusion of Si away from cementite cannot play a major role. Moreover, Si does not affect the growth of carbides during IBT. Therefore, the only mechanism by which Si can possibly restrict carbide precipitation is by hampering the nucleation process itself. Formation of a cementite nucleus in the presence of Si requires the diffusion-controlled ejection of Si at the transformation front, causing a build-up in the concentration of Si around the cementite nucleus. This locally increases the activity of carbon resulting in lowering of the flux of carbon and hence inhibition in development of cementite embryos [45]. Inhibition of carbide formation allows austenite to inherit a large amount of carbon, which decreases M_S to low values and thus a considerable amount of metastable austenite is retained at room temperature.

However, during casting, hot rolling and annealing, the high Si content results in the formation of a very strong oxide layer (red scale defect), which gets easily rolled into the surface during hot rolling. This oxide is not only difficult to remove during pickling but it also causes problems during galvanizing due to poor coatability, resulting in uncoated regions on the steel surface called bare spots [46–48]. Moreover, Si reduces the bainite transformation kinetics appreciably, which makes it imperative to have a long over-aging section so that a long isothermal treatment can be carried out. Both these factors have led to the reduction in the amount of Si used in present TRIP aided steels from 1.5 to 2.0 wt.% used earlier to around 0–1 wt.%; the deficit is compensated by other alloying elements.

In CP steels wherein retained austenite is not required in its final microstructure, the Si content is restricted to <0.15 wt.% [49]. This also ensures relatively faster bainite transformation kinetics.

8.3.4 Aluminum

Aluminum addition is typical to TRIP aided steels in which it is mainly added as an alternative to Si addition. This is because, not only is Al a ferrite stabilizer like Si, but it is also insoluble in cementite and retards its formation [50]. At the same time, unlike Si it does not affect the coatability adversely [51]. In fact, Al results in the formation of an inhibition layer on the steel surface, which in turn prevents the formation of Si/Mn oxides and bare spots [52,53]. Another advantage of Al over Si is that it increases the driving force for austenite to bainite transformation [54], which accelerates the bainite transformation kinetics resulting from an increased nucleation rate. At IBT temperatures varying between 300°C and 500°C, bainite transformation kinetics is fastest for Al containing steels. This becomes especially advantageous for industrial production in conventional galvanizing lines where large over-aging sections are not available. Aluminum also remarkably increases the maximum carbon content of austenite by shifting the T_0 line to the right [55].

Nevertheless, Al addition is associated with its own set of disadvantages. Firstly, it is a weak solid solution strengthening element and so Al alloyed steels cannot reach the strength level possible with Si alloyed steels [56]. Besides, Al, due to its ferrite stabilizing ability, restricts the austenite region of the steel and additions of above 1.2 wt.% make complete austenitization impossible. Further, it increases the M_S temperature thereby destabilizing the austenite. Carbon addition can solve these problems [56], but high carbon content causes problems during welding. In addition, steels with 0.1–0.2 wt.% C and high Al levels are very difficult to continuously cast. To be able to cast these alloys, a small amount of Ti is usually added to fix the free N and prevent AlN formation. Therefore Si is never completely replaced by Al and a partial replacement is generally used [46]. Typically TRIP aided steels contain Al in the range of 0.5–1.0 wt.% (in Al + Si chemistries).

8.3.5 Phosphorus

Like Al, P addition is typical to TRIP aided steels in which they are added to replace Si and limit the usage of Al. Phosphorus is less effective in preventing carbide formation than Al or Si [57]. But, it is a very effective solid solution hardening element and P addition of up to ~0.1 wt.% does not pose any problems during galvanizing or galvannealing [2].

It has been found that steels containing only P and no Si have significantly lower amount of retained austenite than steels with Si and P both, which in turn have lower amount of retained austenite than steels with only Si [58]. However, Chen et al. [18] have found that that steels containing around 0.07 wt.% P in combination with 0.5 wt.% Si have the best mechanical properties. Increasing the P content from 0.07 to 0.2 wt.% results in the change in morphology of the retained austenite from coarse to extremely fine. However, the fine retained austenite in 0.2 wt.% P steels is so stable that it does not undergo deformation-induced transformation and hence does not contribute to enhancing the ductility. On the contrary high P addition leads to segregation of P at the grain boundaries which causes very rapid decrease in the strain hardening rate leading to deterioration of ductility [18].

8.3.6 Niobium

Microalloying additions are incorporated in AHSS to enhance the strength properties, primarily through grain refinement [15] and precipitation strengthening [5,59]. The grain refining effect of Nb results mainly from the prevention of recrystallization in the last stands of the rolling mill. Flattened or pancaked austenite grains with a high dislocation density enhance the nucleation rate of ferrite [60,61]. Additionally, Nb in solid solution lowers the $\gamma \rightarrow \alpha$ transformation temperature [62]. Due to its large atomic size, solute Nb segregates to the austenite grain boundaries and lowers its energy, resulting in the lowering of the activation energy of ferrite formation [62].

In TRIP aided steels, Nb addition has been found to result in higher amount of retained austenite, which can also be attributed to the grain refining effect of Nb [1a,14,15]. The finer grain size of austenite formed during the IA stage accelerates the formation of pro-eutectoid ferrite on cooling from IA temperature to the IBT temperature resulting in higher carbon content of austenite at the beginning of the bainite transformation treatment. This retards the bainite formation kinetics and hence lowers the amount of bainite formed and increases the amount of retained austenite.

However, there are some conflicting reports on the effect of Nb on the stability of retained austenite. Some reports suggest that Nb addition refines the austenite grain size to such an extent that its M_S temperature gets lowered and hence its stability increases. Others also suggest that Nb in solid solution hinders carbide precipitation in the temperature region of bainite formation, which results in an increase in both volume fraction and carbon content of retained austenite [50,63,64]. On the contrary, there is a considerable amount of work to show that Nb addition in fact lowers the stability of retained austenite mainly due to its lower carbon content [58,65]. While some researchers believe this to be due to the tying up of carbon by NbC precipitates [5], others [55,65] attribute it to the lack of the second carbon enrichment during bainite reaction. Timokhinaet al. [65] have argued that in Nb containing TRIP aided steels, the formation of fine bainitic ferrite laths from fine grain sized parent austenite (refined by NbC precipitation) by the displacive bainite reaction leads to an increase in the internal stress. This makes further bainitic transformation and the consequent carbon enrichment difficult.

In order to fully utilize the strengthening contribution of Nb, it is essential that the soaking temperature prior to hot rolling is higher than the dissolution temperature of NbC [66], otherwise the undissolved precipitates only coarsen during the subsequent heat treatments and have negligible effect on strengthening. Therefore, the thermomechanical processing strategy should be such that some amount of Nb remains in solution up to the transformation temperature [62]. The amount of Nb which remains in solid solution during finish rolling precipitates in the ferrite leading to additional strength via precipitation hardening.

8.3.7 Titanium

Like Nb, Ti is added to AHSS in order to increase strength of the steel by grain refinement and precipitation strengthening. However, in TRIP aided steels, Ti addition results in lower ductility than in Nb added steels. This has been attributed to the higher precipitation strengthening effect of TiC and the lower volume fraction of retained austenite in Ti microalloyed steels [66]. Addition of Ti in combination with Nb has been found to have a better strength/ductility balance [1a].

8.3.8 Vanadium

Like Nb and Ti, V also contributes to the strengthening of AHSS by precipitation strengthening as well as grain refinement. The contribution from precipitation strengthening in V microalloyed steels could be over 30% of the total strength in hot rolled steels whereas that from grain refinement could be as high as over 40% in hot rolled and normalized steels [67]. The solubility of V in austenite is more than that of other microalloying elements and so it is more likely to remain in solution prior to transformation. During phase transformation, vanadium precipitates as carbides and/or nitrides, depending on the relative carbon and nitrogen contents. Vanadium is also a ferrite stabilizer and thus increases the driving force for ferrite nucleation from austenite [68]. This results in an increase in the ferrite start temperature and the transformation starts earlier. Vanadium also suppresses the bainite and pearlite transformations, thereby facilitating the formation of martensite/austenite from the carbon enriched austenite remaining after ferrite transformation [68]. In its precipitate form, V retards recovery and softening of the lower transformation products like bainite during prolonged holding at relatively higher temperatures after coiling [68]. The toughness of V-microalloyed grades is usually not adequate and so V is generally used in conjunction with other microalloying elements.

In TRIP aided steels, V also plays an important role in refining the austenite grains and stabilizing it [69]. The retained austenite in V containing steel has been found to be present in the form of tiny particles of size less than 1 μm with film morphology [70]. Hence the stability of retained austenite in these steels is quite high and so gradual transformation of austenite takes place predominately at higher strains.

8.3.9 Chromium

Chromium addition delays the initiation of the $\gamma \rightarrow \alpha$ transformation and thus increases the hardenability of steels. It promotes the formation of lower temperature transformation products, which is why it is usually added in complex phase steels [49]. In TRIP aided steels, though addition of Cr results in retention of higher amount of retained austenite, this retained austenite is generally found to be coarse and unstable [69,70].

8.3.10 Molybdenum

Though Mo does not lower the $\gamma \rightarrow \alpha$ transformation temperature markedly, it severely retards the transformation kinetics, especially in the temperature region of 600–700°C [68]. In fact, it is more effective than Cr in suppressing the pearlite formation. According to the Grossman hardenability calculations, Mo has more potential for increasing hardenability than many other alloying elements [71]. In complex phase steels, usually a combination of Cr + Mo is added [49]. Addition of Mo in the range of 0.1–0.2 wt.% is especially beneficial for the production of galvannealed TRIP aided steels. Galvannealing usually involves annealing at relatively high temperatures of 500–520°C. In this temperature range, which is close to the bainite start temperature, Bs, Mo addition is very effective in suppressing pearlite and/or bainite transformation. Suppression of carbide formation even at these high temperatures enables the retention of metastable austenite with adequate stability in the final microstructure, which is not possible when other alloying elements are used [14].

8.4 Microstructure modeling

The challenge for producing AHSS via the industrial hot rolling/cold rolling routes is that the entire phase transformation sequence has to be completed in a very short time. It has also to be ensured that undesired phases like pearlite etc. are not formed. These stringent requirements necessitate the development of mathematical models for microstructure prediction in these steels in order to ensure proper design and optimization of process parameters.

The models available in literature can be broadly classified into three groups: (a) models for predicting phase transformation during Run-Out-Table (ROT) cooling after hot rolling [34,72,73], (b) models for predicting phase transformation in the CAL after cold rolling [72,74–76], and (c) empirical models [77,78]. These three approaches are discussed briefly in this section.

8.4.1 Phase transformation on ROT

Models for predicting phase transformation from austenite to ferrite during ROT cooling, in general, are based on the Johnson–Mehl–Avrami–Kolmogorov (JMAK) [79–81] model. In order to accommodate the non-isothermal nature of the transformations, the equations are modified according to the Scheil's additivity theory, [82] according to which, the kinetics of any non-isothermal transformation can be considered to be a sum of a series of small isothermal steps. Thus, if a cooling profile is divided into a number of steps, n the fraction transformed at the ith step, X_i is given by [73]

$$\frac{X_i}{X_i^e} = 1 - \exp[-k_i(t' + \Delta t)^n] \tag{8.1a}$$

$$t' = \left[-\frac{1}{k_i}\ln\left(1 - \frac{X_{i-1}}{X_i^e}\right)\right]^{1/n} \tag{8.1b}$$

In the above equations, X_i^e is the phase fraction at thermodynamic equilibrium at the ith step, t' is the equivalent transformation time needed to transform into the fraction X_{i-1} at the temperature of the ith step, Δt is the time step corresponding to the ith step, k_i is the rate constant and n is the time exponent. Whereas the rate constant depends on the temperature and transformation mechanism, the time exponent is assumed to be constant over the temperature range when a unique transformation mechanism operates.

In a model developed by Han and Park [73] to predict the phase transformations during ROT cooling for a TRIP aided steel, the values for k_i and n of the JMAK equation were determined from experimental continuous cooling transformation data and used to predict the phase fractions as a function of carbon content of austenite and prior austenite grain size. Though the model was shown to predict the phase fractions quite accurately, the effect of other alloying elements was not incorporated.

In another approach to predict the microstructure evolution in CP steels processed through hot rolling and ROT cooling [34,72], a sequential transformation model was developed that consisted of sub models for (i) ferrite start, (ii) ferrite growth, (iii) bainite start, (iv) bainite transformation, and (v) formation of martensite and/or retained austenite. This approach was also based on the JMAK approach after incorporating additivity. However, this approach was more detailed than the Han and Park model [73]. The effect of pancaking on austenite grain size was incorporated using an empirical relationship between austenite grain size and the effective strain. Ferrite transformation was modeled considering carbon diffusion controlled early growth of ferrite nucleated at austenite grain corners. In order to predict bainite transformation start, the concept of the critical driving force for the onset of the bainite transformation as proposed by Ali and Bhadeshia [83] was used. In this method, during the ferrite transformation, the driving force for the remaining austenite was calculated continuously and compared with the critical deriving force to determine the bainite start. In order to determine martensite start, the M_S temperature was calculated for the remaining austenite (after accounting for carbon enrichment) using empirical relationships [84]. However, in this approach too, the solute drag effect of other alloying elements was not considered.

8.4.2 Phase transformation during continuous annealing

As discussed earlier, in order to produce AHSS like TRIP aided steels and CP steels via the cold rolling route, the cold rolled sheets are subjected to a continuous annealing cycle which involves an IA step followed by IBT. Though there are a number of models available for bainite transformation [17,85], models for IA are limited. This is probably because of the complex nature of interacting phenomena that take place during IA namely, recrystallization of the cold rolled structure, dissolution of pearlite and carbide particles and formation of austenite, all of which may either occur consecutively or simultaneously. In order to model these complex interactions during IA either cellular automata [75,76] or phase field approach [72] is used. There are also some models which only cater to the phase transformation part of the annealing treatment [74], i.e., pearlite dissolution to form austenite and subsequent growth of austenite into the ferrite, where the phase transformation calculations are performed assuming para-equilibrium conditions. However, since these models do not take care of the ferrite recrystallization part, important effects of heating rate and annealing time cannot be adequately captured by these models.

8.4.3 Empirical models

For most general-purpose applications, empirical models are sufficient to predict the phase fractions of different phases in the microstructure. These models as described in literature [77,78] are primarily based on mass balance equations and dilatometry data and fairly accurate estimates of the phase fractions can be made using these models.

8.5 Deformation-induced transformation of retained austenite

Martensitic transformation is a first order phase transformation and occurs by nucleation and growth [86,87]. For the transformation to occur it is necessary that (i) the free energy of the system decreases and (ii) heterogeneous nucleation sites

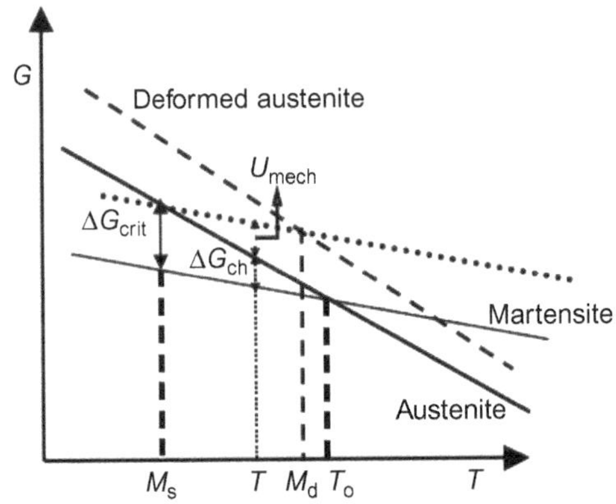

Figure 8.10 Schematic representation of the energetics of martensite transformation [89].

are present [88]. The energetics of transformation can be understood from Fig. 8.10. At the T_0 temperature, the chemical free energy of austenite is equal to that of martensite and they are in thermodynamic equilibrium. But the transformation from austenite to martensite can occur only when the free energy change of the system is large enough to compensate for the surface energy and elastic strain energy associated with martensite. Thus martensite cannot form spontaneously at T_0. It forms only after a certain amount of undercooling to M_S ($<T_0$) at which a critical chemical driving force, $|\Delta G_{crit}|$ becomes available for initiating the transformation [90]. In the absence of any other energy term, the magnitude of the chemical free energy difference, $|\Delta G_{ch}|$ must be greater than or equal to $|\Delta G_{crit}|$ for the transformation to take place.

Martensite transformation involves an invariant-plane strain shape deformation with a large shear component. The transformation can therefore also be regarded as a mode of deformation with a change in crystal structure. It follows that the transformation can be influenced by external deformation [87]. The deformation or strain energy can be regarded as an additional mechanical driving force U_{mech}, which can aid the chemical driving force in accomplishing the transformation [89,91,92]. External deformation and the associated mechanical driving force can thus trigger martensite transformation above the M_S temperature even though chemical driving force is not large enough to initiate the transformation at such temperatures ($>M_S$). In other words, deformation-induced martensitic transformation can be viewed as dissipation of strain energy accumulated during deformation [89].

The magnitude of chemical driving force decreases linearly with an increase in temperature above M_S (Fig. 8.10). Therefore, a proportionally higher mechanical driving force is required to trigger martensite transformation as the temperature is increased beyond M_S. But this is possible only up to a temperature M_d beyond which the chemical driving force is so small that $U_{mech} < |\Delta G_{crit}| - |\Delta G_{ch}|$ and deformation is no longer able to trigger martensitic transformation. At temperatures intermediate between M_S and M_d deformation-induced transformation of austenite

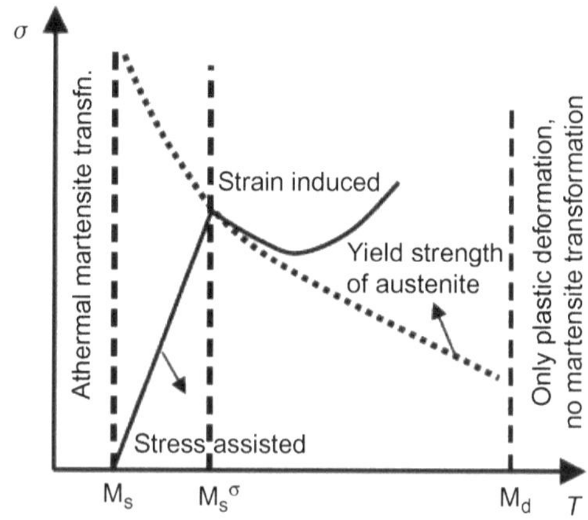

Figure 8.11 Schematic illustration of the influence of temperature and applied stress on martensite transformation mechanisms [93,94]. The solid line indicates the critical stress required to initiate martensite transformation at temperatures above M_S.

to martensite can occur via two modes. These phenomena are illustrated in Fig. 8.11 which shows the variation of critical stress required for initiating martensitic transformation as a function of temperature. The variation of yield strength (YS) of stable austenite with temperature is also superimposed on the same plot. It can be observed from the figure that the critical stress required to cause transformation increases linearly with temperature in the temperature range of M_S to M_S^{σ}. Moreover, the critical stress in this temperature region is lower than the yield stress (for slip) of stable austenite and the transformation occurs by heterogeneous nucleation on the same nucleation sites that are responsible for transformation on cooling below the M_S temperature. This mode of transformation is called *stress assisted transformation* [95]. The maximum temperature, up to which austenite can transform to martensite by the stress assisted mechanism is M_S^{σ} [96]. At this temperature the critical applied stress reaches the YS of stable austenite and above this temperature, the stress needed to initiate martensite transformation is large enough to cause plastic deformation (by slip) in austenite [96]. At temperatures higher than M_S^{σ}, martensite transformation takes places in a plastically deformed austenite on new nucleation sites generated by plastic deformation and at stresses that are considerably lower than is expected from a simple extrapolation of the critical stress-temperature line between M_S and M_S^{σ} (Fig. 8.11). This mode of transformation is termed as *strain-induced transformation* [96]. The above mechanisms are illustrated schematically in Fig. 8.12.

The excellent mechanical properties exhibited by the TRIP aided steels are mainly due to deformation-induced transformation of metastable retained austenite to martensite as described above. However, the beneficial effect of this transformation depends to a great extent on the resistance of retained austenite to the same or the "stability" of austenite which can be understood as the lack of activation energy needed for transformation. If stability is too low or too high the beneficial

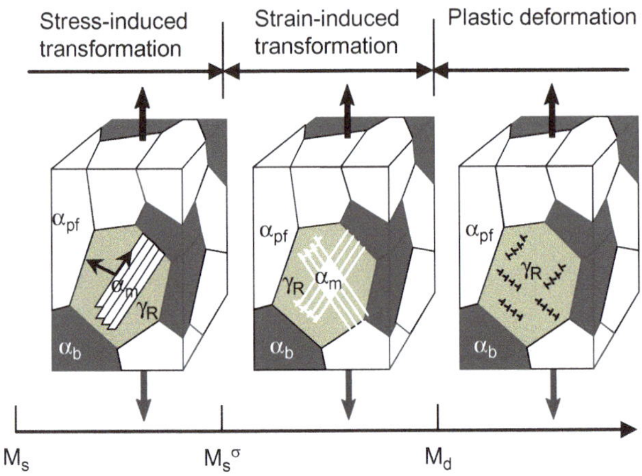

Figure 8.12 Schematic illustration of dominant deformation mechanism of retained austenite in different temperature ranges in TRIP aided steel: (from left to right) stress-induced plasticity, strain-induced plasticity and dislocation glide plasticity. α_{pf}: polygonal ferrite, α_b: carbide free bainite, α_m: martensite, γ_R: retained austenite [97].

effect of steady transformation during forming cannot be utilized [4]. Therefore it is not only the amount of retained austenite, but also its stability on plastic deformation, which is responsible for good formability [12].

Understandably, it is important to develop a mathematical model for the transformation of retained austenite. This would not only enable better understanding of the deformation-induced transformation behavior of retained austenite but also aid in a better design of TRIP aided steels. A number of models have been developed to describe and predict the variation of retained austenite with strain, some of which are described below.

8.5.1 Models for deformation-induced transformation of austenite to martensite

8.5.1.1 Angel [88]/Ludwigson and Berger [98]

A kinetic model to describe deformation-induced transformation of metastable austenite was proposed for austenitic stainless steels.

$$\log\frac{f}{1-f} = p_a \log\varepsilon + k_{ALB} \quad (8.2)$$

In the above equation, f is the volume fraction of martensite, ε is the true strain, p_a is the autocatalytic factor and is equal to 3 for metastable austenitic steels, and k_{ALB} is a constant. The same equation can be modified and applied to TRIP aided steels by expressing the equation in terms of $f_{\gamma\varepsilon}$, the volume fraction of austenite remaining untransformed after application of a true strain, ε, and the initial volume fraction of retained austenite, $f_{\gamma o}$, as:

$$\log\frac{f_{\gamma o}-f_{\gamma\varepsilon}}{f_{\gamma o}}=p_{\mathrm{a}}\log\varepsilon+k_{\mathrm{ALB}} \tag{8.3}$$

However, in this case, since the autocatalytic effect is negligible for TRIP aided steels, as it is suppressed by the presence of a large number of microconstituents other than retained austenite which act as barriers to the autocatalytic propagation of martensitic transformation [89], the factor p_{a} is set equal to 1.

8.5.1.2 Olson and Cohen [99]

The quantitative model for strain-induced martensitic transformation developed by Olson and Cohen [99] assumes that shear band intersections in metastable austenite constitute the primary site for strain-induced nucleation of martensite. According to this model the volume fraction of martensite formed after application of a true strain, ε can be determined as:

$$f=1-\exp\{-\beta_0[1-\exp(-\alpha_0\varepsilon)]^{n_1}\} \tag{8.4}$$

In the above equation, β_0 is related to the probability that a shear band intersection will lead to the formation of martensite nuclei, which in turn depends on the driving force for austenite to martensite transformation, $\Delta G^{\gamma\rightarrow\alpha'}$; α_0 is a strain independent constant and represents the rate of shear band formation; and n_1 is a constant with a value ≥ 2.

8.5.1.3 Matsumura et al. [100]

An empirical formulation to describe strain-induced transformation was developed by Matsumura et al. [100], which incorporated a term, k_{M} to describe the stability of retained austenite:

$$f_{\gamma\varepsilon}=\frac{f_{\gamma o}}{1+\left(\dfrac{k_{\mathrm{M}}}{q_{\mathrm{M}}}\right)f_{\gamma o}\varepsilon^{q_{\mathrm{M}}}} \tag{8.5}$$

In the above equation, q_{M} is the autocatalytic factor and is equal to 1 for TRIP aided steels. The form of this equation is similar to that developed by Angel [88]/Ludwigson and Berger [98].

8.5.1.4 Sugimoto et al. [4]

An empirical model was formulated by determining the best fit equation for a set of experimental data. This equation as denoted below defines the retained austenite stability by a strain-induced transformation factor "k":

$$\ln f_{\gamma\varepsilon} = \ln f_{\gamma o} - k\varepsilon \tag{8.6}$$

Lower value of k indicates higher stability of austenite. In various studies it has been found that k value decreases with increase in temperature and reaches a minimum at about 100–150 °C. According to Sugimoto et al. [4], below this temperature range, deformation-induced martensite transformation (Fig. 8.13) is the dominant mechanism controlling retained austenite stability. On the other hand, at temperatures above this temperature range, the decrease in stability has been attributed to deformation-induced bainite transformation. This phenomenon of a change in the nature of the transformation product with temperature has also been reported by Tsukataniet al. [20]. Apart from temperature, the k value is also affected by other forming conditions like strain rate and state of stress. While increasing strain rate lowers the k value, increasing the mean normal stress, e.g., by changing the state of stress from uniaxial tension to equi-biaxial raises the k value. These effects are illustrated in Fig. 8.13 and will be discussed in more detail in the next section.

8.5.1.5 Shin et al. [89]

This model is based on the internal variable theory of plastic deformation. Here, deformation-induced martensitic transformation is viewed as a continuous relaxation process of internal strain energy accumulated during plastic deformation. Assuming the incremental formation ratio of the number of nucleation sites per unit volume to be proportional to the internal strain energy, a rate parameter was defined to characterize how fast the nucleation sites are formed during plastic deformation and the following equation was derived to describe the deformation-induced martensitic transformation

$$f = f_S\{1 - \exp[-\beta_S \exp(\varepsilon - \varepsilon_0)^{n_2}]\} \tag{8.7}$$

In the above equation, f_S is the saturation value of the fraction of martensite formed, β_S represents the stability of austenite and depends on the chemical

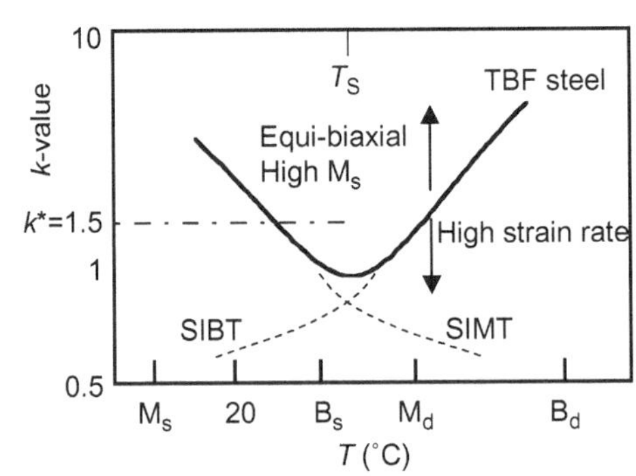

Figure 8.13 Illustration of forming temperature dependence of k value of TRIP aided steel with bainitic ferrite matrix (TBF) [25,101].

composition and temperature, ε_0 is a critical strain below which nucleation sites cannot form and n_2 is the deformation mode parameter which depends on the phase composition and straining route.

8.5.1.6 Sherif et al. [102]

In this model, the dependence of the progress of austenite transformation on the plastic strain, chemical composition and the temperature at which the deformation is carried out, has been elucidated by expressing the factor "k" in Eq. (8.6) as a linear function of chemical driving force for transformation at the temperature of deformation, $\Delta G^{\gamma\rightarrow\alpha'}$, i.e.,:

$$k = k'\Delta G^{\gamma\rightarrow\alpha'} \tag{8.8}$$

Hence the Eq. (8.6) can be rewritten as:

$$\ln f_{\gamma\varepsilon} = \ln f_{\gamma o} - k'\Delta G^{\gamma\rightarrow\alpha'}\varepsilon \tag{8.9}$$

The value of k' in the above equations has been empirically determined to be 0.00446 mol J^{-1}.

8.5.1.7 Mukherjee et al. [103]

This model is based on the thermodynamics of deformation-induced transformation of retained austenite to martensite, according to which $f_{\gamma\varepsilon}$ can be determined as a function of the initial retained austenite volume fraction, $f_{\gamma o}$; the chemical driving force for austenite to martensite transformation, $\Delta G^{\gamma\rightarrow\alpha'}$; and the strain, ε, in the following manner:

$$f_{\gamma\varepsilon} = f_{\gamma o}\left[\frac{\Delta G^{\gamma\rightarrow\alpha'} + A'}{\Delta G^{\gamma\rightarrow\alpha'} + A' - k_1\varepsilon^q}\right] \tag{8.10}$$

In the above equation, A' is elastic strain energy per unit volume of martensite due to shape deformation and $k_1\varepsilon^q$ represents the mechanical driving force or the energy contribution of the deformation strain, wherein k_1 is a proportionality constant and q is the strain exponent.

A comparative study of the different models described above was carried out by Mukherjee et al. [104] in which it was found that among all models, the model proposed by Sheriff et al. [102] gives the best predictions of $f_{\gamma\varepsilon}$. In the case of the other models, though fitting individual data sets resulted in good agreement between experimental and calculated values, the model parameters for these models could not be generalized over all data sets due to their dependence on properties that are specific to individual alloy systems and deformation conditions.

8.5.2 Additional factors affecting the stability of retained austenite

Apart from the effect of deformation conditions like temperature, strain, strain rate, and state of stress, there are a host of other factors as well which have significant influence on the stability of retained austenite and some of these are discussed below.

Among chemical components, the most potent alloying element in increasing the stability of retained austenite is carbon. However, the austenite carbon concentration after IBT is only around 1 wt.% and the corresponding M_S temperature is higher than room temperature, which implies that a certain amount of austenite most certainly transforms to martensite during the final cooling stage. Therefore, chemical enrichment alone cannot account for the retention of 0.1−0.2 volume fraction of austenite in the final microstructure. This incongruity can be explained by considering the stabilizing effects of the size of the austenite grains and also their position relative to their surroundings [105].

Wang and Zwaag [106] have explained the effect of grain size on M_S temperature and hence on the stability of retained austenite, by introducing an interfacial energy term to the total energy change involved in the transformation of austenite to martensite. In their work it was shown that finer austenite grains are associated with higher interfacial energy and hence require larger driving force for transformation, which implies that M_S temperature of fine grain austenite is lower (more stable) than that of coarse grains. However, the increase in interfacial energy was found to be significant only when the austenite grain size is reduced below 1 μm, and for such cases the M_S temperature was found to decrease exponentially with decrease in the size of austenite, signifying that the stabilizing effect of austenite grain size plays a role only when the grain size is less than 1 μm.

However, extremely small retained austenite grains of size smaller than 0.02 μm are also useless for TRIP as M_S is reduced to a very great extent and hence they do not undergo martensitic transformation at all [106].

Based on their work Sugimoto et al. [107] classified the retained austenite morphology into two groups on the basis of the type of phases surrounding it: (i) isolated retained austenite islands lying in a soft ferrite matrix or on the grain boundary adjacent to or away from the hard phases such as bainite or martensite, and (ii) thin retained austenite films existing along martensite or bainite lath boundary or blocky retained austenite in these hard phases. The stability of retained austenite belonging to the second group was found to be higher and this was attributed to its higher average carbon content (as illustrated in Fig. 8.13) and higher hydrostatic pressure of the surrounding phases, both of which lower the M_S temperature. The hydrostatic pressure offers high resistance against the volume extension accompanying martensitic transformation, which reduces the M_S temperature [15,108,109].

The morphology of the matrix and other second phases also has a significant influence on the stability of retained austenite [41,107,110]. Microstructure consisting of granular bainite and isolated regions of retained austenite and martensite has been found to provide optimum stability to retained austenite (Fig. 8.14) and hence maximum contribution toward elongation [65]. In steels which have some martensite in the microstructure due to transformation of austenite during the final cooling step, the martensite present in the microstructure imparts a shielding effect on the

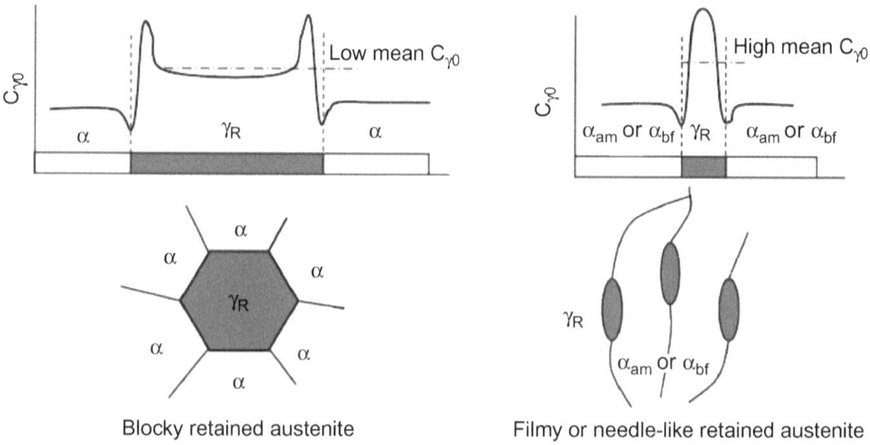

Figure 8.14 Distribution of solute carbon in retained austenite (γ_R) of TRIP aided steels [113], K. Sugimoto, Microalloyed ultra high- and high-strength TRIP-aided sheet steels with different matrix structure, in D. Bhattacharjee, U. K. Chatterjee, S. Chatterjee, M. Dutta and K. K. Ray, eds., Microalloyed Steels: Emerging Technologies and Applications, BESU, IIM, TMS, Kolkata, 2007, pp. 128–138, with permission from Vitasta Publishing [113a].
α, ferrite; α_{am}, annealed martensite; α_{bf}, bainitic ferrite.

retained austenite by reducing the stress increase in the austenite during external deformation. Solid solution strengthening of ferrite by the alloying elements may also produce a similar shielding effect [111].

The morphology can also be controlled by deformation in the two phase $\alpha + \gamma$ region (thermomechanical treatment). Deformation in this region at high strain rate and up to high strains has been found to result in higher volume fraction of retained austenite and general microstructural refinement [112].

8.6 Mechanical properties

The typical flow behaviors of 780 MPa grade TRIP aided and CP steels are compared in Fig. 8.15A. The flow behavior of a 800 MPa DP steel grade is also included in the same figure for the sake of comparison [30,114]. It is observed that while the CP steel has the highest YS, its ductility is much lower than both DP and TRIP aided steels. The TRIP aided steel has the highest ductility among all the three grades.

This difference in the flow behaviors of the three steels can be understood from their corresponding change in differential n values with strain as shown in Fig. 8.15B. It is observed that compared to the moderate strain hardening of CP steel which gradually decreases with strain, the differential n value of TRIP aided steel increases significantly with increasing strain. In the case of DP steel, on the other hand, the initial strain hardening is quite high but it gradually deteriorates with increasing strain. While the high initial strain hardening in DP steel can be attributed to the presence of large number of geometrically necessary dislocations present in the initial microstructure that are generated during the formation of these steels in order to accommodate the volume

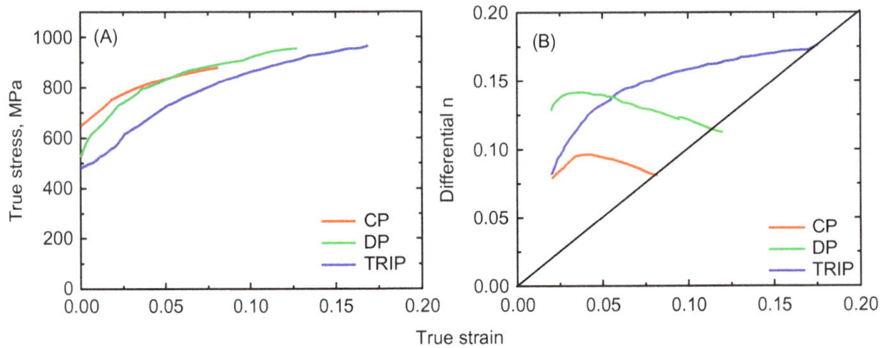

Figure 8.15 (A) Typical flow behaviors [36], Data from B. Ehrhardt, T. Gerber and T.W. Schaumann, Approaches to Microstructural Design of TRIP and TRIP-Aided Cold Rolled High Strength Steels, International Confernce on Advanced High Strength Sheet Steels for Automotive Applications AIST, Winter Park, Colorado, 2004, pp. 39–50 [36a] and (B) differential n values [30] of 780 MPa grade CP, DP, and TRIP aided steels. Data from M. Geiger, M. Merklein and M. Kaupper, Investigation of the mechanical behaviour of advanced high strength steels under various loading conditions, International Journal of Material Forming, 1 (2008), pp. 225–228 [114].

expansion associated with martensitic transformation, the increasing strain hardening capacity with strain in TRIP aided steels can be attributed to the deformation-induced transformation of retained austenite that causes gradual formation of martensite throughout the deformation regime. Numerous dislocations are generated to accommodate the volume expansion and shear deformation accompanying the transformation. Moreover, geometrically necessary dislocations are also generated to accommodate the gradient of deformation between the soft ferritic matrix and hard inclusions (bainite/martensite) [55,115,116]. When progressive transformation of retained austenite takes place, the associated dislocations are introduced gradually and this allows sustaining the strain hardening in a steady manner. The enhancement of strain hardening causes the conditions of plastic instability to be met at higher strains and this Postpones the onset of necking leading to large deformations and high uniform elongation. [116]. This phenomenon is illustrated schematically in Fig. 8.16. The overall properties of the three varieties of AHSS are summarized in Fig. 8.17.

In order to corroborate the extra hardening produced by the deformation-induced transformation of retained austenite, Jacques et al. [115] used neutron diffraction to measure the stress levels of the BCC and FCC phase in a TRIP aided steel and then these were used to calculate the theoretical stress of the steel, $\sigma(\varepsilon)$, by taking into account the contribution from ferrite, bainite and retained austenite but not martensite, and applying the law of mixtures [115]:

$$\sigma(\varepsilon) = f_{\alpha+\alpha_B}\sigma_{\alpha+\alpha_B}(\varepsilon) + f_{\gamma\varepsilon}\sigma_{\gamma}(\varepsilon) \quad (8.11)$$

where $f_{\alpha+\alpha_B}$ is the volume fraction of ferrite (α) + bainite (α_B), $\sigma_{\alpha+\alpha_B}$ is the stress level in ferrite and bainite (i.e., the BCC phases) and σ_{γ} is the stress level in austenite (the FCC phase).

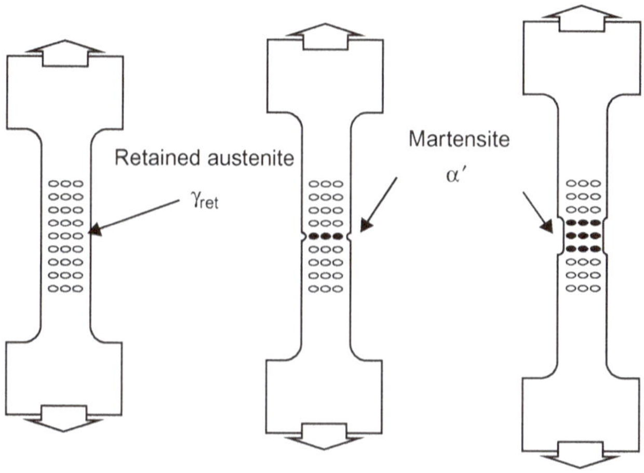

Figure 8.16 Schematic illustration of transformation-induced plasticity mechanism in low alloy TRIP aided steel [97].

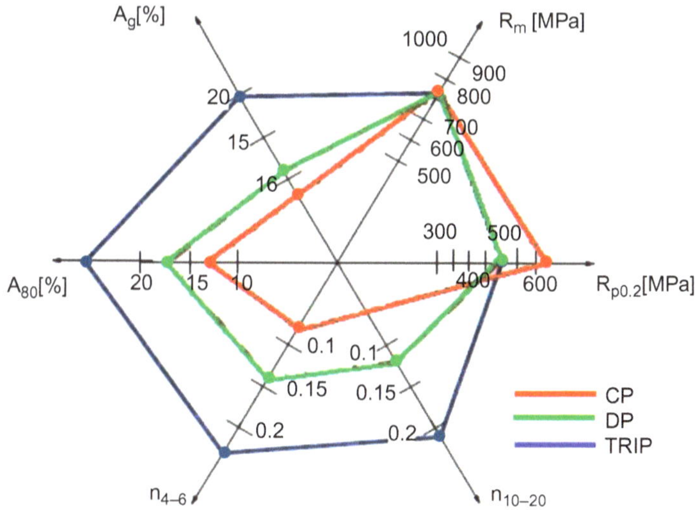

Figure 8.17 Tensile properties of 780 MPa grade CP, DP, and TRIP aided steels [30]. ($R_{p0.2}$ represents the 0.2% offset yield strength; R_m represents the ultimate tensile strength; A_g represents the uniform elongation; A_{80} represents the total elongation for a gage length of 80 mm; n_{4-6} and n_{10-20} represent the strain hardening exponent in the strain ranges of 4–6% and 10–20%, respectively).

Fig. 8.18 compares the theoretical stress calculated thus, with the measured macroscopic stress. The higher values of measured stress and the increasing difference between the measured and calculated stresses were ascribed to extra hardening by progressive martensitic transformation [115].

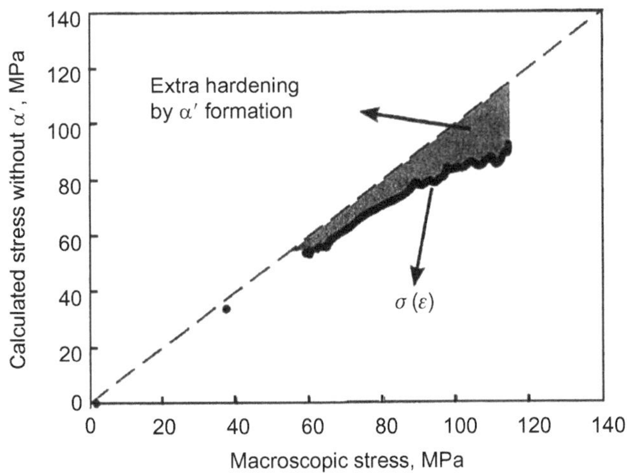

Figure 8.18 Comparison of actual macroscopic stress with the theoretical stress calculated [$\sigma(\varepsilon)$] without taking into account the martensite phase [115].

Apart from the stability of retained austenite, which has a major influence on the mechanical properties of TRIP aided steels, there are a number of other factors as well, which influence the mechanical properties of AHSS. Alloying elements, for instance, have a significant influence because they not only alter the phase transformations occurring during the heat treatments and the resulting microstructure, but also they affect the properties of the microstructural constituents themselves by mechanisms of solid solution strengthening, precipitation strengthening or grain refinement.

The morphology and distribution of the various phases in the microstructure also influence the mechanical properties of CP and TRIP aided steels. In CP steels, e.g., the low strength difference between constituent phases results in a very high resistance to local damage [30]. On the other hand, though TRIP aided steels containing large amounts of retained austenite have improved plastic properties, they exhibit lower fracture toughness owing to the presence of a nearly continuous network of high strength bainite and brittle martensite in the fracture process zone [117]. Premature cracking of martensite brings about faster crack initiation. Moreover, TRIP aided steels which have block-like retained austenite in an annealed microstructure, void nucleation occurs at the boundaries between retained austenite and annealed matrix resulting in lower elongation values [118], whereas fine, homogeneous retained austenite has been found to be most resistant to void formation [119]. However, due to the large strain hardening capacity, these steels exhibit a higher resistance to crack propagation. Therefore, to strike a balance between resistance to plastic instability and resistance to cracking in TRIP aided steels, the amount and morphology of retained austenite require to be controlled in order to avoid premature damage due to a “brittle network effect”[117].

Due to their effect on the stability of retained austenite, forming conditions like temperature and strain rate also have a significant influence on the tensile properties of TRIP aided steels and this is discussed in some detail here.

8.6.1 Effects of temperature

As described earlier, extent of deformation-induced transformation decreases with increase in testing temperature above M_S and is completely suppressed above M_d. As an example, variation in mechanical properties at different test temperatures for two varieties of TRIP aided steels as reported by Tsukatani et al. [20] is shown in Fig. 8.19.

Deformation studies conducted at temperatures varying over wide ranges [20,59] show that at low temperatures, the TS is pretty high and it keeps on increasing on lowering the temperature further. The increase in TS at lower temperatures is much sharper in TRIP aided steels than in conventional steels and is mainly due to the deformation-induced formation of martensite, which increases with reduction of temperature [59]. In general, the YS also shows a similar trend but to a very small extent, which may be due to the fact that at low stresses considerable deformation-induced martensitic transformation does not take place [59]. As the test temperature is increased, TS decreases and reaches a minimum at around 100–250°C after

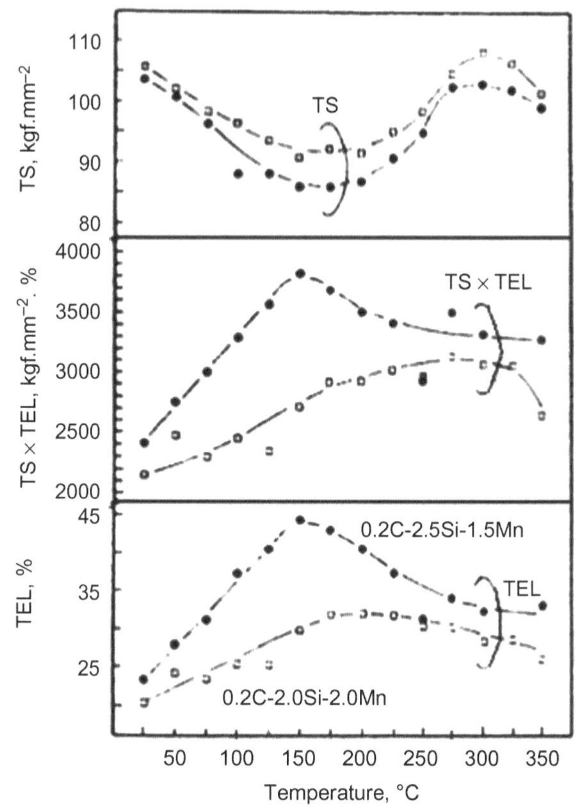

Figure 8.19 Variations in mechanical properties with test temperature in TRIP aided steels [20].*TS*: tensile strength; *TEL*: total elongation, TS × TEL: product of TS and TEL.

which it increases again [19,20,101,120]. The temperature at which minimum TS is obtained roughly coincides with the temperature at which deformation induced martensitic transformation is retarded [4,19]. Tsukatani et al. [20] have attributed the increase in TS at temperatures beyond 200°C to the dynamic strain aging effect and the change in morphology of the deformation-induced transformation product at high deformation temperatures [20].

Ductility as measured by the uniform elongation (UEL) and total elongation (TEL) is also greatly influenced by the test temperature. In conventional TRIP aided steels with polygonal ferrite matrix, elongation is maximum at temperatures at which the retained austenite has optimum stability and is able to undergo slow and progressive deformation-induced transformation to martensite up to very high strains [4,59]. Therefore, at these temperatures the strain hardening rate decreases very gradually with strain over a large strain range and as a result necking is delayed. Moreover, martensitic transformation causes relaxation of severe local stresses, which results in suppression of void formation at the matrix/second phase interface [4]. All these factors lead to high elongation.

8.6.2 Effects of strain rate

Choi et al. [121] observed that increasing the strain rate in the range of 10^{-3} s^{-1} to 10^{2} s^{-1} during quasi-static tensile tests results in an increase in YS and TS but reduces the TEL. Sugimoto et al. [4] have also found that at high strain rates of the order of 10^{-2} s^{-1} the TEL drops drastically and Mukherjee et al. [122] have found this effect to be more significant at higher temperatures. This can be attributed to the excess stability of the retained austenite at such high strain rates due to the concomitant adiabatic heating.

Dynamic tests at high strain rates, on the other hand, can result in significant increase in the YS, TS, UEL, and TEL values in comparison to static tests, as is shown for CMnAl TRIP aided steels in Fig. 8.20. At such high strain rates the

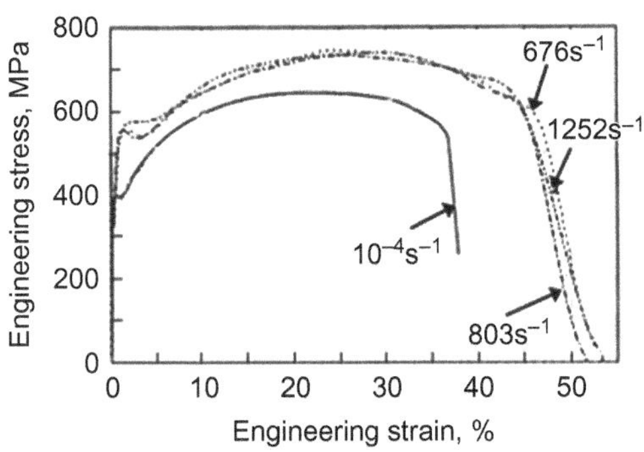

Figure 8.20 Effect of strain rate on stress–strain curves of CMnAl TRIP aided steel subjected to static and dynamic tensile tests [123].

deformation behavior is controlled mainly by two phenomena occurring simultaneously: the work hardening process [124] and the softening caused by adiabatic heating [125]. While the first results in an increase in YS and TS, the second results in an increase in the stability of the retained austenite that slows down its rate of transformation to martensite ensuring more progressive transformation and hence higher UEL and TEL [123].

8.7 Press formability

According to Hilsenet al. [126], bulk formability can be defined as the "ability of the steel to be stamped into a part without failure occurring within the body of the part from exhaustion of ductility." Complex automotive parts may require different forming methods to generate the geometrical shapes associated with them. Each forming method in turn requires different material properties for optimum formability. The product of TS and TEL, i.e., the strength/ductility balance, e.g., gives a fair idea about the press formability of steels and a high strength/ductility balance is an important pre-requisite for press forming. Compared to CP steels which possess low ductility and hence low strength/ductility balance, TRIP aided steels have very good ductility and hence the press formability of these steels is far superior. Nevertheless, it is essential to utilize the merits of high ductility effectively during press forming operations and hence understand the contribution of the TRIP effect under actual press forming conditions.

In case of TRIP aided steels, probably the most important parameter which decides the formability of the steels under various forming conditions is the stress state and its influence on the stability of the retained austenite. The state of applied stress influences the driving force for martensite transformation and hence depending on it, the M_S temperature either increases or decreases. According to Radcliffe and Schatz [108], M_S temperature of plain carbon steels with carbon content varying between 0.3 and 1.2 wt.%, increases by 6.7 °C per 100 MPa increase in the mean normal stress. This increase in M_S temperature implies a decrease in the stability of retained austenite. So the higher mean normal stress during equi-biaxial tensile stretching should correspond to a higher M_S temperature and hence lower stability of retained austenite. Sugimoto et al. [101] have indeed demonstrated that during room temperature stretch forming tests, the retained austenite in TRIP aided steels transforms at a faster rate than during uniaxial tensile tests conducted at the same temperature. They have also observed that the optimum temperature corresponding to maximum stretch formability is higher than the optimum temperature for maximum ductility under uniaxial tension. This higher optimum temperature under stretch forming has been correlated with the higher M_S temperature resulting from the increased normal stress in equi-biaxial tension [101].

Forming of automotive underbody parts, inner panels and suspension parts involves stretch flanging and bending deformations. In the stretch flanging operation, the cut edge of the material experiences a large tensile deformation in a manner similar to expansion of a pierced hole. Hence, the stretch flangeability of a material is measured by a parameter known as the hole expansion ratio or HER, which is nothing but the percentage change in the diameter of a punched hole on being expanded by a conical punch till failure occurs. Stretch flangeability or edge formability can therefore be defined as "the ability of the sheet to be stamped into a part without failure by excessive thinning at a sheared edge or a hole"[126].

The HER and minimum bending radius for 780 MPa grade TRIP aided, CP, and DP steels, are compared in Fig. 8.21. From the figure it is clear that HER and bendability are much better in the case of CP steels.

Both hole expansion and bending involve very high local strains and the final failure in such operations is controlled by void nucleation and growth. Hence for such operations the materials should possess very high resistance against damage [30]. Plastic deformation in a multi-phase microstructure results in strain partitioning between the soft and hard phases, with the strain in the softer phases being much higher than that in the harder phases. Eventually this leads to strain localization in the matrix region close to the hard phases, resulting in nucleation and growth of voids in these regions. Higher strength difference between the constituent phases of the microstructure accelerates this process. Hence, in order to enhance the resistance against damage, the hardness difference between the phases in the material should be low. This explains the better performance of CP grades in which a clear distinction between matrix and second phase is not possible and the hardness difference between the constituent phases is quite low. On the other hand, a high strength difference between the phases in TRIP aided and DP steels, results in very high strains locally, which results in easy void nucleation at the interface between hard martensite and soft ferrite. Therefore, in these materials damage induced failure is observed at relatively low global strains.

However, stretch flangeability of TRIP aided steels has been found to improve remarkably under warm forming conditions. The optimum temperature for maximum stretch flangeability increases linearly with M_S temperature as in forming operations and tensile tests [120]. Also, this temperature is found to be intermediate between those for uniaxial tension and stretch forming which correlates well with the fact that the mean normal stress and hence the M_S temperature in this case are intermediate between those in uniaxial tension and equi-biaxial tension [120].

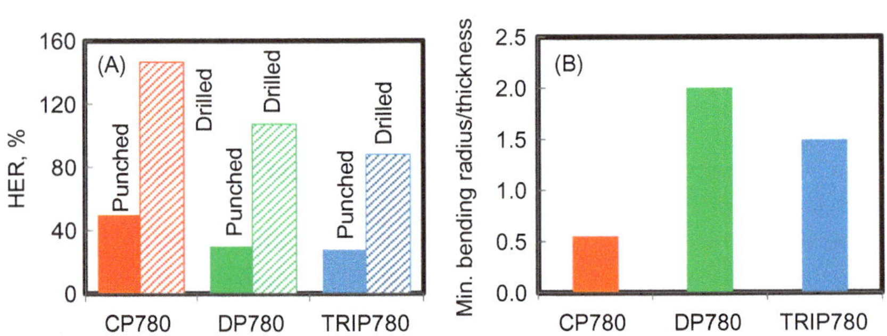

Figure 8.21 (A) Hole expansion ratio (HER) and (B) Ratio of minimum bending radius to thickness of 780 MPa grade CP, DP, and TRIP aided steels [30].

Another factor which has significant influence on the stretch flangeability in TRIP aided steels is the morphology of retained austenite and the surrounding matrix, which according to Kashima et al. [118], is much more than the influence of the stability of retained austenite to deformation-induced transformation.

Kashima et al. [118] have demonstrated that stretch flangeability of TRIP aided steels can be improved by changing the initial hot rolled microstructure, which in turn affects the morphology of the retained austenite and the matrix in the final microstructure. When the initial hot rolled microstructure comprises of ferrite and pearlite (as in conventional TRIP aided steels) blocky retained austenite is formed which easily transforms to martensite enabling easy void nucleation at the ferrite/martensite boundaries during the punching stage itself. The morphology of retained austenite can be modified to a lath-like structure by changing the initial hot rolled microstructure to bainite, martensite or ferrite and martensite. Lath-like retained austenite decreases the stress concentration and restricts void nucleation during the punching operation, which improves stretch flangeability [118]. However, compared to steels with initial hot rolled microstructure of bainite or martensite, steels with initial hot rolled microstructure of ferrite and martensite have lower stretch flangeability, even though both varieties have lath-like retained austenite. It is believed that the presence of soft polygonal ferrite is responsible for lower stretch flangeability. Void nucleation becomes easier due to the high hardness ratio between the annealed martensite (or bainite) and soft polygonal ferrite. When the initial hot rolled microstructure consists of martensite or bainite only, the lath-like morphology of retained austenite mainly prevents void nucleation and crack propagation during the hole punching and subsequent hole expansion stages [118].

8.8 Other mechanical properties

8.8.1 Crashworthiness

The high ductility of TRIP aided steels makes them very strong contenders for application in crash absorbing components of automobiles. The stability and amount of retained austenite in the TRIP aided steel are of utmost importance to achieve this high ductility to ensure that a high amount of energy is absorbed in the event of a crash. It is evident that if all the retained austenite transforms to martensite during the press forming stage itself, the crashworthiness cannot be good. This is mainly because in TRIP aided steels, high carbon brittle martensite is formed which makes the steel prone to failure during a crash. For maximum benefit therefore, the retained austenite should be reasonably resistant to deformation-induced transformation during the press forming stage, but in the event of a crash, the majority of the retained austenite should transform to martensite on impact and absorb a large amount of the impact energy [2].

8.8.2 Fatigue strength

Fatigue behavior of CP and TRIP aided steels has not been studied in great detail. However, from the limited literature available, it appears that TRIP aided steels

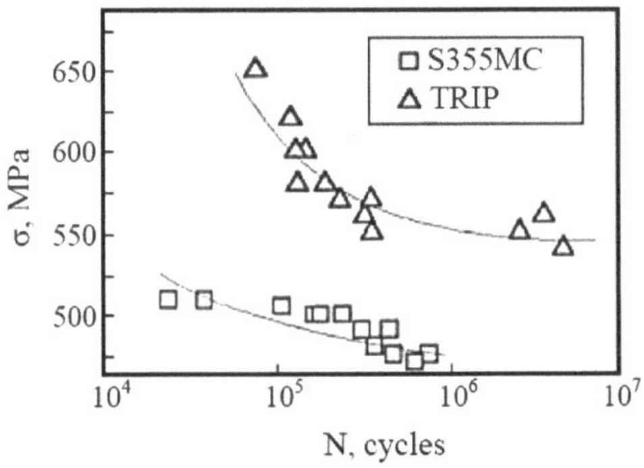

Figure 8.22 S–N curves of TRIP aided and S355MC microalloyed steels [21].

have superior fatigue resistance. For example, the S–N curves of a TRIP aided steel are compared with that of a conventional microalloyed steel in Fig. 8.22. The figure clearly shows the better performance of the TRIP aided steel for which a fatigue limit of around 550 MPa could be achieved. Eberle et al. [127] have also reported similar results.

Song et al. [128] studied the effect of prestraining on the high cycle fatigue strength of TRIP aided steels with different matrix microstructures and different retained austenite characteristics. They found that a tensile prestrain of 10% increased the fatigue limit of the TRIP aided steels, especially in the steel with polygonal ferrite matrix. This was attributed to the effective TRIP of the retained austenite and the high compressive internal stress in the matrix, resulting from the presence of hard second phase. On the other hand, the TRIP aided steel with bainitic ferrite lath matrix exhibited only a small increment in fatigue limit after prestraining. This was believed to be due to the formation of a large amount of strain-induced martensite even during prestraining and low compressive internal stress in the matrix. Of all the parameters examined, the internal stress was found to be most effective in increasing the fatigue strength of the low alloy TRIP aided steels [128].

8.9 Summary

The application of AHSS, which exhibit attractive combinations of high strength and excellent formability, has become a growing trend in the automotive industry due to the possibility of combining weight reduction with improved passenger safety, optimized environmental performance and manufacturing feasibility at an affordable cost.

This chapter reviews the basic concepts of two steels of the AHSS family namely TRIP aided steels and CP steels and presents a detailed description of the processing routes adopted for these steels, their alloy design concepts and details of their microstructure and mechanical properties.

Further, the various mathematical modeling approaches used for predicting the microstructure of these steels are discussed. The importance of retained austenite and its stability against deformation-induced transformation to martensite for achieving the excellent mechanical properties of TRIP aided steels is emphasized. The various empirical models that have been developed to predict the variation of the amount of retained austenite with strain are also presented. Also, the effect of various parameters like chemical composition, morphology and size of retained austenite and forming conditions (test temperature, strain, strain rate, and stress state) on the stability of retained austenite is discussed. The inherent differences in the microstructures of the two varieties of steels and their consequent effect on their respective formabilities are also discussed which is essential for the selection and application of the most appropriate steels for different automotive parts.

As is evident, a significant amount of research has been devoted toward the design and development of these grades of steels. However, the continual emergence of novel design concepts for auto-body components with complex shapes and thinner walls that require more complex forming operations, emphasizes the necessity of further improving the formability of these steels while maintaining the required high strength levels.

References

[1] T. Heller, A. Nuss, Effect of alloying elements on microstructure and mechanical properties of hot rolled multiphase steels, Ironmaking & Steelmaking 32 (4) (2005) 303–308. Available from: http://dx.doi.org/10.1179/174328105X47981.

[1a] T. Heller, A. Nuss, Mechanical properties and behavior of hot-rolled retained-austenite (TRIP)- and dual-phase steels, in: M. Militzer, W.J. Poole, E. Essadiqi (Eds.), International Symposium on Transformation and Deformation Mechanisms in Advanced High Strength Steels, Canadian Institute of Mining, Metallurgy and Petroleum, Vancouver, British Columbia, Canada, 2003, pp. 7–20.

[2] B. Mintz, Hot dip galvanising of transformation induced plasticity and other intercritically annealed steels, Int. Mater. Rev. 46 (2001) 169–197.

[3] A. Sprock, M. Peretic, J. Speer, Compact cooling as an alternative to alloying for production of DP/TRIP steel grades, in: J.G. Speer, B. Nelson, R. Pradhan (Eds.), International Conference on New Developments in Advanced High-Strength Sheet Steels, AIST, Orlando, Florida, 2008, pp. 237–247.

[4] K. Sugimoto, M. Kobayashi, S. Hashimoto, Ductility and strain-induced transformation in a high-strength transformation-induced plasticity-aided dual-phase steel, Metall. Trans. A 23A (1992) 3085–3091.

[5] D. Krizan, B.C.D. Cooman, J. Antonissen, Retained Austenite Stability in Cold Rolled CMnAlSiP Microalloyed TRIP Steels, International Conference on Advanced High Strength Sheet Steels for Automotive Applications, AIST, Winter Park, Colorado, 2004, pp. 205–216.

[6] V.F. Zackay, E.R. Parker, D. Fahr, R. Busch, The enhancement of ductility in high-strength steels, Trans. Am. Soc. Met. 60 (1967) 252–259.

[7] O. Matsumura, Y. Sakuma, H. Takechi, Enhancement of elongation by retained austenite in intercritical annealed 0.4C-1.5Si-0.8Mn steel, Trans. Iron Steel Inst. Jpn. 21 (1987) 570–579.

[8] C. Krempaszky, J. Ocenášek, V. Espinoza, P. Tsipou, Micromechanical modeling of the formability of advanced high strength steels, in: J.G. Speer, B. Nelson, R. Pradhan (Eds.), International Conference on New Developments in Advanced High-Strength Sheet Steels, AIST, Orlando, Florida, 2008, pp. 305–314.
[9] M.H. Demeri, Advanced High-Strength Steels-Science, Technology and Applications, ASM International, Materials Park, Ohio, 2013.
[10] K. Sugimoto, M. Murata, S. Song, Formability of Al-Nb Bearing ultra high-strength TRIP-aided sheet steels with bainitic ferrite and/or martensite matrix, ISIJ Int. 50 (2010) 162–168.
[11] D.V. Pham, J. Kobayashi, K. Sugimoto, Effects of microalloying on stretch-flangeability of TRIP-aided martensitic sheet steel, ISIJ Int. 54 (2014) 1943–1951.
[12] *Dual phase and TRIP steels, Niobium information*, CBMM/NPC, Dusseldorf, Germany, 1997.
[13] B. Sundman, B. Jansson, J.O. Andersson, The thermo-calc data bank system, Calphad 9 (1985) 153–190.
[14] S. Hashimoto, S. Ikeda, K. Sugimoto, S. Miyake, Effects of Nb and Mo Addition on the Mechanical Properties of Hot Rolled TRIP-Aided Steel Sheets, International Conference on Advanced High Strength Sheet Steels for Automotive Applications, AIST, Winter Park, Colorado, 2004, pp. 195–204.
[15] K. Hulka, W. Bleck, K. Papamantellos, Relationships Between Heat Treatment Conditions, Microstructure and Properties of Niobium Microalloyed TRIP Steel, 41st Mechanical Working and Steel Processing Conference, Iron and Steel Society, Baltimore, 1999, pp. 67–88.
[16] D. Kalish, M. Cohen, Structural changes and strengthening in the strain tempering of martensite, Mater. Sci. Eng. 6 (1970) 156–166.
[17] H.K.D.H. Bhadeshia, Bainite in Steels, IoM Communications, London, 2001.
[18] H.C. Chen, H. Era, M. Shimizu, Effect of phosphorus on the formation of retained austenite and mechanical properties in Si-containing low-carbon steel sheet, Metall. Mater. Trans. A, 20A, 1989, pp. 437–445.
[19] K. Sugimoto, N. Usui, M. Kobayashi, S. Hashimoto, Effects of volume fraction and stability of retained austenite on ductility of TRIP-aided dual-phase steels, ISIJ Int. 32 (1992) 1311–1318.
[20] I. Tsukatani, S. Hashimoto, T. Inoue, Effects of silicon and manganese addition on mechanical properties of high-strength hot-rolled sheet steel containing retained austenite, ISIJ Int. 31 (1991) 992–1000.
[21] L. Tosal-Martínez, D. Vanderscheuren, S. Jacobs, S. Vandeputte, Development of a hot-rolled Nb-bearing Si-TRIP steel with excellent fatigue behavior for automotive applications, Steel Res. 72 (2001) 412–415.
[22] Y.R. Cho, S.-K. Kim, H.N. Han, Y.-S. Jin, J.-H. Jung, Development of Hot Rolled High Strength TRIP Steel With Tensile Strength of 780MPa Grade, International Conference on Advanced High Strength Sheet Steels for Automotive Applications, AIST, Winter Park, Colorado, 2004, pp. 71–78.
[23] K. Sugimoto, T. Iida, J. Sakaguchi, T. Kashima, Retained austenite characteristics and tensile properties in a TRIP type bainitic sheet steel, ISIJ Int. 40 (2000) 902–908.
[24] K. Sugimoto, A. Kanda, R. Kikuchi, S. Hashimoto, Kashima, S. Ikeda, Ductility and formability of newly developed high strength low alloy TRIP-aided sheet steels with annealed martensite matrix, ISIJ Int. 42 (2002) 910–915.
[25] K. Sugimoto, Advanced high- and ultra high-strength TRIP-aided steels, Trans. Indian Inst. Met. 59 (2006) 725–738.

[26] G.N. Haidemenopoulos, Austenite stabilization from direct cementite conversion in low-alloy steels, Steel Res. 67 (1996) 93–99.
[27] L. Laquerbe, J. Neutjens, P. Harlet, F. Caroff, P. Cantinieaux, New Processing Route for the Production of Silicon-Free, TRIP Assisted Cold-Rolled and Galvanized Steels, 41st Mechanical Working and Steel Processing Conference, Iron and Steel Society, Baltimore, 1999, pp. 89–99.
[28] K. Eberle, P. Cantinieaux, P. Harlet, New thermomechanical strategies for the production of high strength low alloyed multiphase steel showing a transformation induced plasticity effect, Ironmaking Steelmaking 26 (1999) 176–181.
[29] J.G. Speer, A.M. Streicher, D.K. Matlock, F.C. Rizzo, G. Krauss, Quenching and partitioning: a fundamentally new process to create high strength TRIP sheet microstructures, in: F.B. Damm, M. Merwin (Eds.), Austenite Formation and Decomposition, TMS/ISS, Chicago, Illinois, 2003, pp. 505–522.
[30] F. Spenger, T. Hebesberger, A. Pichler, C. Krempaszky, E.A. Werner, AHSS steel grades: strain hardening and damage as material design criteria, in: J.G. Speer, B. Nelson, R. Pradhan (Eds.), International Conference on New Developments in Advanced High-Strength Sheet Steels, AIST, Orlando, Florida, 2008, pp. 39–49.
[31] R. Kuziak, R. Kawalla, S. Waengler, Advanced high strength steels for automotive industry, Arch. Civ. Mech. Eng. 8 (2008) 103–117.
[32] A. Karelova, C. Krempaszky, E. Werner, T. Hebesberger, A. Pichler, Influence of the edge conditions on the hole expansion property of dual-phase and complex-phase steels, in: J. Fekete, I. Garcia, J. Speer, M. Walp (Eds.), *Advanced High-Strength and Other Specialty Sheet Steel Products for the Automotive Industry, Materials Science & Technology 2007*, AIST, Detroit, Michigan, 2007, pp. 159–169.
[33] W. Bleck, K. Phiu-On, Microalloying of cold-formable multi phase steel grades, Mater. Sci. Forum 500–501 (2005) 97–112.
[34] S. Sarkar, M. Militzer, Microstructure evolution model for hot strip rolling of Nb-Mo microalloyed complex phase steel, Mater. Sci. Technol. 25 (2009) 1134–1146.
[35] C. Mesplont, Phase transformations and microstructure-mechanical properties relations in complex phase high strength steels, Faculty of Engineering and Architecture, University of Ghent (2002).
[36] B. Ehrhardt, T. Gerber, H. Hofmann, T.W. Schaumann, Property related design of advanced cold rolled steels with induced plasticity, Steel Grips 2 (2004) 247–255.
[36a] B. Ehrhardt, T. Gerber and T.W. Schaumann, Approaches to microstructural design of TRIP and TRIP-aided cold rolled high strength steels, International Conference on Advanced High Strength Sheet Steels for Automotive Applications AIST, Winter Park, Colorado, 2004, pp. 39–50.
[37] C. Mesplont, T. Waterschoot, S. Vandeputte, D. Vandershueren, B.C.D. Cooman, *Hot Rolled Bainitic Steels, Thermomechanical Processing of Steels (TMP 2000)*, IoM Communications, London, 2000, pp. 495–504.
[38] Y. Sakuma, O. Matsumura, O. Akisu, Influence of C and annealing temperature on microstructure and mechanical properties of 400°C transformed steels containing retained austenite, ISIJ Int. 31 (1991) 1348–1353.
[39] R. Yi, L. Wen-Yue, W. Shuang, Z. Shuai, Effect of structure pattern on the performance of X80 pipeline steel, J. Iron Steel Res. (International) S1 (2011) 874–877.
[40] Y.D. Morozov, M.Y. Matrosov, S.Y. Nastich, A. Arabei, New generation of high--strength tube steels with a ferrite-bainite structure, Metallurgist 52 (2008) 450–456.

[41] S.J. Kim, C.G. Lee, I. Choi, S. Lee, Effects of heat treatment and alloying elements on the microstructures and mechanical properties of 0.15 wt.% C transformation-induced plasticity-aided cold-rolled steel sheets, Metall. Mater. Trans. A 32A (2001) 505–514.
[42] S.W. Thompson, P.R. Howell, Factors influencing ferrite/pearlite banding and origin of large pearlite nodules in a hypoeutectoid plate steel, Mater. Sci. Technol. 8 (1992) 777–784.
[43] M. Cai, H. Ding, Y. Lee, Z. Tang, J. Zhang, Effects of Si on microstructural evolution and mechanical properties of hot-rolled ferrite and bainite dual phase steels, ISIJ Int. 51 (2011) 476–481.
[44] M. Takahashi, H.K.D.H. Bhadeshia, A model for the microstructure of some advanced bainitic steels, Mater. Trans. Jpn. Inst. Met. 32 (1991) 689–696.
[45] W.S. Owen, The effect of Silicon on the kinetics of tempering, Trans. Am. Soc. Met. 46 (1954) 812–829.
[46] M.D. Meyer, D. Vanderschueren, B.C.D. Cooman, The influence of Al on the Properties of Cold Rolled C-Mn-Si TRIP Steels, 41st Mechanical Working and Steel Processing Conference, Iron and Steel Society, Baltimore, 1999, pp. 265–276.
[47] M.F. Gallagher, J.G. Speer, D.K. Matlock, N.M. Fonstein, Microstructure Development in TRIP-Sheet Steels Containing Si, AI, and P, 44th Mechanical Working and Steel Processing Conference, Iron and Steel Society, Orlando, 2002, pp. 153–172.
[48] Y. Wang, C. Thorning, S. Sridhar, D.M. Haezebrouck, T. Simpson, Surface Oxide Evolution on Cold Rolled C-Mn-Si-Al TRIP Steels, International Conference on Advanced High Strength Sheet Steels for Automotive Applications, AIST, Winter Park, Colorado, 2004, pp. 417–424.
[49] A. Karelova, C. Krempaszky, E. Werner, P. Tsipouridis, T. Hebesberger, A. Pichler, Hole expansion of dual-phase and complex-phase AHS steels—effect of edge conditions, Steel Res. Int. 80 (2009) 71–77.
[50] W. Bleck, A. Frehn, J. Ohlert, *Niobium in Dual Phase and TRIP Steels, International Symposium on Niobium*, TMS, Warrendale, PA, 2001, pp. 727–752.
[51] J. Gao, M. Ichikawa, Development of Coated TRIP Steels at Dofasco, International Conference on Advanced High Strength Sheet Steels for Automotive Applications, AIST, Winter Park, Colorado, 2004, pp. 107–116.
[52] S. Dionne, B. Voyzelle, E. Essadiqi, O. Dremailova, E. Baril, J.R. Mcdermid, et al., Galvanizing of a Hot Rolled C-Mn-Si TRIP Steel, International Conference on Advanced High Strength Sheet Steels for Automotive Applications, AIST, Winter Park, Colorado, 2004, pp. 405–415.
[53] B. Sauerhammer, M. Spiegel, M. Stratmann, Surface Conditioning of High Manganese Content Model Alloys, International Conference on Advanced High Strength Sheet Steels for Automotive Applications, AIST, Winter Park, Colorado, 2004, pp. 425–433.
[54] E.J. Melero, N.H.V. Dijk, L. Zhao, J. Sietsma, S.E. Offerman, J.P. Wright, et al., The effect of aluminium and phosphorus on the stability of individual austenite grains in TRIP steels, Acta Mater. 57 (2009) 533–543.
[55] S. Traint, A. Pichler, M. Blaimschein, B. Röthler, C. Krempaszky, E. Werner, Alloy Design, Processing and Properties of TRIP Steels: A Critical Comparison, International Conference on Advanced High Strength Sheet Steels for Automotive Applications, AIST, Winter Park, Colorado, 2004, pp. 79–98.
[56] S. Jiao, J. Penning, F. Leysen, Y. Houbaert, Deformation and Transformation behavior of Mn-Si and Mn-Al TRIP Steels, 41st Mechanical Working and Steel Processing Conference, Iron and Steel Society, Baltimore, 1999, pp. 499–508.

[57] J. Mahieu, D.V. Dooren, B.C.D. Cooman, Influence of Si, Al and P on the thermodynamics and kinetics of the phase transformations related to continuous galvanization of TRIP aided steels, in: W.J.Pa.E.E.M. Militzer (Ed.), International Symposium on Transformation and Deformation Mechanisms in Advanced High Strength Steels, Canadian Institute of Mining, Metallurgy and Petroleum, Vancouver, British Columbia, Canada, 2003, pp. 21–35.
[58] S. Kim, K. Chin, Effect of Alloying Elements and Manufacturing Conditions on the Mechanical Properties of Cold Rolled TRIP Steels, International Conference on Advanced High Strength Sheet Steels for Automotive Applications, AIST, Winter Park, Colorado, 2004, pp. 99–106.
[59] W. Bleck, S. Kranz, J. Ohlert, K. Papamantellos, Effect of the Testing Temperature on the Mechanical Behavior of Low and High Alloyed Steels Showing the TRIP Effect, 41st Mechanical Working and Steel Processing Conference, Iron and Steel Society, Baltimore, 1999, pp. 295–305.
[60] T. El-Bitar, K.-E. Hensger, C. Klinkenberg, T.E. Weirich, G. Megahed, A. El-Kady, et al., Processing of Nb-microalloyed X52 line pipe steel on Ansdk's CSP plant, in: M. Lamberigts (Ed.), 2nd International Conference on Thermomechanical Processing of Steels (TMP 2004), CRM, Liège Belgium, 2004, pp. 187–194.
[61] C. Klinkenberg, K.E. Hensger, Processing of Niobium microalloyed API grade steel on a thin slab plant, Mater. Sci. Forum 500–501 (2005) 253–260.
[62] M. Gräf, J. Schröder, V. Schwinn, K. Hulka, Production of Large Diameter Pipes Grade X70 with High Toughness Using Acicular Ferrite Microstructures, International Conference on Application and Evaluation of High Grade Linepipes in Hostile Environments, EUROPIPE, Yokohama, Japan, 2002, pp. 1–13.
[63] A.Z. Hanzaki, P.D. Hodgson, S. Yue, *The influence of bainite on retained austenite characteristics in Si-Mn TRIP* steels, ISIJ Int. 35 (1995) 79–85.
[64] E.J. Drewes, E.F. Walker, Niobium bearing steels in the automotive industry, International Symposium on Niobium, TMS, Warrendale, PA, Orlando, 2001, pp. 873–888.
[65] I.B. Timokhina, P.D. Hodgson, E.V. Pereloma, Effect of microstructure on the stability of retained austenite in Transformation-Induced-Plasticity steels, Metall. Mater. Trans. A 35A (2004) 2331–2341.
[66] D. Krizan, L. Barbé, B.C.D. Cooman, J. Antonissen, The influence of micro-alloying elements on the mechanical properties of cold rolled C-Mn-Al-Si-P TRIP steels, in: M. Militzer, W.J. Poole, E. Essadiqi (Eds.), International Symposium on Transformation and Deformation Mechanisms in Advanced High Strength Steels, Canadian Institute of Mining, Metallurgy and Petroleum, Vancouver, British Columbia, Canada, 2003, pp. 395–409.
[67] M. Mukhopadhyay, P. Andrea, Controlling properties of HSLA line pipe steels for oil and gas applications, J. Iron Steel Res. (International) S1 (2011) 701–706.
[68] L.E. Collins, M.J. Godden, J.D. Boyd, Microstructures of linepipe steels, Can. Metall. Q. 22 (1983) 169–179.
[69] V.S. Dwivedi, B.K. Jha, R. Avtar, G. Thomas, Retained austenite and its effect on TRIP aided low carbon high strength multiphase steels, Key Eng. Mater. 84–85 (1993) 406–436.
[70] B.K. Jha, S. Jha, Retention and Deformation Characteristics of Austenite in a TRIP Aided Hot Rolled Steel, 44th Mechanical Working and Steel Processing Conference, Iron and Steel Society, Orlando, 2002, pp. 179–185.
[71] Z. Tang, W. Stumpf, The effect of microstructure and processing variables on the yield to ultimate tensile strength ratio in a Nb-Ti and a Nb-Ti-Mo line pipe steel, Mater. Sci. Eng. A 490 (2008) 391–402.

[72] M. Militzer, F. Fazeli, H.A. Alizamini, Modelling of Microstructure Evolution in Advanced High Strength Steels, 2nd International Conference on Super High Strength Steels, AIM, Peschiera del Garda, 2010, pp. 35–41.
[73] H.N. Han, S.-H. Park, Model for cooling and phase transformation behavior of transformation induced plasticity steel on runout table in hot strip mill, Mater. Sci. Technol. 17 (2001) 721–726.
[74] L. Guo, Modelling of Microstructure in TRIP-Assisted Steels, Department of Ferrous Technology, Graduate Institute of Ferrous Technology, Pohang University of Science and Technology, Pohang, Korea, 2007.
[75] C. Bos, M.G. Mecozzi, J. Sietsma, A microstructure model for recrystallisation and phase transformation during the dual-phase steel annealing cycle, Comput. Mater. Sci. 48 (2010) 692–699.
[76] C. Zheng, D. Raabe, Interaction between recrystallization and phase transformation during intercritical annealing in a cold-rolled dual-phase steel: a cellular automaton model, Acta Mater. 61 (2013) 5504–5517.
[77] M.D. Meyer, J. Mahieu, B.C.D. Cooman, Empirical microstructure prediction method for combined intercritical annealing and bainitic transformation of TRIP steel, Mater. Sci. Technol. 18 (2002) 1121–1132.
[78] M. Mukherjee, S.B. Singh, O.N. Mohanty, Microstructural characterization of TRIP-aided steels, Mater. Sci. Eng. A 486 (2008) 32–37.
[79] W.A. Johnson, R.F. Mehl, Reaction kinetics in processes of nucleation and growth, Trans. Am. Inst. Min. Metall. Eng. 135 (1939) 416–458.
[80] M. Avrami, Kinetics of phase change. III: granulation, phase change and microstructure, J. Chem. Phys. 9 (1941) 177–184.
[81] A.N. Kolmogorov, Statistical theory of crystallization of metals. (in Russian), Izvestia Akademia Nauk SSSR Ser. Mathematica (Izv. Akad. Nauk SSSR, Ser. Mat; Bull. Acad. Sci. USSR. Ser. Math) 1 (1937) 355–359.
[82] E. Scheil, Anlaufzeit der austenitumwandlung, Arch. Eisenhüttenwes 8 (1935) 565–567.
[83] A. Ali, H.K.D.H. Bhadeshia, Nucleation of widmanstätten ferrite, Mater. Sci. Technol. 6 (1990) 781–784.
[84] K.W. Andrews, Empirical formulae for the calculation of some transformation temperatures, J. Iron Steel Inst. 203 (1965) 721–727.
[85] M. Hillert, Diffusion in growth of bainite, Metall. Mater. Trans. A 25A (1994) 1957–1966.
[86] J.W. Christian, Thermodynamics and kinetics of martensite, in: G.B. Olson, M. Cohen (Eds.), International Conference on Martensitic Transformations ICOMAT '79, MIT Press, Cambridge, MA, 1979, pp. 220–234.
[87] C.M. Wayman, H.K.D.H. Bhadeshia, Physical Metallurgy, Elsevier Science Publishers, Amsterdam, 1996.
[88] T. Angel, Formation of martensite in austenitic stainless steels, J. Iron Steel Inst. 177 (1954) 165–174.
[89] H.C. Shin, T.K. Ha, Y.W. Chang, Kinetics of deformation-induced martensitic transformation in a 304 stainless steel, Scr. Mater. 45 (2001) 823–829.
[90] M. Cohen, E.S. Machlin, V.G. Paranjpe, Thermodyamics in Physical Metallurgy, ASM, Cleveland, OH, 1950.
[91] J.R. Patel, M. Cohen, Criterion for the action of applied stress in the martensitic transformation, Acta Metall. 1 (1953) 531–538.
[92] G.B. Olson, M. Cohen, Stress-assisted isothermal martensitic transformation application to TRIP steels, Metall. Trans. A 13A (1982) 1907–1914.
[93] G.B. Olson, M. Cohen, A mechanism for the strain-induced nucleation of martensitic transformations, J. Less-Common Met. 28 (1972) 107–118.

[94] M. Azuma, N. Fujita, M. Takahashi, T. Iung, O. Bouaziz, M. Gouné, et al., A Model for Microstructure Development in High Strength Steels, International Conference on Advanced High Strength Sheet Steels for Automotive Applications Proceedings, AIST, Winter Park, Colorado, 2004, pp. 269–276.
[95] G.B. Olson, M.A. Azrin, Transformation behavior of TRIP steels, Metall. Trans. A 9A (1978) 713–721.
[96] G.F. Bolling, R.H. Richman, *The plastic deformation-transformation of paramagnetic f.c.c.* Fe-Ni-C alloys, Acta Metall. 18 (1970) 673–681.
[97] B.C.D. Cooman, Structure-properties relationship in TRIP steels containing carbide-free bainite, Curr. Opn. Solid State Mater. Sci. 8 (2004) 285–303.
[98] D.C. Ludwigson, J.A. Berger, Plastic behavior of metastable austenitic stainless steels, J. Iron Steel Inst. 192 (1969) 63–69.
[99] G.B. Olson, M. Cohen, Kinetics of strain-induced martensitic nucleation, Metall. Trans. A 6A (1975) 791–795.
[100] O. Matsumura, Y. Sakuma, H. Takechi, TRIP and its kinetics aspects in austempered 0.4C-1.5Si-0.8Mn steel, Scr. Mater. 21 (1987) 1301–1306.
[101] K. Sugimoto, M. Kobayashi, A. Nagasaka, S. Hashimoto, Warm stretch-formability of TRIP-aided dual-phase sheet steels, ISIJ Int. 35 (1995) 1407–1414.
[102] M.Y. Sherif, C.G. Mateo, T. Sourmail, H.K.D.H. Bhadeshia, Stability of retained austenite in TRIP-assisted steels, Mater. Sci. Technol. 20 (2004) 319–322.
[103] M. Mukherjee, S.B. Singh, O.N. Mohanty, Deformation-induced transformation of retained austenite in transformation induced plasticity-aided steels: a thermodynamic model, Metall. Mater. Trans. A 39A (2008) 2319–2328.
[104] M. Mukherjee, T. Bhattacharyya, S.B. Singh, Models for austenite to martensite transformation in TRIP aided steels: a comparative study, Mater. Manufact. Process. 25 (2010) 206–210.
[105] T.V. Rompaey, Q. Fornémont, P.J. Jacques, T. Pardoen, B. Blanpain, P. Wollants, Micromechanical modelling of TRIP steels, Steel Res. 74 (2003) 365–369.
[106] J. Wang, S.V.D. Zwaag, Stabilization mechanisms of retained austenite in transformation-induced plasticity steel, Metall. Mater. Trans. A 32A (2001) 1527–1529.
[107] K. Sugimoto, M. Misu, M. Kobayashi, H. Shirasawa, Effects of second phase morphology on retained austenite morphology and tensile properties in a TRIP-aided dual-phase steel sheet, ISIJ Int. 33 (1993) 775–782.
[108] S.V. Radcliffe, M. Schatz, The effect of high pressure on the martensitic reaction in iron-carbon alloys, Acta Metall. 10 (1962) 201–207.
[109] Y. Pyshmintsev, M.D. Meyer, B.C.D. Cooman, R.A. Savray, V.P. Shveykin, M. Vermeulen, The influence of the stress state on the plasticity of transformation induced plasticity-aided steel, Metall. Mater. Trans. A 33A (2002) 1659–1667.
[110] A. Mark, D. Boyd, E. Essadiqi, The Effects of Microstructure on Stability of Retained Austenite in Multiphase Steels, International Conference on Advanced High Strength Sheet Steels for Automotive Applications, AIST, Winter Park, Colorado, 2004, pp. 307–313.
[111] P.J. Jacques, J. Ladrière, F. Delannay, On the influence of interactions between phases on the mechanical stability of retained austenite in Transformation-Induced Plasticity multiphase Steels, Metall. Mater. Trans. A 32A (2001) 2759–2768.
[112] J. Majta, A. Bator, A.K. Zurek, Influence of the Thermomechanical Treatment on the TRIP Effect, International Conference on Advanced High Strength Sheet Steels for Automotive Applications, AIST, Winter Park, Colorado, 2004, pp. 395–401.
[113] K. Sugimoto, Fracture strength and toughness of ultra high strength TRIP aided steels, Mater. Sci. Technol. 25 (2009) 1108–1117.

[113a] K. Sugimoto, Microalloyed ultra high- and high-strength TRIP-aided sheet steels with different matrix structure, in: D. Bhattacharjee, U.K. Chatterjee, S. Chatterjee, M. Dutta, K.K. Ray (Eds.), Microalloyed Steels: Emerging Technologies and Applications, BESU, IIM, TMS, Kolkata, 2007, pp. 128–138. With permission from Vitasta Publishing
[114] M. Geiger, M. Merklein, M. Kaupper, Investigation of the mechanical behavior of advanced high strength steels under various loading conditions, Int. J. Mater. Form. 1 (2008) 225–228.
[115] P.J. Jacques, Transformation-induced plasticity for high strength formable steels, Curr. Opn. Solid State Mater. Sci. 8 (2004) 259–265.
[116] E. Girault, A. Mertens, P. Jacques, Y. Houbaert, B. Verlinden, J.V. Humbeeck, Comparison of the effects of silicon and aluminum on the tensile behavior of multi-phase TRIP-assisted steels, Scr. Mater. 44 (2001) 885–892.
[117] P. Jacques, Q. Furnèmont, T. Pardoen, F. Delannay, On the role of martensitic transformation on damage and cracking resistance in TRIP assisted multiphase steels, Acta Mater. 49 (2001) 139–152.
[118] T. Kashima, Y. Tsukasa, K. Sugimoto, S. Hashimoto, The Effects of Initial Microstructures Before Annealing on Ductility of TRIP-Aided Steel Sheets With Annealed Matrix, International Conference on Advanced High Strength Sheet Steels for Automotive Applications, AIST, Winter Park, Colorado, 2004, pp. 315–323.
[119] S.B. Lee, J.G. Speer, D.K. Matlock, The Influence of Phase Distributions and Interfaces on Fracture and Formability of High Strength Sheet Steels, International Conference on Advanced High Strength Sheet Steels for Automotive Applications, AIST, Winter Park, Colorado, 2004, pp. 383–394.
[120] K. Sugimoto, A. Nagasaka, M. Kobayashi, S. Hashimoto, Effects of retained austenite parameters on warm stretch-flangeability in TRIP-aided dual-phase sheet steels, ISIJ Int. 39 (1999) 56–63.
[121] I.D. Choi, D.M. Bruce, S.J. Kim, C.G. Lee, S.H. Park, D.K. Matlock, et al., Deformation behavior of low carbon TRIP sheet steels at high strain rates, ISIJ Int. 42 (2002) 1483–1489.
[122] M. Mukherjee, O.N. Mohanty, S. Hashimoto, T. Hojo, K. Sugimoto, Mechanical properties of TRIP-aided Steels with different matrix microstructures, Asia Steel International Conference, The Iron and Steel Institute of Japan, Fukuoka, 2006, pp. 662–666.
[123] L. Samek, B.C.D. Cooman, J.V. Slycken, P. Verleysen, J. Degrieck, Quasi-adiabatic effects during the high strain rate deformation of dispersed-phase ferrous alloys with strain-induced martensitic transformation, in: M. Militzer, W.J. Poole, E. Essadiqi (Eds.), International Symposium on Transformation and Deformation Mechanisms in Advanced High Strength Steels, Canadian Institute of Mining, Metallurgy and Petroleum, Vancouver, British Columbia, Canada, 2003, pp. 77–91.
[124] G.T. Hahn, A model for yielding with special reference to the yield-point phenomena of iron and related bcc metals, Acta Metall. 10 (1962) 727–738.
[125] G.E. Dieter, Mechanical Metallurgy, McGraw Hill Book Company, Singapore, 1988.

[126] R.R. Hilsen, T.E. Fine, G.J. Hansen, Stamping potential of hot-rolled, columbium-bearing high-strength steels, in: M. Korchynsky (Ed.), Microalloying '75, Union Carbide Corporation, New York, 1977, pp. 654–664.
[127] K. Eberle, P. Harlet, P. Cantinieaux, M.V. Populitere, New Thermomechanical Strategies for the Realization of Multiphase Steels Showing a Transformation Induced Plasticity (TRIP) Effect, 40th Mechanical Working and Steel Processing Conference, Iron and Steel Society, Pittsburgh, PA, 1998, pp. 251–258.
[128] S.M. Song, K. Sugimoto, K. Kandaka, A. Futamura, M. Kobayashi, S. Masuda, Effects of prestraining on high-cycle fatigue strength of high strength low alloy TRIP-aided steels, Mater. Sci. Res. Int. 9 (2003) 223–229.

Bake hardening of automotive steels

9

E. Pereloma[1] *and I. Timokhina*[2]
[1]University of Wollongong, Wollongong, NSW, Australia, [2]Deakin University, Geelong, VIC, Australia

9.1 Introduction

Bake hardening (BH) steels are characterized by a significant increase (~10%) in yield strength (YS) after pre-formed component is subjected to the paint baking operation during manufacturing of various automotive body parts (outer door panels, outer skin of hood, fenders, and trunk (boot) lids, etc.) [1]. This increase in strength is termed bake hardening response (BHR) and results in improved dent and crash resistance. Fig. 9.1 illustrates schematically the BHR determination from stress–strain curves before and after paint baking operation, as well as corresponding arrangements of dislocations and solute atoms. Paint baking at 150–200°C is the final operation to give the automotive part a nice aesthetic look. The additional strength improvement resulting from this process also assists the manufacturers in reduction of the thickness of the sheet used, thus decreasing the car weight and environmental pollution.

BH is fundamentally a strain-aging phenomenon with nitrogen and carbon atoms playing the key role. However, even the presence of 5–10 ppm of nitrogen could lead to unwanted severe strain-aging effects during room temperature storage of cold-rolled sheet [2–5], thus the low-carbon steels are typically Al-killed to remove N from the solid solution in the form of AlN and BHR is controlled by the amount of solute C present. Conventional BH steels were developed in 1980s following the need for higher strength cold-rolled sheet with a good formability [6]. With development of degassing technologies, the carbon content was reduced from >0.05 wt.% C to 0.01–0.03 wt.% C in extra-low carbon steels to <50 ppm (0.001 ± 0.004%C) in ultra-low carbon (ULC) steels to <30 ppm in super-ultra-low carbon (SULC) steels [1]. Addition of small amounts of carbide-forming elements (Ti,Nb) to SULC steels could further reduce the concentration of C to 15–20 ppm, which opened the possibility of production of superformable, bake-hardenable steels [1]. However, it was shown that carbon content <10 ppm degrades the steel ductility [7] and that the optimum amount of carbon to prevent ambient temperature strain aging and achieve the noticeable BHR is 15–25 ppm. All these steels exhibit a single phase ferritic microstructure and could achieve the maximum tensile strength of < 600 MPa.

Further demand for higher strength steels used in automotive applications led to the development of so called Advanced High Strength Steels (AHSS) among which are dual-phase (DP), transformation-induced plasticity (TRIP), complex phase, and

Automotive Steels. DOI: http://dx.doi.org/10.1016/B978-0-08-100638-2.00009-2

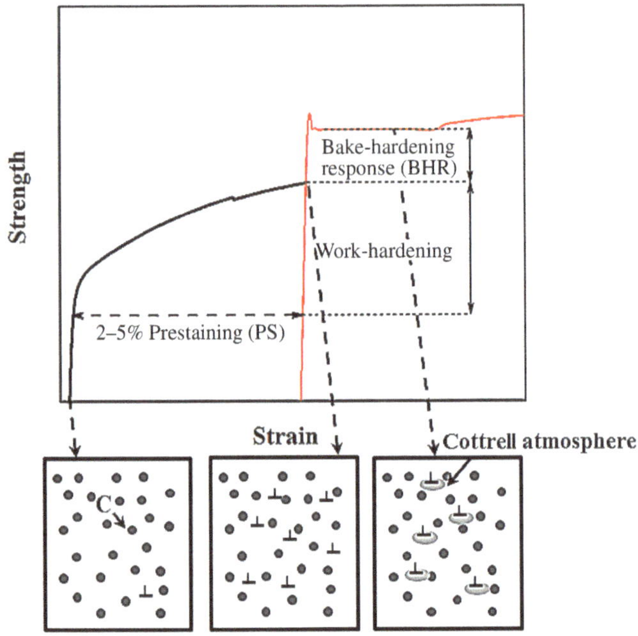

Figure 9.1 Schematic of bake hardening response determination and corresponding distribution of carbon atoms in ultra-low carbon ferritic steels.

martensitic steels. Although these steels are mostly used for structural parts (door beams, bumper reinforcements, etc.) a significant body of research has been carried out to evaluate the BH potential in these steels [8–14]. Up to 100 MPa BHR was reported for AHSS after 2–10% pre-straining [9,10]. Despite the observed strengthening due to BH, there were reports questioning the possible premature failure of BH components made from martensitic steels.

In this overview, the mechanism of BH and factors affecting it will be addressed for both conventional steels and AHSS. Furthermore, the steel processing parameters affecting the BHR will be analyzed and discussed.

9.2 Mechanisms of bake hardening response

During forming of the part, typically ~2–5% strain (locally the amount could be even 10–20%) is accumulated, which is due to the dislocation activity leading to the increased dislocation density arising from the fresh dislocations formation and interaction of existing ones. So-called Snoek effect [15] could lead to the mechanical response of the material. This effect is related to the preferential occupation of those interstitial sites by atoms under elastic stress, which are oriented along stress direction. However, on load removal the atoms are again distributed randomly. Snoek process is a fast one involving atom jumps between the neighboring sites.

Although this stress-induced ordering of interstitial atoms is the origin of internal friction, it cannot be detected at temperatures above −12°C [4] due to the measurement speed limitations of the currently available techniques. Thus, its operation during BHR could only be presumed.

Elsen and Hougardy [16] have suggested two stages of BH occurring in ferrite at elevated temperatures with time:

Stage I—Formation of so-called Cottrell atmospheres at dislocations; and
Stage II—Precipitation of coherent carbides.

During stage I, strain aging mechanism originally proposed by Cottrell and Bilby [17] operates. It involves the diffusion of interstitial atoms to freshly-formed dislocations during previous deformation. These interstitials form atmospheres around the core of dislocations and pin them. According to Cottrell-Bilby model [17], the extent of pinning varies linearly with $t^{2/3}$:

$$\frac{N(t)}{N_s} = n_0 b^3 \left(\frac{\pi}{2}\right)^{1/3} \left(\frac{ADt}{kT}\right)^{2/3} \tag{9.1}$$

where $N(t)/N_s$ is the fraction of pinning complete, n_0 is the matrix solute number density, b is Burgers vector, t is time, k is Boltzmann's constant, T is temperature in Kelvin, D is the carbon diffusivity in the matrix and A is the "interaction parameter," which is a measure of the binding energy of the solute atom to the dislocation. Although in the Cottrell-Bilby model N_s was assumed to be one carbon atom per atomic plane threaded by a dislocation, more recent experimental evidence using Atom Probe Tomography (APT) [18–20] puts this number in the vicinity of 12–15. The discrepancy between the atom probe data and Cottrell-Bilby model predictions is due to the inability of the model to take into account saturation of the dislocation with interstitial atoms and the retardation of diffusion with time due to the depletion of solute in the vicinity of atmosphere and the reduction of the driving force due to the relief of the strain field. Therefore, this model is valid only for the early stages of aging (up to 30% completion). The model proposed by Harper [21] (Eq. 9.2) allows for the saturation effect:

$$W = \frac{n_t}{n_0} = 1 - \exp\left[-3L\left[\frac{\pi}{2}\right]^{1/3}\left(\frac{ADt}{kT}\right)^{2/3}\right] \tag{9.2}$$

where W is the fraction of atoms segregated in time t to the dislocations with the total length L per unit volume. Harper's model found a good agreement with experiment for solute depletion up to 0.9, but could not account for the back diffusion of interstitials from the atmosphere [22]. Its application to experimental data gives much higher values of 50–100 atoms per atomic plane threaded by the dislocation [22], than those assumed by Cottrell and Bilby. Similarly, the modified model of Cochardt et al. [23] predicts that the saturation number of carbon atoms in an atmosphere of radius 30b (b is the Burgers vector) will be 15 per atomic plane. Cochardt et al. [23]

stated that within a distance 1b the carbon atoms are more stable within the atmosphere than in cementite, as their maximum interaction energy is 0.75 eV, whereas the binding energy of C in cementite is only ~0.5 eV [3].

The generalized form of Eq. (9.2) is often used for characterization of aging kinetics [24]:

$$\ln(1 - W) = -\left(\frac{t}{\tau}\right)^n = -(kt)^n \tag{9.3}$$

where τ is a temperature-dependent relaxation constant obeying an Arrhenius-type relation. Using τ, the activation enthalpy of the aging process can be derived. The kinetic parameter n is determined from fitting experimental data to Eq. (9.3). Any deviation in the value of n from 2/3 is generally considered as indication of a change in the mechanism of carbon segregation on dislocations [5,25]. According to De et al. [24], Harper's model could be successfully applied for description of aging kinetics until the completion of atmospheres formation in BH ULC steels, which have a very low carbon content. De et al. [24] also proposed another equation (Eq. 9.4) for aging kinetics determination based on the modified Hartley's model [26]:

$$\frac{\Delta\sigma}{\Delta\sigma_{\text{atm}}} = K_1 + K_2\left(\frac{Dt}{T}\right)^{2/3} \tag{9.4}$$

where $\Delta\sigma$ is the difference between the upper yield stress after pre-straining and aging and the flow stress at the end of pre-straining, $\Delta\sigma_{\text{atm}}$ is the maximum increase in the YS for atmosphere saturation, t is the aging time, T is the aging temperature, D is the diffusion coefficient, while K_1 and K_2 are the constants for the constant test conditions.

It should be noted that C could segregate not only to the statistically-stored dislocations within the grains but also to the dislocations forming the grain boundaries or other planar defects [27–29]. Although the process at grain boundaries is much slower than saturation of atmospheres at dislocations due to planar, rather than cylindrical, source-sink geometry, a rough estimate of the time required to achieve 0.5 carbon atoms per atomic plane threaded by a dislocation in a high angle boundary at 94°C in a low-carbon steel is approximately 6000s, which is sufficient to restrict the generation of new dislocations from the grain boundaries [30]. Taking into account that BH is carried out at nearly twice as high temperature, less time will be required for this process. Saturation of carbon at grain boundaries could be even more feasible during repeated paint baking process commonly used in the production of luxury cars.

If all of the sites allowing strong interaction of dislocations with carbon atoms are occupied and atmosphere is condensed, then according to Refs. [4,16,31–33] stage II begins by nucleation of coherent carbides at vacancies or dislocations. Activation energy for such precipitation was calculated to be $74 \pm 0.9\ \text{kJ mol}^{-1}$ [34], which is lower than the activation energy for the formation of Cottrell atmosphere ($87.1 \pm 10\ \text{kJ mol}^{-1}$) [35]. Taking into account that (i) the activation energy for transition ε-carbide formation is 70.4 ± 2.9 kJ mol^{-1} [36], which is close to the one predicted by De et al. [34]; (ii) it has better lattice matching with ferrite than cementite [5], and (iii) the formation of ε-carbide is preferred at temperatures below 250°C along elastically soft <100> direction having the minimum activation barrier to precipitation [5,37], ε-carbide formation would be expected during baking stage. However, only limited evidence to date exists on the account of carbides precipitation during BH. Only advances in APT allowed elucidate carbon atoms clustering along the dislocation lines and formation of carbides of close to Fe_3C composition (21 at.% < 25 at.%) at dislocation nodes in ferrite of TRIP steel (Fig. 9.2) [38].

Based on the above, the sequence of events taking place during long holding time could be summarized as follows (Fig. 9.3): formation of atmospheres at dislocations → formation of carbon clusters in ferrite matrix → segregation of carbon to planar defects → formation of carbides. However, in order to achieve a required BH response the following conditions should be satisfied [39,40]: (i) although the presence of freshly formed, mobile dislocations is essential, the rate of dislocation recovery during baking should be slow preventing the softening and (ii) a sufficient concentration of *mobile* solutes for pinning of these dislocations.

However, contrarily to conventional high strength steels (HSS), the AHSS contain other than ferrite phases, such as bainite and martensite. Thus, during the BH of these steel grades, changes in bainite and/or martensite occur, which also contribute to BHR. Bleck et al. [41] suggested tempering of martensite in DP and TRIP steels to be another mechanism of BH in these grades. The tempering of bainite with formation of $Fe_{32}C4$ or $Fe_4C_{0.63}$ carbides was reported [42], but no presence

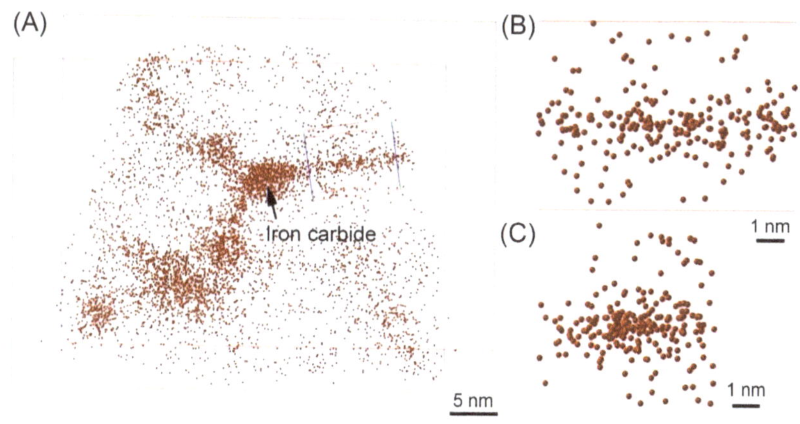

Figure 9.2 Atom probe maps of carbon segregation to dislocations and carbide formation in intercritically annealed TRIP steel with 0.2 wt.% C after 4% pre-straining and 30 min bake hardening at 180°C.

Source: Reprinted from E. Pereloma, H. Beladi, L. Zhang, I. Timokhina, Understanding the behavior of advanced high-strength steels using atom probe tomography. Metall. Mater. Trans. A 43 (2011) 3958–3971, with copyright permission by Springer.

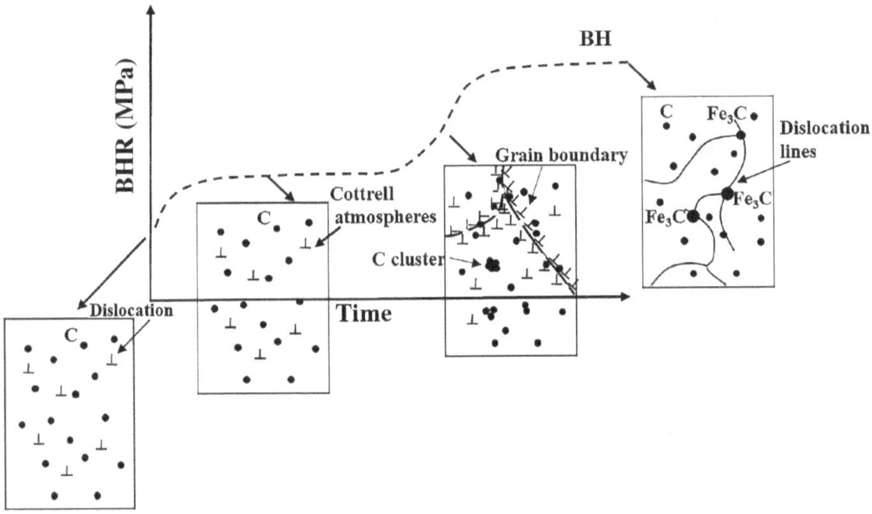

Figure 9.3 Schematic representation of bake hardening mechanisms. BHR is bake hardening response.

of ε-carbide, which however was pronounced in tempered martensite of TRIP steels [13,43]. Three different strengthening stages of BH in the multi-phase steels have been summarized as follows [12]: (i) Cottrell atmosphere formation around mobile dislocations in ferrite, (ii) the precipitation of low temperature carbides in ferrite, and (iii) formation of clusters and particles in martensite/bainite. In the presence of strong carbide-forming elements, such as Nb and Mo, their clustering with C and formation of fine precipitates in ferrite and bainite were also reported [13]. This will be addressed in more detail in Section 9.4 below.

9.3 Factors affecting bake hardening response

9.3.1 Composition

As already mentioned, nitrogen is eliminated to prevent room temperature aging and provide the automotive parts a "shelf life" of three months in storage. Thus, carbon is the main element controlling the BHR. However, there is no direct correlation between the concentration of carbon in steel and BHR, i.e., BHR increases with carbon content and reaches its maximum at a certain carbon level, then there is no further effect of carbon solute increase on BHR (Fig. 9.4). It was reported by Rubianes and Zimmer [44] that steels with <3 ppm carbon, do not show BHR and are stable during prolonged storage. On the other hand, when >7 ppm carbon present there is a good BHR but it is independent of carbon concentration and the time of room temperature storage is significantly reduced. The range of 3–7 ppm carbon provides a good combination of BHR and "shelf life". Wilson and Russel [4] and Hanai et al. [45] concluded that 5 ppm of carbon are required to complete the

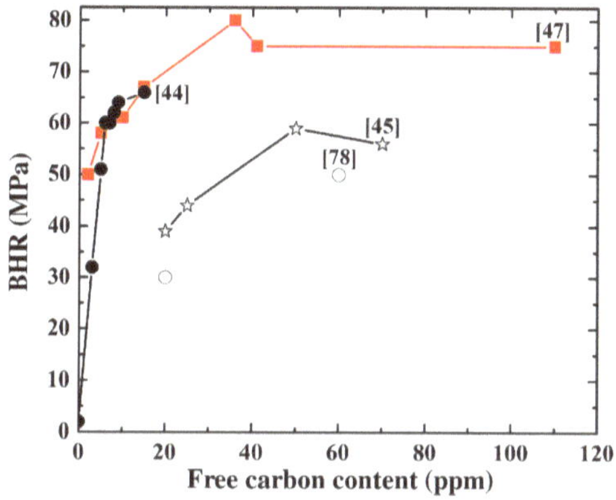

Figure 9.4 Effect of carbon content on bake hardening response (BHR).

formation of Cottrell atmospheres. On the other hand, De et al. [24] assumed the atmosphere saturation when one carbon atom is present per plane threaded by the dislocation and showed that 1 ppm of carbon is enough to saturate all dislocations in ferrite with dislocation density of 10^{14} m^{-2}. Earlier Kalish and Cohen [46] showed that in a steel containing 10 ppm all carbon could be trapped at dislocations with density 10^{14} m^{-2}. There have been different reports on much higher amounts of carbon corresponding to the maximum BHR: (i) Hanai et al. [45] have found that in Al-killed low carbon (0.04 wt.%) steel the BH response reaches its maximum at 20 ppm solute carbon; (ii) for a constant grain size of 17 μm, increase from 0 to 35 ppm results in BHR raising from 40 to 80 MPa with BHR remaining independent of carbon content at 70 MPa with $\geq$40 ppm carbon [47] (Fig. 9.4). The discrepancy in these amounts with the ones reported by Rubianes and Zimmer [44] lies in the measurement techniques used; the latter utilized internal friction which is unable to account for the carbon located at grain boundaries [48]. It is generally an agreement that the saturation stage in BHR is associated with the completion of Cottrell atmospheres formation and the absence of stage II [49], which is typical for the steels with a very low carbon content, e.g., ULC-BH steels.

Presence of Nb, Ti, Mo, and V could have effect on dissolution of carbides at annealing temperature, thus the amount of carbon in solute. However, each carbide-forming element will exhibit different degrees of the effect depending on the carbide dissolution temperatures and the strength of binding energy to carbon. For example, after annealing at 750°C in steels alloyed with vanadium, 20 ppm carbon could be released into solid solution. In contrast, in the presence of titanium and niobium the amount of free carbon in solution will be increased only by 1–2 ppm [50]. It was suggested that in order for any carbon to escape into solid solution during annealing the Nb/C ratio should exceed 0.3 [51].

In addition, re-precipitation of carbides during thermo-mechanical processing will influence the amount of available carbon for formation of atmospheres at dislocations. Contrarily to iron carbides which have a lower binding energy between Fe and C (~0.5 eV) than that of carbon to dislocations (0.75 eV), affinity between carbon and these elements is strong, with for example the binding energy between carbon and Nb in NbC being 2.3 eV [52]. This explains the findings that with increasing Nb/C ratio [53] and Ti content [54], the BHR in ULC steels decreases.

In Ti-stabilized steels, the amounts of S and Mn need to be controlled in order to prevent formation of $Ti_4C_2S_2$ and TiS (the latter serves as heterogeneous nucleation sites for TiC formation, which should be avoided) assisting the required level of carbon remain in solid solution leading to the BHR [1,55,56]. Furthermore, the deterioration of BHR with increasing Mn content [45,57] is linked with the preferential formation of Mn-C dipoles [58–60], which reduces the amount of free carbon in solution. Similarly, the deterioration of BHR with increasing Mo and Cr content would also be expected due to the same reason of dipoles formation, which makes the diffusion of C atoms more difficult.

On the other hand, an increase in BHR is observed with addition of P [57,61], which could also participate in locking the dislocations as shown by APT studies [62,63]. Similarly, the addition of Si also increases BHR via increase in available solute carbon [45].

9.3.2 Effect of processing parameters

The main features of the steel having the pronounced effect on BHR are the amount of solute carbon present and the density of dislocations. These are affected by the prior to BH processing stages, which are discussed below.

9.3.2.1 Annealing temperature

While some researchers found that increasing the annealing temperature reduces the corresponding strength increment after BH [45,64], others found that an increase in the annealing temperature effectively increases the BHR [53,54,65,66]. Fig. 9.5 shows some data from the literature for different steel grades. Unfortunately, the values for austenite grain sizes after annealing at various temperatures were not provided in any of the papers; the effect was mostly considered based on the amount of soluble carbon present. As was shown in Refs. [54,66], the amount of carbon solute in ferrite increases with annealing temperature until the temperature for the full carbides or cementite dissolution is reached. An increase in BHR by ~20 MPa with an increase of temperature from 750°C to 850°C is seen for interstitial free (IF) steel, whereas for extra deep drawable (EDD) steel increase in temperature from 500°C to 750°C resulted in >45 MPa BHR rise (Fig. 9.5). However, with increase in annealing temperature the grain size also increases and the Sv (the ratio of grain boundaries area to the total area) decreases, which may affect the BHR as discussed in Section 9.3.3. This may explain the contradiction in the data with respect to the effect of annealing temperature on BHR.

In addition to the effect of annealing temperature on grain size, the chemical composition of steel also affects the choice of annealing temperature for the achievement of the required BHR. It was shown that in Nb-stabilized steels with Nb/C ratio in the range of 0.7–2.0, an increase in annealing temperature from

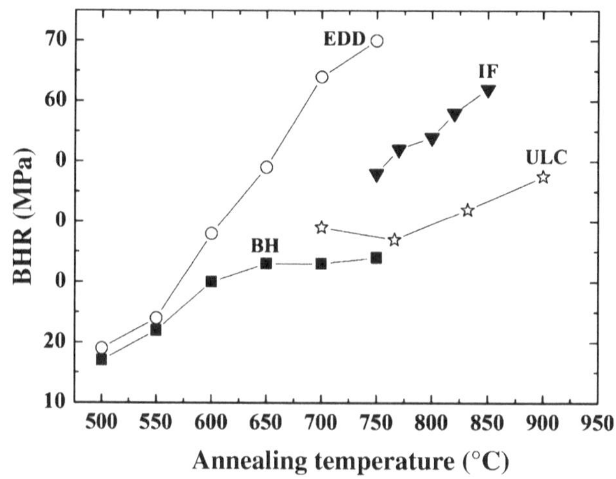

Figure 9.5 Effect of annealing temperature on bake hardening response (BHR) for different steel grades.
Source: Data is taken for EDD, BH, and IF steels from S. Das, Bake Hardening in Low and Medium Carbon Steels. Indian Institute of Technology, Kharagpur 2012 and for ULC steel from L.-J. Chiang, K.-C. Yang, I.-C. Hsiao, Effects of annealing conditions on bake hardenability for ULC steels. 2011: China Steel Technical Report. 1–6.

750°C to 950°C resulted in further BHR increase by ~30–50 MPa [51,67]. Contrarily, in understabilized steel there was no effect on BHR after increase of soaking temperature to 900°C and a significant effect was observed only when the Nb/C ratio was ~2.1 [68]. In Ti-stabilized steel without any excess of Ti, the BHR increased from ~10 MPa to 40 MPa as a result of annealing temperature increase from 800°C to 900°C [67]. This BHR is lower than that in Nb-stabilized steels. However, in fully Ti-Nb-stabilized steels, no effect of annealing temperature was detected, whereas a 10 MPa BHR increase was achieved with Nb/C ratio of 0.55 [69]. A considerable decrease in BHR was recorded after holding Nb-containing steels during processing at 700–800° or 300°C. The former was explained by the formation of NbC, whereas the latter one by Fe_3C, both reducing the amount of carbon available during BH [70]. As was shown by Storozheva et al. [71] the increase in the annealing temperature from 770°C to 890°C is accompanied by a decline in the YS for the steel with a hypostoichiometric composition (Nb/C ~0.5), whereas for the steel with a hyperstoichiometric composition (Nb/C ~ 1.3) the YS rises.

9.3.2.2 Cooling rate

Cooling rate after annealing affects: (i) the ability of carbides to re-precipitate; (ii) carbon diffusion into grain boundaries, and (iii) further austenite grain growth and ferrite grain size. In multi-phase steels, the volume fractions of phases present are also controlled by the cooling rates. As was shown by Chiang et al. [66], the variation in slow cooling rates ($<20°Cs^{-1}$) has a marginal effect on BHR due to a very small difference in solute carbon content. However, still the trend indicated that an increase in cooling rate results in a higher BHR due to more carbon trapped in the solid solution. When a larger variation in cooling rates (10–550°Cs^{-1}) was applied to ULC steel [24,34], a ~20 MPa higher BHR was achieved after cooling at 550°Cs^{-1}compared to the one after cooling at 10°Cs^{-1}. Furthermore, the BHR and completion of stage I were noticeable at shorter baking times. These observations correlate with more solute carbon available for dislocations locking during paint baking after faster cooling rates from annealing temperatures.

9.3.2.3 Effect of pre-straining

During forming, different autoparts undergo various amounts of deformation, however the level of straining is typically low. Furthermore, even in one part this strain may vary (2–20%), which is common for the stamped parts [72]. Typically, 2–5% pre-straining by tensile test is used to simulate the effect of pre-forming on BHR. However, in practice the state of deformation (tensile, compression, shear) will depend on the application and shape of the component. The amount of pre-straining has a direct effect on the number density of dislocations, thus influencing the BHR for the same steel composition. It should be also noted that the effect of work hardening from forming operations may be more pronounced than the effect from the subsequent BH, as was found by Durrenberger et al. [73] for a top-hat section manufactured from two different AHSS subjected to the BH.

The degree of pre-straining exerts different influence on stage I and II of BH, which is also somewhat dependent not only on steel composition but also on the BH temperature due to the diffusion-controlled nature of the process. In this regard, temperature of pre-straining also affects the subsequent BHR [74]; the higher the pre-straining temperature, the more carbon will diffuse to dislocations resulting in solute depletion and reduced BHR.

As seen from Fig. 9.6, for the same steel composition and constant BH temperature and time, the BHR is either independent of pre-straining [49,75] or decreases monotonically with strain [9,16,47,74]. These trends are applicable to both single and multi-phase steels. When the maximum BHR corresponds to the completion of carbon atmosphere formation at dislocations, then the BHR is independent of strain [75]. This is because stage I is independent of strain, if there is a sufficient carbon content to lock all the dislocations [16]. However, if stage II contributes to BHR, the higher amount of carbon is consumed in stage I, the less amount will be available for stage II, which may lead to the reduction of BHR associated with stage II (and as a result the total BHR) if there is insufficient total carbon concentration for completion of both stages. If all carbon is consumed for completion of both stages, then further increase in pre-straining will have no effect on the BHR. This was clearly demonstrated by Elsen and Hougardy [16] for the steel with 5 ppm carbon (Fig. 9.6). However, if there is a sufficient concentration of carbon present to lock all the dislocations formed after different pre-strains and time is enough to accomplish this, then BHR will increase with an increase in pre-strain [76,77].

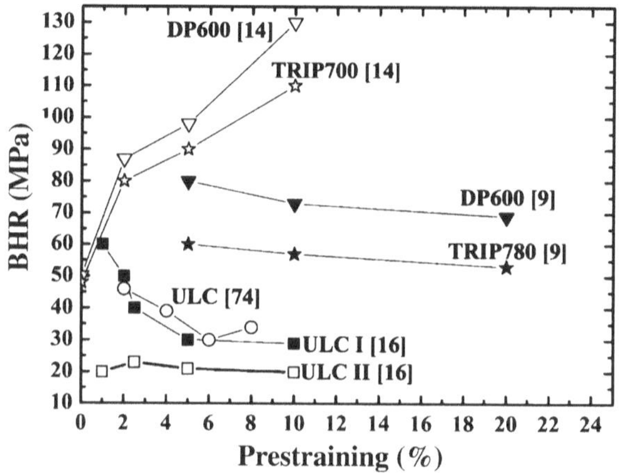

Figure 9.6 Effect of pre-straining on bake hardening response (BHR). ULC-I refers to completion of stage I of bake hardening (BH) only, whereas ULC-II refers to the BH with occurrence of stage II according to Elsen and Hougardy [16].

9.3.2.4 Effect of strain path

Commonly, temper rolling is the last operation before the BH of steels. However, it may produce high local strains exceeding the macroscopic extensions applied. In these local areas the degree of stress relaxation is expected to be much higher than that in other areas of sheet, thus leading to the reduction in BHR [1]. For the same macroscopic strain, temper rolling should produce a larger decrease in BH than tensile strain due to this strain inhomogeneity. This needs to be kept in mind when tensile pre-straining is used in laboratory simulations of BHR.

The similarities (>5% strain) and differences (<5% strain) in the deformation substructure formation and relaxation as a function of strain after tensile loading and stamping were also highlighted for DP and TRIP steels by Timokhina et al. [72] (Fig. 9.7). In this regard, it is important to remember that the change in strain path before and after BH could lead to the different effect on BHR. As was shown by Ballarin et al. [78] for two BH steels: (i) the pinning of dislocations by carbon atoms is efficient and leads to the YS increase when both strains are applied in the same direction; (ii) when the second strain is applied in the reverse direction, the slip is reactivated in the opposite direction on the same slip planes, but the movement of dislocations is restricted due to the pinning leading to the YS increase; and (iii) when the second strain is applied in approximately orthogonal direction to the first one, the new slip systems are activated, which were latent during the pre-straining and thus are not subjected to carbon pinning. In this case, there is no influence on YS, but the increase in the flow stress soon after yielding was observed due to the strengthening effect from atmospheres formation at forest dislocations.

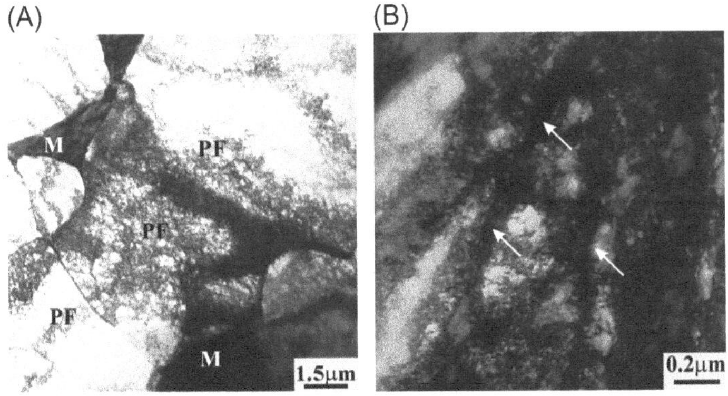

Figure 9.7 Difference in deformation substructures formed in DP steel subjected to (A) Five percent equivalent tensile strain during stamping and (B) After five percent uniaxial tensile loading. Arrows indicate the microband walls. M is martensite and PF is polygonal ferrite. *Source*: Reprinted from I. Timokhina, E. Pereloma, P. Hodgson, The formation of complex microstructures after different deformation modes in advanced high-strength steels. Metall. Mater. Trans. A 45(10) (2014) 4247–4256, with copyright permission from Springer.

9.3.3 Effect of grain size

In order to evaluate the effect of grain size on BHR, two main characteristics need to be considered: (i) the distributions of carbon atoms and dislocations in grain interior and (ii) near/at grain boundaries. To date, the data reported in the literature are controversial. On one hand, it was concluded that an increase in grain size is accompanied by the increase in carbon content intragranularly resulting in cementite formation during cooling and diminished BHR [47,67,79]. Similarly, it was found in Refs. [40,45,80] that for a given carbon content the BHR increases with a decrease in grain size (Fig. 9.8) and that the effect becomes more pronounced with an increase in carbon content.

On the other hand, Takahashi et al. [81] did not find any dependency of YS after BH on the grain size, whereas De et al. [82] showed that a decrease in grain size leads to a decrease in the matrix carbon content and hence BHR in the ULC-BH steels after continuous annealing. This behavior was attributed to the role of grain boundaries as sinks for carbon atoms, which results in reduction of free carbon available intragranularly for dislocations locking. However, De et al. [82] found that for ULC-BH steels there are a critical grain size and a specific pre-straining for which BHR increases with the grain size reduction. This effect could be explained, as suggested by Sakata et al. [67], by the release of carbon atoms from the boundaries and their contribution to the BH. However, it is not clear what the driving force would be for carbon release, unless strong carbide-forming elements with high affinity to carbon, like Nb, are present in the matrix. Although even in this case, carbon would first preferentially react with such elements and not segregate to the dislocations and grain boundaries [52]. Takahashi et al. [81] also argued that due to the strong binding energy of carbon atoms to grain boundaries, the BH

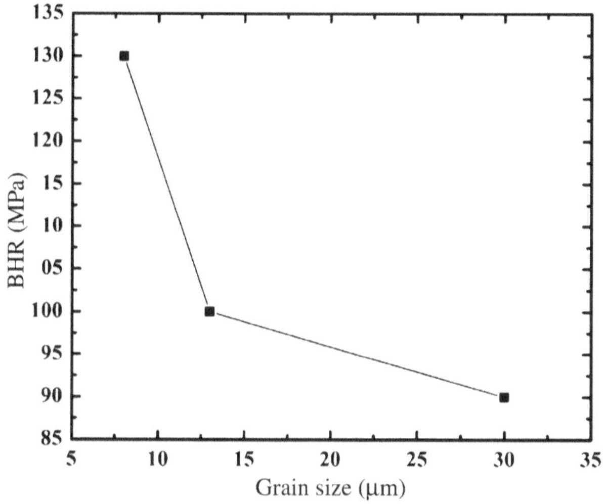

Figure 9.8 Effect of grain size on the yield strength increment after bake hardening. *Source*: Data is taken from V. Ballarin, M. Soler, A. Perlade, X. Lemoine, S. Forest, Mechanisms and modeling of bake-hardening steels: part I. Uniaxial tension. Metall. Mater. Trans. A 40 (6) (2009) 1367–1374.

temperatures of ~170°C is too low for their escape. Similarly, Storojeva et al. [71] stated that the carbon atoms segregated to the grain boundaries are more stable and require higher temperatures for strain aging compared to those inside the grain interior. As was already pointed out in Section 9.2, the time required to saturate grain boundaries with carbon is much longer than that for the formation of Cottrell atmospheres in the grain interior. This is due to the planar geometry of the boundaries. In addition, carbon will be required to diffuse from larger distances in order to achieve the saturation level at grain boundaries comparable to that at statistically-stored dislocations within grain interior. With grain size reduction two opposite factors operate: the area fraction of grain boundaries increases and thus requires more carbon for its saturation, but the diffusion distances from the center to the boundaries are reduced. In the case of sufficiently high BH temperature and sufficient time, and when there is enough carbon available for grain boundaries saturation, BHR should increase with decrease in grain size. The reason for this is inhibited generation of dislocations from grain boundaries due to pinning by carbon segregation, and the increased stress necessary to eject dislocations from the boundary [30].

9.3.4 Effect of bake hardening temperature and time

Temperature at which BH is carried out affects both the rate of dislocations re-arrangement (recovery) and the diffusivity of carbon. Generally, with an increase in BH temperature, the BHR increases (Fig. 9.9). At low temperatures (<120°C), both these processes are slow and it was reported that incubation time up to 30 minutes

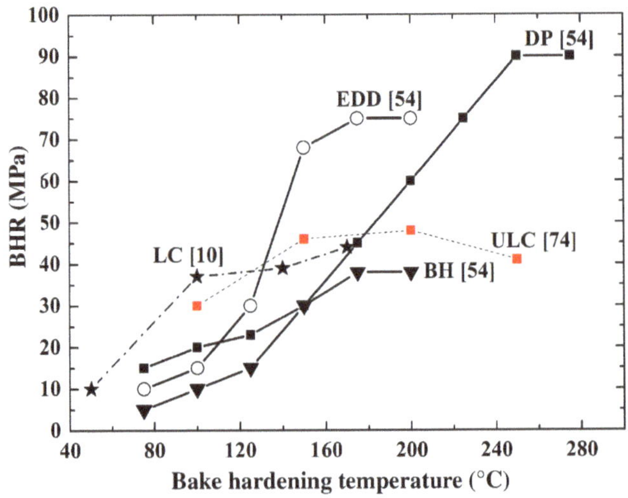

Figure 9.9 Effect of bake hardening temperature on yield strength increment.

might be required for obtaining of noticeable BHR in Ti-stabilized ULC steels [49,75], which is also associated with the time dependence for completion of stage I as was found for BH and ULC steels [16]. At higher temperatures, due to high carbon atoms diffusion rates the stage I is completed in a very short time, which is becoming shorter with BH temperature increase [16,49,83,84]. If holding time is shorter than that required for the completion of stage II, then the BHR will be reduced and the lower BHR will be the later stage II starts, i.e., at lower BH temperature [41,80]. However, if time is long enough for completion of precipitation in stage II, then the maximum BHR is independent of temperature [14,41,49]. Although, further coarsening of precipitates and their number density decrease at higher temperatures or longer times may lead to the lowering of BHR. It should be also noted that longer times will be required to achieve a similar BHR in the presence of alloying elements, which could form dipoles with carbon in ferrite (such as Mn, Mo, Cr), thus increasing the activation energy for carbon diffusion.

9.4 Bake hardening of multi-phase steels

9.4.1 DP and TRIP steels

As mentioned in Sections 9.1 and 9.2, BH could be also applied to DP and TRIP steels for their YS improvement while maintaining the thickness of steel sheet used in the final body structure or providing the pathway for thickness reduction. The microstructure of DP steels contains 40–80 vol.% of soft and ductile ferrite phase with dispersed second phase regions of predominantly hard martensite. TRIP steels microstructure also contains a large fraction of soft ferrite in addition to carbide-free

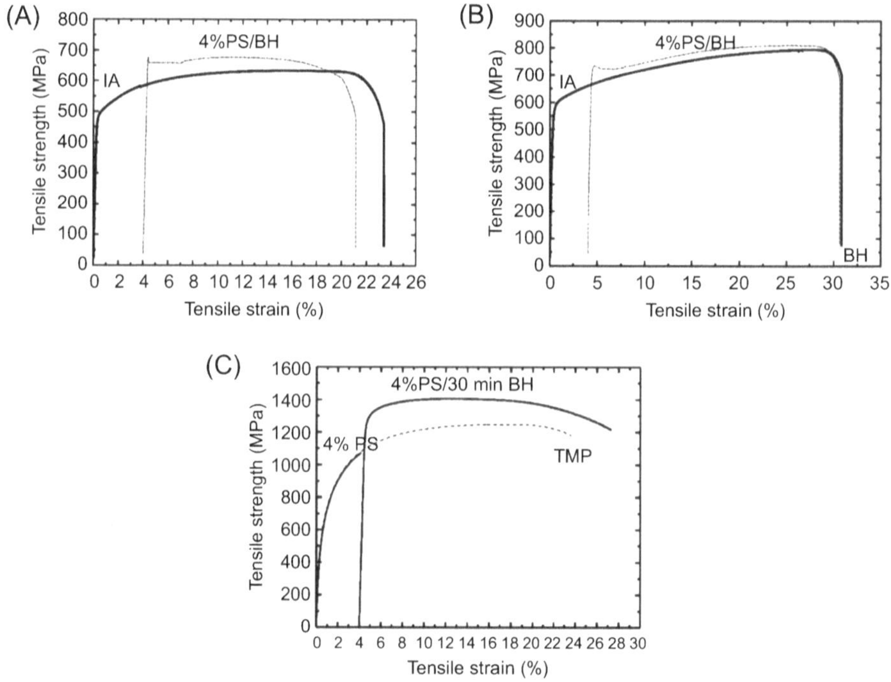

Figure 9.10 Representative stress–strain curves of (A) DP, (B) TRIP after IA, and (C) TRIP after TMP steels.
Source: Reprinted from I.B. Timokhina, P.D. Hodgson, S.P. Ringer, R.K. Zheng, E.V. Pereloma, Effect of bake hardening on the structure-property relationship of multiphase steels for the automotive industry. Steel Research Int. 80 (7) (2009) 507–514, with copyright permission by Springer.

bainite and small amounts of retained austenite (5–20 vol.%) and martensite. Both steel grades are relatively low cost as their compositions are low carbon, low alloyed; however the typical carbon content in TRIP steels (0.15–0.2 wt.%) is more than double that in DP ones (0.04–0.09 wt.%). One of the important characteristics of these steels is high rate of work hardening. Their complex microstructure can increase the work-hardening rate in the additional ways to the single phase ferrite behavior, such as through the formation of mobile dislocations in the soft ferrite matrix in the vicinity of hard phase in DP steels and the transformation of retained austenite to martensite during forming in TRIP steels [85–87].

There are two main routes for the production of DP and TRIP steels: (i) an intercritical annealing after cold rolling (IA) and (ii) thermomechanical processing (TMP). The maximum BHR has been observed in the intercritically annealed TRIP and DP steels after 2–10% of pre-straining in an interrupted tensile test [1,86,88]. The YS increase after pre-straining and 30 minutes holding at 175°C was accompanied by the return of the yield point, a slight increase in the tensile strength and a decrease in the elongation for both DP600 and TRIP780 IA steels (Fig. 9.10).

As seen from Fig. 9.6, the BHR for both steels decreased with increase in the amount of pre-straining. However, for all levels of pre-strain (5–20%), the DP steel showed a higher BHR compared to the TRIP steel. Similarly, Ramazani et al. [14] reported an increase in BHR for both DP600 and TRIP700 steels after baking treatment for 20 minutes at 170 or 220°C after pre-straining to 2–10% with DP600 achieving a higher BHR than TRIP steel (Fig. 9.6). It is also worth noting that TRIP700 steel maintains continuous increase in BHR during long holding times, whereas BHR decreases in DP600 steel due to over-aging. The differences in the observed DP and TRIP steels behaviors in Refs. [14] and [9] are due to: (i) a higher carbon content of 0.22 wt.% in TRIP700 steel compared to 0.12 wt.% in TRIP780 and (ii) the 30 and 20 minutes baking times used, respectively, resulting in the attaining of different strain-aging stages.

In comparison with ferritic steels, DP steels portray a higher BHR. An earlier study of the BH mechanism in the DP steel has proposed [89] that the increase in the YS after treatment was caused by the presence of mobile dislocations formed in ferrite around the dispersed martensite islands during their formation named as geometrically necessary inhomogeneous dislocations (GNDs) (Fig. 9.11A). However later research [9,12,14,42,90] highlighted a more complex interplay of BH mechanisms in AHSS, as mentioned in Section 9.2. Although, two other mechanisms operating in ferrite are the same as in the IF, ULC, and LC steels; namely formation of Cottrell atmospheres (Fig. 9.11B) and precipitation of fine iron carbides, predominantly on defects (Fig. 9.11C).

The significant BHR in DP steels is not only due to the mechanisms operating in ferrite, but is also associated with the response of martensite to BH treatment. Although the BH temperatures are not high enough and at times are shorter than those required to achieve the complete decomposition of martensite, the presence of excess of carbon and high density of dislocations promote fast kinetics of carbon atmospheres formation at dislocations, segregation of carbon at lath boundaries, which restricts ejection of new dislocations during plastic deformation, and precipitation of intermediate carbides and ε carbides (Fig. 9.11D). All these early stages of martensite tempering contribute to the operation of strengthening mechanisms resulting in YS increment in both DP and TRIP steels.

Furthermore, the BH of AHSS appeared to be even more complex. Firstly, the formation of mobile dislocations in ferrite as a result of martensite transformation during quenching of the DP steel and the microstructural changes such as TRIP effect in the TRIP steels during pre-straining lead to the formation of complex dislocation substructure, such as cells, microbands and shear bands in ferrite matrix (Fig. 9.12A–C), depending on the level of pre-strain [9]. The formation of this dislocation substructure in ferrite further promotes the formation of Cottrell atmospheres, Fe-C clusters, intermediate and equilibrium carbides (Fig. 9.12E).

Secondly, the differences in the BH kinetics of ferrite, bainite, retained austenite, and martensite as well as the effect of strain partitioning between soft and hard phases during straining could also affect the BH behavior [9,34]. It was also found that the presence of bainite in the microstructure of the TRIP steel contributes significantly to the BHR, as supersaturated carbon content and high dislocation density

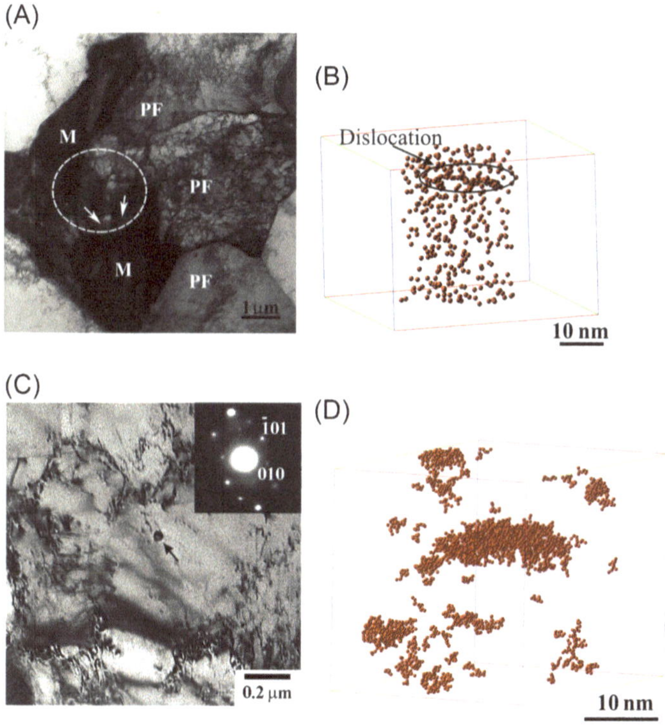

Figure 9.11 TEM images (A, C) and APT maps (B, D) of DP steel showing: (A) the local increase in the dislocation density of ferrite near martensite after intercritical annealing shown by *arrows*, (B) carbon segregation to dislocation in ferrite, (C) formation of carbides in ferrite (shown by *arrow*) after bake hardening (BH) (zone axis in inset is ($[101]\alpha//[101]_c$)), and (D) formation of carbides in martensite after BH.
Source: Reprinted from I.B. Timokhina, P.D. Hodgson, S.P. Ringer, R.K. Zheng, E.V. Pereloma, Effect of bake hardening on the structure-property relationship of multiphase steels for the automotive industry. Steel Res. Int. 80 (7) (2009) 507–514, with copyright permission by Springer.

in bainitic ferrite accelerates the dislocation pinning by carbon atmospheres, as well as clusters and carbides formation (Fig. 9.12G) [9,42,91]. It is interesting to note that if to target both the TRIP-assisted steel composition and the temperature of carbide-free bainite formation in such a way as to form a large fraction of bainite and prevent a significant recovery of the dislocation structure in bainitic ferrite resulted from the shear mechanism of its formation, then a significant BHR could be realized without any pre-straining [92].

The recent APT study was able to prove the segregation of carbon during baking to the retained austenite twin planes formed after pre-straining in the TRIP steels

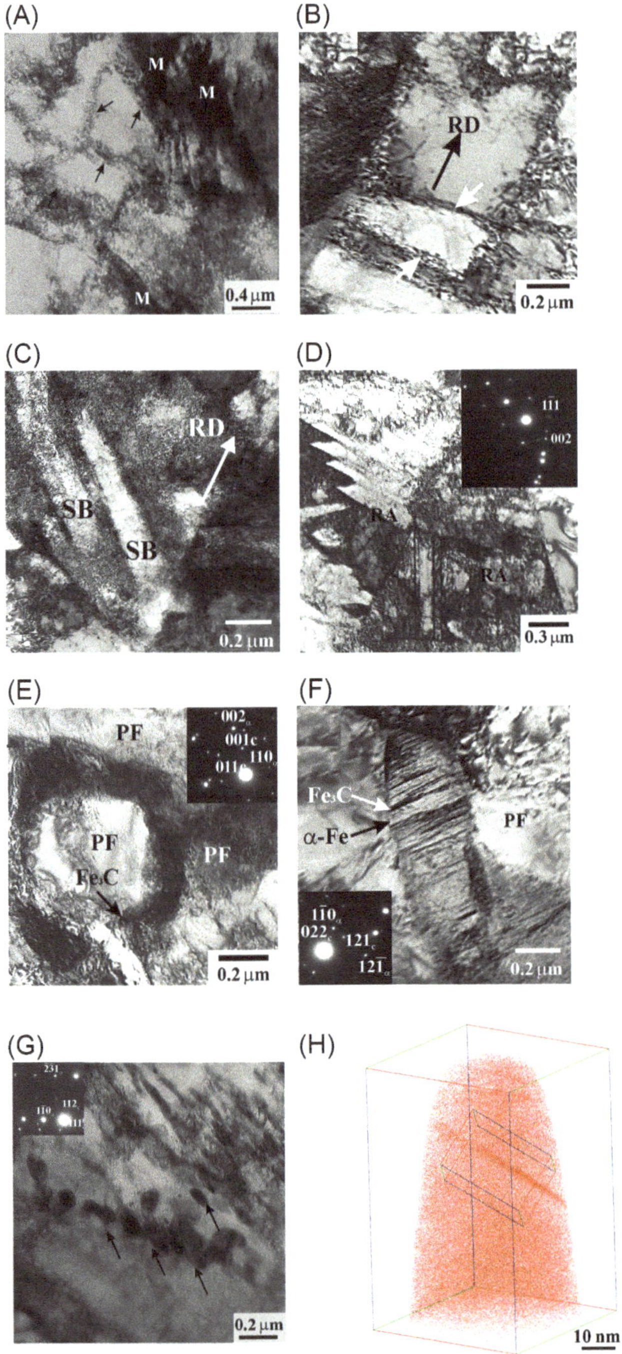

Figure 9.12 (Continued)

that could also affect the BHR (Fig. 9.12H) [42]. The decomposition of the retained austenite during BH is possible (Fig. 9.12F), however it could only be observed after either a very high pre-strain (20%) or long BH treatment at high temperature, which is not applicable for industry [9,14].

The study of BH behavior of the TRIP steel produced by TMP has shown the difference from that observed for TRIP steel after IA. The conventional composition of TRIP steels produced by TMP is Fe: 0.15–0.2 wt.% C: 1.5 wt.% Mn: 1.5 wt.% Si with possible substitution of Si by Al and/or addition of Mo, Cu, Nb, and P [93,94]. The presence of carbide-forming elements in the steel composition of this TRIP steel and complexity of microstructure with ferrite, bainite, retained austenite, and martensite make the understanding of the BH in this type of steels even more complex. Contrarily to IA TRIP steels, the continuous yielding behavior and much higher BHR have been observed for TMP TRIP steels (Fig. 9.10C) [13,52]. The main difference in yielding behavior of these two steels is related to the dominant weak or strong pinning of the dislocations. In IA TRIP steel, the pronounced yield point on stress–strain curve is due to the unlocking of dislocations from Cottrell atmospheres [38,95]. On the other hand, in TMP TRIP steels alloyed with strong carbide-forming elements (Nb, Mo), a strong pinning of dislocations by alloy carbides present in both before and after BH treatment conditions (Fig. 9.13), causes formation of new mobile dislocations during deformation and continuous yielding. It has also been believed that the high volume fraction of the bainitic ferrite in TMP TRIP steels, where there is no effect of dislocation unlocking, is responsible for the observed continuous yielding.

Moreover, not only Fe_3C precipitated in ferrite and bainitic ferrite after prestraining and baking stage (Fig. 9.13C), an increase in the number density of the Nb and Mo carbides, their coarsening and formation of new Nb/Mo clusters were found in both the bainitic ferrite and ferrite using APT (Fig. 9.13A, B). It appeared

◀ **Figure 9.12** TEM micrographs (A–F) and carbon atom map (H) of intercritically annealed TRIP steel showing: (A) dislocation cell substructure in ferrite after 5% pre-strain (*arrows* indicate cell walls), (B) formation of microbands (shown by *arrows*) in ferrite in the 10% pre-strain, (C) formation of shear bands (SB), and (D) retained austenite twinning after 20% pre-strain, zone axis is $[110]\gamma$, (E) formation of fine carbides in polygonal ferrite after bake hardening (BH), zone axis is $[110]\alpha//[100]c$, (F) decomposition of retained austenite ($[113]\alpha//[111]c$), (G) formation of carbides after BH in bainitic ferrite, zone axis in inset is $[115]\alpha//[\bar{1}10]c$, and (H) carbon segregation to the twin plane in the retained austenite after BH. BF is bainitic ferrite, M is martensite. PF is polygonal ferrite, RA is retained austenite, RD is rolling direction.
Source: Reprinted from I.B. Timokhina, P.D. Hodgson, E.V. Pereloma, Transmission electron microscopy characterization of the bake-hardening behavior of transformation-induced plasticity and dual-phase steels. Metall. Mater. Trans. A 38 (10) (2007) 2442–2454; and I.B. Timokhina, P.D. Hodgson, S.P. Ringer, R.K. Zheng, E.V. Pereloma, Effect of bake hardening on the structure-property relationship of multiphase steels for the automotive industry. Steel Res. Int. 80 (7) (2009) 507–514, with copyright permission by Springer.

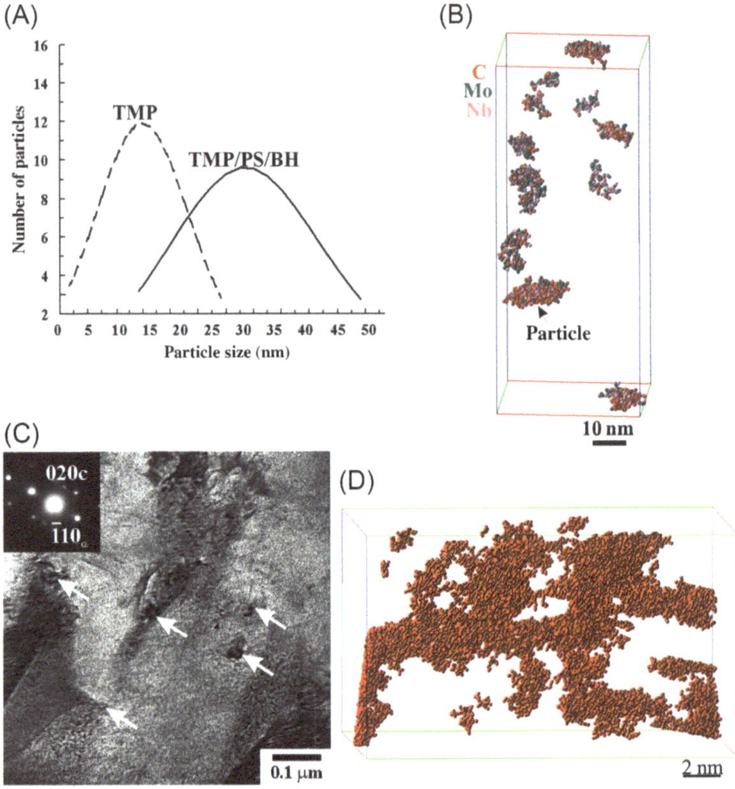

Figure 9.13 (A) Particle size distribution in the TRIP steel after TMP and after 4% pre-straining and 20 min bake hardening at 180°C (PS/BH), (B) atom map showing the formation of C-Mo-Nb clusters and particles in the bainitic ferrite, (C) bright field TEM showing the formation of Fe_3C in bainitic ferrite after PS/BH (zone axis is $[\bar{1}00]_C$), and (D) carbon atom map showing the decomposition of martensite/retained austenite after PS/BH of TRIP steel (matrix atoms were removed for clarity in B and D).
Source: Reprinted from I.B. Timokhina, M. Enomoto, M.K. Miller, E.V. Pereloma, Microstructure-property relationship in the thermomechanically processed C-Mn-Si-Nb-Al-(Mo) transformation-induced plasticity steels before and after prestraining and bake hardening treatment. Metall. Mater. Trans. A 43 (7) (2012) 2473–2483, with permission from Springer.

that the high dislocation density and high carbon content of bainitic ferrite and formation of complex dislocation substructure as a result of pre-straining allowed considering the pipe diffusion of Nb and Mo along dislocation cores during BH treatment. However, even the rate of pipe diffusion extrapolated from the Arrhenius plot for high-temperature diffusivity is insufficient to account for the coarsening of alloy carbides at such a low temperature. Only when a much higher pipe diffusivity of Nb and Mo atoms was assumed, similar to that proposed by Gjostein [96] for surface diffusivity, the possible diffusion of these atoms on interparticle distances during BH was corroborated [13].

The formation of Fe-C carbides in ferrite/bainitic ferrite, and in martensite as a result of martensite decomposition was also observed in the TMP TRIP steel (Fig. 9.13C and D) [13,43]. This was a continuous process starting from the formation of C-clusters as precursors to carbide formation, followed by a compositional evolution and an increase in the size of carbides [13,52]. The contribution of new carbides and clusters formation after pre-straining and BH treatment in the TMP TRIP steel to the YS increase could exceed 300 MPa [13]. The retained austenite in the TRIP steel after TMP was found to be more stable to the pre-straining than in the IA TRIP steel due to the higher carbon content and the constraining effect of the hard bainitic ferrite laths [95].

9.4.2 Bainitic steels

Recently the BH after pre-straining has been studied [72] in the steel with fully bainitic microstructure produced by the TMP. The microstructure consists of fine layers of bainitic ferrite and retained austenite (Fig. 9.14A). A significant BHR of ~222 MPa and an increase in YS of ~790 MPa was reached due to the presence of the bainitic ferrite in bainite with high dislocation density and high carbon content (Fig. 9.14B). The microstructural events such as segregation of C, Mn, and Mo at the γ/α interface, segregation of carbon at the dislocations with formation of Cottrell atmospheres, the formation of transition and equilibrium carbides in bainitic ferrite and retained austenite were observed during BH process (Fig. 9.14C–F).

9.4.3 Martensitic steels

As mentioned earlier, BH steels initially have a low proof stress and good formability, making them suitable for the more complex forming operations encountered during fabrication of outer body parts [1]. The BH of martensitic steels utilizes the phenomenon of strain aging reported earlier [46,98]; an increase in YS is achieved by migrating solute carbon atoms during low temperature aging (70–200°C) to the dislocations produced during straining, which results in dislocations locking. The process is similar to the one observed in martensite of multi-phase steels during baking treatment (Figs. 9.11D and 9.13D). An increase in the YS of martensitic steels mainly depends on martensite dislocation density and carbon level in solution. An initial dislocation density of martensite reaches the order of 10^{15} m^{-2} and further pre-straining creates even more dislocations which then are stabilized by carbon pinning [99]. As was reported by Kalish and Cohen [46], the most rapid increase in YS occurs within the first several percent of deformation (up to 10%) and then it continues to increase linearly. The strengthening due to the deformation is more evident in the medium carbon (0.25–0.45 wt.% C) steels, and lower carbon martensitic steels are found to undergo less hardening [46]. The morphology of martensite appeared to have no effect on the bake-hardening behavior of martensitic steels [98].

A more recent study [8] showed the BHR of 40–50 MPa in the Docol 1200 M martensitic steel, and the more pronounced effect of BH temperature on the BHR

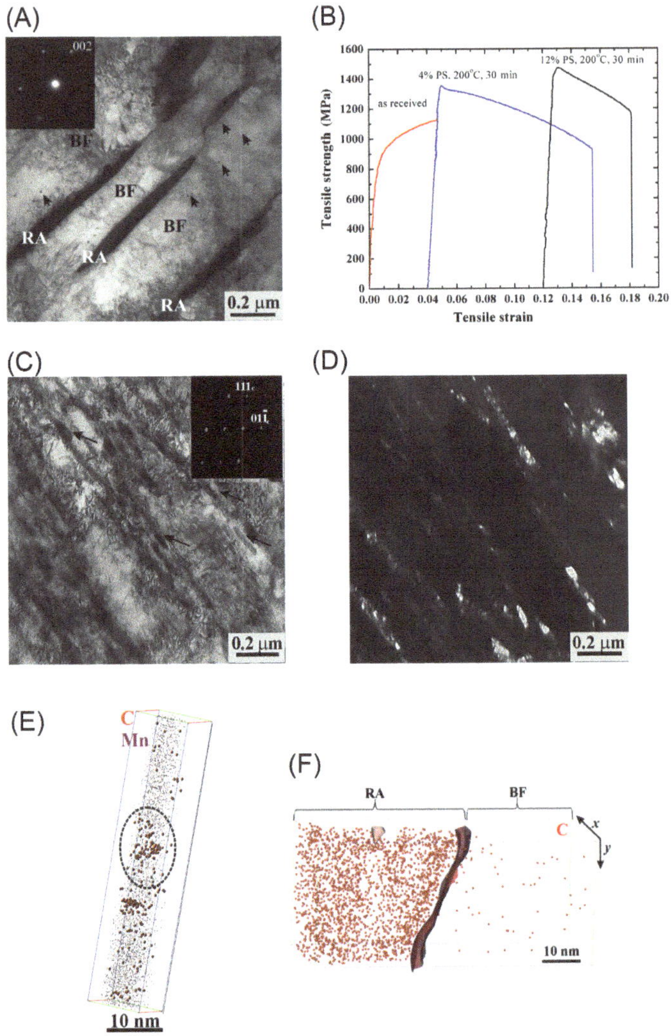

Figure 9.14 (A) TEM micrograph of the bainite after TMP, (B) mechanical properties after TMP and PS/BH, (C, D) formation of carbides in the bainitic ferrite after PS/BH (zone axis in inset in c is $[2\bar{1}\bar{1}]$), (E) segregation of carbon at the dislocations in the bainitic ferrite after PS/BH, and (F) segregation of carbon at the retained austenite/bainitic ferrite interface after PS/BH. *Source*: Reprinted from I. Timokhina, H. Beladi, X.-Y. Xiong, P.D. Hodgson, On the low temperature strain aging of bainite in the TRIP steel. Metall. Mater. Trans. A 44 (11) (2013) 5177–5191, with copyright permission by Springer.

rather than time. The increase in the YS of the martensitic steels was accompanied by the return of the yield point and drastic decrease in total elongation in both plane stress and plane strain tensile conditions. However, less rapid failure of Docol 1200 M steel was encountered during bending tests [100]. It appeared that after BH

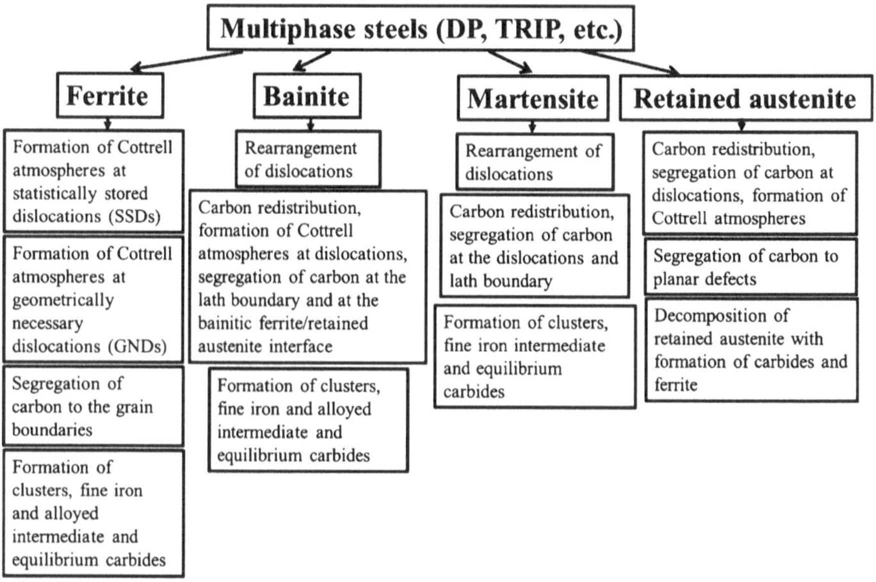

Figure 9.15 Summary of bake hardening mechanisms operating in multi-phase steels.

treatment, the stress level is raised so much by the Cottrell locking that once the atmospheres are unpinned, a localized necking is initiated as there is no support from work hardening and/or precipitate strengthening to withstand this high load [100]. The bake hardened martensitic steels could be used in components where pre-straining from forming operations can be located in regions of subsequent uniform plane strain or uniaxial tension [6]. However, it was pointed out that any conditions of yielding should be avoided during such parts service life [100]. It has also been reported for grades (013–0.14C–0.1Si – 1.7Mn – 0.2–0.4Cr (wt.%)) of martensitic steels produced by Salzgitter Flachstahl GmbH that it is possible to attain in restricted areas 130–220 MPa BHR depending on C and Mn content. This can potentially be used for local strengthening of the steel without significant degradation of overall ductility [101].

The summary of the operating mechanisms during BH of multi-phase steels is given in Fig. 9.15. As can be seen, there are both similarities and differences, which are predominantly dictated by the initial carbon content, carbon solubility, dislocation density, and diffusion distances in each phase.

9.5 Modeling

Prediction of BHR is important for both a better understanding of the phenomena and utilization in the industry. Ability to easily manipulate the steel compositions and processing variables and estimate the resulting properties could save significant costs associated with laboratory and industrial trials. Many theoretical models were developed to address the different aspects of the BH phenomenon (kinetics, effects of carbon segregation to grain boundaries or pre-existing carbides) and the effects of selected processing variables, as well as the problem as a whole. Among these models are those that use a phenomenological approach based on physical phenomena [16,21,24,26,102] including the application of Monte Carlo algorithm [83,103] and those that utilize empirical modeling based on data-driven artificial neural network analysis [54,77]. Some of the former models were addressed in Section 9.2. A good description of different types of the models could be found in Refs. [54,104]. However, due to the complexity of BH mechanisms and a large number of factors affecting the kinetics, there is yet to appear a model accounting for all these variables, especially in application to multi-phase steels.

9.6 Effect of bake hardening on the performance of automotive steels

Due to a variety of requirements, different automotive parts utilise various steel grades to achieve the required properties. For example, outer body panels should possess good surface quality, dent resistance and also a good hemming ability when used in closure panels. On other hand, inner body panels should display good formability, weldability and superior strength. The selection of an appropriate steel candidate is based on both performance and manufacturing criteria.

Bake hardenable steels are the material of choice for many applications as significant strength increases are achieved after forming operations followed by paint bake cycle. LC and ULC-BH steels were designed as dent-resistant steels, as they perform better than IF steels [105].

It was reported [106] that paint baking operation applied to tubes made of Docol 600 and Docol 800 steels led to an increase in YS by 250 and 120 MPa, respectively. BH also had a positive effect on the fatigue strength, but more so on smooth surface samples performance than of the notched ones. Although BH did not influence the energy absorptions, the peak load was slightly higher than that of non-treated materials.

Durrenger et al. [73] evaluated the effect of pre-strain and BH on the crash properties of a top-hat section made of TRIP780 steel produced by ArcelorMittal. They concluded that although the absorption capacity increases after pre-straining, there is a limit in thickness reduction above which it is no longer compensated by the material hardening, thus leading to a slightly lower crash force. Baking after pre-straining demonstrated a positive effect on the absorbed energy.

As BH increases the strength of the steel, its bending force and springback increase [107,108]. Recent observations also point to the improved bending angle for multiphase steels due to the reduced difference in strength between the phases. Thinner gauges of high strength steel together promote higher springback. However, if sheet is formed before paint bake, then the springback is reduced, as the strength is gained at a later processing stage.

There is still a limited information on in-service performance of parts made of bake hardened multi-phase AHSS. With further developments targeting the third generation of AHSS, more research and design efforts are required for evaluation of the effects of paint baking operations on the properties of the parts produced using these novel steels.

9.7 Summary

During production of automotive parts, the phenomenon of BH is widely employed. The steel composition and processing parameters play the defining role in attaining the required BHR for a particular application. The availability of free carbon to participate in both SSD and GND dislocations locking, and then in clustering and fine precipitation, is the essential requirement for the occurrence of BHR. However, the mechanisms are complex with different metallurgical variables, such as dislocation density, grain size, presence of iron or alloy carbides, as well as solute segregation at the boundaries and interfaces, which are able to exert a significant influence on the BHR. It also should be kept in mind that the amount of free carbon needs to be controlled in order to prevent the strain-aging of parts during storage. Furthermore, this strain-aging process occurs differently in multi-phase steels compared with conventional ferritic steels not only due simply to the presence of these phases, but because of their interactions during both pre-straining and baking stages.

Further explorations of the effects of carbide-forming microalloying additions, complex strain state before paint baking stage, volume fractions of phases present in order to optimize both the steel compositions and processing parameters, are required. Development of comprehensive models will assist the progress of these issues.

References

[1] L. Baker, S. Daniel, J. Parker, Metallurgy and processing of ultralow carbon bake hardening steels, Mater. Sci. Technol. 18 (4) (2002) 355–368.

[2] J.D. Baird, Strain aging of steel—A critical review-part I: practical aspects, J. Iron Steel 5 (1963) 186–192.

[3] J.D. Baird, Strain aging of steel—A critical review-part I: practical aspects (cont), J. Iron Steel 6 (1963) 326–334.

[4] D. Wilson, B. Russel, The contribution of atmosphere locking to the strain-aging of low carbon steels, Acta Metall. 8 (1960) 36–45.

[5] W.C. Leslie, The Physical Metallurgy of Steels, McGraw Hill Book Company, New York, 1981.

[6] H. Kantereit, Bake hardening behavior of advanced high strength steels under manufacturing conditions. 2011.

[7] W.B. Hutchinson, K.I. Nilson, J. Hirsch, Metallurgy of Vacuum Degassed Products, TMS, Warrendale, PA, 1990.

[8] J. Hagström, L. Ryde, Bake Hardening in Dual Phase and Martensitic High Strength Steels, Swerea KIMAB, 2008.
[9] I.B. Timokhina, P.D. Hodgson, E.V. Pereloma, Transmission electron microscopy characterization of the bake-hardening behavior of transformation-induced plasticity and dual-phase steels, Metall. Mater. Trans. A 38 (10) (2007) 2442–2454.
[10] M. De Meyer, A.K. Del, L. Tosal-Martinez, B.C. De Cooman, The bake hardening and aging behavior of cold rolled TRIP steels, in 43rd mechanical working and steel processing conference 2001: Charlotte, USA 349–358.
[11] M. Krieger, Y. Estrin, M. Janecek, M. Niemeyer, S. Paul, Mechanical properties and bake hardening behavior of two cold rolled multiphase sheet steels subjected to CGL heat treatment simulation, Steel Res. Int. 77 (2006) 668–674.
[12] T. Waterschoot, A.K. De, S. Vandeputte, B.C. Cooman, Static strain aging phenomena in cold-rolled dual-phase steels, Metall. Mater. Trans. A 34 (13) (2003) 781–791.
[13] I.B. Timokhina, M. Enomoto, M.K. Miller, E.V. Pereloma, Microstructure-property relationship in the thermomechanically processed C-Mn-Si-Nb-Al-(Mo) transformation-induced plasticity steels before and after prestraining and bake hardening treatment, Metall. Mater. Trans. A 43 (7) (2012) 2473–2483.
[14] A. Ramazani, S. Bruehl, T. Gerber, W. Bleck, U. Prahl, Quantification of bake hardening effect in DP600 and TRIP700 steels, Mater. Des. 57 (2014) 479–486.
[15] J.L. Snoek, Effect of small quantities of carbon and nitrogen on the elastic and plastic properties of iron, Physica 8 (7) (1941) 711–733.
[16] P. Elsen, H.P. Hougardy, On the mechanism of bake hardening, Steel Res. 64 (1993) 431–436.
[17] A.H. Cottrell, B.A. Bilby, Dislocation theory of yielding and strain aging of iron, Proc. Phys. Soc. 62A (1949) 49–62.
[18] J. Wilde, A. Cerezo, G.D.W. Smith, Three-dimensional atomic-scale mapping of a cottrell atmosphere around a dislocation in iron, Scr. Mater. 43 (1) (2000) 39–48.
[19] E.V. Pereloma, M.K. Miller, Atomic observations of warm rolled low carbon steels alloyed with Cr and P, Microsc. Microanal. 11 (Suppl. S02) (2005) 876–877.
[20] E.V. Pereloma, I.B. Timokhina, J.J. Jonas, M.K. Miller, Fine-scale microstructural investigations of warm rolled low-carbon steels with and without Cr, P, and B additions, Acta Mater. 54 (17) (2006) 4539–4551.
[21] S. Harper, Precipitation of carbon and nitrogen in cold-worked alpha-iron, Phys. Rev. 83 (4) (1951) 709–712.
[22] J.D. Baird, Strain aging of steel—A critical review-part II: the theory of strain aging, J. Iron Steel 7 (1963) 368–374.
[23] A.W. Cochardt, G. Schoek, H. Wiedersich, Interaction between dislocations and interstitial atoms in body-centered cubic metals, Acta Metall. 3 (6) (1955) 533–537.
[24] A.K. De, S. Vandeputte, B.C. De Cooman, Kinetics of strain aging in bake hardening ultra low carbon steel—a comparison with low carbon steel, J. Mater. Eng. Perform. 10 (2001) 567–575.
[25] S.I. Neife, E. Pink, H.P. Stüwe, The isothermal precipitation behavior in a low-carbon steel at low temperatures investigated by internal friction, Scr. Metall. et Mater. 30 (3) (1994) 361–366.
[26] S. Hartley, Strain-aging in tantalum, Acta Metall. 14 (10) (1966) 1237–1246.
[27] D. McLean, Grain Boundaries in Metals, Clarendon Press, London, UK, 1957.
[28] D.V. Wilson, Role of grain boundaries in the discontinuous yielding of low-carbon steels, Met. Sci. J. 1 (1) (1967) 40–47.

[29] S. Ohsaki, K. Hono, H. Hidaka, S. Takaki, Characterization of nanocrystalline ferrite produced by mechanical milling of pearlitic steel, Scr. Mater. 52 (4) (2005) 271–276.
[30] E.V. Pereloma, V. Bata, R.I. Scott, R.M. Smith, Effect of Cr and Mo on strain aging behavior of low carbon steel, Mater. Sci. Eng. A 527 (10–11) (2010) 2538–2546.
[31] J.W. Cahn, Nucleation on dislocations, Acta Metall. 5 (3) (1957) 169–172.
[32] C.C. Dollins, Nucleation on dislocations, Acta Metall. 18 (11) (1970) 1209–1215.
[33] D.M. Barnett, On nucleation of coherent precipitates near edge dislocations, Scr. Metall. 5 (4) (1971) 261–266.
[34] A.K. De, S. Vandeputte, B.C. De Cooman, Kinetics of low temperature precipitation in a ULC-bake hardening steel, Scr. Mater. 44 (4) (2001) 695–700.
[35] R.K. Seal, New Ultra- Low Carbon High Strength Steels with Improved Bake Hardenability for Enhanced Stretch Formability and Dent Resistance, University of Pittsburgh, Pittsburgh, 2006.
[36] H. Abe, T. Suzuki, Thermoelectric power versus electrical conductivity plot for quench aging of low carbon aluminium killed steel, Trans. ISIJ 20 (1980) 690–695.
[37] H.K.D.H. Bhadeshia, R.W.K. Honeycombe, Steels: Microstructure and Properties, 3rd ed, Butterworth-Heinemann, Oxford, UK, 2006.
[38] E. Pereloma, H. Beladi, L. Zhang, I. Timokhina, Understanding the behavior of advanced high-strength steels using atom probe tomography, Metall. Mater. Trans. A 43 (11) (2011) 3958–3971.
[39] W.C. Jeong, Effect of prestrain on aging and bake hardening of cold-rolled, continuously annealed steel sheets, Metall. Mater. Trans. A 29A (1998) 463–467.
[40] D.K. Matlock, B.J. Allen, J.G. Speer, Aging behavior and properties of ultra-low carbon bake-hardening steels, in: W. Bleck (Ed.), Modern LC and ULC Sheet Steels for Cold Forming: Processing and Properties, Aachen, Germany, Verlag Mainz, 1998, pp. 265–276.
[41] W. Bleck, S. Brühl, T. Gerber, A. Katsamas, R. Ottaviani, L. Samek, S. Traint, Control and exploitation of the bake-hardening effect in multi-phase high-strength steels. 2007.
[42] I.B. Timokhina, E.V. Pereloma, S.P. Ringer, R.K. Zheng, P.D. Hodgson, Characterization of the bake-hardening behavior of transformation induced plasticity and dual-phase steels using advanced analytical techniques, ISIJ Int. 50 (4) (2010) 574–582.
[43] E.V. Pereloma, M.K. Miller, I.B. Timokhina, On the decomposition of martensite during bake hardening of thermomechanically processed transformation-induced plasticity steels, Metall. Mater. Trans. A 39 (13) (2008) 3210–3216.
[44] J.M. Rubianes, P. Zimmer, The bake hardenable steels, Rev. Metall. Cah. Int. Technol. (1996) 99–109.
[45] S. Hanai, N. Takemoto, Y. Tokunaga, Y. Mizuyama, Effect of grain size and solid solution strengthening elements on the bake hardenability of low carbon aluminum-killed steel, Trans. Iron Steel Inst. Jan. 24 (1) (1984) 17–23.
[46] D. Kalish, M. Cohen, Structural changes and strengthening in the strain tempering of martensite, Mater. Sci. Eng. 6 (1970) 156–166.
[47] A. Van Snick, K. Lips, S. Vandeputte, B.C. De Cooman, J. Dilewijns, Effect of carbon content, dislocation density and carbon mobility on bake hardening, in: W. Bleck (Ed.), Modern LC and ULC Sheet Steels for Cold Forming: Processing and Properties, Aachen, Germany, Verlag Mainz, 1998, pp. 413–424.
[48] V. Massardier, N. Lavaire, M. Soler, J. Merlin, Comparison of the evaluation of the carbon content in solid solution in extra-mild steels by thermoelectric power and by internal friction, Scr. Mater. 50 (12) (2004) 1435–1439.

[49] A. De, S. Vandeputte, B. De Cooman, Static strain aging behavior of ultra low carbon bake hardening steel, Scr. Mater. 41 (8) (1999) 831–837.
[50] K.A. Taylor, J.G. Speer, Development of vanadium alloyed, bake hardenable sheet steels for hot dip coated applications, 39th MWSP, ISS, Indianapolis, IN, 1998, pp. 49–61.
[51] K. Lips, X. Yang, K. Mols, The effect of coiling temperature and continuous annealing on the properties of bake hardenable IF steels, Steel Res. 67 (9) (1996) 357–363.
[52] E.V. Pereloma, K.F. Russell, M.K. Miller, I.B. Timokhina, Effect of pre-straining and bake hardening on the microstructure of thermomechanically processed CMnSi TRIP steels with and without Nb and Mo additions, Scr. Mater. 58 (12) (2008) 1078–1081.
[53] J.N.A. Starling, The Evolution of solute Carbon during Continuous Annealing of Interstitial Free Steel and Its Effects on Bake Hardenability, Queen's University, Kingston, Canada, 1998.
[54] S. Das, Bake Hardening in Low and Medium Carbon Steels, Indian Institute of Technology, Kharagpur, India, 2012.
[55] K. Tsunoyama, K. Satoh, T. Obara, K. Sakata, K. Hasigusi, T. Irie, Hot-and Cold-Rolled Sheet Steels, TMS-AIME, 1988, pp. 155–164.
[56] K. Kawasaki, T. Senuma, S. Sanagi, Processing, Microstructure and Properties of Microalloyed and Other Modern HSLA Steels, ISS, 1991, pp. 133–144.
[57] H. Taniguchi, T. Kiyokawa, M. Mizutani, R. Okamoto, Effect of Mn on strain aging behavior of ultra low carbon steel sheets, Tetsu-to-Hagane 91 (2005) 662–669.
[58] C.C. Li, W.C. Leslie, Effect of dynamic strain aging on subsequent mechanical properties of carbon steels, Metall. Trans. A 9A (1978) 1765–1775.
[59] H. Abe, T. Suzuki, S. Okada, Decomposition of Mn-C dipoles during quench-aging in low-carbon aluminium-killed steels, Trans. Jan. Inst. Met. 25 (4) (1984) 215–225.
[60] K. Tagashira, N. Sumita, H. Hayakawa, Effect of Si on the formation of Mn-C dipoles in Fe-Si-Mn-C alloys, Mater. Trans., JIM 30 (1989) 10–18.
[61] A. Okamoto, M. Takahashi, T. Hino, Development of bake-hardenable al-killed steel sheet by box-annealing process, Trans. Iron Steel Inst. Jan. 21 (11) (1981) 802–811.
[62] M.K. Miller, APT characterization of solute segregation to individual dislocations, TMS Lett. 1 (2004) 19–20.
[63] M.K. Miller, Atom probe tomography characterization of solute segregation to dislocations, Microsc. Res. Tech. 69 (5) (2006) 359–365.
[64] P. Meissen, V. Leroy, Scavenging additions of boron in low C low Al steels, Steel Res. 60 (1989) 320–328.
[65] W. Al-Shalfan, J.G. Speer, D.K. Matlock, K. Findley, Effect of annealing time on solute carbon in ultralow-carbon Ti-V and Ti-Nb steels, Metall. Mater. Trans. A 37 (1) (2006) 207–216.
[66] L.-J. Chiang, K.-C. Yang, I.-C. Hsiao, Effects of annealing conditions on bake hardenability for ULC steels. 2011: China Steel Technical Report. 1–6.
[67] K. Sakata, S. Satoh, T. Kato, O. Hashimoto, Metallurgical principles and applications for producing extra-low carbon IF steels with deep drawability and bake hardenability, International Forum for Physical Metallurgy of IF Steels, ISIJ, Tokyo, Japan, 1994, pp. 279–288.
[68] A. Itami, K. Ushioda, N. Kimura, H. Asano, Y. Kimura, K. Koyama, Development of new formable cold-rolled sheet steels for automotive body panels, Nippon Steel Technical Reports, Nippon Steel, 1995, pp. 26–32.
[69] Y.-S. Hwang, H.-C. Chen, T.-S. Chou, High Strength Steels for the Automotive Industry, ISS, 1994, pp. 187–193.

[70] S. Satoh, T. Irie, O. Hashimoto, Tetsu-to-Hagane 68 (1983) 1362.
[71] L.M. Storozheva, K. Éscher, R. Bode, K. Hulka, D.A. Burko, Effect of niobium and coiling and annealing temperatures on the microstructure, mechanical properties, and bake hardening of ultralow-carbon steels for the automotive industry, Met. Sci. Heat Treat. 44 (3) (2000) 102–107.
[72] I. Timokhina, E. Pereloma, P. Hodgson, The formation of complex microstructures after different deformation modes in advanced high-strength steels, Metall. Mater. Trans. A 45 (10) (2014) 4247–4256.
[73] L. Durrenberger, X. Lemoine, A. Molinari, Effects of pre-strain and bake-hardening on the crash properties of a top-hat section, J. Mater. Process. Technol. 211 (12) (2011) 1937–1947.
[74] K. Dehghani, J.J. Jonas, Dynamic bake hardening of interstitial-free steels, Metall. Mater. Trans. A 31A (5) (2000) 1375–1384.
[75] A.K. De, K. De Blauwe, S. Vandeputte, B.C. De Cooman, Effect of dislocation density on the low temperature aging behavior of an ultra low carbon bake hardening steel, J. Alloys Compd. 310 (1–2) (2000) 405–410.
[76] S. Das, S.B. Singh, O.N. Mohanty, Effect of free carbon and aging condition on bake hardening, Ironmaking Steelmaking 38 (2) (2011) 139–143.
[77] S. Das, S.B. Singh, O.N. Mohanty, H.K.D.H. Bhadeshia, Understanding the complexities of bake hardening, Mater. Sci. Technol. 24 (1) (2008) 107–111.
[78] V. Ballarin, A. Perlade, X. Lemoine, O. Bouaziz, S. Forest, Mechanisms and modeling of bake-hardening steels: part II. Complex loading paths, Metall. Mater. Trans. A 40 (6) (2009) 1375–1382.
[79] T. Obara, K. Sakata, M. Nishida, T. Irie, Effects of heat cycle and carbon content on the mechanical properties of continuous-annealed low carbon steel sheets, Kawasaki Steel Technical Report, Kawasaki Steel Japan, 1985, pp. 25–35.
[80] V. Ballarin, M. Soler, A. Perlade, X. Lemoine, S. Forest, Mechanisms and modeling of bake-hardening steels: part I. Uniaxial tension, Metall. Mater. Trans. A 40 (6) (2009) 1367–1374.
[81] J. Takahashi, M. Sugiyama, N. Maruyama, Quantitative observation of grain boundary carbon segregation in bake hardening steels. in Nippon Steel Technical Report. 2005: Japan. 28–33.
[82] A.K. De, S. Vandeputte, B. Soenen, B.C. De Cooman, Effect of grain size on the static strain aging of a ULC bake hardening steel, Z. Metallkunde 95 (2004) 713–717.
[83] J.Z. Zhao, A.K. De, B.C. Cooman, Formation of the cottrell atmosphere during strain aging of bake-hardenable steels, Metall. Mater. Trans. A 32 (2) (2001) 417–423.
[84] J.Z. Zhao, A.K. De, B.C. De Cooman, Kinetics of Cottrell atmosphere formation during strain aging of ultra-low carbon steels, Mater. Lett. 44 (6) (2000) 374–378.
[85] O. Matsumura, Y. Sakuma, H. Takechi, TRIP and its kinetic aspects in austempered 0.4C-1.5Si-0.8Mn steel, Scr. Metall. 21 (1987) 1301–1306.
[86] M. Kinoshita, A. Nishimoto, Microstructure and properties of ferrite + martensite dual phase stainless steel sheet, CAMP-ISIJ 3 (1990) 1780–1785.
[87] V.F. Zackay, E.R. Parker, D. Fahr, R. Bush, The enhancement of ductility in high strength steels, Trans. ASM 60 (1967) 252–257.
[88] B.C. De Cooman, Structure-properties relationship in TRIP steels containing carbide-free bainite, Curr. Opin. Solid State Mater. Sci. 8 (3–4) (2004) 285–303.
[89] T. Furukawa, H. Mrikawa, M. Endo, H. Takechi, M. Koyama, O. Akisue, T. Yamada, Process factors for cold-rolled dual phase sheet steels, Trans. ISIJ 21 (1981) 812–818.

[90] T. Huper, S. Endo, N. Ishikawa, K. Osawa, Effect of volume fraction of constituent phases on the stress–strain relationship of dual phase steels, ISIJ Int. 39 (3) (1999) 288–294.
[91] L.C. Zhang, I.B. Timokhina, A. La Fontaine, S.P. Ringer, P.D. Hodgson, E.V. Pereloma, Effect of pre-straining and bake hardening on the microstructure and mechanical properties of CMnSi TRIP steels, La Metallurgia Italiana 101 (6) (2009) 49–55.
[92] S. Das, I. Timokhina, S.B. Singh, E. Pereloma, O.N. Mohanty, Effect of bainitic transformation on bake hardening in TRIP assisted steel, Mater. Sci. Eng. A 534 (2012) 485–494.
[93] Y. Sakuma, O. Matsumura, H. Takechi, Mechanical properties and retained Austenite in intercritically heat treated bainite-transformed steel and their variations with Si and Mn additions, Metall. Trans. A 22A (1991), 489-498
[94] H.K.D.H. Bhadeshia, D. Edmonds, The bainite transformation in a silicon steel, Metall. Trans. A 10A (1979) 895–907.
[95] I.B. Timokhina, P.D. Hodgson, S.P. Ringer, R.K. Zheng, E.V. Pereloma, Effect of bake-hardening on the structure-property relationship of multiphase steels for the automotive industry, Steel Res. Int. 80 (7) (2009) 507–514.
[96] N.A. Gjostein, Met. Surf. ASM (1963) 99–103.
[97] I. Timokhina, H. Beladi, X.-Y. Xiong, P.D. Hodgson, On the low temperature strain aging of bainite in the TRIP steel, Metall. Mater. Trans. A 44 (11) (2013) 5177–5191.
[98] A. Sherman, G. Eldis, M. Cohen, The aging and tempering of Iron-Nickel-Carbon martensites, Metall. Mater. Trans. A 14 (1983) 995–1005.
[99] A. Keh, S. Weissman, Deformation substructure in body-centered cubic metals, Electron Microscopy and Strength of Crystals, Interscience, New York, 1963.
[100] K. Lindqvist, Bake Hardening Effect in Advanced High Strength Steels, Chalmers University of Technology, Sweden, 2013, p. 90.
[101] M. Asadi, H. Palkowski, Influence of thermo-mechanical processing parameters and chemical composition on bake hardening ability of hot rolled martensitic steels, Steel Res. Int. 80 (2009) 499–506.
[102] S. Berbenni, V. Favier, X. Lemoine, M. Berveiller, A micromechanical approach to model the bake hardening effect for low carbon steels, Scr. Mater. 51 (4) (2004) 303–308.
[103] B. Soenen, A.K. De, S. Vandeputte, B.C. De Cooman, Competition between grain boundary segregation and Cottrell atmosphere formation during static strain aging in ultra low carbon bake hardening steels, Acta Mater. 52 (12) (2004) 3483–3492.
[104] S. Das, S.B. Singh, O.N. Mohanty, Bake hardening, Encyclopedia of Iron, Steel, and Their Alloys, Taylor & Francis, 2016, pp. 306–319.
[105] G. Manikandan, R. Verma, A. Raj, Effect of yield strength, pre-strain and curvature on stiffness and static dent resistance of formed panel, 5th International and 26th All India Manufacturing Technology, Design and Research Conference, IIT Guwahati: IIT Guwahati, Assam, India, 2014, 353-1-7
[106] J.-O. Sperle, K. Olsson, High strength automotive steel sheets for weight reduction and safety applications, High-Strength Steels for Automotive Industry, AIME, Baltimore, MD, USA, 1994.
[107] E. Wold, High strength low alloy (HSLA) cold forming steels, Mater. Design 11 (1990) 243–246.
[108] E. Billur and T. Altan, Challenges in forming advanced high strength steels, in: Proceedings of New Developments in Sheet Metal Forming, Stuttgart, Germany, 2010, pp. 285–304.

Bainitic and quenching and partitioning steels

10

E. De Moor and J.G. Speer
Colorado School of Mines, Golden, CO, United States

10.1 Introduction

Substantial research efforts by academic and industrial research groups, as well as product application engineering groups, to develop next generation advanced high strength steel (AHSS) sheets are ongoing. Development of novel families of steels exhibiting combinations of properties superior to currently available products such as Dual Phase (DP) or Transformation Induced Plasticity (TRIP) aided steels is of particular interest, preferably using lean or fairly lean alloying approaches. As indicated by an iso-strain modeling approach, very fine high strength ferritic microstructures and/or martensitic steels containing substantial retained austenite fractions are promising candidates to generate innovative properties beyond tensile property envelopes exhibited by various so-called first and second generation AHSS [1–4]. In this present chapter bainitic steels and quenching and partitioning will be discussed, as novel properties have been obtained for these steels, and both approaches are being implemented by industry.

10.2 Bainitic steels

Substantial austenite retention can be obtained by carbon enrichment during a bainitic transformation if abundant cementite formation is prevented by judicious alloying. The bainite reaction is effectively employed to stabilize austenite in low carbon TRIP aided steels where the reaction most often takes place isothermally at, e.g., 400–450°C following intercritical annealing. In order for austenite to be retained, competing reactions such as cementite formation need to be suppressed so that the carbon is available for austenite enrichment and stabilization through suppression of the martensite start (M_s) temperature. As the overall fineness of the microstructure is related to the isothermal bainitic transformation temperature, substantial research has also been conducted to decrease bainite start temperatures and very fine high strength microstructures have been obtained [5–7]. Given the high strength, these microstructures are sometimes referred to as "superbainite." In order to suppress the bainite transformation temperatures, substantial alloying needs to be conducted and example chemistries are shown in Table 10.1. Involved transformation times however are beyond typical sheet steel processing windows. The acceleration of the bainite

Automotive Steels. DOI: http://dx.doi.org/10.1016/B978-0-08-100638-2.00010-9

Table 10.1 Chemical compositions of steels used for bainitic heat treatments

	Ref.
0.79C-1.94Mn-1.59Si-1.33Cr-0.30Mo-0.11 V	[5]
0.98C-1.89Mn-1.46Si-1.26Cr-0.26Mo-0.09 V	[5–7]
0.83C-1.98Mn-1.57Si-1.02Cr-0.24Mo-1.54Co	[5,7]
0.78C-1.95Mn-1.49Si-0.97Cr-0.24Mo-1.60Co-0.99Al	[5,7]
0.80C-2.01Mn-1.59Si-1Cr-0.24Mo-1.51Co	[8,12]
0.79C-1.98Mn-1.56Si-1.01Cr-0.24Mo-1.51Co-1.01Al	[8,9]
0.43C-3.00Mn-2.02Si	[10,11]
0.19C-2.90Mn-2.13Si	[10]
1.1C-3.00Mn-2.02Si	[10]
0.22C-3.00Mn-2.03Si	[11]
0.39C-2.05Si-4.08Ni	[11]
0.32C-1.97Mn-1.45Si-1.26Cr-0.26Mo-0.10 V	[12]
0.31C-1.51Si-1.44Cr-0.25Mo-3.52Ni-0.10 V	[12]
0.30C-1.51Si-1.42Cr-0.25Mo-3.53Ni	[12]
0.31C-1.51Si-1.44Cr-3.52Ni-0.25Mo-0.10 V	[13]
0.30C-1.51Si-1.42Cr-3.53Ni-0.25Mo	[13]
0.29C-2.25Mn-1.50Si-0.26Mo	[13]
0.29C-1.97Mn-1.46Si-0.46Cr-0.25Mo	[13]
0.29C-1.56Mn-1.49Si-1.47Cr-0.25Mo	[13]
0.27C-1.53Mn-1.71Si-1.47Ni-0.17Cr-0.24Mo	[13]
0.22C-2.23Mn-1.50Si-0.25Mo	[13]
0.21C-1.56Mn-1.46Si-1.49Cr-0.24Mo	[13]

Notes:
All the chemical compositions in this table and henceforth in this chapter are in wt pct unless otherwise mentioned.

reaction and austenite decomposition has also been studied in the presence of some initial martensite [14–16].

Although intercritical annealing is traditionally employed to generate TRIP steel microstructures, increased annealing temperatures resulting in reduced or absent intercritical ferrite fractions have received significant attention more recently and obtained carbide-free bainitic microstructures are frequently referred to as TRIP-aided bainitic ferrite (TBF). The holding temperature for the isothermal transformation can be in the bainitic regime or slightly below the M_S temperature of the austenite present at the soaking temperature [17]. Example heat treating schedules and resulting tensile properties are shown in Figs. 10.1 and 10.2 respectively. Continuously decreasing strength levels, both yield and tensile, are observed with increased holding temperatures whereas greater elongation levels are obtained for holding temperatures above the M_S temperature of 417°C for the alloy studied in Ref. [17].

The evolution of the engineering stress-strain curves with austempering temperature is illustrated in Fig. 10.3 [18]. It should be noted that the M_S temperature is reported to be 423°C for the 0.2C-1.5Mn-1.5Si steel and 416°C for the niobium microalloyed steel [18]. Therefore, the isothermal holding temperatures span a wide

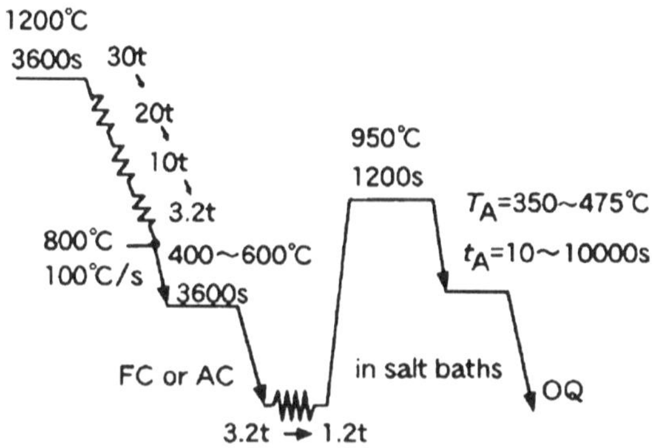

Figure 10.1 Processing schematic detailing hot rolling, cold rolling, and annealing for a 0.20C-1.51Si-1.51Mn TRIP type bainitic steel [17].
Source: K.-I. Sugimoto, T. Iida, J. Sakaguchi, T. Kashima, ISIJ Intl. 40 (9) (2000) 902–908.

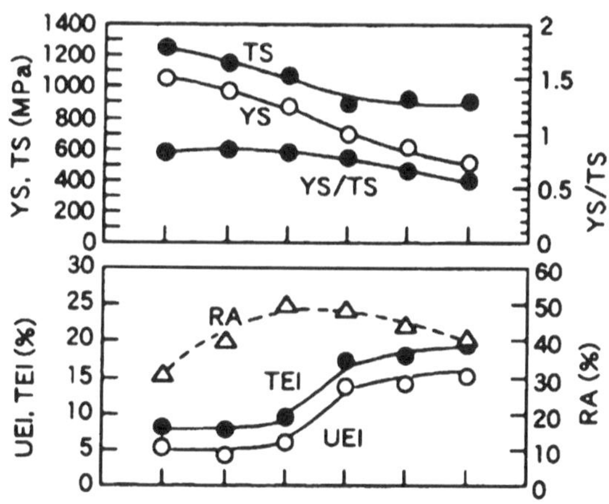

Figure 10.2 Yield strength (YS), tensile strength (TS), uniform (UEl) and total elongation (TEl), left axis and ratio of yield to tensile strength (YS/TS) and reduction in area (RA), right axis as a function of isothermal annealing temperature T_A shown in Fig. 10.1 and a holding time of 200 s [17].
Source: K.-I. Sugimoto, T. Iida, J. Sakaguchi, T. Kashima, ISIJ Intl. 40 (9) (2000) 902–908.

range above and below the M_s temperature. Increased strength and reduced ductility are observed with decreasing isothermal holding temperature. Niobium microalloying of TBF steels has also been considered in more recent work [19].

The presence of retained austenite in the fine microstructure is important to obtain next generation properties and austenite is effectively stabilized to room temperature following isothermal holding. A strong correlation between the product of ultimate tensile strength and total elongation and retained austenite volume fraction was obtained as shown in Fig. 10.4 [18].

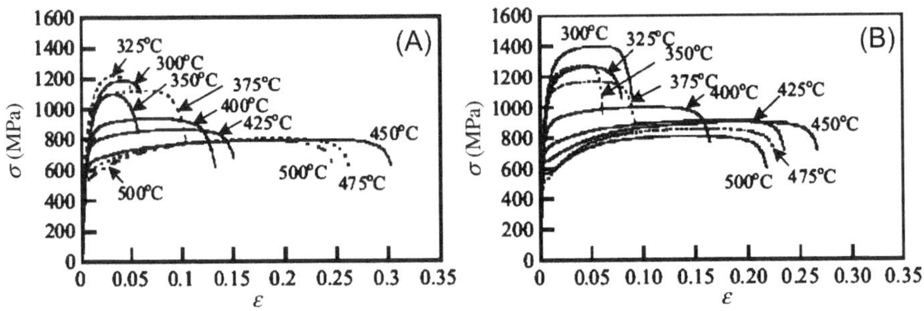

Figure 10.3 Engineering stress-strain curves evolution with isothermal heat treating temperature for (A) 0.2C-1.5Mn-1.5Si and (B) 0.2C-1.5Mn-1.5Si-0.048 Nb steels [18].
Source: K.-I. Sugimoto, M. Murata, Y. Mukai, Proc. Mater. Sci. Technol. (2007) 15–26.

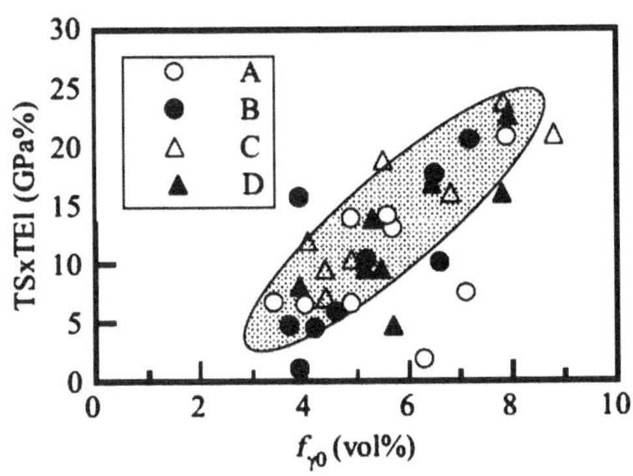

Figure 10.4 Product of tensile strength and total elongation as a function of retained austenite denoted as $f_{\gamma 0}$ volume fraction [18].
Source: K.-I. Sugimoto, M. Murata, Y. Mukai, Proc. Mater. Sci. Technol. (2007) 15–26.

10.3 Quenching & partitioning

Quenching and Partitioning (Q&P) steels are receiving substantial attention as possible third generation AHSS candidates and substantial development work is underway. The present Chapter will highlight some of the more recent work on Q&P heat treating response. A schematic of the Q&P heat treating profile is shown in Fig. 10.5 [20,21] and the procedure consists of reheating a steel to a temperature within the intercritical region or full austenitization, quenching to a predetermined, so-called quenching temperature (QT), between the martensite start and finish temperatures, M_s and M_f respectively, followed by holding at that temperature or at a higher temperature. The latter step is referred to as the partitioning step and holding at the initial quenching temperature is referred to as one-step Q&P, whereas a partitioning treatment at a temperature in excess of the QT is referred to as two-step Q&P. Martensite forms during the initial quenching step and a partially martensitic, partially austenitic microstructure is present at the QT if sufficiently high cooling rates are employed and the alloy exhibits sufficient hardenability to avoid austenite decomposition during the initial quenching step. The aim of the partitioning

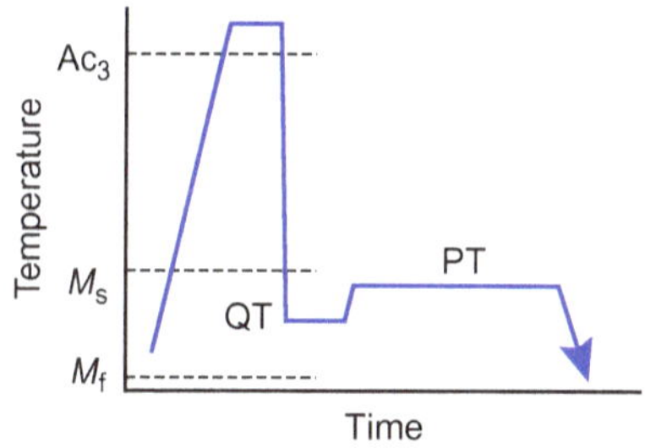

Process Step	Microstructure
Reheating (> A_{c3})	Austenite $C_\gamma = C_i$
Quenching (QT)	Austenite + Martensite $C_\gamma = C_i$, $C_M = C_i$
Partitioning (PT)	Carbon-enriched Austenite + Carbon-depleted Martensite $C_\gamma > C_i$, $C_M < C_i$
Post-partitioning cooling to RT	Carbon-enriched Austenite + Carbon-depleted Martensite + Fresh Martensite

Figure 10.5 Schematic illustrating the quenching and partitioning heat treating schedule with full austenitization. QT is the quenching temperature, PT is the partitioning temperature, RT refers to room temperature and M_s and M_f are the martensite start and finish temperatures, respectively. C_i, C_γ, and C_M refer to the carbon contents of the initial alloy, austenite, and martensite, respectively [20].
Source: Adapted from D.K. Matlock, V.E. Brautigam, J.G. Speer, Mater. Sci. Forum 426–432 (2003) 1089–1094.

treatment is to stabilize austenite present in the microstructure by carbon enrichment via carbon depletion of the martensite and transport to the austenite. As the austenite enriches in carbon, the M_s temperature decreases and once the M_s temperature reaches room temperature, all of the austenite present at the QT can be retained at room temperature. The QT is a critical heat treating parameter for Q&P, as this temperature governs the amount of initial martensite formed, and therefore the volume fraction of austenite remaining in the microstructure, in addition to indirectly governing the potential for austenite retention, as the amount of martensite also sets the amount of carbon available for austenite enrichment and stabilization. A straightforward calculation consisting of the application of the Koistinen-Marburger equation to predict initial and final martensite fractions can be conducted to determine the temperature at which the greatest austenite fractions may be obtained for a given alloy, assuming full carbon depletion of martensite and enrichment in austenite [22].

In order to achieve the most effective partitioning and associated austenite retention by carbon partitioning from martensite into the remaining austenite, competing reactions must be limited. Sufficient retardation of cementite precipitation can be achieved by alloying additions of silicon, aluminum, and/or phosphorus [23] in austempered bainitic [24] and Q&P martensitic microstructures [25–27]. Early Q&P material response studies have employed low carbon TRIP steel type compositions [25,28]. More recently, richer alloy compositions have been investigated and a sampling of employed chemical compositions is shown in Table 10.2. A number of alloying approaches

Table 10.2 Example chemical compositions of steels employed in recent Q&P studies

Ref.	
Steel A [29]	1.07C-2.9Mn-2.2Si-0.048Al
Steel B [30]	0.59C-2.9Mn-2Si-0.038Al
0.2C-3Mn-1.6Si [31]	0.2C-3Mn-1.6Si-0.06Al
0.3C-3Mn-1.6Si [32]	0.29C-2.95Mn-1.59Si-0.06Al
0.3C-5Mn-1.6Si [32]	0.28C-4.96Mn-1.64Si-0.06Al
H0 [33]	0.43C-0.59Mn-2.03Si-0.008Al-0.03Mo-1.33Cr
H1 [29]	0.43C-0.59Mn-2.6Si-0.008Al-0.03Mo-1.33Cr
H2 [29]	0.43C-1.17Mn-2.6Si-0.008Al-0.03Mo-1.33Cr
CMnSi [26]	0.2C-1.63Mn-1.63Si
CMnSi [26]	0.24C-1.61Mn-1.45Si-0.3Al
CMnAlSi [26]	0.25C-1.7Mn-0.55Si-0.69Al
MoCMnAl [26]	0.24C-1.6Mn-0.12Si-1.41Al-0.17Mo
MoCMnSi [26]	0.21C-1.96Mn-1.49Si-0.25Mo
HiSi [34]	0.195C-3.5Mn-1.54Si-0.006Al
LSi [30]	0.198C-3.5Mn-0.45Si-0.22Al
De Knijf et al. [35,36]	0.25C-3Mn-1.5Si
de Diego-Calderon et al. [37,38]	0.25C-3Mn-1.5Si-0.023Al
Seo et al. [39]	0.21C-4Mn-1.6Si-1Cr
Arlazarov et al. [40]	0.29C-3Mn-1.4Si
H_CL_{Ni} [41,42]	0.37C-2.27Mn-2.45Si-0.58Mo-0.8Cr
L_CL_{Ni} [39]	0.22C-2.42Mn-2.49Si-0.49Mo-0.72Cr
H_CH_{Ni} [38,39]	0.39C-2.39Mn-2.64Si-0.5Mo-0.78Cr
L_CH_{Ni} [38,39]	0.28C-2.37Mn-3Si-0.6Mo-0.71Cr
CMnSi [28]	0.19C-1.59Mn-1.63Si-0.036Al
Cheng et al. [43]	0.19C-1.53Mn-1.55Si-0.025Al-0.45Mo-1.01Cr-1.01Cu
40SiMnNiCr [44]	0.41C-1.30Mn-1.27Si-1.01Ni-0.56Cr
Steel N [26]	0.32C-0.07Mn-1.91Si-13.95Ni-0.25Mo
Steel M [42]	0.27C-8.23Mn-1.94Si-0.24Mo

significantly impact retained austenite levels and/or mechanical properties and some trends are highlighted below. Recent studies have particularly expanded the range of Mn and C concentrations compared to early work.

10.4 Substitution of silicon by aluminum

TRIP steel studies have shown that silicon alloying may result in poor surface quality following hot dip galvanizing as the formation of silicon-containing surface oxides may result in poor zinc wettability and therefore corrosion resistance may be impaired as bare spots may be present on the galvanized strip [45]. Partial or full replacement of silicon by aluminum has proved successful for hot dip galvanized

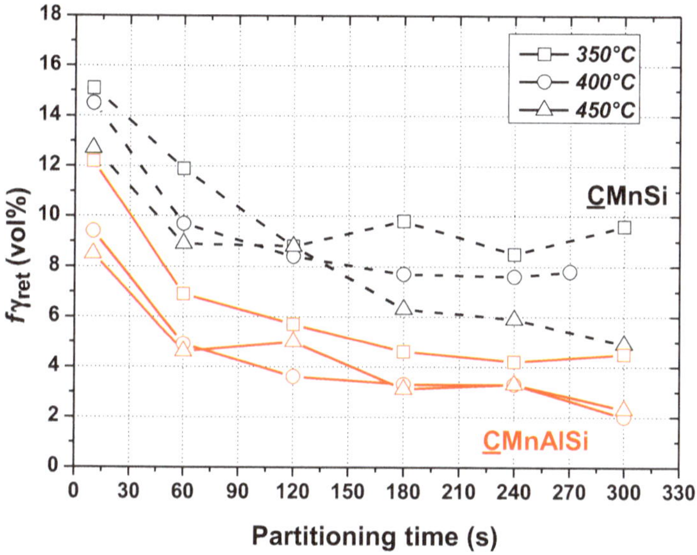

Figure 10.6 Retained austenite fractions ($f\gamma_{ret}$) stabilized by Q&P processing of a 0.24C-1.61Mn-1.45Si (CMnSi) and a 0.25C-1.70Mn-0.69Al-0.55Si (CMnAlSi) steel as a function of partitioning time for partitioning temperatures shown in the legend [26].
Source: E. De Moor, J.G. Speer, D.K. Matlock, C. Föjer, J. Penning, Materials Science and Technology (MS&T), October 25–29, 2009, Pittsburgh, PA, 2009, pp. 1554–1563.

TRIP steels with good zinc coverage [46]. In addition, the austempering treatment can be performed at temperatures compatible with the hot dip galvanizing process [47] and increased bainite transformation kinetics have been observed for Al alloyed TRIP steels compared to CMnSi alloys [46], and one benefit of Q&P processing is its insensitivity to bainite transformation kinetics.

More recent research has indicated that the development and prevalence of surface oxides in CMnSi alloys may relate to the employed soaking temperature and a reduction in oxides has been observed for higher reheating temperatures exceeding the A_3 temperature [48,49]. Limited surface oxides developed and good zinc wettability was obtained. This observation may be relevant as full austenitization leads to the greatest strength levels in Q&P studies, although extensive efforts have been undertaken to develop hot-dip coating practices for higher-Si steels.

Hot dip galvanizing of Q&P steels is of interest [50,51] and the effect of partial or full substitution of silicon by aluminum on the Q&P response is therefore important and has been studied [32]. Fig. 10.6 shows corresponding austenite fractions for a 0.24C-1.61Mn-1.45Si (CMnSi) and a 0.25C-1.70Mn-0.69Al-0.55Si (CMnAlSi) steel as a function of partitioning time for a range of partitioning temperatures and times. Thus, Al is not considered as effective in Q&P steels as in earlier TRIP steel grades.

It is of interest to note that the austenite fractions obtained in the CMnAlSi alloy are lower than in the base CMnSi alloy despite a higher bulk carbon content of 0.25 compared to 0.20 wt pct [46].

10.5 Manganese alloying

As manganese is an effective austenite stabilizing element in addition to providing solid solution strengthening, elevated manganese levels have been examined, with considerable success [30–32,36,42,52].

It is also of interest to note that other automotive steel concepts under consideration at present use substantial manganese levels (e.g., 7 to 10 wt pct) and are frequently referred to as "medium manganese steels." As these concepts are being explored, some emphasis is placed on reducing alloying levels, and alloys containing approximately 5 wt pct manganese are under consideration [53,54]. This level has also been employed in Q&P studies [32,55]. Fig. 10.7 shows engineering stress-strain curves obtained for alloys containing 3 wt pct manganese and carbon levels of 0.2 or 0.3 wt pct, partitioned using the indicated conditions. Tensile strength levels approaching 1500 MPa with substantial elongation levels are observed at the higher bulk carbon level resulting in greater strength and improved elongation in some cases.

Fig. 10.8 shows the effect of elevated manganese levels on the tensile properties of 0.3C-Mn-1.6Si steels. Manganese is observed to increase tensile ductility and ultimate tensile strength levels. It is interesting to note that an increase in manganese levels from 1.5 to 3 wt pct appears to result in more significant effects than a further increase to 5 wt pct manganese. Increased partitioning times result in decreasing ultimate tensile strength levels for all alloys whereas different trends are observed for tensile elongation. Namely, the leanest alloy exhibits decreasing ductility with increasing partitioning time whereas the alloys with elevated manganese levels show

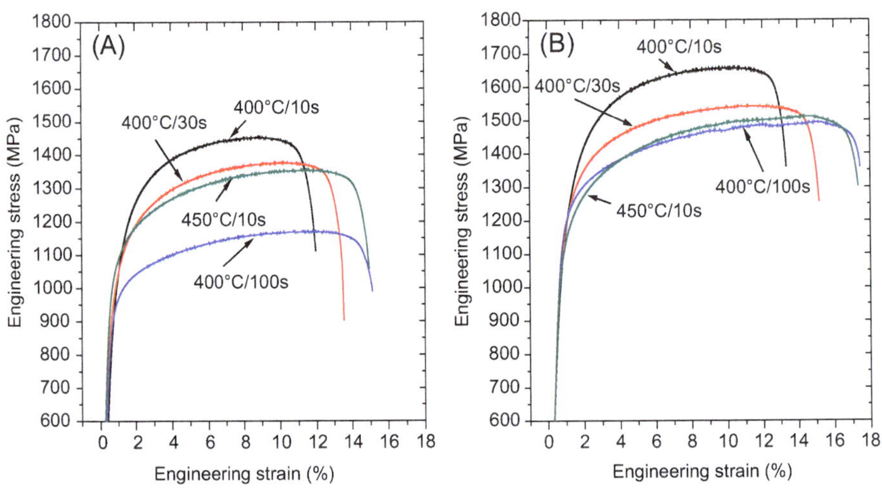

Figure 10.7 Engineering stress-strain curves obtained in (A) 0.2C-3Mn-1.6Si and (B) 0.3C-3Mn-1.6Si steels Q&P heat treated following full austenitization using the indicated partitioning conditions [32].
Source: E. De Moor, J.G. Speer, D.K. Matlock, J.H. Kwak, S.B. Lee, ISIJ Intl. 51 (2011) 137–144.

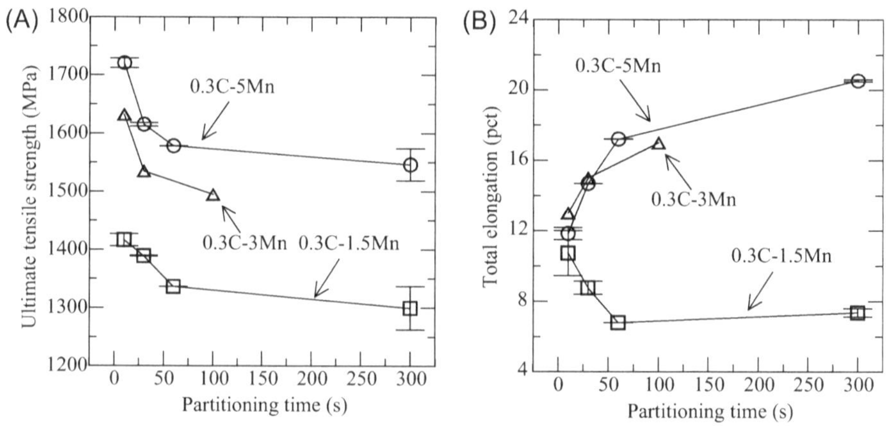

Figure 10.8 Effect of manganese content on (A) ultimate tensile strength and (B) total elongation shown as a function of partitioning time for partitioning at 400°C [55,56].
Source: E. De Moor, J. Kähkönen, J.G. Speer, D.K. Matlock, Proceedings of the Pacific Rim International Conference, Kyoto, Japan, August 2016, in press.

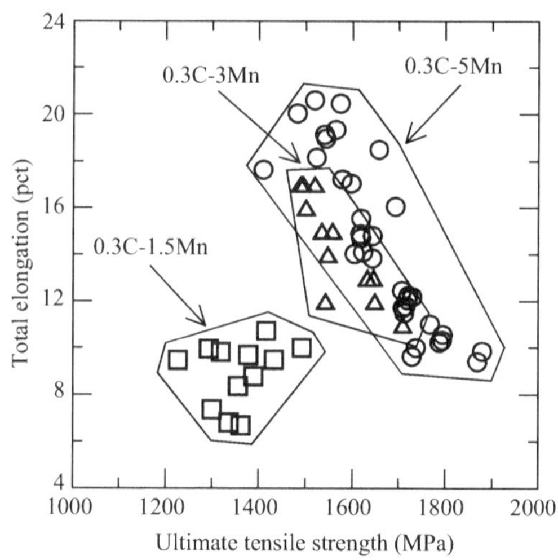

Figure 10.9 Total elongation versus ultimate tensile strength for 0.3C-Mn-1.6Si steels containing various levels of manganese [55,56].
Source: E. De Moor, J. Kähkönen, J.G. Speer, D.K. Matlock, Proceedings of the Pacific Rim International Conference, Kyoto, Japan, August 2016, in press.

continuously increasing tensile elongation with partitioning time. Longer partitioning times beyond the scope of the study may even further increase elongation levels. Additional data for the alloys are summarized in Fig. 10.9 and it is of interest to note that a number of partitioning conditions led to ultimate tensile strength and total elongation combinations of approximately 1500 MPa or greater and 20 pct respectively.

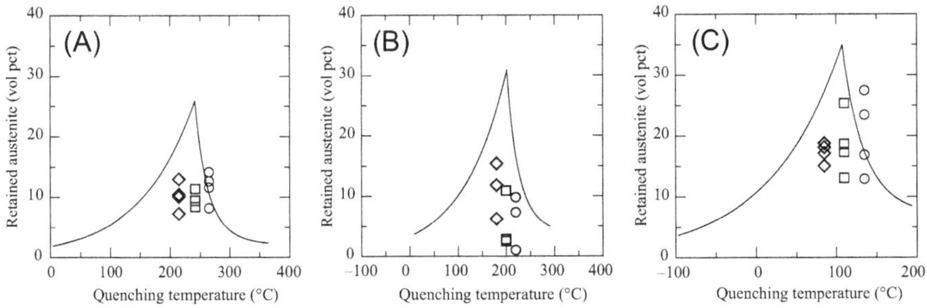

Figure 10.10 Calculated based on idealized carbon partitioning (solid line) and experimentally obtained retained austenite levels (discrete data points) as a function of quenching temperature after Q&P processing of (A) 0.3C-1.5Mn-1.6Si (B) 0.3C-3Mn-1.6Si and (C) 0.3C-5Mn-1.6Si steels using full austenitization and partitioning at 400°C for various times [55].
Source: J. Kähkönen, Master of Science thesis, Colorado School of Mines, Golden, CO, 2016.

The corresponding austenite fractions are plotted as a function of quenching temperature in Fig. 10.10 for all three manganese levels. Manganese is an austenite stabilizing element, and the highest bulk manganese level resulted in the greatest austenite fractions following Q&P heat treating with values close to the levels predicted using idealized carbon partitioning, where all the carbon is assumed to partition from martensite into austenite. It should be noted that the intermediate manganese steel showed lower fractions than the leanest alloy in some cases.

10.6 Carbon alloying

Martensite hardness and strength are known to correlate with its carbon content [57]. Kähkönen *et al.* (2015) studied the effect of carbon on tensile properties and austenite retention in 0.3 and 0.4 wt pct carbon containing CMnSi steels [58] and compared the results to literature data for 0.2 wt pct Q&P steels. Resulting stress-strain curves are shown in Fig. 10.11 for two partitioning conditions, 400°C for 10 s and 300 s, respectively. It is apparent that increased bulk carbon levels substantially increase tensile strength levels whereas tensile elongation levels overall seem less affected. Fig. 10.12 shows corresponding austenite levels for the three bulk alloy carbon levels as a function of quenching temperature. Both calculated fractions assuming "idealized" carbon partitioning conditions and experimental fractions obtained following partitioning at 400°C for various times are shown. An increase in the maximum calculated fractions is obtained with increased bulk carbon content, with optimum quenching temperatures (i.e., those temperatures resulting in the greatest austenite fractions) showing slight increases. The experimental austenite fractions increase with increasing bulk carbon content and the

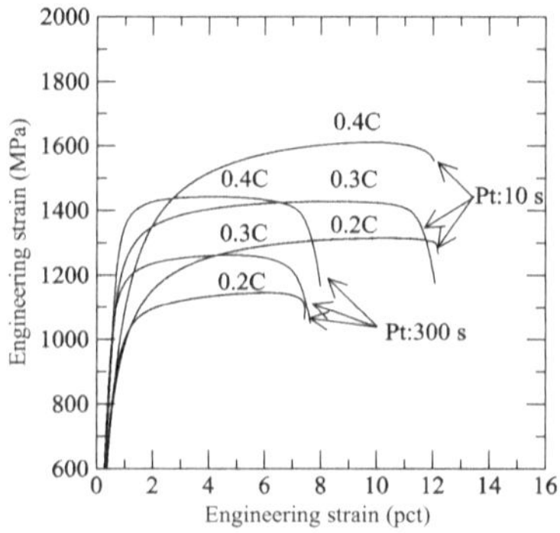

Figure 10.11 Engineering stress-strain curves obtained in 0.2, 0.3, and 0.4C-1.5Mn-1.6Si Q&P processed steels with partitioning conducted at 400°C for 10 and 300 s respectively [58].
Source: J. Kähkönen, E. De Moor, G.A. Thomas, J.G. Speer, SAE Intl. J. Mater. Manuf. 8 (2) (2015) 419–424.

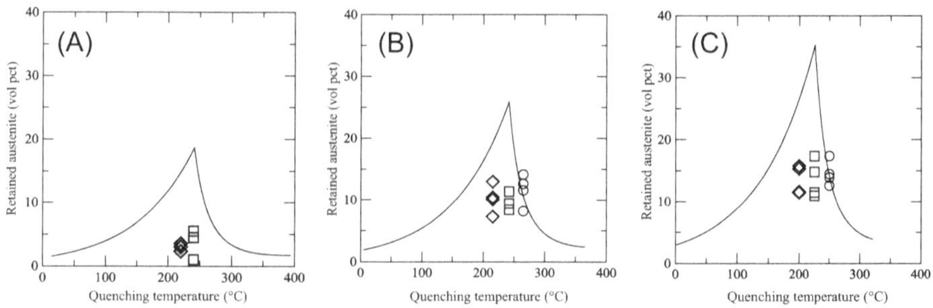

Figure 10.12 Calculated (solid line) and experimentally obtained retained austenite fractions (discrete data points) as a function of quenching temperature obtained by Q&P processing of (A) 0.2C-1.5Mn-1.6Si (B) 0.3C-3Mn-1.6Si and (C) 0.4C-5Mn-1.6Si steels using full austenitization and partitioning at 400°C for various times [32,58].
Source: J. Kähkönen, E. De Moor, G.A. Thomas, J.G. Speer, SAE Intl. J. Mater. Manuf. 8 (2) (2015) 419–424.

fractions approximate calculated levels better. Fig. 10.13 shows the evolution of retained austenite carbon contents as a function of partitioning time for partitioning at 400°C. Increased carbon concentrations are obtained with longer partitioning time. Bulk carbon alloy content appears to only have modest effects on austenite carbon levels.

Figs. 10.12 and 10.13 presented results obtained from X-ray diffraction (XRD) analysis with austenite carbon contents determined from lattice parameter measurements and correlation with carbon content via the Cullity method [59]. More recently, Mössbauer spectroscopy has been employed to characterize Q&P microstructures [27]. Direct correlations between Mössbauer absorption spectra

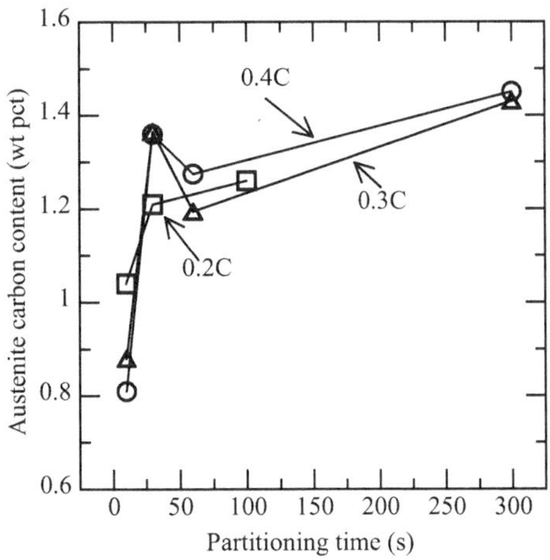

Figure 10.13 Retained austenite carbon content as a function of partitioning time following partitioning at 400°C for the three indicated bulk alloy carbon levels in C-1.5Mn-1.5Si steels [32,58].
Source: J. Kähkönen, E. De Moor, G.A. Thomas, J.G. Speer, SAE Intl. J. Mater. Manuf. 8 (2) (2015) 419–424.

and the amount and location of carbon with respect to iron atoms can be made and therefore Mössbauer spectroscopy provides a more direct measure of carbon content compared to inference via lattice parameter changes. Analysis of Mössbauer spectra enables determination of austenite carbon levels, carbon levels present as a carbide, carbide identification, and phase fraction determination. A mass balance calculation can be conducted to determine the carbon content of the martensite and example results are shown in Fig. 10.14A where the amount of carbon in the various phases is plotted as a fraction of the bulk alloy carbon. The carbon content of the austenite initially increases with partitioning time (Pt) whereas a reduction is observed at the longest Pt of 300 s. η carbides were detected and the fraction of carbon consumed by the formation of these transition carbides is also included in the figure. Increased η carbon levels are observed with partitioning time. The results of the mass balance calculation show continuously decreasing carbon levels in the martensite with increased partitioning time. It is also interesting to note that the ultimate tensile strength levels of the Q&P steels decrease with increasing partitioning time as shown in Fig. 10.14B for the corresponding tensile samples. This may suggest that ultimate strength levels may be directly correlated to the martensite strength and carbon content. It should be recognized, however, that no detailed study of austenite mechanical stability and potential transformation induced plasticity contributions to strain hardening were pursued.

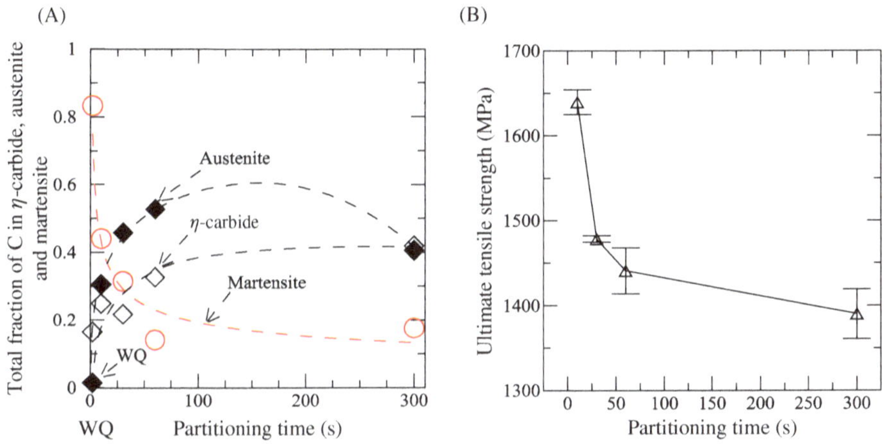

Figure 10.14 (A) Mössbauer spectroscopy results showing the distribution of carbon between austenite, η carbide and martensite via mass balance calculation in 0.4C-1.5Mn-1.5Si Q&P samples as a function of partitioning time at 400°C. WQ refers to a fully austenitized and water quenched sample. (B) Corresponding ultimate tensile strength as a function of partitioning time for Q&P samples [27,58].
Source: (A) D. Pierce, D.R. Coughlin, D.L. Williamson, K.D. Clarke, A.J. Clarke, J.G. Speer, et al., Acta Mater. 90 (2015) 417–430; (B) J. Kähkönen, E. De Moor, G.A. Thomas, J.G. Speer, SAE Intl. J. Mater. Manuf. 8 (2) (2015) 419–424.

10.7 Molybdenum additions

As illustrated by Fig. 10.6, decreasing austenite fractions with increased partitioning time are frequently observed in CMnSi alloys [26] and attempts have been made by alloying to reduce this sensitivity of the austenite fractions to prolonged partitioning. The effect of a molybdenum addition was investigated in Ref. [60] and resulting austenite fractions are shown in Fig. 10.15 as a function of partitioning time for the three indicated partitioning temperatures. Austenite fractions stabilized in a base CMnSi steel are also provided as a reference, from Fig. 10.6. In general, the molybdenum containing alloy exhibits greater austenite fractions than the base alloy and the sensitivity to partitioning time is reduced. The highest partitioning temperature shows the greatest austenite levels in the Mo added alloy. Corresponding tensile properties are shown in Fig. 10.16 on a total elongation versus tensile strength diagram. The molybdenum modified alloy shows the best strength/ductility combinations. In general, partitioning at an increased temperature results in reduced strength levels and similar or improved tensile elongation levels. Also included in the figure are mechanical property data for a 0.17Mo-0.24C-1.6Mn-0.12Si-1.41Al steel (labeled MoCMnAl in the figure). This alloy was studied to investigate whether

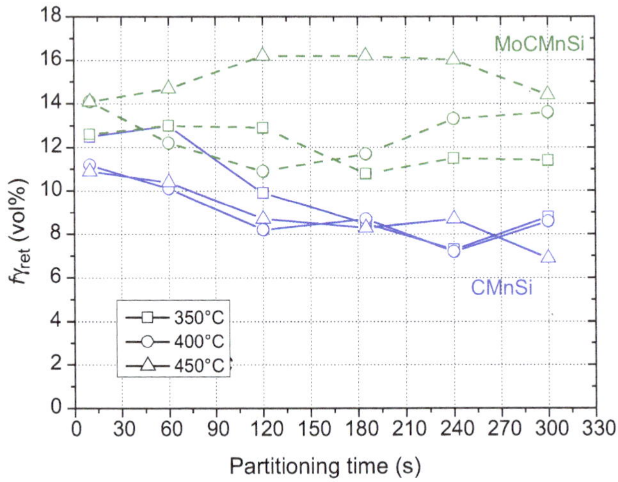

Figure 10.15 Retained austenite fractions stabilized in Q&P heat treated 0.2C-1.63Mn-1.63Si and 0.25Mo-0.21C-1.96Mn-1.49Si steels as a function of partitioning time [26]. *Source*: E. De Moor, J.G. Speer, D.K. Matlock, C. Föjer, J. Penning, Materials Science and Technology (MS&T), October 25–29, 2009, Pittsburgh, PA, 2009, pp. 1554–1563.

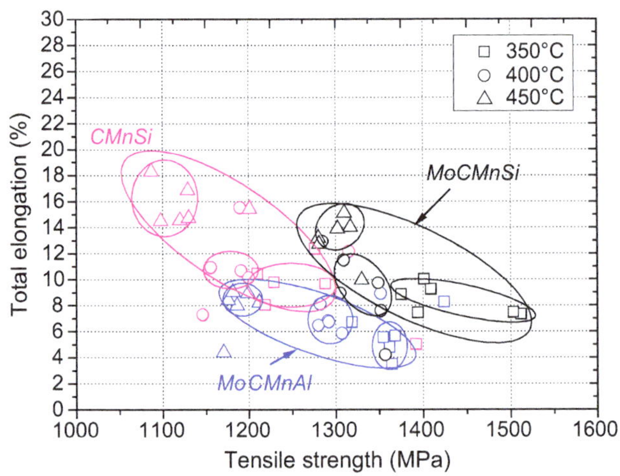

Figure 10.16 Tensile properties plotted on a total elongation versus tensile strength plot for 0.2C-1.63Mn-1.63Si, 0.25Mo-0.21C-1.96Mn-1.49Si, and 0.17Mo-0.24C-1.6Mn-1.41Al steels partitioned at the indicated partitioning temperatures for times ranging from 10 to 300 s [26]. *Source*: E. De Moor, J.G. Speer, D.K. Matlock, C. Föjer, J. Penning, Materials Science and Technology (MS&T), October 25–29, 2009, Pittsburgh, PA, 2009, pp. 1554–1563.

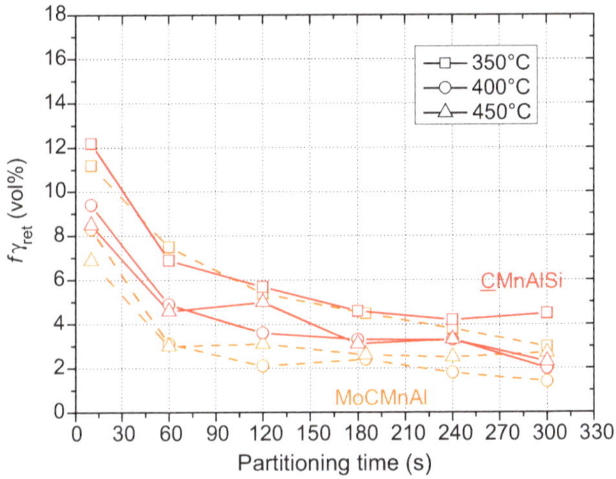

Figure 10.17 Retained austenite fractions obtained in 0.25C-1.7Mn-0.69Al-0.55Si and 0.17Mo-0.24C-1.6Mn-1.41Al-0.12Si steels as a function of partitioning time for the indicated partitioning temperatures [26].
Source: E. De Moor, J.G. Speer, D.K. Matlock, C. Föjer, J. Penning, Materials Science and Technology (MS&T), October 25–29, 2009, Pittsburgh, PA, 2009, pp. 1554–1563.

additions of molybdenum to a CMnAl alloy would be able to counteract the effect of aluminum. As discussed in the context of Fig. 10.6, the partial substitution of silicon by aluminum results in reduced austenite retention following Q&P. The austenite fractions obtained in the molybdenum added CMnAl steel are shown in Fig. 10.17 and compared to the fractions stabilized in a CMnSiAl steel; it can be noted that the same data are presented and compared to the austenite fractions stabilized in a CMnSi base alloy in Fig. 10.6. The investigated molybdenum level did not seem to beneficially affect the austenite levels and thus was not able to compensate for the influence of aluminum substitution. It can be noted that the aluminum level in the molybdenum containing alloy is higher than in the molybdenum free alloy, which may reduce austenite retention, whereas manganese levels were greater in the molybdenum containing alloy. Corresponding tensile property combinations for the MoCMnAl alloy are also included in Fig. 10.16. The lowest combinations of ultimate tensile strength and total elongation were observed for the Mo-Al containing alloy which may relate to the limited austenite retention and perhaps reduced contributions from transformation induced plasticity. Example engineering stress-strain curves are provided in Fig. 10.18 for the three alloys discussed. The MoCMnAl alloy clearly shows the most limited amount of strain hardening in addition to the lowest retained austenite fraction. The CMnSi and MoCMnSi alloys show significant strain hardening and it is noteworthy that the stress-strain curve obtained for the MoCMnSi steel appears to exhibit similar (initial) strain hardening

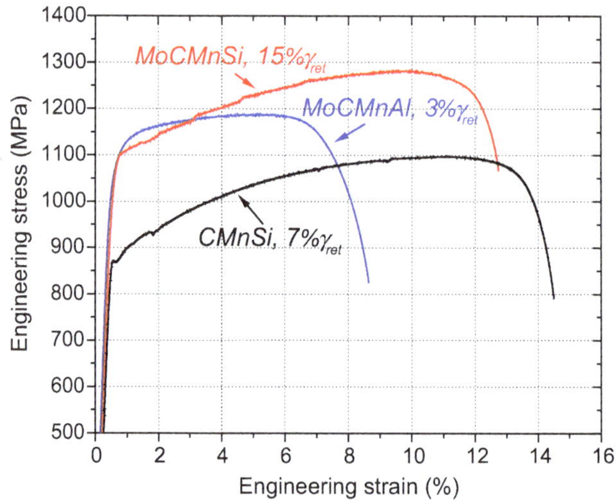

Figure 10.18 Engineering stress-strain curves for samples partitioned at 450°C for 300 s. Austenite fractions are included in the labels [26].
Source: E. De Moor, J.G. Speer, D.K. Matlock, C. Föjer, J. Penning, Materials Science and Technology (MS&T), October 25–29, 2009, Pittsburgh, PA, 2009, pp. 1554–1563.

albeit at greater strength levels, and also contains approximately twice the amount of retained austenite.

As indicated by the chemical compositions list in Table 10.2, other alloying approaches exist and richer alloys are being investigated. A detailed discussion of the various approaches is beyond the scope of the current review.

10.8 Competing reactions during partitioning

As indicated by Fig. 10.6, austenite decomposition may be occurring during the partitioning treatment in addition to mechanisms competing for available carbon such as (transition) carbide formation. Austenite levels may decrease by movement of the martensite/austenite boundary [61] as evidenced by recent experimental observations [62,63], decomposition into ferrite and cementite and/or bainite formation type mechanisms in particular if quench temperatures only moderately lower than the M_s temperature are used. It should be recognized that, although these phenomena may occur and reduce austenite fractions, substantial experimental work supports austenite stabilization by carbon enrichment from martensite carbon depletion as an active and/or predominant mechanism [30,64–66]. A few highlights of these experimental observations are presented here with no intent to be comprehensive.

Fig. 10.19 shows nano-secondary ion mass spectroscopy (SIMS) results of a Q&P heat treated 0.29C-2.95Mn-1.59Si steel quenched to 180°C and partitioned at 400°C for 100 s [67]. The investigated area is imaged in Fig. 10.19A by scanning electron microscopy (SEM) and electron backscatter diffraction (EBSD) with phase

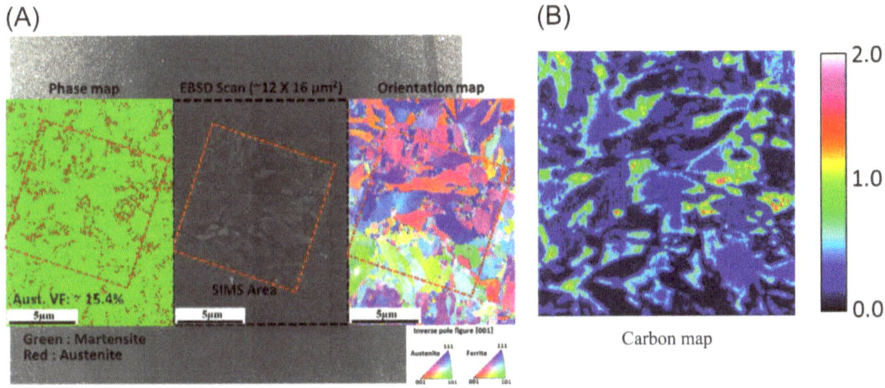

Figure 10.19 Nano-secondary ion mass spectroscopy of a 0.29C-2.95Mn-1.59Si steel sample quenched to 180°C and partitioned at 400°C for 100 s. Investigated area shown by (A) secondary electron imaging and electron back scattering diffraction including a phase map with green regions corresponding to martensite and red regions to austenite and inverse pool figure map and (B) resulting nano-SIMS color coded carbon signal with the legend in wt pct [67].
Source: K.S. Choi, Z. Zhu, X. Sun, E. De Moor, M.D. Taylor, J.G. Speer, et al., Scripta Mater. 104 (2015) 79–82.

maps and inverse pole figure maps shown. The corresponding nano-SIMS carbon signal obtained for this area is provided in Fig. 10.19B. Substantial carbon enrichment of austenite adjacent to martensitic constituents is clearly observed in the nano-SIMS map. Bainitic laths are not present in the microstructure. Carbon levels lower than the bulk carbon content are observed in the martensitic regions. It should be noted that identification of any (transition) carbides in the martensite was not possible at the available resolution and potential carbon present in carbides is therefore averaged with the carbon levels of the martensitic features.

A different study employed significant alloying for increased hardenability and to retard the bainitic transformation [42,63]. The rich alloying substantially depressed the M_s temperature so that austenite was stable at room temperature following water quenching. Subsequent partitioning behavior was then assessed by XRD, transmission electron microscopy (TEM), and atomprobe tomography at various stages. Fig. 10.20 shows austenite lattice parameter variations with partitioning time in a 0.31C-13.95Ni-1.91Si-0.24Mo steel [63]. Partitioning at 300°C shows a continuously increasing lattice parameter indicative of carbon enrichment and the increase occurs earlier for partitioning at 400°C whereas limited change is observed at a PT of 200°C.

EBSD was also employed to identify martensite and austenite phases and corresponding interphase interfaces. Focused ion beam (FIB) liftouts were then taken to include both phases and the boundary. The resulting compositional profile generated by atomprobe tomography is shown in Fig. 10.21. Substantial carbon depletion of martensite and enrichment of austenite are apparent, further supporting carbon partitioning from martensite to austenite as an effective mechanism for austenite stabilization and retention.

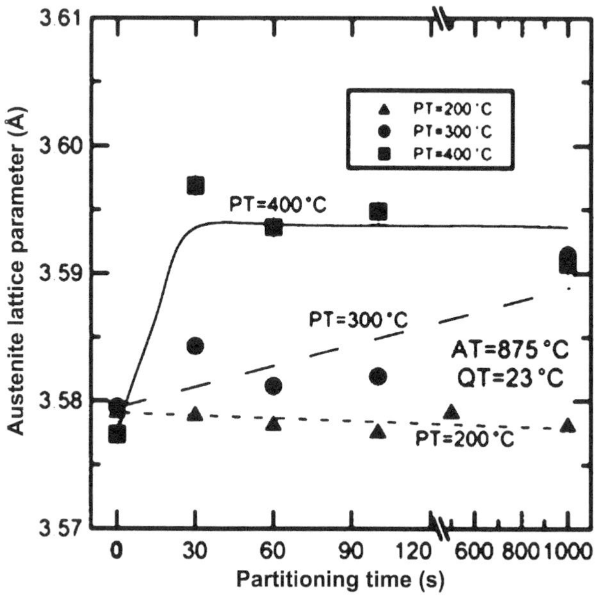

Figure 10.20 Austenite lattice parameter change as a function of partitioning time for the indicated partitioning temperatures in a 0.31C-13.95Ni-1.91Si-0.24Mo steel [63].
Source: G.A. Thomas, F. Danoix, J.G. Speer, S.W. Thompson, F. Cuvilly, ISIJ Intl. 54 (12) (2014) 2900–2906.

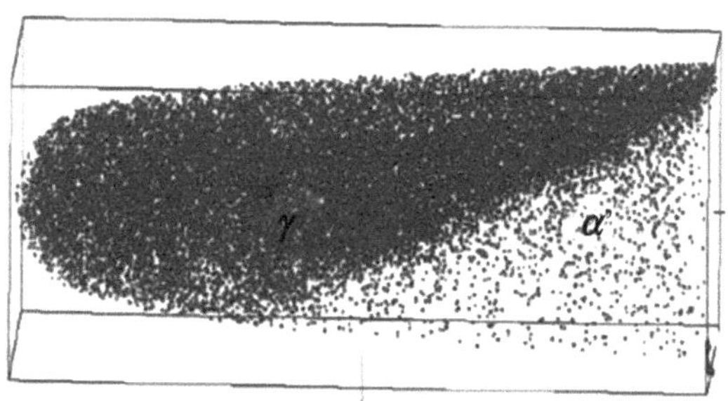

Figure 10.21 Carbon distribution map of a 0.31C-13.95Ni-1.91Si-0.24Mo Q&P sample partitioned at 400°C for 100 s. Analyzed volume was $25 \times 25 \times 63\ nm^3$ and contained 1.5×10^6 ions. Austenite and martensite were identified by EBSD prior to FIB liftout generation and atomprobe analysis [63].
Source: G.A. Thomas, F. Danoix, J.G. Speer, S.W. Thompson, F. Cuvilly, ISIJ Intl. 54 (12) (2014) 2900–2906.

Although carbon partitioning from martensite is a substantial contributor to carbon enrichment and stabilization of austenite, competing reactions have also been confirmed, such as the formation of transition carbides. Fig. 10.22 shows a transmission electron micrograph taken from a 9260 steel (with composition

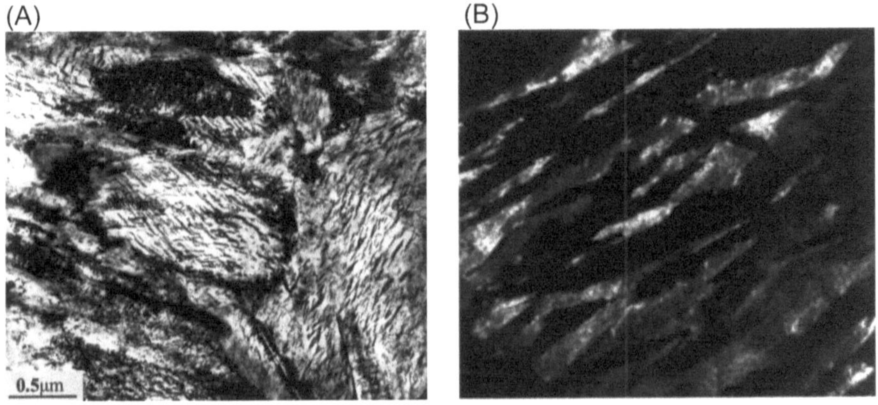

Figure 10.22 Transmission electron micrographs of 9260 Q&P microstructures (A) bright field image of a sample quenched to 150°C and partitioned at 250°C and (B) dark field image generated using an austenite reflection of a sample quenched to 190°C and partitioned at 400°C [68].
Source: D.V. Edmonds, K. He, F.C. Rizzo, B.C. De Cooman, D.K. Matlock, J.G. Speer, Mater. Sci. Eng. A438–440 (2006) 25–34.

0.60C–0.95Mn–1.96Si) Q&P processed using various partitioning conditions [68]. Fig. 10.22A shows abundant transition carbides identified as ε carbides by selected area diffraction work. Other researchers have also reported ε carbides in some alloys following Q&P processing [69] in addition to η carbides in other studies including the Mössbauer spectroscopy results in Fig. 10.14 [27,70]. Tempering reactions leading to transition carbide formation therefore warrant special attention to understand the competition with martensite carbon depletion. Fig. 10.22B shows a dark field image using an austenite reflection indicating that in this condition, substantial austenite fractions were present in the microstructure. It should be noted that the corresponding sample was partitioned at a higher temperature and fewer transition carbides were apparent. Clearly the various competing mechanisms are composition and processing dependent, so it is difficult to capture their importance.

Mössbauer spectroscopy has been conducted more recently to identify and quantify transition carbide formation and map the distribution of carbon levels in austenite, carbides, and martensite in a 0.4C-1.5Mn-1.5Si steel [27,71]. η transition carbides were detected and confirmed by electron diffraction and the corresponding amount of η carbides determined by Mössbauer spectroscopy is shown in Fig. 10.23 for a variety of heat treating conditions including full austenitization and water quenching (WQ), quenching and tempering (Q&T), and Q&P. One-step Q&P, where partitioning is conducted by extended holding at the quenching temperature, and two-step Q&P with partitioning at 400°C, were both considered. It is of particular interest to note that the amount of η carbides observed in the Q&T sample is significantly greater than the amounts observed in the Q&P samples. Some transition carbides formed through autotempering were also detected in the WQ samples.

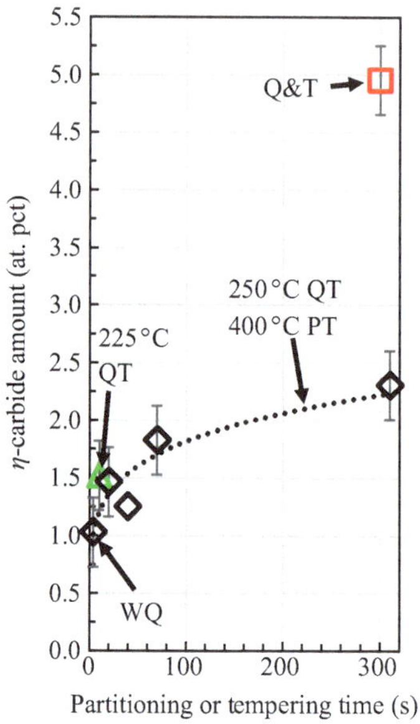

Figure 10.23 Amount of η carbides determined by Mössbauer spectroscopy in a 0.4C-1.5Mn-1.5Si steel heat treated using full austenitization and water quenching (WQ), quenching and tempering (Q&T) at 400°C for 300 s, one-step Q&P processing by holding at a quenching temperature of 225°C for 10 s and two-step Q&P heat treating with partitioning at 400°C for the indicated times [27].
Source: D. Pierce, D.R. Coughlin, D.L. Williamson, K.D. Clarke, A.J. Clarke, J.G. Speer, et al., Acta Mater. 90 (2015) 417–430.

Carbide fractions in the one-step and two-step Q&P samples were greater than in the WQ samples and fractions increase with increasing partitioning time.

The distribution of carbon among the various phases present in these samples was presented earlier in Fig. 10.14A. The carbon fractions contained in austenite and η carbide were determined directly from the Mössbauer absorption spectra whereas the martensite carbon fraction was determined by a mass balance calculation. As pointed out earlier, continuously decreasing levels of carbon solute with time are observed in the martensite whereas the carbon tied up in η carbides increases with time. At the longest partitioning time, approximately 20 pct of the total bulk carbon content (or 0.1 wt pct) remains in martensite whereas 40 pct (or 0.2 wt pct) is tied up in a carbide; the remainder having partitioned into austenite.

10.9 Local formability of bainitic and Q&P steels

In addition to tensile ductility and "global" formability, "local" formability, in particular near sheared edges, is of growing concern and interest. Failures have been

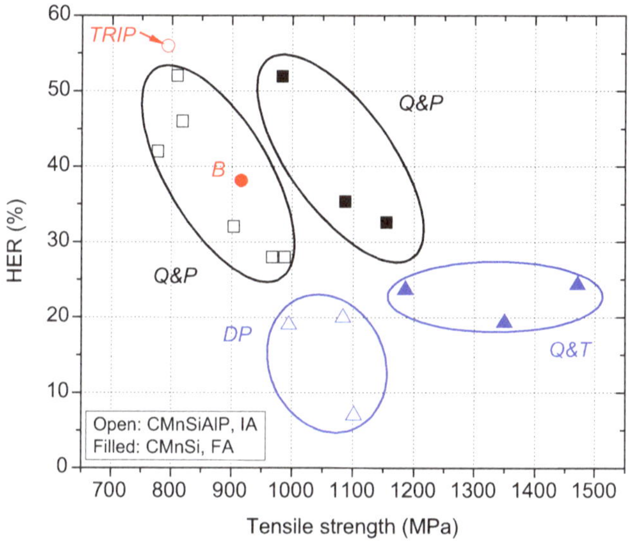

Figure 10.24 Hole expansion ratio (HER) as a function of tensile strength for intercritically annealed (IA, open symbols) and fully austenitized (FA, filled symbols) microstructures including Q&P, Q&T, tempered dual phase, bainitic and TRIP microstructures [72].
Source: E. De Moor, D.K. Matlock, J.G. Speer, C. Föjer, J. Penning, SAE Tech. Publ. #2012-01-0530, SAE, Warrendale, PA, 2012. Reprinted with Permission from SAE International.

reported to initiate near sheared edges, and "damage" and/or strain hardening near the sheared edge is believed to play a crucial role in understanding resistance to local sheared edge failure.

A few studies have investigated the local formability of Q&P and bainitic steels and Fig. 10.24 shows hole expansion ratios (HER) obtained in fully austenitized (closed symbols) and intercritically annealed microstructures (open symbols) plotted as a function of tensile strength [72]. Heat treating following soaking consisted of two-step Q&P, austempering resulting in intercritically austenitized TRIP or fully bainitic ("B") microstructures as well as water quenching and tempering to generate Q&T and tempered DP microstructures. In general the fully austenitized materials showed greater HER values at a given strength level than the intercritically annealed microstructures due to reduced hardness differences between individual phases [73]. The fully austenitized Q&P steels showed greater HER and tensile strength combinations than the Q&T and bainitic materials. The tempered DP steels showed comparable or greater tensile strength and lower HER than the intercritically annealed Q&P steels. The TRIP microstructure showed greater HER than the intercritically annealed Q&P steels at comparable tensile strength levels. It should be noted that hole preparation in this study was performed by drilling, and a flat die punch was used during expansion.

Other work employed punching as a hole preparation method and a conical die punch [74]. A systematic variation in microstructures was obtained using the heat

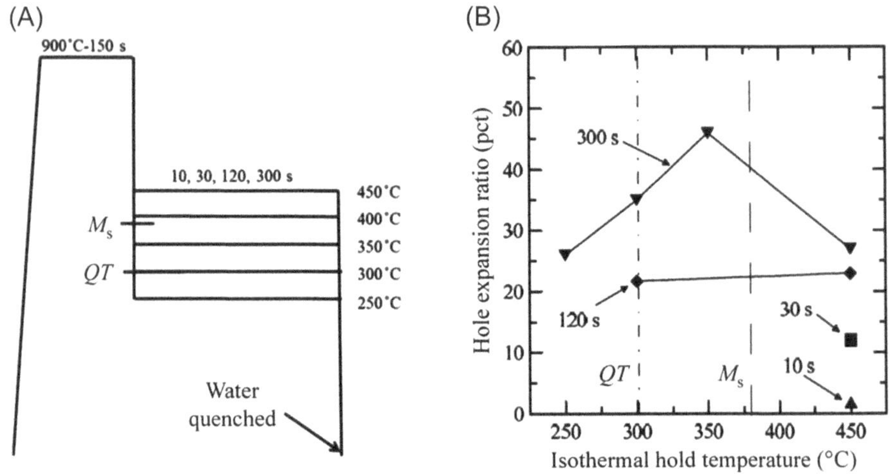

Figure 10.25 (A) Heat treating schedule employed in Ref. [74] on a 0.25C-2.05Mn-1.49Si steel and (B) associated hole expansion ratios as a function of isothermal hold temperature for the indicated hold times.
Source: R.J. Johnson, E. De Moor, N. Fonstein, D.N. Hanlon, A. Pichler, in: E. De Moor, H.J. Jun, J.G. Speer, M. Merwin (Eds.), Proceedings of the International Symposiu on New Developments in Advanced High-Strength Sheet Steels, AIST, Warrendale, PA, 2013, pp. 71–83.

treating schedule shown in Fig. 10.25A for a 0.25C-2.05Mn-1.49Si steel. Samples isothermally held in the bainitic transformation regime above M_s, TBF type microstructures and one-step Q&P steels were generated using this heat treating matrix. The M_s temperature was calculated to be 380°C and the optimum QT 301°C. The corresponding hole expansion ratios are shown in Fig. 10.25B for the indicated hold times. It should be recognized that the heat treatments resulted in strength and total elongation variations as shown in Fig. 10.26A and the hole expansion data as plotted as a function of ultimate tensile strength in Fig. 10.26B. Superior tensile ductility over a wide range of tensile strength levels was obtained for a hold at 450°C, resulting in bainitic microstructures containing fairly coarse retained austenite islands and laths as shown in Fig. 10.27A. Some fresh martensite, formed during final water quenching to room temperature, was also observed. It is interesting to note that the corresponding hole expansion values are lower than the values obtained for the other heat treatments as shown in Fig. 10.25B. Q&P microstructures showed the greatest HER at ultimate tensile strength levels of 1600 MPa. An example Q&P microstructure is provided in Fig. 10.27B showing a very fine microstructure containing retained austenite as fine films.

Another study detailing hole expansion behavior of samples held isothermally at a variety of temperatures above and below the M_s temperature (reported to range from 416 to 433°C) following full austenitization was conducted by

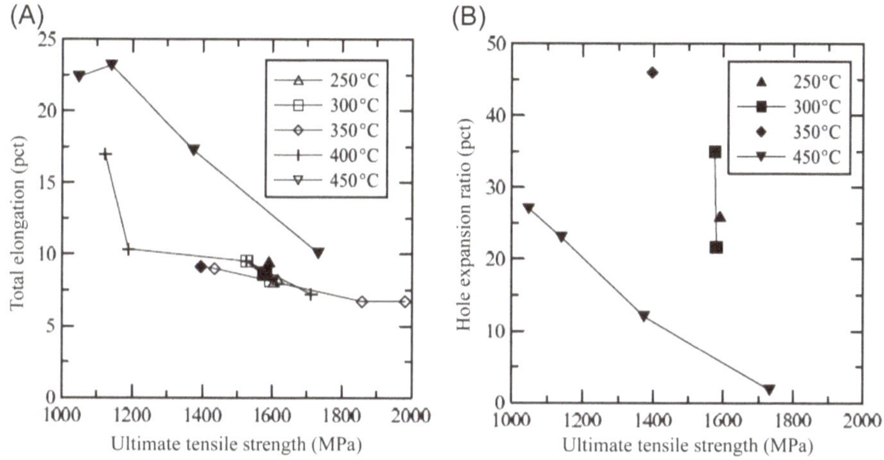

Figure 10.26 (A) Total elongation as a function of ultimate tensile strength for the heat treatments shown in Fig. 10.25 and (B) hole expansion ratio as a function of ultimate tensile strength for select heat treating conditions [74].
Source: R.J. Johnson, E. De Moor, N. Fonstein, D.N. Hanlon, A. Pichler, in: E. De Moor, H. J. Jun, J.G. Speer, M. Merwin (Eds.), Proceedings of the International Symposiu on New Developments in Advanced High-Strength Sheet Steels, AIST, Warrendale, PA, 2013, pp. 71–83.

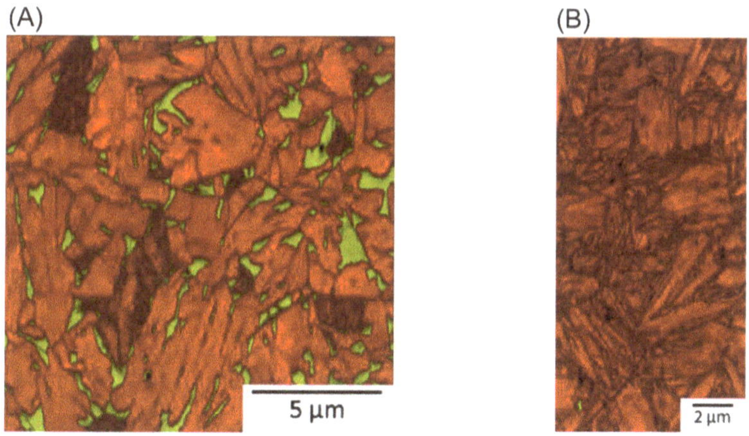

Figure 10.27 Combined phase and image quality map for 0.25C-2.05Mn-1.49Si steel samples fully austentized and isothermally held at (A) 450°C for 120 s and (B) 300°C for 300 s [74].
Source: R.J. Johnson, E. De Moor, N. Fonstein, D.N. Hanlon, A. Pichler, in: E. De Moor, H. J. Jun, J.G. Speer, M. Merwin (Eds.), Proceedings of the International Symposiu on New Developments in Advanced High-Strength Sheet Steels, AIST, Warrendale, PA, 2013, pp. 71–83.

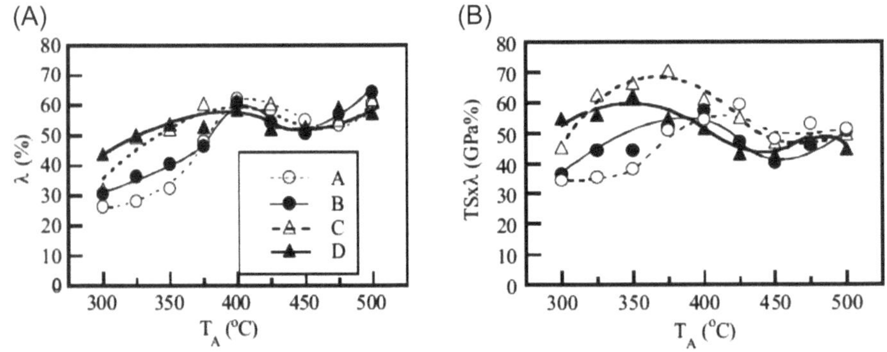

Figure 10.28 (A) Hole expansion ratio (λ) and (B) its product with tensile strength (TS) plotted as a function of isothermal hold temperature (T_A) for TBF steel samples isothermally held above and below the M_s temperature ranging from 416 to 433°C. Symbols in the legend correspond to composition: open circle-A: 0.19C-1.54Si-1.51Mn, closed circle-B: 0.20C-0.99Si-1.51Mn, open triangle-C: 0.21C-1.50Si-1.50Mn-0.048 Nb, triangle-D: 0.20C-1.00Si-1.50Mn-0.049 Nb [18].
Source: K.-I. Sugimoto, M. Murata, Y. Mukai, Proc. Mater. Sci. Technol. (2007) 15–26.

Sugimoto et al. [18]. In addition, the effect of niobium microalloying was studied. Resulting hole expansion ratios (λ) are plotted as a function of isothermal hold temperature (T_A) in Fig. 10.28 [18] along with the product of the hole expansion ratio and the obtained tensile strengths. Lower hole expansion ratios are obtained below M_s albeit at higher strength levels typically. Niobium microalloying has a beneficial effect on hole expansion and its product with tensile strength in particular for isothermal holding below the M_s temperature.

10.10 Conclusions

This chapter reviewed selected aspects of bainitic and Q&P steels as potential third generation AHSS candidates due to their fine high strength microstructure containing retained austenite. Heat treating and alloying strategies were reviewed in addition to tensile and hole expansion properties. The competition between carbon partitioning and martensite tempering reactions such as transition carbide formation was discussed.

Acknowledgments

The authors gratefully acknowledge the support of the sponsors of the Advanced Steel Processing and Products Research Center, an industry-university cooperative research center at Colorado School of Mines. This work was partially supported by the Advanced Manufacturing Office of the Department of Energy under award number DE-EE0005765.

Disclaimer

This report was prepared as an account of work sponsored by an agency of the United States Government. Neither the United States Government nor any agency thereof, nor any of their employees, makes any warranty, express or implied, or assumes any legal liability or responsibility for the accuracy, completeness, or usefulness of any information, apparatus, product, or process disclosed, or represents that its use would not infringe privately owned rights. Reference herein to any specific commercial product, process, or service by trade name, trademark, manufacturer, or otherwise does not necessarily constitute or imply its endorsement, recommendation, or favoring by the United States Government or any agency thereof. The views and opinions of authors expressed herein do not necessarily state or reflect those of the United States Government or any agency thereof.

References

[1] D.K. Matlock, J.G. Speer, in: H.C. Lee (Ed.), Proceedings of the The 3rd International Conference on Structural Steels, The Korean Institute of Metals and Materials, Seoul, Korea, 2006, pp. 774–781.
[2] D.K. Matlock, J.G. Speer, in: A. Haldar, S. Suwas, D. Bhattacharjee (Eds.), Microstructure and Texture in Steels and Other Materials, Springer, London, 2009, pp. 185–205.
[3] S. Kang, J.G. Speer, D. Krizan, D.K. Matlock, E. De Moor, Mater. Des. 97 (2016) 138–146.
[4] K.S. Choi, X. Hu, X. Sun, M. Taylor, E. De Moor, J. Speer, et al., SAE Technical Publication #2014-01-0812, SAE, Warrendale, PA, 2014, 9 p.
[5] F.G. Caballero, H.K.D.H. Bhadeshia, Curr. Opin. Solid State Mater. Sci. 8 (2004) 251–257.
[6] C. García-Mateo, F.G. Caballero, H.K.D.H. Bhadeshia, ISIJ Intl. 43 (8) (2003) 1238–1243.
[7] C. García-Mateo, F.G. Caballero, H.K.D.H. Bhadeshia, ISIJ Intl. 43 (11) (2003) 1821–1825.
[8] C. Garcia-Mateo, F.G. Caballero, Mater. Trans. 46 (8) (2005) 1839–1846.
[9] C. Garcia-Mateo, F.G. Caballero, ISIJ Intl. 45 (11) (2005) 1736–1740.
[10] H.K.D.H. Bhadeshia, D.V. Edmonds, Metal Sci. 17, 1983, pp. 411–419.
[11] H.K.D.H. Bhadeshia, D.V. Edmonds, Metal Sci. 17, 1983, pp. 420-425.
[12] F.G. Caballero, H.K.D.H. Bhadeshia, K.J.A. Mawella, D.G. Jones, P. Brown, Mater. Sci. Tech. 17 (May 2001) 517–522.
[13] F.G. Caballero, M.J. Santofimia, C. García-Mateo, J. Chao, C. García de Andrés, Mater. Des. 30 (6) (2009) 2077–2083.
[14] S.M.C. Van Bohemen, M.J. Santofimia, J. Sietsma, Scripta Mater. 58 (2008) 488–491.
[15] A. Navarro-López, J. Sietsma, M.J. Santofimia, Metall. Mater. Trans. A 47 (2016) 1028–1039.
[16] D.H. Kim, J.G. Speer, B.C. De Cooman, Metall. Mater. Trans. A 40 (2009) 2048–2060.
[17] Koh-ichi Sugimoto, T. Iida, J. Sakaguchi, T. Kashima, ISIJ Intl. 40 (9) (2000) 902–908.

[18] K.-I. Sugimoto, M. Murata, Y. Mukai, Proc. of Materials Science and Technology (2007) 15–26.
[19] K. Hausmann, D. Krizan, K. Spiradek-Hahn, A. Pichler, E. Werner, Mater. Sci. Eng. A 588 (2013) 142–150.
[20] D.K. Matlock, V.E. Bräutigam, J.G. Speer, Mater. Sci. Forum 426–432 (2003) 1089–1094.
[21] J.G. Speer, D.K. Matlock, B.C. De Cooman, J.G. Schroth, Acta Mater. 51 (2003) 2611–2622.
[22] J.G. Speer, A.M. Streicher, D.K. Matlock, F. Rizzo, G. Krauss, Austenite Formation and Decomposition, ISS/TMS, Warrendale, PA, 2003, pp. 505–522.
[23] W.S. Owen, Trans. ASM 46 (1954) 812.
[24] B.C. De Cooman, Curr. Opin. Solid State Mater. Sci. 8 (2004) 285–303.
[25] E. De Moor, S. Lacroix, A.J. Clarke, J. Penning, J.G. Speer, Metall. Trans. A 39A (2008) 2586–2595.
[26] G.A. Thomas, E. De Moor, J.G. Speer, in: E. De Moor, H.J. Jun, J.G. Speer, M. Merwin (Eds.), Proceedings of the Internationl Symposium on New Developments in Advanced High-Strength Sheet Steels, AIST, Warrendale, PA, 2013, pp. 153–165.
[27] D. Pierce, D.R. Coughlin, D.L. Williamson, K.D. Clarke, A.J. Clarke, J.G. Speer, et al., Acta Mater. 90 (2015) 417–430.
[28] A.M. Streicher, J.G. Speer, D.K. Matlock, B.C. De Cooman, Proceedings of the International Conference on Advanced High Strength Sheet Steels for Automotive Applications, AIST, Warrendale, PA, June 6–9, 2004, Winter Park, CO, pp. 51–62.
[29] Y. Toji, G. Miyamoto, D. Raabe, Acta Mater. 86 (2015) 137–147.
[30] Y. Toji, H. Matsuda, M. Herbig, P.-P. Choi, D. Raabe, Acta Mater. 65 (2014) 215–228.
[31] E. De Moor, J.G. Speer, D.K. Matlock, J.H. Kwak, S.B. Lee, ISIJ Intl. 51 (2011) 137–144.
[32] E. De Moor, J.G. Speer, D.K. Matlock, C. Föjer, J. Penning, Materials Science and Technology (MS&T) 2009, October 25–29, 2009, Pittsburgh, PA, pp. 1554–1563.
[33] B. Mašek, H. Jirková, D. Hauserová, L. Kučerová, D. Klauberová, Mater. Sci. Forum 654–656 (2010) 94–97.
[34] M.J. Santofimia, T. Nguyen-Minh, L. Zhao, R. Petrov, I. Sabirov, J. Sietsma, Mater. Sci. Eng. A 527 (23) (2010) 6429–6439.
[35] D. De Knijf, C. Föjer, L.A.I. Kestens, R. Petrov, Mater. Sci. Eng. A 638 (2015) 219–227.
[36] D. De Knijf, E.P. Da Silva, C. Föjer, R. Petrov, Mater. Sci. Technol. 31 (7) (2015) 817–828.
[37] I. De Diego-Calderón, D. De Knijf, J.M. Molina-Aldareguia, Rev. Metal. 51 (2015) 1–12.
[38] I. De Diego-Calderón, D. De Knijf, M.A. Monclús, J.M. Molina-Aldareguia, I. Sabirov, C. Föjer, et al., Mater. Sci. Eng. A 630 (2015) 27–35.
[39] E.J. Seo, L. Cho, B.C. De Cooman, Metall Mater. Trans. A 46 (1) (2015) 27–31.
[40] A. Arlazarov, O. Bouaziz, J.P. Masse, F. Kegel, Mater. Sci. Eng. A 620 (2014) 293–300.
[41] A.R. Martins, F. Rizzo, D. Coelho, J.G. Speer, D. Matlock, K. He, et al., Proc. Mater. Sci. Technol. (2009) 1564–1574.
[42] F.C. Rizzo, A.R. Martins, J.G. Speer, D.K. Matlock, A. Clarke, B.C. De Cooman, Mater. Sci. Forum 539–543 (2007) 4476–4481.

[43] M. Cheng, H.W. Song, X. Li, S.H. Zhang, T. Lin, Mater. Sci. Forum 788 (2014) 340–345.
[44] H.Y. Li, X.W. Lu, W.J. Li, X.J. Jin, Metall. Mater. Trans. A 41 (2010) 1284–1300.
[45] J. Mahieu, S. Claessens, B.C. De Cooman, Metall. Trans. A 32 (2001) 2905–2908.
[46] J. Mahieu, J. Maki, B.C. De Cooman, S. Claessens, Metall. Trans. A 33 (2002) 2573–2580.
[47] M. De Meyer, D. Vanderschueren, B.C. De Cooman, ISIJ Intl. 39 (8) (1999) 813–822.
[48] T. Van De Putte, D. Borissov, D. Loison, J. Penning, M. Rohwerder, S. Claessens, Proceedings of the International Conference on New Developments in Advanced High-Strength Sheet Steels, AIST, June 15–18, 2008, Orlando, FL, pp. 69–80.
[49] T. Van De Putte, D. Loison, J. Penning, S. Claessens, Metall. Trans. A 39 (2008) 2875–2884.
[50] G.A. Thomas, J.M. Losz, United States Patent Application, US 2014/0338797 A1.
[51] G.A. Thomas, L.G. Garza-Martinez, United States Patent Application, US 2014/0338798 A1.
[52] M.J. Santofimia, T. Nguyen-Minh, L. Zhao, D.N. Hanlon, T.A. Kop, J. Sietsma, in: J. G. Speer, B. Nelson, R. Pradhan (Eds.), Proceedings of the International Conference on New Developments in Advanced High-Strength Sheet Steels, AIST, Warrendale, PA, 2008, pp. 191–198.
[53] S. Kang, E. De Moor, J.G. Speer, Metall. Trans. A 46 (3) (2015) 1005–1011.
[54] M.J. Merwin, Proceedings of Materials Science and Technology 2007, September 16–20, 2007, Detroit, MI, pp. 515–536.
[55] J. Kähkönen, Master of Science thesis, Colorado School of Mines, Golden, CO, 2016.
[56] E. De Moor, J. Kähkönen, J.G. Speer, D.K. Matlock, Proceedings of the Pacific Rim International Conference, Kyoto, Japan, August 2016, in press.
[57] R.A. Grange, C.R. Hribal, L.F. Porter, Metall. Trans. A 8 (1977) 1775–1785.
[58] J. Kähkönen, E. De Moor, G.A. Thomas, J.G. Speer, SAE Intl. J. Mater. Manuf 8 (2) (2015) 419–424.
[59] B.D. Cullity, Elements of X-Ray Diffraction, 3rd ed., Prentice-Hall, New Jersey, 2001, p. 640.
[60] E. De Moor, C. Föjer, A.J. Clarke, J. Penning, J.G. Speer, Proceedings of the Conference on New Developments on Metallurgy and Applications of High Strength Steels, Buenos Aires, Argentine, May 26–28, 2008.
[61] J.G. Speer, R.E. Hackenberg, B.C. De Cooman, D.K. Matlock, Phil. Mag. Lett. 87 (6) (2007) 370–382.
[62] D. De Knijf, M.J. Santofimia, H. Shi, V. Bliznuk, C. Föjer, R. Petrov, et al., Acta Mater. 90 (2015) 161–168.
[63] G.A. Thomas, F. Danoix, J.G. Speer, S.W. Thompson, F. Cuvilly, ISIJ Intl. 54 (12) (2014) 2900–2906.
[64] T.D. Bigg, D.V. Edmonds, E.S. Eardley, J. Alloys Comp. 577S (2013) S695–S698.
[65] T.D. Bigg, D.K. Matlock, J.G. Speer, D.V. Edmonds, Solid St. Phen. 172-174 (2011) 827–832.
[66] G.A. Thomas, J.G. Speer, Mater. Sci. Technol. 30 (9) (2014) 998–1007.
[67] K.S. Choi, Z. Zhu, X. Sun, E. De Moor, M.D. Taylor, J.G. Speer, et al., Scripta Mater. 104 (2015) 79–82.
[68] D.V. Edmonds, K. He, F.C. Rizzo, B.C. De Cooman, D.K. Matlock, J.G. Speer, Mater. Sci. Eng. A438-440 (2006) 25–34.
[69] F. HajyAkbary, J. Sietsma, G. Miyamoto, T. Furuhara, M.J. Santofimia, Acta Mater. 104 (2016) 72–83.

[70] D.R. Coughlin, D.T. Pierce, J.G. Speer, A.J. Clarke, P.D. Jablonski, Ő.N. Doğan, V.K. Judge, E. De Moor, K.D. Clarke, in: M. Militzer, G. Botton, L.-Q. Chen, J. Howe, C. Sinclair, H. Zurob (Eds.), Proceedings of the International Conference on Solid-Solid Phase Transformations in Inorganic Materials, 2015, pp. 1051–1058.
[71] D.T. Pierce, D.R. Coughlin, D.L. Williamson, J. Kähkönen, A.J. Clarke, K.D. Clarke, et al., Scripta Mater. 121 (2016) 5–9.
[72] E. De Moor, D.K. Matlock, J.G. Speer, C. Föjer, J. Penning, SAE Tech. Publ. #2012-01-0530, SAE, Warrendale, PA, 2012.
[73] K. Hasegawa, K. Kawamura, T. Urabe, Y. Hosoya, ISIJ Intl. 44 (3) (2004) 603–609.
[74] R.J. Johnson, E. De Moor, N. Fonstein, D.N. Hanlon, A. Pichler, in: E. De Moor, H.J. Jun, J.G. Speer, M. Merwin (Eds.), Proceedings of the International Symposiu on New Developments in Advanced High-Strength Sheet Steels, AIST, Warrendale, PA, 2013, pp. 71–83.

High Mn TWIP steel and medium Mn steel

11

B.C. De Cooman
Graduate Institute of Ferrous Technology, POSTECH, Pohang, South Korea

11.1 Introduction

Steels have plasticity-enhancing mechanisms which are not yet fully exploited in automotive applications. These mechanisms are not yet fully understood and their activation during straining requires a fundamental redesign of the way in which the composition, processing, and microstructure of automotive steel grades are selected. Most formable ferritic steels used in the automotive industry have uniform engineering elongations less than 25%, and relatively low ultimate tensile strengths (<< 1 GPa). The formability of these steels is based on the control of their crystallographic texture, rather than their stain hardening. As a consequence, higher strengths are usually achieved at the cost of plasticity. Fully austenitic steels and multi-phase steels which contain austenite can be designed to achieve an enhanced strain hardening rate, making it possible to achieve both strength and high formability. High Mn twinning-induced plasticity (TWIP) steels and medium Mn steels belong to this category of ferrous alloys. The microstructure of these two types of steels are compared in Fig. 11.1.

Medium Mn (5–12 wt.-% Mn) steel and high Mn (12–25 wt.-% Mn) TWIP steel have been shown to achieve superior combinations of strength and ductility [1–3]. High Mn TWIP steels in particular have been the subject of intense research and materials development efforts aimed at developing this new class of ultra-high strength steels for automotive applications. In terms of plasticity-enhancing mechanisms, high manganese steels can be divided into three types based on the active plasticity-enhancing mechanism (Table 11.1): TWIP steel [4,5], transformation-induced plasticity (TRIP) or TWIP + TRIP steel [6–8], and shear band or micro band induced plasticity (SBIP/MBIP) steel [9–11]. The mechanical properties of these types of steels are critically dependent on the steel composition, the grain size and the stacking fault energy (SFE) of the austenite phase.

In fully austenitic 12–25 wt.-% Mn TWIP steel deformation twinning is caused by the passage of Shockley partials on every successive {111} austenite plane. The stability of the austenite is high, i.e., it has a sub-zero martensite start (Ms) temperature, and the room temperature stacking fault energy (SFE) is typically in the range of 15–45 mJ/m^2. Whereas deformation twinning occurs when the austenite room temperature SFE is higher than 15 mJ/m^2, strain-induced ε-martensite is formed when the SFE is less than 15 mJ/m^2. The strain-dependent density of deformation twins results in a dynamic Hall-Petch effect which leads to an excellent

Automotive Steels. DOI: http://dx.doi.org/10.1016/B978-0-08-100638-2.00011-0

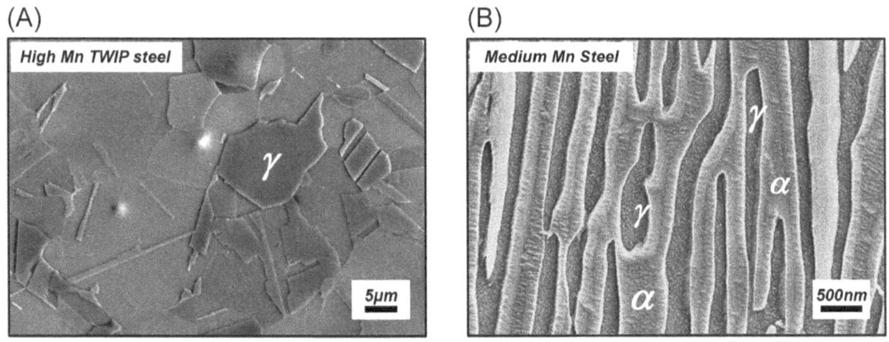

Figure 11.1 Comparison of the microstructure of high Mn TWIP steel (A) and medium Mn steel (B). The TWIP steel has a coarser grained (grain size >2 μm) fully austenitic microstructure. Its high ductility is achieved by the activation of the plasticity-enhancing mechanical twinning mechanism (TWIP effect) during straining. Medium Mn steel has an ultra-fine grained (grain size <500 nm) austenitic-ferritic microstructure. Its ductility is controlled by the activation of the plasticity-enhancing strain-induced transformation mechanism (TRIP effect) or a combination of the TWIP and the TRIP effect occurring in succession (TWIP + TRIP effect).

Table 11.1 Overview of the main properties of the different types of high and medium Mn steels

	High Mn		Medium Mn	
Wt.-% Mn	>25	22–15	12–8	7–4
Processing	Conventional Annealing		Intercritical annealing	Quenching and Partitioning
Cold rolled	Shear bands	Deformed γ	Deformed α'	
After annealing	γ		UFG $\gamma + \alpha + \alpha'$	
Plasticity	SBIP	TWIP	TWIP + TRIP→TRIP	
γ_{isf} (mJ/m^2)	$\gamma_{isf} > 75$	$15 < \gamma_{isf} < 45$	$15 < \gamma_{isf} < 45$	$\gamma_{isf} < 15$
γ-stability	γ-composition		γ-composition and size	
Role of Mn	γ-stability/SFE		γ-stability/SFE/Hardenability	

combination of strength and ductility. Both the TRIP effect and the TWIP effect are dependent on the austenite grain size, as the austenite grain refinement enhances the austenite stability and results in a decrease of TRIP and TWIP kinetics.

Medium Mn multi-phase steels exhibiting the TWIP + TRIP plasticity-enhancing mechanisms have a low content of the austenite stabilizing elements

Mn (<12 wt.-%) and C (<0.4 wt.-%) and a slightly higher content of the ferrite stabilizing element Al (~3 wt.-%) as compared to conventional TWIP steel. The medium Mn TWIP + TRIP steels have a multi-phase microstructure consisting of ferrite and austenite, obtained by intercritical annealing. The alloy composition and the intercritical annealing conditions are chosen in such a manner that the room temperature austenite SFE is in the range of 15–45 mJ/m^2. Medium Mn steels exhibit a very high tensile strength in the range of 1.0–1.5 GPa after intercritical annealing. The presence of both ferrite and austenite during intercritical annealing results in the partitioning of C and Mn to austenite and an increase of the austenite stability. During straining deformation twinning (TWIP effect) and strain-induced transformation (TRIP effect) are activated successively in the austenite. The $\gamma \rightarrow \gamma_{Twin}$ TWIP effect is activated soon after plastic deformation starts. After the volume fraction of deformation twins is saturated, the $\gamma \rightarrow \alpha'$ TRIP effect is initiated at twin-twin intersections. The succession of the TWIP and TRIP plasticity-enhancing mechanisms results in a high work hardening rate and tensile elongations can reach 65%.

Strain-induced phase transformations, mechanical twinning and shear band-induced plasticity are known to substantially enhance both strength and ductility. Materials which use these concepts have extraordinary mechanical properties which make them attractive in many structural applications. Since the groundbreaking research contributions of Grässel et al. and Frommeyer et al. [12–16] on Fe-Mn-Si-Al TWIP steel with a very low C content (<100ppm), there has been a global research effort aimed at developing a better understanding of the plasticity-enhancing mechanisms which are activated during the plastic deformation of these and similar C-alloyed high Mn face-centered cubic (fcc) γ-Fe.

Grässel and Frommeyer [12–16] observed that large scale deformation twinning occurred in the microstructure of TWIP steel when the Mn content was larger than 25 wt.-%, the Al content was larger than 3 wt.-% and when the Si content was in the range of 2–3 wt.-%. The reported high strength-ductility balances, i.e., the product of the ultimate tensile strength and the total elongation (UTS × TE-values or $R_m \times$ A-values, in European publications) of over 50,000 MPa × %, resulted mainly from an exceptionally high total elongation value of 80%. Current TWIP steels exhibit both high ultimate tensile strengths and extremely large elongations (60%) over a wide strain rate range ($10^{-3} - 10^{+3}$ s^{-1}). Their importance as breakthrough structural materials cannot be understated as the TWIP steel alloy concept has also opened the possibility of creating new steel grades with a very wide range of properties, which have already been reviewed in the literature [17–19]. Fig 11.2 illustrates the exceptional strength-ductility combination exhibited by TWIP steel, by comparing the properties of Ti-stabilized interstitials-free (Ti-IF) ferritic steel with those for a 1000 MPa UTS Fe-18%Mn-0.6%C-1.5%Al (in wt.-%) TWIP steel. The Ti-IF steel is considered one of the most formable industrial steel grades, although at much lower strength levels. Its formability is based on a strict control of the crystallographic texture development during the cold rolling and the subsequent recrystallization annealing. As compared to Ti-IF steel, TWIP steel is characterized by a high strain hardening, a large uniform elongation and a high

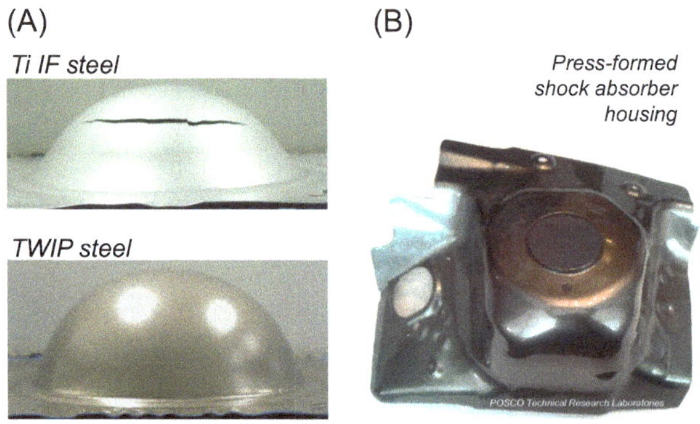

Figure 11.2 (A) Comparison of a bulge test carried out on a Ti-stabilized interstitial-free ferritic steel and an austenitic Fe-18%Mn-0.6%C-1.5%Al (in wt.-%) TWIP steel. (B) Press-formed automotive shock absorber made of Fe-18%Mn-0.6%C-1.5%Al (in wt.-%) TWIP steel. *Source*: courtesy of the POSCO Technical Research Laboratories, Gwangyang, Korea.

ultimate tensile strength. It also has a relatively low yield strength and virtually no post-uniform elongation. The superior forming properties of specific TWIP steels, such as the Fe-18%Mn-0.5%C-1.5%Al (in wt.-%) TWIP steel, have also already been reviewed in great detail [20]. In fact, this class of highly ductile, high, and ultra-high strength steels has been the subject of a truly international research and development effort, in both industrial and academic laboratories, aimed at their industrial development for large scale use in the automotive industry, LNG-shipbuilding, oil-and-gas exploration and transport, construction industry, and non-magnetic structural applications. In the case of automotive materials, the main motivation for the current interest in Fe-Mn-Al-C TWIP steel for use in automotive application is related to the fact that they can provide materials with an ultra-high strength for structural reinforcement, a superior ductility for the ease of press forming, and a large energy absorption capability for improved vehicle crashworthiness. Their large scale use of TWIP steel could potentially result in considerable progress in achieving a reduction of passenger vehicle mass, a lowering of their greenhouse gas emissions, a drastic increase of their gas mileage and an improvement of passenger safety. Medium Mn steels have been shown to offer similar advantages at lower cost. The medium Mn steels have a more complex physical metallurgy as compared to high Mn TWIP steels and their industrial development is currently less advanced.

11.2 High Mn TWIP steel

11.2.1 Principle of high Mn TWIP steel alloy design

The analysis of single crystal of alloys with a low intrinsic stacking fault energy γ_{isf} prompted Venables [21] to conclude that there was a parabolic relationship between the stress required to initiate deformation twinning and γ_{isf}. Remy and Pineau [22] also emphasized the role of γ_{isf} by arguing that the occurrence of deformation twinning required γ_{isf} to be within a specific range: 20 mJ/m^2 < γ_{isf} < 50 mJ/m^2. In the case of high Mn TWIP steel, various γ_{isf}-ranges have been proposed in which the TWIP mechanism is activated: 12–35 mJ/m^2 [23], 25–60 mJ/m^2 [24], 25–80 mJ/m^2 [25]. The occurrence of strain-induced transformations has also been linked to the γ_{isf} [26]. Although these observations are largely empirical, they have contributed to the fact that Venables' opinion has remained a widely accepted view on the subject and that major efforts have been made to compute the temperature and composition dependence of γ_{isf} for TWIP steels in order to predict their mechanical properties. The empirically determined γ_{isf}-ranges in which specific deformation modes are activated are shown in Fig. 11.3. A low γ_{isf} (γ_{isf} < 15 mJ/m^2) results in stress or strain-induced solid-solid phase transformation (TRIP-effect), an intermediate γ_{isf} (15 mJ/m^2 < γ_{isf} < 45 mJ/m^2) enables mechanical twinning, and a high γ_{isf} (γ_{isf} > 45 mJ/m^2) leads to perfect dislocation glide and shear banding. The temperature plays a key role in the relation between γ_{isf}, deformation-twinning and the plasticity of TWIP steel, as the value of γ_{isf} increases with increasing temperature. At high temperature, perfect dislocation glide with frequent cross-slip is expected when γ_{isf} becomes larger than 45 mJ/m^2. At lower temperatures, i.e., when 15 mJ/m^2 < γ_{isf} < 45 mJ/m^2, large dislocation dissociation

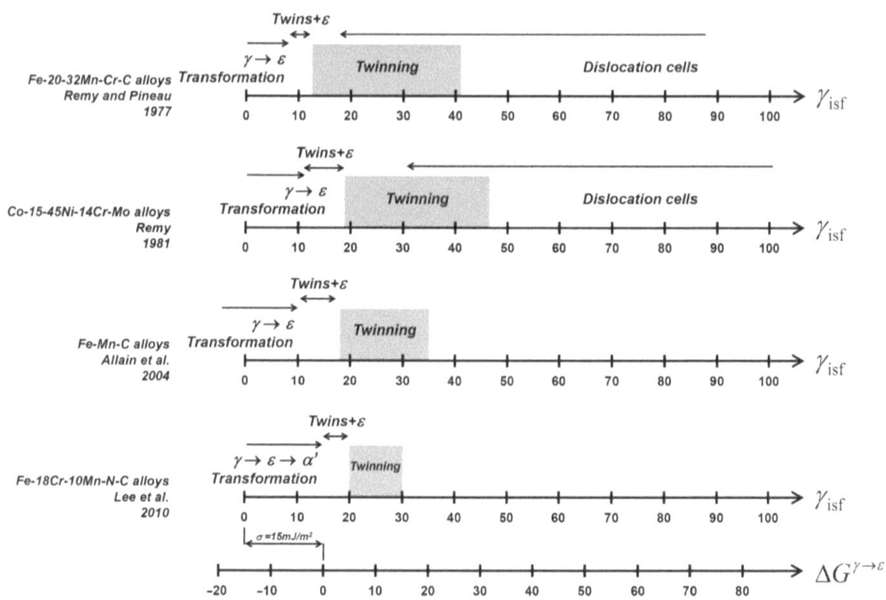

Figure 11.3 Intrinsic stacking fault γ_{isfe} energy ranges for transformation, deformation twinning and dislocation glide in fcc γ-Fe alloys reported in the literature for Fe-(20–32)% Mn-Cr-C (in wt.-%) alloys (data originally published in [27]), Co-(15–45)%Ni-14%Cr-Mo (in wt.-%) alloys (data originally published in [28]), TWIP steel (data originally published in [29]) and Fe-18%Cr-10%Ni-C-N (in wt.-%) alloys (data originally published in [30]). The relation between the γ_{isf}-axis and $\Delta G^{\gamma\rightarrow\varepsilon}$ is also indicated, assuming that $2 \times \sigma^{\gamma/\varepsilon} = 15$ mJ/m^2.

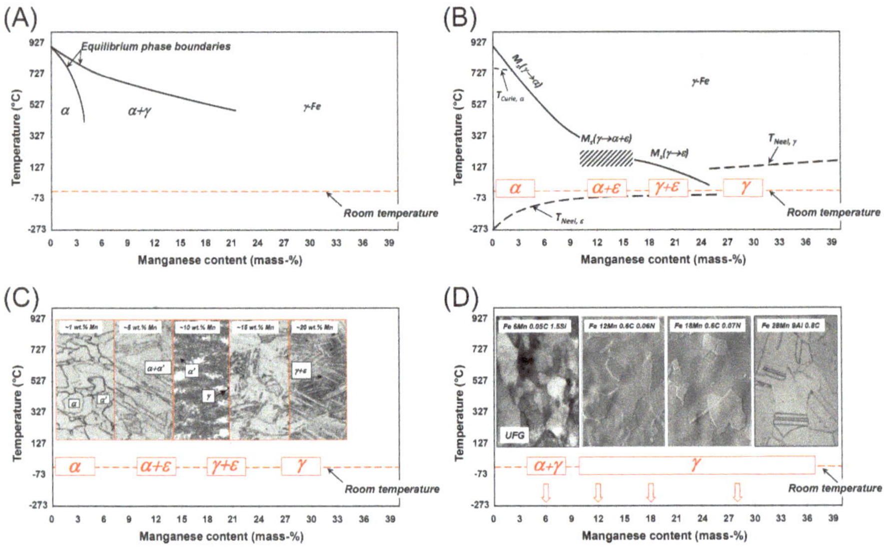

Figure 11.4 (A) Equilibrium binary Fe-Mn phase diagram. (B) Non-equilibrium phase transformation diagram for Fe-Mn binary alloys. (C) Room temperature microstructure evolution corresponding to (B). (D) Room temperature microstructure evolution in the Fe-Mn-C alloy system.

widths lead to the nucleation of twins and their growth by thickening. As the temperature is further reduced deformation-induced ε-martensite formation gradually replaces deformation twinning during straining. The interaction between dislocations, grain boundaries, stacking faults, and deformation twins results in a sustained high strain hardening during deformation.

The microstructure of Fe-Mn, Fe-Mn-C, Fe-Mn-C-Al and more complex multi-component ferrous alloys such as Fe-Mn-C-Al-Si alloys in which deformation-induced twinning can occur, can best be understood by consideration of non-equilibrium diagrams rather than equilibrium phase diagrams. This is illustrated in Fig. 11.4 for binary Fe-Mn and ternary Fe-Mn-C alloys systems. In the Fe-Mn alloys system close to 30 wt.-% Mn is needed to obtain meta-stable austenite phase at room temperature. The addition of 0.6 wt.-% C can stabilize the austenite phase for Mn contents as low as 12 wt.-%.

Fig. 11.5 indicates that the maximum C content is limited to approximately 0.6 wt.-% by the requirement to avoid the precipitation of carbides and obtain a homogeneous austenitic microstructure at room temperature. The additional requirement that the stacking fault energy γ_{isf} should be at least 15 mJ/m^2, in order to initiate deformation twinning during deformation at room temperature, clearly limits the minimum Mn content to about 18 wt.-%.

In the seminal work of Schumann [31] on the microstructure and mechanical properties of Fe-Mn-C alloys, the composition was mainly considered in terms of a means to achieve room temperature phase stability for the fcc crystal structure.

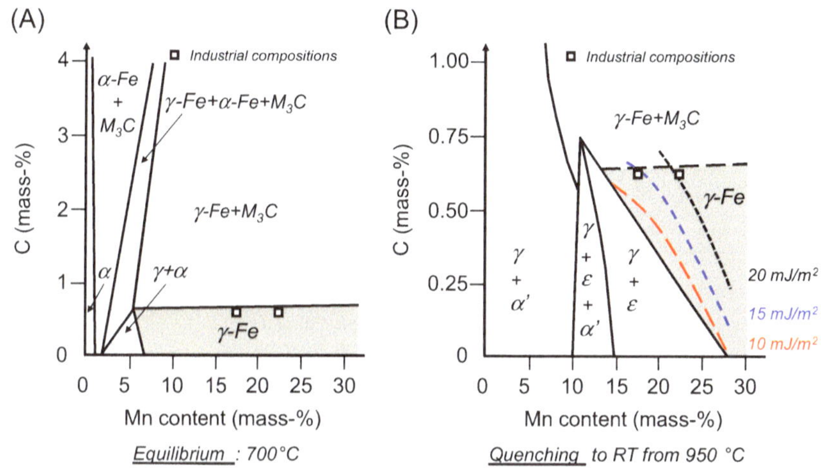

Figure 11.5 (A) Isothermal section of the Fe-rich corner of the equilibrium Fe-Mn-C phase diagram at 700°C. (B) Room temperature microstructure diagram after quenching Fe-Mn-C alloys from 950°C. The austenite phase stability range is indicated in gray in (A) and (B).

The observation of the deformation microstructure also allowed for the determination of a characteristic boundary line in the Mn-C plane, between a region of stable compositions deforming by dislocation glide and twinning, and a region of unstable compositions, i.e., composition leading to the formation of the hexagonal close-packed (hcp) ε-phase during straining:

$$\%\mathrm{Mn} = 32 - 20 \times \%\mathrm{C} \tag{11.1}$$

Here %C and %Mn are the bulk wt.-% of C and Mn in the alloy, respectively. In the range of (15–25) wt.-%Mn and (0.4–0.8) wt.-%C, this equation corresponds approximately to alloy compositions with an iso-γ_{isf} value of approximately 30 mJ/m^2 according to Nakano and Jacques [32] and 20 mJ/m^2 according to Saeed-Akbari et al. [33].

There are two basic computational approaches to determine γ_{isf}: (1) the thermo-dynamic approach and (2) the ab initio approach. In the thermodynamic approach, the intrinsic stacking fault energy is computed by the means of the CALPHAD method used for the modeling of equilibrium thermodynamic properties of multi-component alloys. The CALPHAD method uses experimental and theoretical information to obtain a complete thermodynamic description of the phase equilibria and the thermo-chemical properties of alloy systems. This thermodynamic description consists of the Gibbs energy of each phase, expressed as a function of composition and temperature. Extrapolation makes it possible to compute thermodynamic properties when no experimental data is available. In the thermodynamic approach, γ_{isf} is calculated using an equilibrium thermodynamic formalism originally proposed by Olson and Cohen [34], which

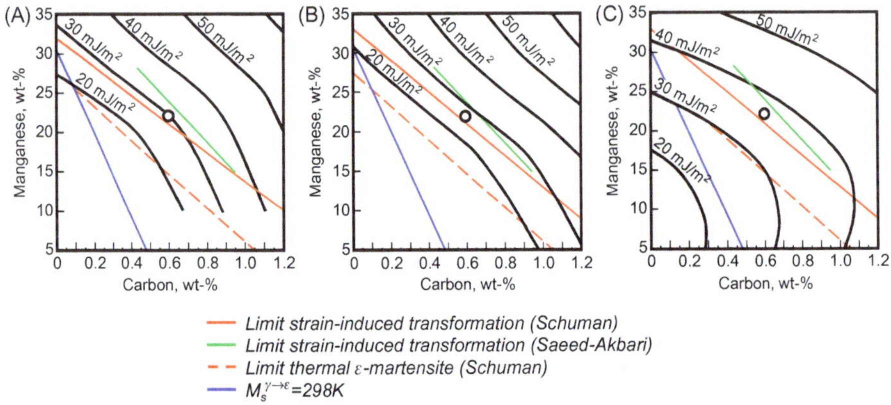

Figure 11.6 Mn and C content dependence of the stacking fault energy of austenitic ternary Fe-Mn-C alloys according to (A) Allain et al. (data originally published in [23]), (B) Saeed-Akbari et al. (data originally published in [33]), and (C) Nakano and Jacques (data originally published in [32]).

separates the stacking fault formation energy into contributions from the Gibbs free energy difference between the hcp ε-phase and fcc γ-phase, based on Hirth's definition of an intrinsic stacking fault as a thin plate of the hcp ε-phase [35], as extended by Olson and Cohen [34]:

$$\gamma_{\text{isf}} = 2 \cdot \frac{4}{\sqrt{3} \cdot a^2 \cdot N_{\text{A}}} \cdot \Delta G_{\text{C}}^{\gamma\rightarrow\varepsilon} + 2 \cdot \sigma^{\gamma/\varepsilon} \tag{11.2}$$

Here $\Delta G_{\text{C}}^{\gamma\rightarrow\varepsilon}$ is the chemical driving force for the $\gamma\rightarrow\varepsilon$ transformation, i.e., the difference between the Gibbs free energy of the γ-phase and the ε-phase. $\sigma^{\gamma/\varepsilon}$ is the γ/ε interface energy. Pierce et al. reported values for $\sigma^{\gamma/\varepsilon}$ in the range of 8–12 mJ/m^2 for TWIP steels [36]. A value of 10 mJ/m^2 applies in the case of Fe-Mn-C and Fe-Mn-Al-C TWIP steel [23,37]. Typical results of the thermodynamic computation of the Mn and C content dependence of the stacking fault energy of austenitic ternary Fe-Mn-C alloys are compared in Fig. 11.6.

The ab initio approach is based on the computation of the generalized stacking fault energy (GSFE) surface of a TWIP steel. The GSFE surface represents the energy required to rigidly shear two parts of a supercell with the fcc crystal structure on a {111} glide plane in a <112> direction. The GSFE surface can be used to understand the role of characteristic energy barriers in deformation twinning and predict the twinnability of TWIP steel. A number of computer codes are available for the ab initio computation of the electronic structure of solids. When applied to the problem of deformation twinning in TWIP steel, these simulation programs use specific theoretical approaches, such as density functional theory, to compute from first principles the correlation between specific rigid lattice shear displacements

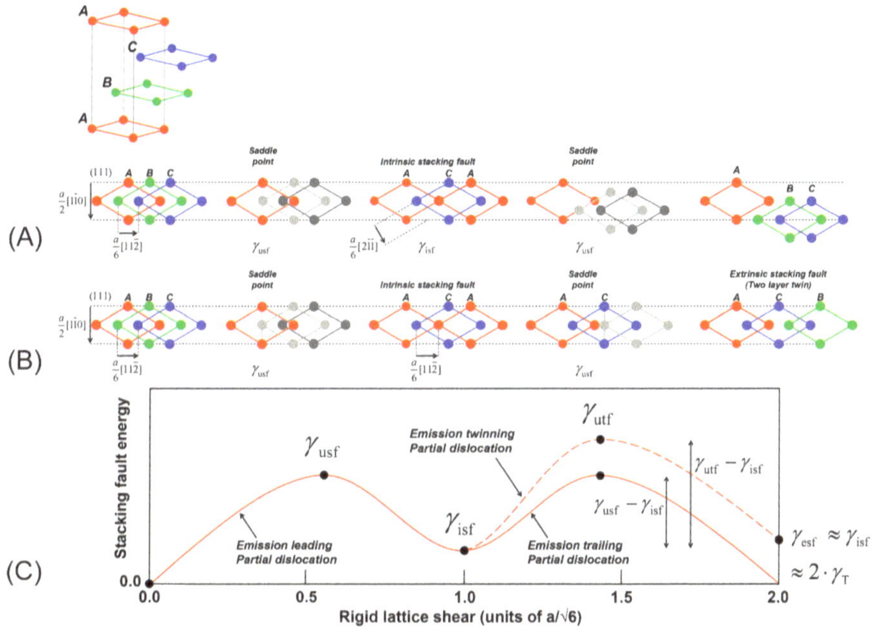

Figure 11.7 Schematic of the supercell method, which consists of sliding part of the supercell by a distance a/6 < 211 >, i.e., one partial dislocation Burgers vector, on a {111} plane, to create an intrinsic stacking fault. A two-layer twin is obtained by shifting the next plane by the same distance. (A) Rigid lattice shear corresponding to the emission of a leading partial dislocation followed by the emission of the trailing partial dislocation, which restores the perfect lattice. (B) Rigid lattice shear corresponding to the passage of a partial dislocation followed by the passage of a twinning partial dislocation, which results in the formation of a two-layer twin. (C) Schematic of the generalized planar fault energy for the shear in (A) and (B). The solid generalized stacking fault energy curve corresponds to the energy landscape for slip by a dissociated dislocation. The dotted line is the energy landscape for the formation of a two layer twin by shearing on the adjacent plane.

and the corresponding free energy changes associated with the nucleation of an intrinsic stacking fault and a two-layer twin. The energy barriers associated with this sequence of shears is illustrated in the schematic GSFE curve shown in Fig. 11.7.

In the GSFE, the first maximum encountered is the unstable stacking fault energy, γ_{usf}, which corresponds to the energy required to nucleate a partial dislocation [38]. A key parameter for mechanical twinning is the unstable twinning fault energy γ_{utf}, the saddle point energy associated with a rigid {111} < 112 > shift in the presence of an intrinsic stacking fault on an adjacent {111} plane [39]. The leading partial must overcome an energy barrier equal to γ_{usf}. The trailing partial requires overcoming an energy barrier equal to $\gamma_{usf} - \gamma_{isf}$. A large difference between γ_{usf} and γ_{isf} will therefore promote the formation of wide stacking faults.

Assuming that a two layer twin acts as an embryo for a full size deformation, the tendency for twinning by nucleation a partial dislocation on a slip plane adjacent to a stacking fault is predicted by considering the ratios $\frac{\gamma_{isf}}{\gamma_{usf}}$ and $\frac{\gamma_{utf}}{\gamma_{usf}}$. The $\frac{\gamma_{isf}}{\gamma_{usf}}$ ratio characterizes the energy barrier for slip. When the $\frac{\gamma_{utf}}{\gamma_{usf}}$ ratio is close to one, the barrier for the nucleation of a partial dislocation and the barrier for the nucleation of a two layer mechanical twin are similar. Twinning will be observed when $\frac{\gamma_{usf}}{\gamma_{utf}}$ is large, i.e., the energy barrier for the emission of the twinning partial is low, and $\frac{\gamma_{isf}}{\gamma_{usf}}$ is small.

The alloy content of TWIP steel is typically (15−30) wt.-% Mn, (0−1) wt.-% C, (0−3) wt.-% Al, and (0−3) wt.-% Si. Secondary alloying additions include Cr, Cu, N, Nb, Ti and/or V. The steel composition plays a key role in the determining the value of γ_{isf}. In the following paragraphs the influence of the most important alloying elements on the mechanical properties of TWIP steel is therefore reviewed in detail.

11.2.2 TWIP steel composition

11.2.2.1 Carbon

C plays a central role in the properties of TWIP steel as: (1) C increases the lattice parameter, (2) C increases the stability of the austenite relative to ε-martensite (i.e., C increases $\Delta G^{\gamma\rightarrow\varepsilon}$), (3) C promotes paramagnetism (i.e., C decreases the Néel temperature T_N^{γ}), (4) C substantially increases γ_{isf}, with a larger effect at lower Mn content in Fe-Mn-C TWIP steel, and (5) C has a strong solid solution hardening effect. Solute C occupies octahedral interstitial sites in γ-Fe and alloys. The small lattice distortion caused by solute C is isotropic. The nearest neighbors (NN) and next nearest neighbor (NNN) interactions between C-C pairs in austenite have been studied by Mössbauer Spectroscopy (MS), thermodynamic estimations and Monte-Carlo simulations [40,41]. The NN and NNN C-C interactions have been found to be repulsive in γ-Fe-C alloys. This results in a reduction of the probability that an interstitial octahedral site is occupied in the vicinity of an octahedral site already occupied by a C atom, leading to an increased C diffusivity with increasing C/Fe-atomic ratio [42,43]. The ab initio analysis of Fe-Mn-Al-C alloys has also shown that there is a pronounced repulsion between interstitial C atoms in TWIP steel [44−47]. The preferred location of C atoms is away from the stacking fault plane. The stacking fault energy is highest when the C atom is in the stacking fault plane. The effect is very short range, and the stacking fault energy rapidly decreases when the C atom is positioned further from the fault plane. This dependence of γ_{isf} on the distance of the C from the stacking fault plane is shown in Fig. 11.8 based on the work of Gholizadeh [48]: the further C atoms are from the stacking fault plane, the smaller their influence on γ_{isf}. Only a few atomic hops of the C atom away from the stacking fault are required to obtain a decrease of the γ_{isf} to its C-free stacking fault value. Experimental evidence for this nano-diffusion of C away from the stacking fault has been observed by Hickel et al. [50].

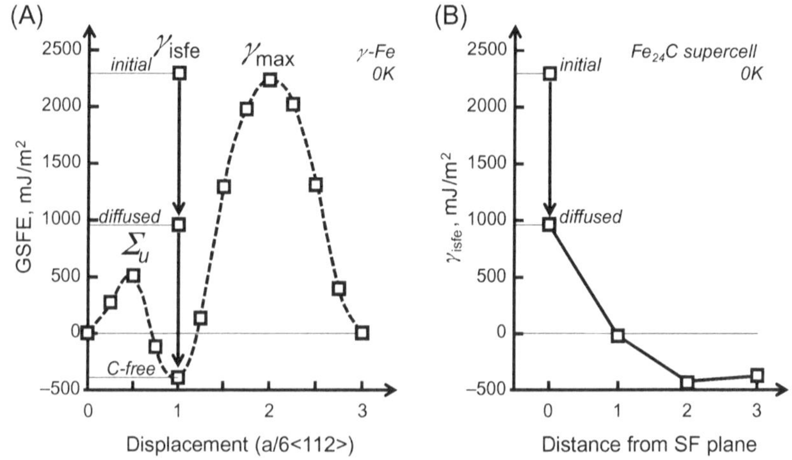

Figure 11.8 (A) Ab initio calculation of the generalized stacking fault energy for γ-Fe at 0 K, illustrating the effect of C on the value of γ_{isf}. γ_{isf} is very high when a C atom is in its initial state, i.e., positioned on the stacking fault plane during the passage of a dissociated dislocation. As the C diffuses to a more favorable position inside the stacking fault γ_{isf} is reduced but γ_{isf} is still high compared to γ_{isf} for a C-free stacking fault. (B) Effect of the distance of solute C from the stacking fault plane on γ_{isf}. Data redrawn from the work of Gholizadeh and Limmer [48,49].

According to Choi et al. [51] the solid solution strengthening effect of C is +279 MPa/wt.-% C. Allain reported a similar solid solution hardening effect of +250 MPa/wt.-% C [52].

11.2.2.2 Manganese

As Mn increases the $\varepsilon \rightarrow \gamma$ transformation temperature, it is expected that Mn-additions result in a decrease of γ_{isf}. The Mn-content dependence of the γ_{isf} is non-linear (Fig. 11.9). In binary Fe-Mn alloys, Mn decreases γ_{isf} at low Mn contents. The lowest γ_{isf} value is obtained in the range of 10–16 at % Mn. At higher concentrations, Mn additions increase γ_{isf}. The parabolic dependence of γ_{isf} is due to the effect of Mn on the γ-ε energy difference: an increase of the Mn content suppresses the formation of the ε-phase at high Mn content. The experimentally observed non-linear dependence has been confirmed by ab initio calculations for disordered Fe-Mn alloys. The strong attractive NN C-Mn (+26 kJ/mol) and NNN C-Mn (+13 kJ/mol) interaction between Mn and C atoms in TWIP steels result in the formation of Mn-C pairs [45,46]. Kang et al. [53] carried out ab initio calculations of the energy change related to the addition of Mn atoms to Fe_{6-x}-Mn_x-C octahedra, in which the carbon atoms occupy the central octahedral interstitial position. They reported an energy decrease of −0.0528 eV per Mn atom (−5.09 kJ/mol), i.e., the C-Mn interaction

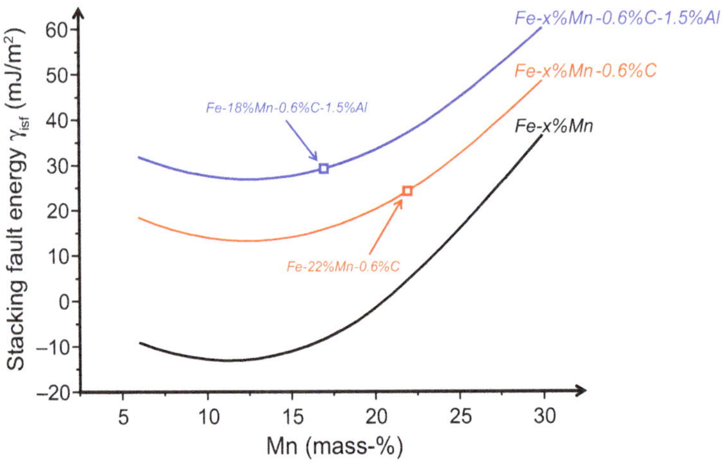

Figure 11.9 Mn content dependence of the intrinsic stacking fault energy of Fe-Mn, Fe-Mn-C and Fe-Mn-C-Al TWIP steels (data originally published in [33]).

is attractive. The formation of Mn-C pairs reduces the C diffusivity. The attractive C-Mn interaction also leads to an increase of the solubility limit of C in γ-Fe when Mn is added.

11.2.2.3 Aluminium

The role of Al in TWIP steel is complex. Al strongly stabilizes γ-Fe–Mn random alloys. Al also increases the γ_{isf} [54]. Jin and Lee [55] have studied the influence of Al-additions on γ_{isf} for a Fe-18%Mn-0.6%C-x%Al (in wt.-%) TWIP steel by X-ray diffraction (XRD) and transmission electron microscopy (TEM). They report the following linear Al-content dependence:

$$\gamma_{isf}(\mathrm{mJ/m^2}) = 20 + 7.8 \cdot \%\mathrm{Al(wt.\text{-}\%)} \tag{11.3}$$

Kang et al. reported a similar increase of γ_{isf} by $+10\ \mathrm{mJ/m^2}$ per wt.-% Al, i.e., from 17 mJ/m^2 for Fe-18%Mn-0.6%C to 37 mJ/m^2 for Fe-18%Mn-0.6%C-2% Al (in wt.-%) [56]. These results are in agreement with the increase of approximately $+8\ \mathrm{mJ/m^2}$ per wt.-% Al predicted by thermodynamic calculation [33]. Al-additions suppress the dynamic strain hardening in C-alloyed TWIP steels. Based on the study of the temperature dependent properties of austenitic Fe-30% Mn-x%Al-1.0%C (in wt.-%) alloys, Shun et al. [57] reported that Al additions increase the apparent activation energy for the onset of serration from 14.4 to 22.3 kcal/mol due to the addition of 2.7 wt.-% Al, suggesting that Al-additions reduce the C diffusivity within the core of dislocations. The strong attractive NNN C-Al interaction (+25 kJ/mol) results in an increase of the solubility limit of carbon in γ-Fe when Al is added. As the NN C-Al interaction (−10 kJ/mol, −0.104 eV/at) is repulsive, the formation of local order resembling the crystal structure of perovskite Fe_3AlC κ-carbide is favored [40]. This compound has been experimentally identified by Park et al. in Fe−22%Mn−0.6%C-(0-3-6)%Al (in wt.-%) TWIP steels [58]. Al is also a solid solution strengthening element, and Al additions result in a significant increase of the yield strength.

11.2.2.4 Silicon

Si is a strong solid solution strengthening element in γ-Fe alloys. The solute Si atoms have a repulsive interaction. In the presence of C, the NN Si-C interaction is strongly repulsive. The NNN Si-C interaction is attractive. In the presence of both C and Mn, i.e., as in an actual TWIP steel, there is a repulsive Si-Mn interaction. According to Tian and Zhang [59], an increase of the Si content up to 8 at.-%Si in a decrease of γ_{isf} in Fe−Mn−Si−C alloys. This was confirmed by Jeong et al. [60] who measured experimentally a decrease of γ_{isf} of -4 mJ/m^2 per wt.-% of Si, for Si-additions up to 1.5 wt.-%. They also reported that the decrease of γ_{isf} in a 1.5% Si-added TWIP steel resulted in a higher strain hardening, which they attributed to the increased activation twinning.

11.2.2.5 Nitrogen

Lee et al. [61] have shown that the N-content dependence of γ_{isf} was approximately $+100$ mJ/m^2/%N for a Fe-15%Mn-2%Cr-0.6%C-x%N (in wt.-%) TWIP steel. They explained the observed retardation of twinning kinetics and reduction of the thickening of the twins to this increase of γ_{isf} with increasing N content. Huang et al. reported that N increases γ_{isf} in Fe-(20−22)%Mn-0.01%C-(0.7−2.5)%Al -(2−3)%Si -(0−0.05)%N (in wt.-%) TWIP steel [62]. Similar results were obtained by Jiang et al. [63] for Fe − 30%Mn − 6%Si − (0−0.047%) N (in wt.-%) steel.

11.2.2.6 Copper, Nickel

Both Cu and Ni stabilize the γ-phase relative to the ε-phase. Cu and Ni atoms cluster in austenite, i.e., they favor Cu-Cu and Ni-Ni bonds over Cu-Fe and Ni-Fe bonds. Lee et al. studied the effect of Cu additions on the mechanical behavior of TWIP steel [64]. They reported that the addition of 2 wt.-% Cu to a Fe-Mn-C-Al TWIP steel increases γ_{isf} from 22.8 mJ/m^2 to 26.0 mJ/m^2. They also reported that Cu additions retard the kinetics of twin formation and influence the type of serrations on the stress-strain curve. The effect of Cu on the dynamic strain aging (DSA) of Fe-Mn-C-Al TWIP steels is very similar to the effect of N addition in altering the type of serration and increasing the number of Portevin-Le Châtelier (PLC) bands. Cu additions also result in an increase in total elongation without a loss of strength. Ab initio calculations of Limmer [49] suggest an important influence of Cu and Ni on γ_{isf}. When Cu and Ni are located in the stacking fault plane, they increase γ_{isf} by 9.9 mJ/m^2 per atom-% Cu and 10.0 mJ/m^2 per atom-% Ni.

11.2.2.7 Chromium

While Cu and Ni atoms prefer to cluster in austenite, the interaction between Cr atoms appears to point to a planar segregation preference, otherwise Cr atoms prefer to segregate, favoring Cr-Fe bonds over Cr-Cr bonds. In the presence of C, the C-Cr interaction is always attractive. The attractive C-Cr is maintained in the presence of both C and Mn. The preferred position is the 180 degree-oriented Cr-C-Mn octahedral point defect complex, due to the weak Cr-Mn repulsion. Ab initio computations for binary γ-Fe-Cr alloys predict that the effect of Cr is similar to the effect of Mn. Both decrease γ_{isf}. When located at the stacking fault Cr decreases γ_{isf} by -1.7 mJ/m^2 per atom-% Cr for Cr contents up to 5.6 atom-% Cr [46]. Thermodynamic stacking fault energy models also predict that Cr additions decrease γ_{isf}.

11.2.3 Mechanical properties of TWIP steel

11.2.3.1 Elastic properties of TWIP steel: modulus anomaly

The magneto-elastic properties of anti-ferromagnetic high Mn Fe-alloys are characterized by a pronounced modulus anomaly. The modulus is higher in the high-temperature paramagnetic state than in the low-temperature magnetically ordered anti-ferromagnetic state. The origin of this "ΔE"-effect is a spontaneous volume magneto-striction, i.e., a volume expansion caused by the anti-ferromagnetic ordering when the alloy is cooled below the Néel temperature, which leads to a deviation from the normal temperature dependence of the modulus [65,66]. The decrease in the modulus is due to the fact that the increase in the interatomic distances associated with the volume expansion also leads to a weaker interatomic bonding, i.e., lattice softening results in a stiffness reduction. A systematic study of Young's modulus of relatively highly Mn-alloyed ternary face-centered cubic Fe-(28.0-37.5) at.% Mn-(1.5-3.0)at.% C alloys was carried out by Reeh et al. by means of nano-indentation [67]. The measured Young's modulus values were in the range of 185 ($\pm$ 12)–220($\pm$ 20) GPa for bulk samples. This corresponds to a room temperature shear modulus in the range of 72–85($\pm$ 20) GPa. These values are in broad agreement with those reported by Jung and De Cooman for the temperature and Al-content dependence of the shear modulus values of Fe-18%Mn-0.6%C-x%Al (in wt.-%) TWIP steels (Fig. 11.10) [68,69]. They reported a room temperature shear modulus in the range of 76.5–79.0 GPa. Allain [52] points to a substantially lower shear modulus if the Néel temperature for the anti-ferromagnetic transition occurs close to room temperature or at higher temperatures. The Néel temperature of Al-free Fe-22%Mn-0.6%C (in wt.-%) TWIP steel is 321 K (+48°C). The shear modulus is reduced from a value of 72 GPa at +107°C to a minimum value of 60.5 GPa at -74°C. The room temperature shear modulus of Fe-22%Mn-0.6%C (in wt.-%) alloy is 62 GPa [52]. Steinmetz et al. reported a shear modulus value of 52.5 GPa for Fe-22%Mn-0.6%C (in wt.-%) TWIP steel [69]. Table 11.2 gives an overview of the reported shear modulus of various TWIP steels.

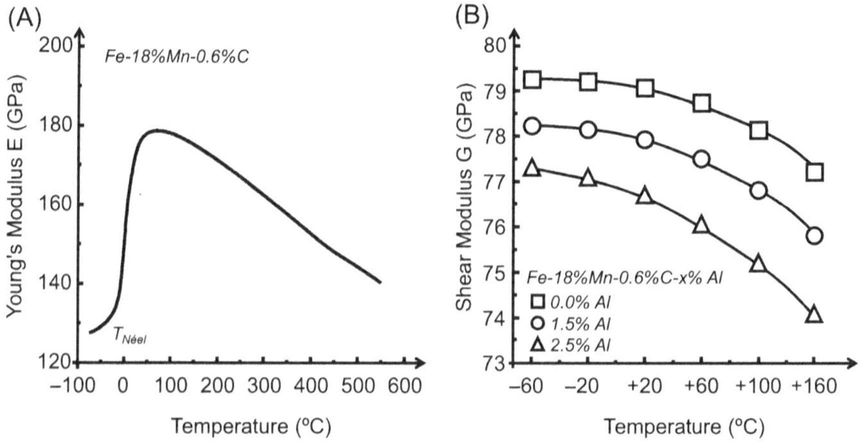

Figure 11.10 (A) Temperature dependence of Young's modulus of Fe-18%Mn-0.6%C (in wt.-%) TWIP steel. The anti-ferromagnetic ordering results in a reduction of the modulus below the Néel temperature. (B) Effect of Al on the shear modulus of Fe-18%Mn-0.6%C-x% Al (in wt.-%) TWIP steel (data originally published in [68,69]).

Table 11.2 Room temperature shear modulus of selected TWIP steels

Composition (in wt.-%)	*G* (GPa)	Ref.
Fe-24%Mn-0.7%C*	62.4	[70]
Fe-22%Mn-0.5%C*	63.6	[70]
Fe-18%Mn-0.5%C*	73.2	[70]
Fe-22%Mn-0.6%C*		[71]
[5 4 22] single crystal	67	[71]
[16 6 23] single crystal	77	[71]
Fe-22%Mn-0.6%C*,*	62	[17]
Fe-18%Mn-0.6%C**	79.0	[68,69]
Fe-18%Mn-0.6%C-1,5%Al**	78.0	[68,69]
Fe-18%Mn-0.6%C-2,5%Al**	76.7	[68,69]

*, Anti-ferromagnetic; **, Polycrystal.

Plastic deformation takes place when the critical shear stress for dislocation glide is reached. Reported values for the critical shear stress for dislocation glide and deformation twinning are listed in Table 11.3 for various TWIP steels.

11.2.3.2 Twinning mechanisms in TWIP Steel

Deformation twinning results from a homogeneous shearing of the matrix by a highly coordinated glide of partial dislocations with the same Burgers vector,

Table 11.3 Overview of the reported critical shear stress values for dislocation glide and twinning in TWIP steels

Composition (in wt.-%)	Compression axis	Sample type	Deformation	τ_{CRSS} (MPa)	Ref.
Fe-22%Mn-0.6%C	[0 0 1]	Micro-pillar (0.7 μm)	Twinning	488	[72]
Fe-22%Mn-0.6%C	[16 6 23]	Micro-pillar (2.0 μm)	Dislocation glide	206	[71]
Fe-22%Mn-0.6%C	[1 3 6]	Micro-pillar (3.9 μm)	Twinning	188	[72]
Fe-22%Mn-0.6%C	–	Polycrystal (EN10002)	–	160	[71]
Fe-22%Mn-0.6%C	[5 4 22]	Micro-pillar (2.0 μm)	Twinning	154	[71]
Fe-18%Mn-0.6%C	–	Polycrystal (ASTM E8)	–	144	[73]
Fe-22%Mn-0.6%C	[0 1 1]	Single crystal (4.0 mm)	Dislocation glide	76	[71]
Fe-22%Mn-0.6%C	[0 0 1]	Single crystal (4.0 mm)	Twinning	72	[71]

with exactly one dislocation gliding on successive {111}-type twinning planes. In polycrystalline materials twin formation by {111}<112> slip results in a volume orientation change of a certain volume fraction of the twinned grains which will affect the crystallographic texture. {111}<112> slip requires the complete separation of partial dislocations, or the existence of partial dislocation sources. The spontaneous formation of twins is very unlikely as it would require very high stresses. In the generally accepted model for deformation twinning wide stacking faults are formed first. Overlapping stacking faults then generate twin embryos and twin growth occurs by the addition of stacking fault on the twin plane. The wide stacking faults which form the twin embryos can be (1) the result of suitable dislocation reactions or (2) they can be emitted from interfaces such as grain boundaries. The conditions for twin nucleation from stacking faults emitted at grain boundaries are not well defined.

Deformation twins are formed soon after yielding in TWIP steel. Twins are mainly formed on the primary twin system and the contribution of secondary twins systems to deformation twinning is limited. The twinned region consists of very thin twin plates. Fig. 11.11 illustrates the evolution of the twin structure.

The various models for deformation twinning have in common that deformation twinning requires achieving a high enough dislocation density and a high local stress to trigger the twin nucleation and growth. The schematics in Fig. 11.12 illustrate the key aspects of the deformation twinning mechanisms which have been reported in relation to TWIP steel in the literature. It is generally accepted

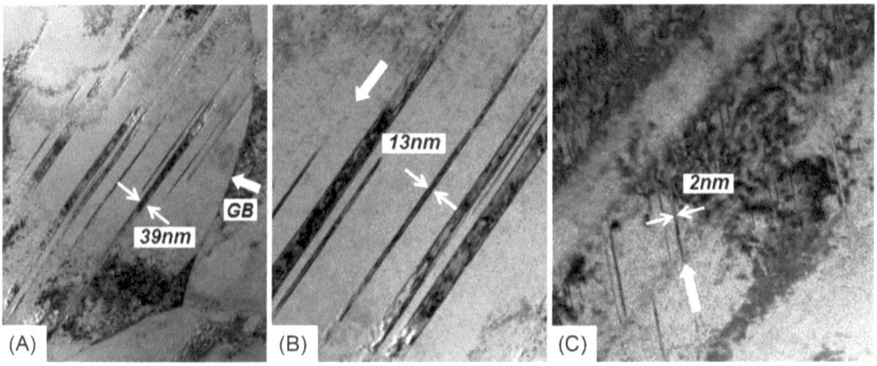

Figure 11.11 (A) Deformation twins in a Fe-18%Mn-0.6%C-1.5%Al (in wt.-%) TWIP steel deformed to an engineering strain of 10%. The micrograph illustrates that the deformation twins cross the entire length of the grain. (B) Higher magnification micrograph illustrating the nano-scale thickness of the deformation twins. (C) Observation of planar faults in the region between deformation twins at high strains.

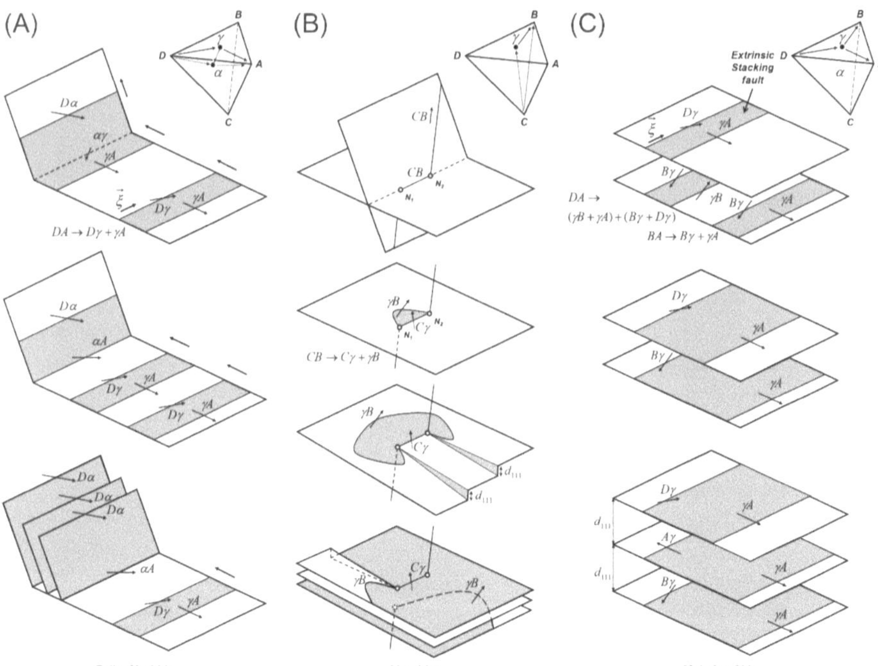

Figure 11.12 Schematic of (A) the Fujita-Mori-Liu stair-rod cross slip twinning model [81, 82], (B) the Venables pole dislocation mechanism model [21] and (C) the Mahajan-Chin three layer twin nucleus model [77].

that twinning requires the presence of dissociated dislocations, i.e., a low γ_{isf}, and that the nucleation of the twins requires dislocation reactions leading to the formation of twinning partial dislocations trailing an intrinsic stacking fault, and a process which suppresses the formation or the movement of the corresponding trailing partial dislocation. In the Venables model [21], a twinning partial, formed on a long jog on the primary slip plane, rotates around a specific pole dislocation. The model requires the formation of a large enough jog, which implies that multiple intersections must have taken place in multi-slip conditions [21,74]. Glide-type twin nuclei models are characterized by an intrinsic simplicity. Pure glide type models for twin formation do not require multi-slip conditions. The simplest glide-type twin nuclei model is the Copley-Kear-Buyn's dislocation break-away mechanism [75,76]. In the Copley-Kear-Byun model twinning is assumed to initiate by partial dislocation breakaway, solely based on straightforward application of the Peach-Kohler equation on the partial dislocation on either sides of a dissociated primary dislocation. The twin thickening mechanism is based on the chance overlap of wide stacking fault generated by the breakaway process. The Mahajan-Chin model [77] for the formation of a three layer twin nucleus is also a glide-type source. In this model two dislocations with coplanar Burgers vectors form a set of three Shockley partials on three consecutive slip planes, resulting in an extrinsic fault configuration. This three layer extrinsic stacking fault is considered to act as a twin nucleus. Steinmetz et al. [78] considered a strongly pinned segment of a pre-existing three-layer Mahajan-Chin stacking fault as the twin nucleus in their model for the TWIP effect. Mahato et al. [79] observed extrinsic stacking faults by TEM analysis of TWIP steel. Taking the viewpoint that extrinsic stacking faults serve as precursors to the formation of twin nuclei, they concluded that the Mahajan-Chin mechanism operates in their case. Kibey et al. [80] also favored the Mahajan-Chin model for deformation twinning.

Based on the direct observation of twin formation, Liu et al. [81] concluded that a slightly revised version of the Fujita-Mori model [82] was compatible with their in situ observations made in a Fe-24%Mn-0.5%C (in wt.-%) TWIP steel. The Fujita-Mori-Liu model requires prior activation of slip on two slip systems. In the model, the leading partial dislocation of a primary slip dislocation splits into a stair-rod dislocation and a twinning partial dislocation on the conjugate slip plane. A key feature of the Fujita-Mori model involves the "stair-rod cross slip" mechanisms, which allows for cross slip of partial dislocations. The model is based on a dislocation reaction originally proposed by Cohen and Weertman [83], whereby a perfect a/2 $<110>$-type dislocation at the head of a dislocation pileup in front of a strong barrier dissociates in a sessile a/3 $<111>$-type Frank dislocation and a a/6 $<112>$-type partial dislocation. The partial dislocation can glide away from the Frank dislocation on the conjugate slip plane trailing a wide stacking fault. Idrissi et al. [84] have studied the twin nucleation mechanism in Fe-20%Mn-1.2%C (in wt.-%) TWIP steel by TEM. Their results appear support the Cohen-Weertman or the Miura-Takamura-Narita [85] nucleation mechanisms, whereas twin thickening appears to be due to a pole mechanism as proposed by Venables [21].

Considering the large variety of possible mechanisms and in the absence of clear microscopic evidence for the proposed models in many reports, it remains very plausible that grain boundaries may be playing a key role in the nucleation of deformation twins by emission of twinning partials from grain boundaries. According to de las Cuevas et al. [86], the deformation twins are nucleated at high-angle boundaries. The emitted twins have a sub-micrometer or nanometer thickness and they cross the grain until they hit a grain boundary, another twin, or a strong obstacle.

11.2.3.3 Critical stress for twinning

The formation of deformation twins is generally considered to be preceded by dislocation activity which results in the formation of a dislocation mediated twin nucleating process, such as the Venables pole mechanism [21,87], the Cohen-Weertman deviation process [83] or the Mahajan-Chin stacking fault process [77]. While various equations are available to determine the twinning stress theoretically for specific deformation twinning models, the experimental determination of the twinning stress of TWIP steel is notoriously difficult. Values for experimentally measured critical deformation twinning stress are listed in Table 11.4. In tensile deformation, deformation winning appears to initiate in $<111>$/td (td: tensile direction) oriented grains. At room temperature, deformation twinning in TWIP steels is initiated at the flow stress corresponding to the stress at the onset of stage B in the strain hardening curves which correlates with primary twinning. There is only a small difference between the yield stress and the twinning stress of TWIP steel. Based on their study of the effect of grain size on yield strength of Fe-22%Mn-0.6% C (in wt.-%) TWIP steel, Bouaziz et al. [3] reported that the twin nucleation stress was independent of the grain size. It was approximately 550 MPa, i.e. $\tau_T \approx 180$ MPa,

Table 11.4 Reported critical twinning shear stress value for TWIP steel

Composition (in wt.-%)	Twinning stress (MPa) σ_T $(\tau_T)^*$	Grain size (μm)	γ_{isf} (mJ/m^2)	Ref.
Fe-22%Mn-0.5%C	272 (89)	–	–	[88]
Fe-22%Mn-0.6%C	220 (72)	–	23	[71]
Fe-22%Mn-0.6%C**	220 (72)	1.3–25	23	[17]
Fe-22%Mn-0.1%C-1.2Al	251 (82)	15	16	Present work
Fe-17%Mn-0.4%C-1.3%Al	312 (102)	15	17	idem
Fe-18%Mn-0.6%C-1.5%Al	422 (138)	5	25	idem
Fe-15%Mn-0.7%C-2%Al-2%Si	190 (62)	84	30	[89]
Fe-15%Mn-0.7%C-2%Al-2%Si	352 (115)	10	30	[89]
Fe-15%Mn-0.7%C-2%Al-2%Si	563 (184)	4.3	30	[89]
Fe-27%Mn-2.5%Si-3.5%Al	272 (89)	NA	18	[79]

*Assuming $\sigma_T = 3.06\ \tau_T$; **, [001] single crystal in compression (twinning favored).

for a grain size in the range of 1.3–25 μm. By studying the accumulation of permanent strain during cyclic tests below the yield point, Rahman et al. [89] estimated the twinning stress and its grain size dependence. They reported a single crystal critical twinning stress of 67 MPa [89]. This value is close to the critical resolved shear stress of 72 MPa reported by Choi et al. for a Fe-22%Mn-0.6%C (in wt.-%) single crystal oriented to favor deformation twinning, i.e., in compression along the [001]-axis [71].

Classical twinning theories generally predict that the critical resolved shear stress for twinning is proportional to γ_{isf}:

$$\tau_T \cdot b_{112} \sim \gamma_{isf} \tag{11.4}$$

Narita and Takamura [90] reported the following twinning stress equation:

$$\tau_T = \frac{\gamma_{isf}}{2 \cdot b_P} \tag{11.5}$$

Application of Eq. (11.5) to Fe-22%Mn-0.6%C (Composition in wt.-%, $\gamma_{isf} = 23\ \mathrm{mJ/m^2}$, $b_P = 0.147 \times 10^{-9}$ m) and Fe-18%Mn-0.6%C-1.5%Al (Composition in wt.-%, $\gamma_{isf} = 30\ \mathrm{mJ/m^2}$, $b_P = 0.147 \times 10^{-9}$ m) yields τ_T as 78 MPa and 102 MPa, respectively. Suzuki and Barrett [91] assumed that a small segment of dislocation consisting of a sessile partial and a twinning partial pinned between two pinning points act as twin nucleus. They defined the twinning stress as the critical stress for the formation of an unstable bow-out of the twinning partial:

$$\tau_T = \frac{\gamma_{isf}}{2 \cdot b_P} + \frac{G \cdot b_P}{L_0} \tag{11.6}$$

Here L_0 is the length of the sessile partial dislocation of the twin nucleus. Assuming a pole mechanism is operating, grain size reduction has been reported to make deformation twinning more difficult [92]. Gutierrez-Urrutia et al. [93] also observed that, in the grain size range of 3–50 μm, a grain size reduction resulted in an increase of the twinning stress in a Fe-22%Mn-0.6%C (in wt.-%) TWIP steel. They proposed the following grain size dependent twinning stress equation:

$$\tau_T = \frac{\gamma_{isf}}{b_p} + \frac{G \cdot b_p}{D} \tag{11.7}$$

D is the grain diameter. The first term in the equation is the stress required to achieve partial dislocation separation to create a twin and the second term is the stress needed to nucleate a twin at a grain boundary. Using $\tau_T = 0.447 \times \sigma_T$, $\gamma_{isf} = 22\ \mathrm{mJ/m^2}$, $G = 62$ MPa and $b_P = 0.147$ nm, the twinning stress for Fe-22% Mn-0.6%C (in wt.-%) steel according to Gutierrez-Urrutia et al. [93] is given by:

$$\begin{gathered}\tau_T = \frac{\gamma_{isf}}{b_p} + \frac{G \cdot b_p}{D} = \frac{22 \cdot 10^{-3}\,\mathrm{Nm/m^2}}{0.147 \cdot 10^{-9}\,\mathrm{m}} + \frac{62 \cdot 10^{9}\,\mathrm{Pa} \cdot 0.147 \cdot 10^{-9}\,\mathrm{m}}{2 \cdot 10^{-6}\,\mathrm{m}} = 154.3\,\mathrm{MPa} \\ \sigma_T = 2.238 \cdot \tau_T = 345.3\,\mathrm{MPa}\end{gathered} \tag{11.8}$$

Steinmetz et al. [78] assumed that a small segment of a Mahajan-Chin three layer twin between two pinning points act as twin nucleus. They defined the twinning stress as the critical stress for the formation of an unstable bow-out of the three twinning partials:

$$\tau_T = \frac{\gamma_{isf}}{3 \cdot b_P} + \frac{G \cdot 3 \cdot b_P}{L_0} \tag{11.9}$$

Here L_0 is the length of the sessile partial dislocation in the twin embryo. According to Steinmetz et al. $L_0 = 261$ nm for Fe-22%Mn-0.6%C (in wt.-%) TWIP steel. According to Venables the twinning stress τ_T consists of a nucleation stress and a twin growth stress [94]:

$$(1 - \frac{2 \cdot \theta}{3 \cdot \lambda} + \frac{(1-\nu) \cdot L}{1.84 \cdot G \cdot b} \cdot \theta^2 \cdot \tau_T) \cdot \tau_T = \frac{\gamma_{isf}}{b_P} \tag{11.10}$$

Here θ, is an orientation-dependent term. It is the ratio of the Schmid factors for slip and twinning. b_P is the Burgers vector of a partial a/6 < 112 > -type dislocation, and b is the Burgers vector of an un-dissociated a/2 < 110 > -type dislocation. λ, is a constant related to the forest dislocation density and the critical resolved shear stress for slip. In practice, $\theta \approx \lambda \approx 1$. Calculated twinning stress values based on the previous equations are compared in Fig. 11.13 for a Fe-18%Mn-0.6%C-1.5%Al (in wt.-%) TWIP steel. According to Buyn [76] twinning is activated when the applied stress is high enough to initiate partial dislocation breakaway in straight screw dislocation segments. The twinning stress is defined as the critical stress at which the partial dislocation separation on a screw dislocation diverges:

$$\tau_T = \frac{2 \cdot \gamma_{isf}}{b_P} \tag{11.11}$$

Fig. 11.13 also illustrates the effect of an externally imposed shear stress on the separation of the two partial dislocations in a TWIP steel with $\gamma_{isf} = 30$ mJ/m^2 based on Buyn's equation:

$$d = \frac{(2 - 3 \cdot \nu)}{8 \cdot \pi \cdot (1 - \nu)} \cdot \frac{G \cdot b_P^2}{\left(\gamma_{isf} - \frac{\tau \cdot b_P}{2}\right)} \tag{11.12}$$

The figure shows that the Buyn model predicts twinning stress that is much larger than the experimentally measured critical twinning stresses. The Steinmetz model is closest to the experimentally determined values. This is an important observation as it implies that (1) the critical dislocation density required for twin-nucleating dislocation-dislocation interactions is achieved after small amounts of strain, and (2) twinning is very easily initiated in TWIP steel. Alternatively, the result can also be interpreted as an indication that there are pre-existing multi-layer

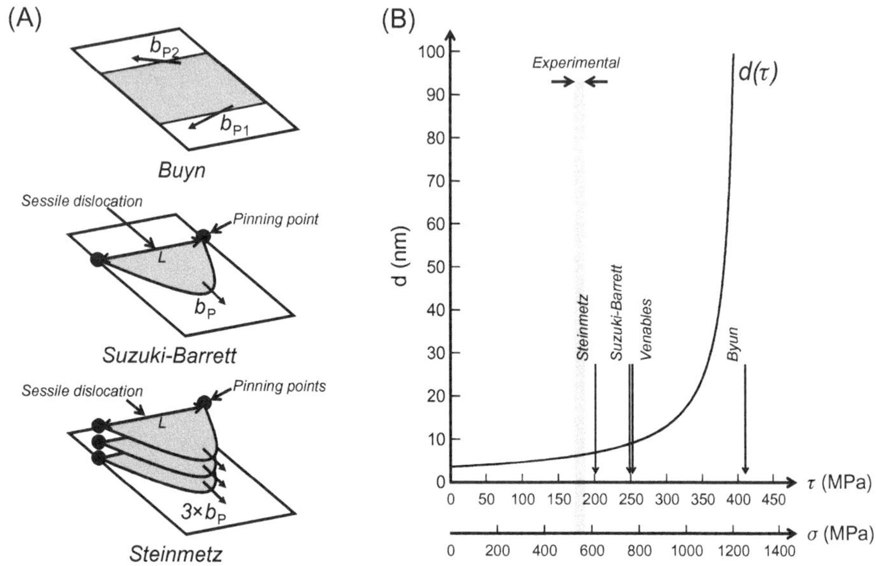

Figure 11.13 (A) Schematic of the Buyn, Suzuki-Barrett, and Steinmetz twin nucleation models. (B) Stress dependence of the partial dislocation separation according to Buyn [76]. The values for the twinning stress according to the Buyn [76], Suzuki-Barrett [91], Venables [94] and Steinmetz [78] models are compared with the experimentally measured value for a Fe-18%Mn-0.6%C-1.5%Al (in wt.-%) TWIP steel. The following parameter values were used for the computed twinning shear stress: $\gamma_{isf} = 30$ mJ/m^2, $G = 79$ GPa, $b_P = 0.147$ nm, $\nu = 0.312$, $L = 261$ nm [78], $L = 135$ [94].

Table 11.5 Reported twinning shear stress values of TWIP steels based on Steinmetz's equation

Composition (in wt.-%)	τ_T MPa	σ_T $2.24\times\tau_T$ MPa	σ_T $3.06\times\tau_T$ Mpa	γ_{isfe} mJ/m^2	G GPa	L_0 nm	Ref.
Fe-22%Mn-0.5%C	160	358	490	23	63.6	260	[70]
Fe-18%Mn-0.5%C	160	358	490	16	73.2	260	[70]
Fe-24%Mn-0.7%C	183	410	560	24	62.4	260	[70]
Fe-22%Mn-0.6%C	192	430	588	29	52.5	260	[78]
Fe−27Mn−2.5Si−3.5Al	201	450	615	18.8	66.0	200	[79]

twin nuclei in the recrystallized microstructure. An obvious choice location for these pre-existing nuclei could be grain boundaries where Mahajan-Chin type three layer twin nuclei could be part of the grain boundary defect structure, as suggested by Steinmetz (Table 11.5).

More recent deformation twinning theories have proposed that it was γ_{isf}, the unstable twinning fault energy, rather than γ_{isf}, which controlled twinning nucleation:

$$\tau_T \cdot b_{112} \sim \gamma_{utf} \tag{11.13}$$

This relation has been verified for pure metals [95] and for Cu-Al alloys by Kibey et al. [95,96]. According to Ogata et al. the ideal twining stress, i.e., the activation energy required for the homogeneous nucleation of a deformation twin in a defect-free crystal, i.e., not involving a heterogeneous twin nucleus model such as a three layer twin nucleus, can be computed using energy barrier values of calculated generalized planar fault energy (GPFE) curves [80,97]:

$$\tau_T = \frac{\pi}{b_P} \cdot (\gamma_{utf} - 2 \cdot \gamma_{tsf}) \approx \frac{\pi}{b_P} \cdot \left(\gamma_{usf} - \frac{\gamma_{isf}}{2}\right) \tag{11.14}$$

Here γ_{utf} is the unstable twin fault energy and $2 \times \gamma_{tsf}$ is the stable twin fault energy. As $\gamma_{isf} \approx 2 \cdot \gamma_{tsf}$ and $\gamma_{utf} = \gamma_{usf} + \frac{\gamma_{isf}}{2}$. Using the 0 K values available for critical energy barriers in TWIP steels one obtains: $\tau_T = 14$ GPa [48,49]. This value is considerably larger than the experimental value for the critical twin nucleation stress and indicates that twin formation requires a heterogeneous nucleation at pre--existing lattice defects.

11.2.3.4 Twinning kinetics and twinning saturation

Deformation twinning in TWIP steel saturates at relatively low volume fractions. The deformation twins in TWIP steel do not thicken appreciably and the mean twin thickness is in the range of 20–30 nm. The volume fraction of deformation twins in TWIP steel remains small and ε_T, the contribution of the twins to the macroscopic strain, is very small. Assuming a twin volume fraction $f_T \approx 0.1$:

$$\varepsilon_T = \frac{1}{M} \cdot \frac{\sqrt{2}}{2} \cdot f_T \approx 0.023 \tag{11.15}$$

Here M is the Taylor orientation factor and $\sqrt{2}/2$ is the twinning strain. The main contribution of deformation twinning to strain hardening is therefore due to its influence on the dislocation density evolution by the promotion of planar glide, and the reduction of the dislocation annihilation rate. Deformation twinning does also not contribute directly to the evolution of the crystallographic texture during deformation. Literature data shows that the total volume fraction of deformation twins is close to 10% [98–101]. Haase et al. [102] reported even lower volume fractions of deformation twins in cold rolled TWIP steel. They studied the influence of the cold rolling reduction on the measured volume fraction of the Cu-Twin texture component and calculated the value of the corresponding deformation volume fraction. They reported a saturation volume fraction of deformation twins of about 5%. According to Kim et al. 20% of the grains do not twin during deformation in uni-

Table 11.6 Typical parameter values for the strain dependence of the twin volume fraction in Fe- TWIP steel at 293 K. Steel compositions are in wt.-%

Parameter	Fe-18%Mn-0.6%C	Fe-18%Mn-0.6% C-1.5%Al	Fe-22%Mn-0.6%C
$F_{T,sat}$	0.15	0.14	0.15
β	7.50	8.00	3.00
ε_i	0.04	0.03	0.06–0.12
m	1	1	2
Ref.	[100]	[100]	[3]

axial tensile tests [103]. The twinning kinetics in the grains which exhibit deformation twinning saturate at a volume percentage of approximately 15%. The strain-dependent kinetics for the evolution of the twinning volume fraction inside the grains which undergo deformation twinning, $F_T(\varepsilon)$, has a characteristic S-shape dependence on the applied strain. $F_T(\varepsilon)$ has usually been described by means of the following empirical equation:

$$F_T(\varepsilon) = F_{T,sat} \cdot \left(1 - e^{-\beta \cdot (\varepsilon - \varepsilon_i)}\right)^m \tag{11.16}$$

Here $F_{T,sat}$ is the saturation twin volume fraction, and ε_i, is the critical strain for the nucleation of the deformation twinning. Reported values for ε_i are usually small. As for example, Bracke et al. [88] already observed the micro-twins at a true strain of 0.02. β and m are adjustable parameters which need to be determined experimentally. Table 11.6 list typical parameter values for Fe-18%Mn-0.6%C-x%Al (in wt.-%) TWIP steel at 293 K [100]. Fig. 11.14 shows twinning kinetics data reported in the literature for various TWIP steels.

11.2.3.5 Solid solution hardening

The hardening effect of solute atoms in austenitic fcc Fe alloys is known to be much less pronounced than in the case of ferritic bcc Fe alloys. Bouaziz et al. have proposed the following equation for the lattice friction stress, which includes the solid solution hardening effect of C and Mn in Fe-Mn-C TWIP steel [108]:

$$\sigma_0(\text{MPa}) = 228 + 187 \times \%\text{C} - 2.0 \times \%\text{Mn} \tag{11.17}$$

Choi et al. derived an alternative composition-dependent solid solution hardening equation for the crtical resolved shear stress and the lattice friction stress for Fe-Mn-C-Al-Si TWIP steel [71]:

$$\begin{aligned}\tau_{crss}(\text{MPa}) &= 31.7 + 91.3 \times \%\text{C} - 0.49 \times \%\text{Mn} + 16.2 \times \%\text{Si} + 6.7 \times \%\text{Al} \\ \sigma_0(\text{MPa}) &= 97 + 279 \times \%\text{C} - 1.5 \times \%\text{Mn} + 49.6 \times \%\text{Si} + 20.5 \times \%\text{Al}\end{aligned} \tag{11.18}$$

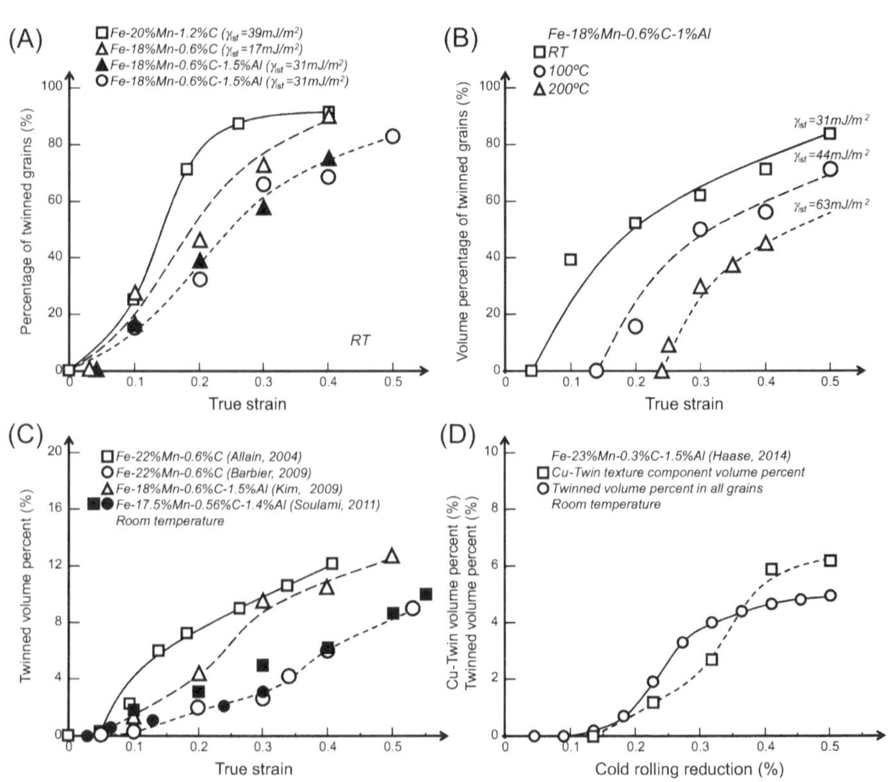

Figure 11.14 (A) Strain dependence of the number fraction of twinned grains in TWIP steel based on data from Renard et al. [104], Beladi et al. [105] and Kim [100]. (B) Strain dependence of the volume fraction of twinned grains in TWIP steel at different temperatures based on data from Shterner et al. [106]. (C) Strain dependence of the volume fraction of twins in TWIP steel based on data from Allain [98], Barbier [99], Kim [100] and Soulami et al. [101]. (D) Influence of the cold rolling reduction on the measured volume fraction of the Cu-Twin texture component and the calculated value of the corresponding deformation volume fraction [107].

Eqs. (11.17) and (11.18) predict that the yield strength of TWIP steel is low, and that a large strain is needed to see the strength benefit associated with the high strain hardening. As relatively large deformations are required to achieve a high flow stress, the use of standard TWIP steel is limited to complex press formed parts requiring a large amount of plastic strain. Additional hardening mechanisms have to be introduced to achieve a high strength at small strains. An obvious option is pre-straining, but it results in considerable reduction of the elongation as illustrated in Fig. 11.15. The alternative strengthening mechanisms, which involve grain size refinement and precipitation hardening by micro-alloying, are reviewed in the following paragraphs.

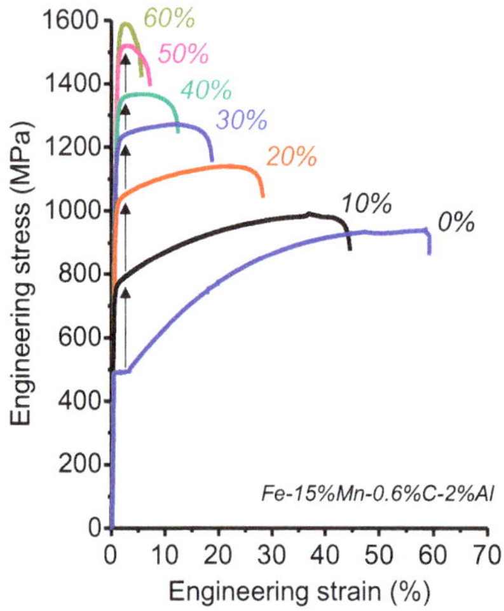

Figure 11.15 Figure illustrating the effect of rolling deformation on the flow curve of Fe-15%Mn-0.6%C-2%Al (in wt.-%) TWIP steel. A considerable decrease in elongation and strain hardening is associated with the increase of the yield strength.

11.2.3.6 Grain size strengthening in TWIP steel

The yield stress, σ_{YS}, of polycrystalline TWIP steel has been shown to follow the Hall-Petch relationship:

$$\sigma_{ys}(\text{MPa}) = \sigma_{ys}^{0} + \frac{k_{ys}^{HP}}{\sqrt{d}} \tag{11.19}$$

Here the parameter σ_{YS}^{0} includes the lattice friction stress, the solid solution strengthening contribution of the alloying elements, and the strain hardening contribution of the initial dislocation density, k_{ys}^{HP} is the dislocation un-pinning parameter at the yield point, and d is the grain size. Table 11.7 reviews the available literature data for the parameters σ_{ys}^{0} and k_{ys}^{HP} for TWIP steels. Fig. 11.16 is a selection of grain size strengthening data of TWIP steels with a similar composition. The least square fit of the Hall-Petch equation to the data yields the following Hall-Petch relationship:

$$\sigma_{ys}(\text{MPa}) = 189 + \frac{413}{\sqrt{d(\mu m)}} \tag{11.20}$$

Kang et al. [117] have analyzed the influence of the C content on the Hall-Petch coefficient k_{ys}^{HP} using a slow cooling procedure to ensure equilibrium C segregation to the available grain boundaries sites. They reported that k_{ys}^{HP} increased from 218 MPa $\mu m^{1/2}$ (6.9 MPa $mm^{1/2}$) to 344 MPa $\mu m^{1/2}$ (11.5 MPa $mm^{1/2}$) when the

Table 11.7 Hall-Petch equation parameters for TWIP steel

Composition (in wt.-%)	σ^0_{ys} (MPa)	k^{HP}_{ys} (MPa μm$^{1/2}$)	Ref.
Fe-22%Mn-0.6%C	137	449	[5]
Fe-22%Mn-0.6%C	170	428	[109]
Fe-22%Mn-0.6%C	157	357	[110]
Fe-20%Mn-0.6%C	158	485	[111]
Fe-22%Mn-0.5%C-0.08%N	219	478	[112]
Fe-15%Mn-0.7%C-2%Al-2%Si	305	330	[89]
Fe-31%Mn-3%Al-3%Si	53	764	[113]

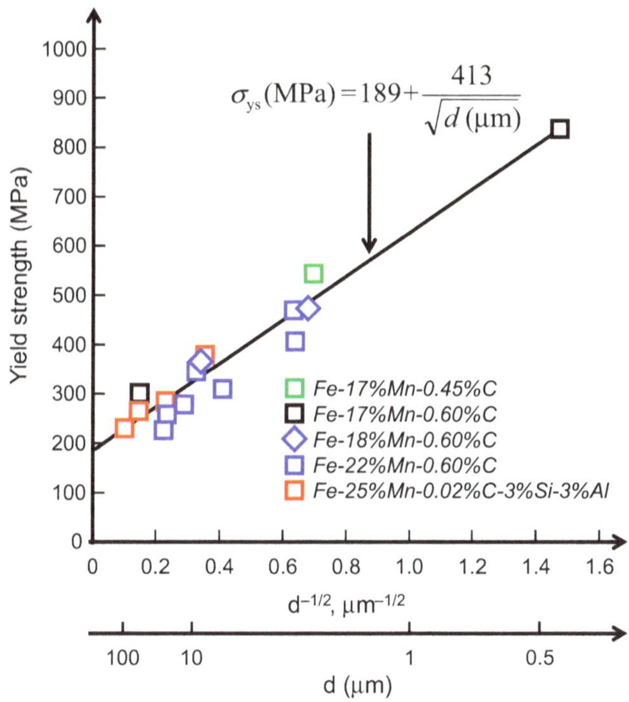

Figure 11.16 Verification of the Hall-Petch relationship for various TWIP steels. Data for Fe-17%Mn-0.45%C (in wt.-%) TWIP steel from Gwon and De Cooman [114]. Data for Fe-17% Mn-0.6%C (in wt.-%) TWIP steel from Lee et al. [115], for Fe-22%Mn-0.6%C (in wt.-%) from Scott et al. [109], and for Fe-25%Mn-3%Si-3%Al-0.02%C (in wt.-%) from Wang et al. [116].

C content was increased from 0.3 wt.-% to 0.6 wt.-%. They suggested two possible explanations for this pronounced increase of k^{HP}_{ys}. This effect could be due to (1) an increased stress for the activation of dislocation sources in the grain matrix when the C content is increased, or (2) an increased density of grain boundary ledges acting as dislocation sources [118].

11.2.3.7 Micro-alloyed TWIP steel

Micro-alloying of TWIP steel has been studied as a method to increase the yield strength by grain size reduction and precipitation hardening. Results are available for low-C Fe-25%Mn-3%Si-3%Al-0.02%C (in wt.-%) TWIP steel [116], and high-C Fe-17%Mn-0.6%C (in wt.-%) [115] and Fe-22%Mn-0.6%C (in wt.-%) TWIP steels [109]. The observed precipitate strengthening has been reported to be due to the Orowan bypassing strengthening mechanism. No equilibrium solubility equations and precipitation kinetics information are currently available for nitrides, carbides or carbo-nitrides of micro-alloying additions of Ti, Nb, or V in TWIP steel. It is therefore possible that some of the observed strength increase is due to solid solution hardening by the micro-alloying additions. V appears to be the most efficient micro-alloying addition, as it both reduces the grain size and leads to the formation of 3.4 nm diameter precipitates according to Scott et al. [109]. The VC precipitation does not significantly affect the work hardening. According to Chateau et al. [119] the VC particles do no change the strain hardening behavior. Based on their TEM analysis they concluded that VC precipitates do not appear to act as strong obstacles to the propagation of the microtwins, as the VC particles do not appear to block microtwins. Fig. 11.17 illustrates the strengthening effect of V additions. The addition of 0.21wt.-% V yields a relatively low strengthening effect of 140 MPa, which leads them to conclude that despite the relatively important amount of added V, a large fraction of it must remain in solid solution. The solid solution strengthening effect of V is approximately 17 MPa/wt.-% V. This conclusion appears to be in contrast to the work reported by Yen et al. [120] on the effect of V_4C_3 carbide precipitates in Fe-21.6%Mn-0.63% C-0.87%V (in wt.-%) TWIP steel. They observed a high density of precipitates formed

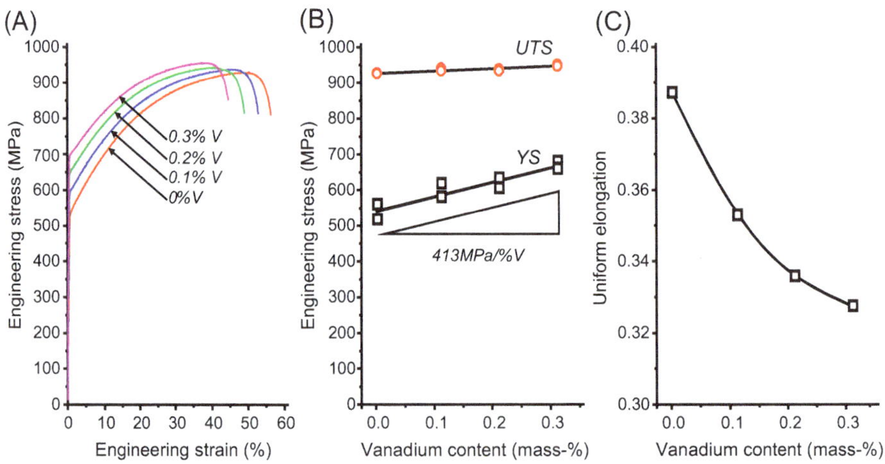

Figure 11.17 Effect of V micro-alloying on (A) the flow curve, (B) the strength, and (C) the uniform elongation of TWIP steel. The strengthening effect of V is about 417 MPa/wt.-% V. The strengthening effect combines the strengthening effects due to the grain size reduction and the precipitation.

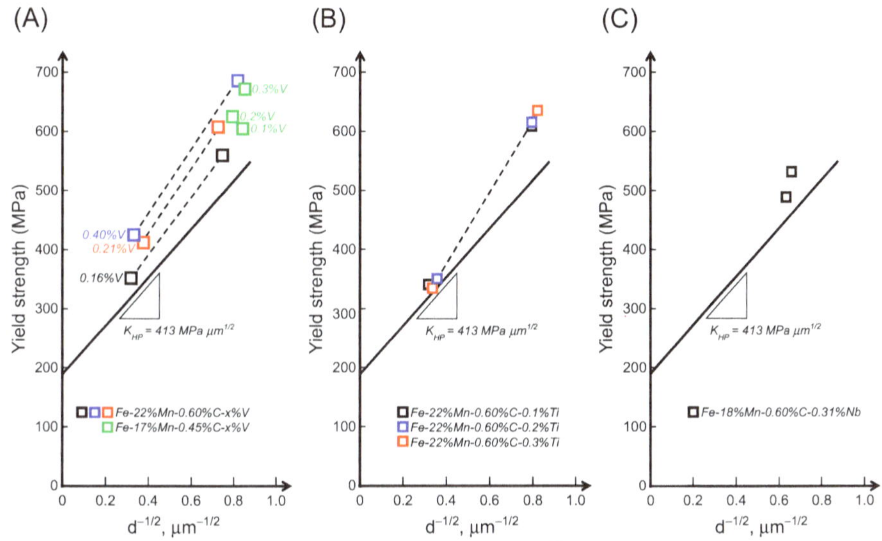

Figure 11.18 Influence of micro-alloying additions of (A) V, (B) Ti and (C) Nb on the Hall-Petch relationship of TWIP steel. The data is for V-alloyed Fe-22%Mn-0.6%C (in wt.-%) [109] and Fe-17%Mn-0.45%C (in wt.-%) TWIP steel [114].

during a recrystallization annealing treatment of 60 s at 850°C. The average precipitate diameter was 13 ± 7.7 nm. They reported a precipitate strengthening effect due to the Orowan bypassing mechanism of 198 MPa. This is a small strengthening effect considering the large V-addition, suggesting that a reduction of the precipitate size to about 5 nm is required to achieve a more pronounced strengthening effect.

Ti micro-alloying results in a high hardening effect (+1380 MPa/wt.-%Ti). This effect saturates at about 0.1 wt.-% Ti, resulting in a maximum strength increase of approximately +150 MPa. The micro-alloying effect of Nb additions to TWIP steel are not well known (Fig. 11.18).

Ueji et al. [121,122] reported that grain size reduction in C-free Fe-31%Mn-3%Al-3%Si (in wt.-%) TWIP steel (γ_{isf} ~42 mJ/m^2) resulted in the strong inhibition of deformation twinning and an important reduction in ductility. Gutierrez-Urrutia et al. reported that a grain refinement did not suppress deformation twinning in a Fe-22%Mn-0.6%C (in wt.-%) TWIP steel (γ_{isf} ~23 mJ/m^2), but that the deformation twinning became more difficult and that a smaller grain size was associated with a reduction of the twin area fraction [123].

11.2.3.8 Strain hardening in TWIP steel

The key reasons for the combination of high strength and plasticity of TWIP steel are (1) the reduction of the dislocation mean free path, especially for dislocation on slip planes intersecting the twinning plane, (2) the size strengthening contribution

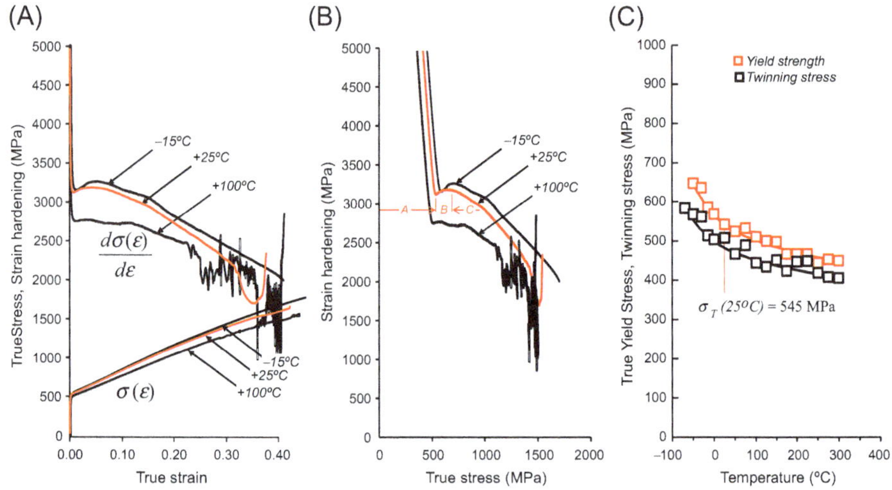

Figure 11.19 (A) True stress-true strain curve and strain hardening curve for Fe-18%-0.6% C-1.5%Al (in wt.-%) TWIP steel at −15°C, 25°C and 100°C. (B) Strain hardening versus stress curve indicating the different hardening stages for the flow curves shown in (A). (C) Temperature dependence of the yield stress and the twinning stress for Fe-18%-0.6%C-1.5%Al (in wt.-%) TWIP steel.

of the very narrow deformation twins, and (3) the high dislocation storage capacity, due to the strong inhibition of cross slip resulting from the low value of γ_{isf}, an increase in the dislocation constriction energy and the separation of dislocations by the deformation twins. According to Kalidindi et al. the strain hardening response of low γ_{isf} fcc alloys which are subject to deformation twinning goes through four distinct stages [124,125]. Fig. 11.19 identifies these four strain hardening stages in the strain hardening curve of Fe-18%Mn-0.6%C-1.5%Al (in wt.-%) TWIP steel. In the initial strain hardening stage A, the strain hardening is controlled by the dislocation density evolution in the absence of twin formation. Stage A is usually observed during a relatively short strain interval. No twinning takes place in stage A and it can be considered as an incubation stage in which dislocations interactions generate twin nuclei. Overlapping stacking faults are usually observed in this stage. Stage B is the primary twinning stage, which is characterized by an increasing strain hardening. It starts when primary deformation twinning is initiated in favorably oriented grains. Stage B typically starts in the true strain range of 0.03–0.04, which indicates that dislocation activity and dislocation-dislocation interactions occur prior to twinning. It also suggests that dislocations play a role in the creation of twin initiation sites. Fig. 11.19 also illustrates that the twinning stress, as defined by the transition from stage A to stage B, is only slightly higher than the yield stress. The highest strain hardening is observed in the transition from stage B to stage C. In stage C the strain hardening decreases gradually with increasing strain. This decreasing strain hardening is attributed to a reduction in the rate of primary twin formation. As the original grain size is reduced by the progressive segmentation of the grains by the deformation twins, higher stresses are required to generate more twins. The slowly decreasing strain hardening observed in stage C is terminated when extensive twinning is initiated on secondary twin systems. Secondary twin systems are activated in stage D. Multiple twin-twin intersections are formed which considerably limit dislocation glide distances. This leads to a reduction of the rate at which the strain hardening decreases with strain, and a more or less constant strain hardening behavior is observed in stage D. Note that the four stages are not always observed.

In its simplest form, the basic model for strain hardening is that the dominant deformation mode in TWIP steel is dislocation glide, i.e., strain hardening is controlled by the material's dislocation storage capacity during plastic deformation. The deformation-induced twinning results in a reduction of the effective dislocation glide distance. This results in a "dynamical Hall-Petch effect", which leads to a pronounced increase of the strain hardening, a high tensile strength and a considerable uniform elongation. This early model has considerably evolved. As the thin deformation twins appear to be impenetrable for dislocation to glide through, they could (1) act as a source of back stress and kinematic, rather than isotropic, hardening, or (2) contribute to strength as a result of their thickness, which is typically in the nanometer range. Three models have been put forward to explain the mechanism of plasticity enhancement due to deformation twinning: (1) the Allain-Bouaziz "dynamic Hall-Petch effect" [3,5,17,23], (2) the Bouaziz backstress effect [3], and (3) the Guttierez "composite effect" [93]. The key aspects of the three models are illustrated in Fig. 11.20. Model calculations based on the Allain-Bouaziz dynamic Hall-Petch concept are shown in Fig. 11.21. The results illustrate the influence of the deformation twinning kinetics on the mechanical properties of a TWIP steel using a two variables constitutive model

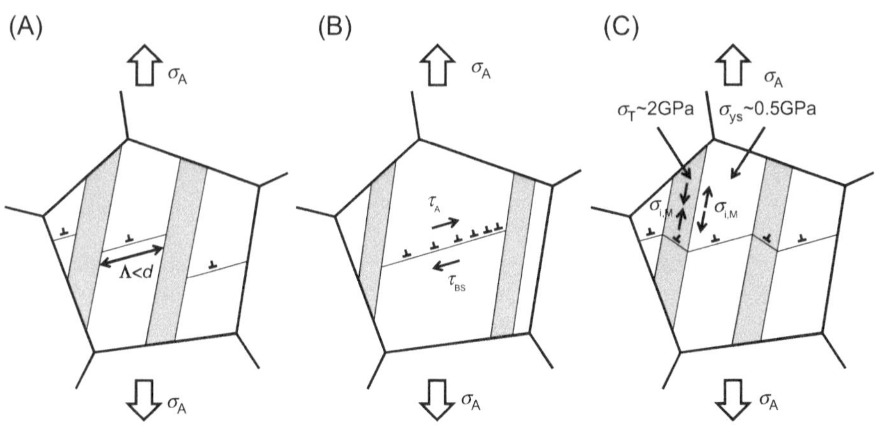

Figure 11.20 Mechanism of plasticity enhancement due to deformation twinning according to (A) the Allain-Bouaziz "dynamic Hall-Petch effect" [3,5,17,23], (B) the Bouaziz backstress effect [3], and (C) the Guttierez "composite effect" [93].

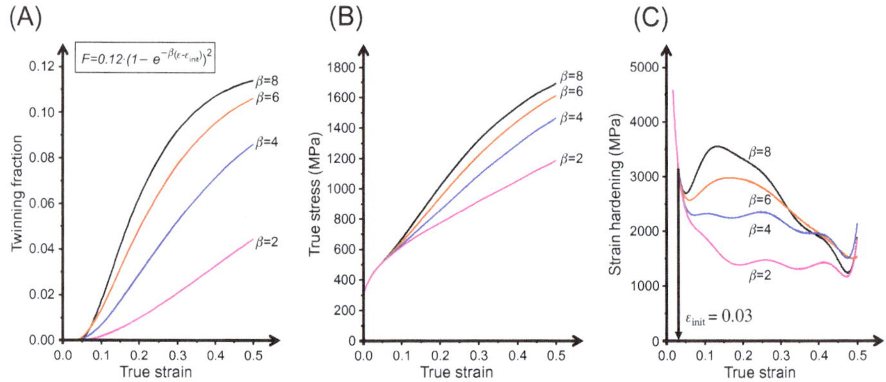

Figure 11.21 Effect of a reduction of the twinning kinetics on (A) the strain dependence of the twin volume fraction, (B) the stress-strain curve, and (C) the strain hardening based on the model developed by Kim et al. [126,127].

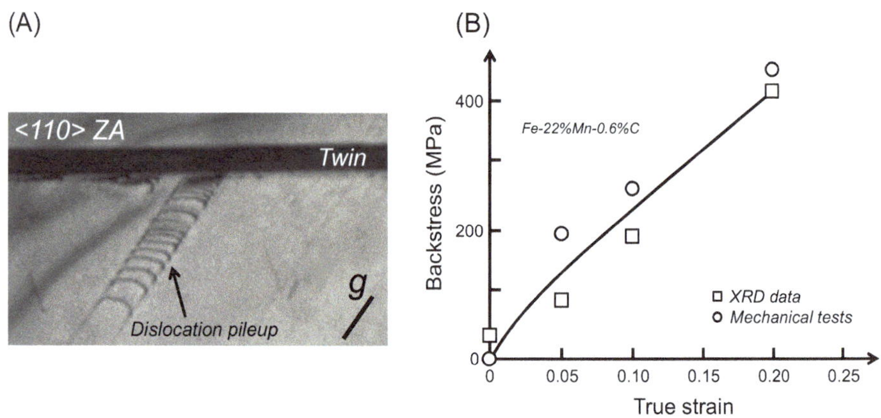

Figure 11.22 (A) Partial dislocation pileup at a deformation twin. (B) Strain dependence of the backstress in TWIP steel [3].

proposed by Kim et al. in which the twins boundaries act as impenetrable obstacles for dislocation glide [126,127].

Using forward-reverse shear tests, Bouaziz et al. [3] were able to detect an important Baushinger effect. They reported that the backstress contribution to hardening is approximately equal to 50% of the flow stress at a strain of 0.2. The results suggest that, in addition to the isotropic hardening, there is a substantial kinematic hardening contribution to strain hardening. According to Bouaziz et al, the backstress is generated by pileups of dislocations arrested at twin boundaries. Fig. 11.22 shows a TEM micrograph of a pileup of identical partial dislocations at a twin. The progressively larger interdislocation distance is a clear indication of the repulsive interaction between the dislocations and the twin boundary.

Sevillano and de las Cuevas also noted that compatibility requires the generation of backward and forward internal stresses [110,128]. The backward stresses, developed in the softer austenite matrix in which the twins are formed, must be balanced by forward stresses in the hard twins. This leads to a composite work hardening effect. The back stress results in an increase of the forward work-hardened stress and a permanent softening when the direction of the stress is inverted. Sevillano and de las Cuevas also proposed an explanation for the considerable strain hardening in TWIP steel. They argued that it is a strengthening effect due to the deformation twins, which contribute to the strength of the twin-austenite aggregate because of their very small thickness. The very hard twins separate regions of softer non--twinned matrix. According to Sevillano there are therefore two contributions to the flow stress: (1) the contribution of the matrix region, which is gradually being refined by an increasing volume fraction of deformation twins, and (2) the contribution from the very thin, high strength deformation twins.

There are many observations which suggest that the relation between deformation twinning and strain hardening may be more complex than the three models developed in the previous paragraphs which emphasize the relation between the evolution of the density of deformation twins and the strain hardening. Most TWIP steels retain a large ductility when deformation twinning is restricted, i.e., by a change in composition, grain refinement, or an increase in temperature. For example, Bouaziz et al. described two TWIP steels with the same γ_{isf}, i.e., the Fe-30%Mn and Fe-22%Mn-0.6%C (in wt.-%) TWIP steels, which develop a very different dislocation microstructure, with extensive twinning in Fe-30%Mn and no twinning in Fe-22%Mn-0.6%C (in wt.-%) [108]. Both steels have a large ductility. Another example suggesting that deformation twinning does not necessarily play a key role in the strain hardening of TWIP steel is shown in Fig. 11.23. It compares the flow curves of Fe-18%Mn-0.6%C-1.5%Al (in wt.-%) TWIP steel at temperatures in the range of $-70°C$ to $+300°C$. Despite the fact that γ_{isf} varies over a wide range, form 17 mJ/m^2, at $-70°C$, to 82 mJ/m^2, at $+300°C$, the plasticity of the material remains high, even when the dynamic strain aging is very pronounced, as can be concluded from the serrated flow curves measured in the high temperature range of 100°C-300°C. As deformation twinning is suppressed when $\gamma_{isf} > 45$ mJ/m^2, i.e. at $+108°C$, these flow curves suggest that deformation twins do not necessarily play the most important role in determining the conditions for a pronounced strain hardening observed for TWIP steel. Other processes leading to a favorable dislocation density evolution may also lead to a large strain-hardening during tensile deformation. A low γ_{isf} and the presence of the deformation twins may only be two of the factors contributing to an intrinsically high strain hardening by a lowering of the rate of dynamic recovery, which increases the strain-hardening due to an enhanced dislocation accumulation. In addition to deformation twinning, other contributions to the strain hardening of TWIP steel have therefore been proposed: (1) short range ordering, (2) dynamic strain aging, i.e., the interaction between mobile dislocations and interstitial C atoms of a point defect complex involving a C atom [129], (3) pseudo-twinning, i.e., the strengthening effect associated with the lattice

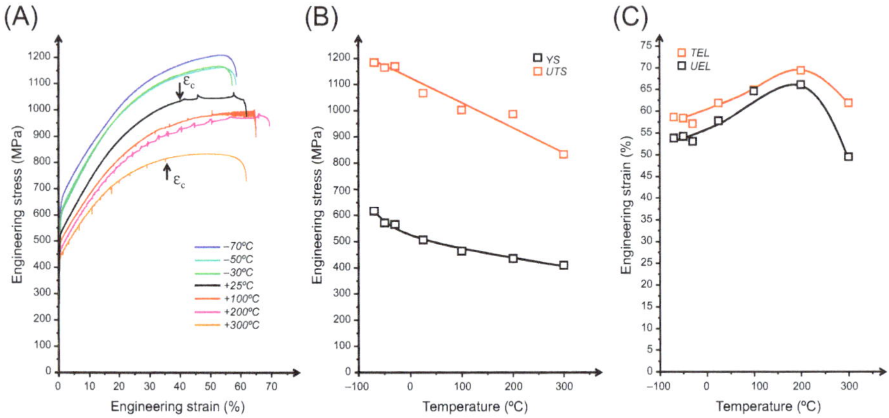

Figure 11.23 (A) Engineering stress-strain curves of Fe-18%Mn-0.6%C-1.5%Al (in wt.-%) TWIP steel in the temperature range of −70°C to +300°C as measured with a strain rate of 10^{-3} s^{-1}. (B) Temperature dependence of the yield strength (YS) and the ultimate tensile strength (UTS). (C) Temperature dependence of the uniform and total elongation (UEL and TEL). Note in (A) that the stress-train curves in the range of 25°C-300°C are serrated. The TWIP steel has an inverse behavior for the occurrence of serrations at high temperature: whereas a critical strain (ε_c) of about 40% must be reached for the initiation of the serrations at 25°C, a critical strain of about 35% must be reached to suppress the serrations at 300°C. Note also in (B) that the uniform elongation peaks at a temperature of 200°C. At this temperature the γ_{isf} is approximately 62 mJ/m^2, i.e., deformation twinning is suppressed.

distortion due to the change of the interstitial C atom position from an octahedral site to a tetrahedral site during the twinning shear when the C is on the glide plane [130], and (4) stacking faults.

11.2.4 Forming properties

TWIP steel is characterized by a high strain hardening, a large uniform elongation and a high ultimate tensile strength. It also has a relatively low yield strength and virtually no post-uniform elongation. In addition, some serrations are visible on the stress-strain curve of C-alloyed TWIP steels at high strain levels. The superior forming properties of specific TWIP steels, such as the Fe-18%Mn-0.5%C-1.5%Al (in wt.-%) TWIP steel, have also been reviewed in detail [20,131]. In fact, this class of highly ductile, high and ultra-high strength steels has been the subject of a truly international research and development effort, in both industrial and academic laboratories, aimed at their industrial development for large scale use in the automotive industry, LNG-shipbuilding, oil-and-gas exploration and transport, and non-magnetic structural applications. In the case of automotive materials, the main motivation for the current interest in Fe-Mn-Al-C TWIP steel for use in automotive application is related to the fact that they can provide materials with an ultra-high strength for structural reinforcement, a superior ductility for the ease of press

Table 11.8 Mechanical properties of Fe-15%Mn-0.7%C-2.5%Al-2.5%Si (in wt.-%) TWIP steel (strain rate $10^{-2}s^{-1}$) [131]

	0° (//RD)	45° to RD	90° to RD (//TD)
0.2% Yield strength, MPa	633	630	652
Tensile strength, MPa	1035	999	1012
r-value	0.787	0.962	0.997
E-modulus, GPa	172	168	175
Uniform elongation, %	49.2	52.3	51.2
Total elongation, %	56.7	59.9	57.1
Strain rate sensitivity	N.A.		

Table 11.9 Mechanical properties of Fe-18%Mn-0.5%C-1.5%Al (in wt.-%) TWIP steel (strain rate 10^{-3} s^{-1}) [20]

	0° (//RD)	45° to RD	90° to RD (//TD)
0.2% Yield strength, MPa	457.2	465.3	481.7
Tensile strength, MPa	960.1	944.2	951.0
r-value	0.816	1.188	1.339
E-modulus, GPa	180	188.2	194.3
Uniform elongation, %	–	–	–
Total elongation, %	68.9	64.9	67.8
Strain rate sensitivity	− 0.007		

forming, and a large energy absorption capability for improved vehicle crashworthiness. In addition, their large-scale use is expected to result in a reduction of vehicle mass, a lowering of greenhouse gas emissions, a drastic increase of gas mileage and an improvement of passenger safety. The mechanical properties of two TWIP steels are listed in Tables 11.8 and 11.9.

11.2.5 The high strain rate behavior of TWIP steel

High strain rate properties are of considerable technical interest, e.g., in automotive application for passenger safety-related parts which are expected to absorb energy in case of collisions. Fig. 11.24 illustrates the result of an axial crush test carried out on various advanced high strength steel grades. The TWIP steel has clearly superior crash properties. Using a split Hopkinson pressure bar Ha et al. have studied the microstructure of Fe-22%Mn-0.4%C (in wt.-%) TWIP steel ($\gamma_{isf} = 35$ mJ/m^2) after high strain rate deformation at 10^3 s^{-1} [132]. Curtze and Kuokkala [133] studied the high strain rate behavior of Fe-(25-28)%Mn-1.6% Al-0.08%C (in wt.-%) TWIP steels at 10^3 s^{-1}. Both reported a considerable increase in tensile strength at high strain rates, e.g., to 2265 MPa for Ha et al, and

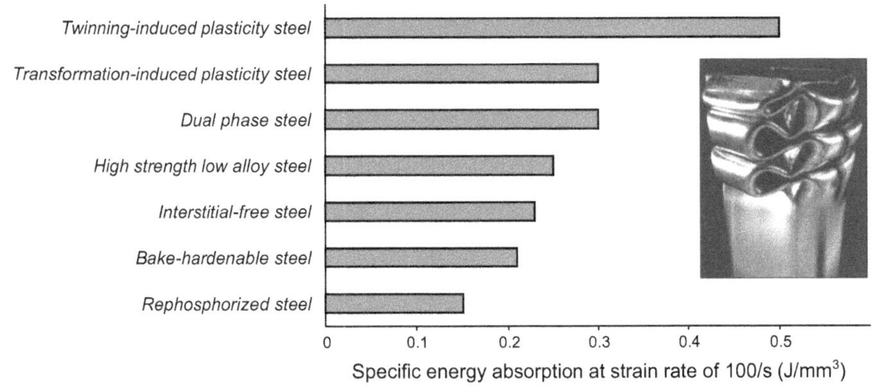

Figure 11.24 Comparison of the crash performance properties of various high strength automotive steel grades obtained by means of the axial crush test.

an increasing strain hardening up to a true strain of approximately 0.19. They reported that the high strain rate deformation leads to nanostructured austenite which appears random at low magnification. When the microstructure is observed at high magnification it consists of a nano-structure of 20–30 nm wide grains which are twin-related to the neighboring grains. According to Ha et al. the formation of twin-matrix lamellas within the grains occurs only in the early stages of the high strain rate deformation. The adiabatic heating of the sample at larger strains has two consequences: (1) an increase of γ_{isf}, and hence the suppression of twin formation, and (2) dynamic recovery. The grain fragmentation resulting from dislocation-twin boundary interactions and dynamic recovery generates the nano-structured austenite.

11.3 Medium Mn TRIP and TWIP + TRIP steel

Whereas TWIP steels have an excellent balance of strength and ductility, the high production cost and the low productivity related to their high Mn content (18–30 wt.-%) have hampered production and industrial utilization of TWIP steel. Efforts have been undertaken to develop steels with lower Mn contents which retain the superior tensile properties achieved by TWIP steel. Steels with a reduced C and Mn content as low as 12 wt.-% can have a homogeneous austenitic microstructure, but their SFE is also considerably less than 20 mJ/m^2. As a consequence, strain-induced ε-martensite transformation takes place during deformation and the tensile properties are poor. This is illustrated in Fig. 11.25. Medium Mn steels with an ultra-fine grained austenitic-ferritic microstructure obtained by intercritical annealing have therefore been developed. The properties of the medium Mn steels and high Mn TWIP steels are compared in Fig. 11.26. While TWIP steels achieve impressive elongations, the medium Mn steel achieves higher strength levels and adequate levels of plasticity. Some medium Mn steel concepts have been shown to outperform TWIP steels in terms of both strength and elongation, as shown in Fig. 11.26.

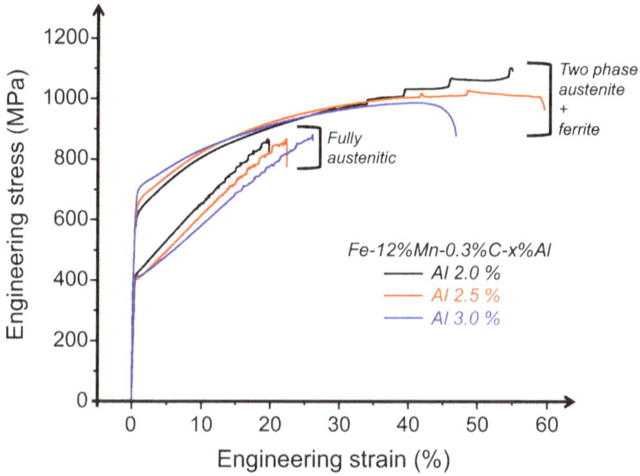

Figure 11.25 Comparison of the stress-strain curves of Fe-12%Mn-0.3%C-x%Al (in wt.-%) steel with a fully austenitic microstructure, obtained by annealing in the austenite stability range, and the stress-strain curves of the same steel with a two-phase, austenite + ferrite, microstructure obtained by intercritical annealing. Both strength and elongation performance are considerably improved when the steel has a two-phase microstructure. The addition of Al gives the alloy concept more flexibility. In the example, Al increases the yield point and suppresses the serrations in the intercritically annealed steel.

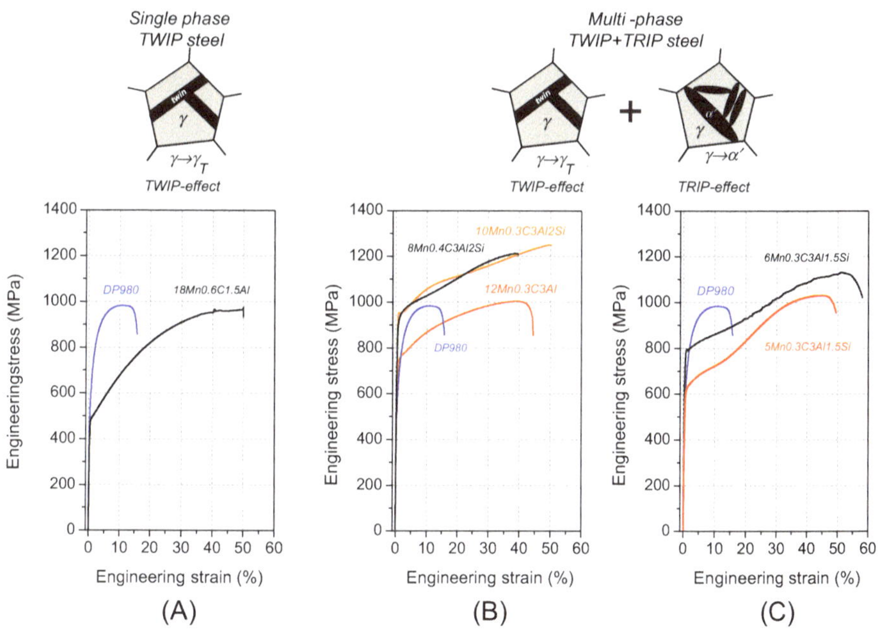

Figure 11.26 Comparison of the mechanical properties of a first generation advanced high strength steel (DP980), a second generation high strength TWIP steel (18%Mn-0.6%C-1.5% Al, in wt.-%), and the intercritically annealed multi-phase TWIP + TRIP steels containing 6–12 wt.-%Mn.

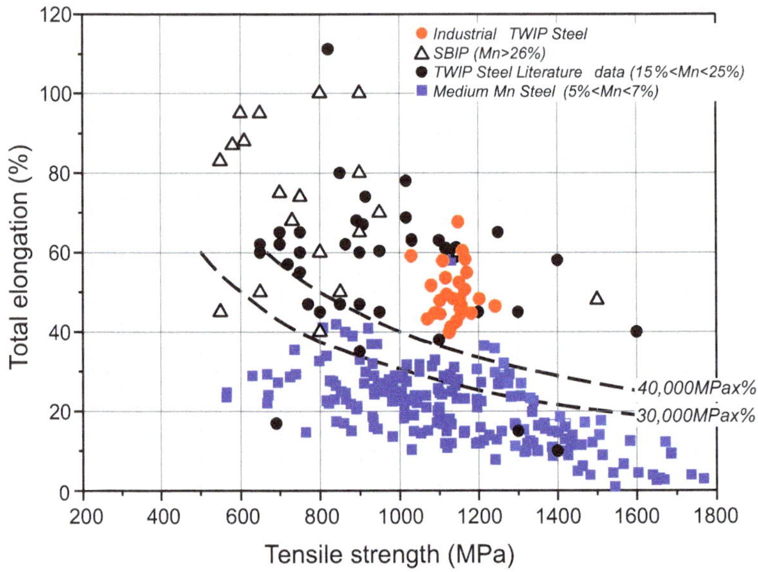

Figure 11.27 Comparison of strength-ductility properties of high Mn TWIP steel, high Mn shear band-induced plasticity steel, and medium Mn steel. Whereas TWIP steels achieve strength-ductility balances >40,000 MPa-%, the strength-ductility balance of high performance medium Mn steels is typically in the range of 30,000 MPa-% to 40,000 MPa-%.

Fig. 11.26 compares the engineering stress-strain curves of the mechanical properties of a first generation advanced high strength steel (DP980), a second generation high strength TWIP steel (18%Mn-0.6%C-1.5%Al, in wt.-%), and the intercritically annealed multi-phase TWIP + TRIP steels containing 6–12 wt.-% Mn. Intercritically annealed medium Mn steels with Mn contents as low as 5 wt.-% also show excellent properties when the TWIP effect is activated at low strain. When the room temperature stacking fault energy of the austenite is in the range of 15–20 mJ/m^2, it is possible to activate both mechanical twinning (TWIP effect) and strain-induced transformation (TRIP effect) in the austenite grains in succession. When secondary twins are formed and intersect with the primary twins during straining, nucleation sites for α'-martensite are created, leading to the subsequent activation of the TRIP effect. The resulting TWIP + TRIP effect, whereby the TWIP effect and the TRIP effect are activated in succession, results in high tensile strengths in the range of 1000–1200 MPa and a ductility characterized by total elongations in the range of 20–30% (Fig. 11.27).

11.3.1 Principles of medium Mn TWIP + TRIP steel alloy design

The design principles for medium Mn ferrous alloys have recently been published by Lee and De Cooman [134–138]. The basic concept is shown in Fig. 11.28. It illustrates the intercritical annealing concept for medium Mn steel. The medium

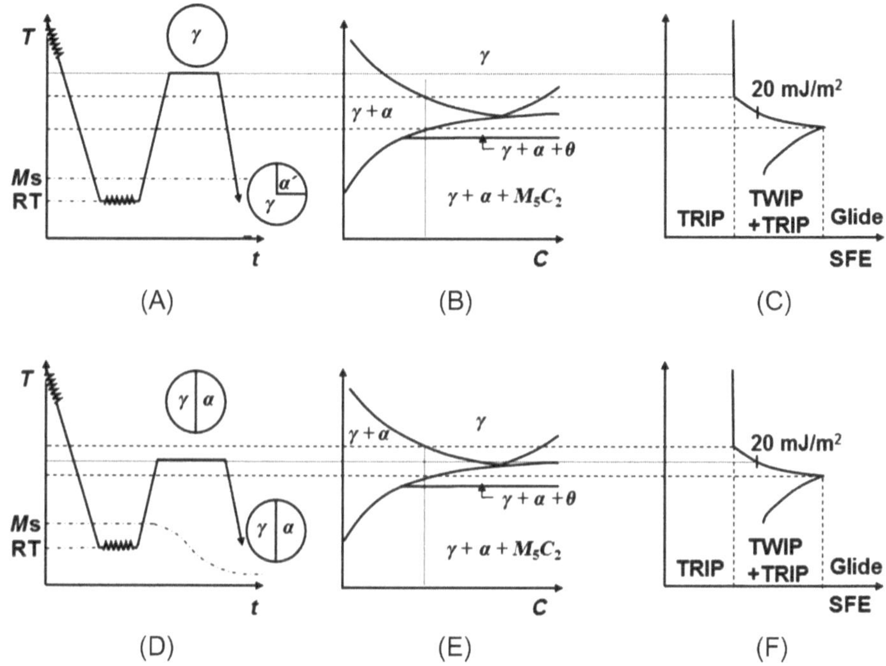

Figure 11.28 Schematic illustrating the intercritically annealed medium Mn steel concept. Annealing in the austenite stability range using the thermal cycle shown in (A) results in a homogeneous austenite phase with the same Mn and C content as the nominal alloy as illustrated in the pseudo-binary Fe-C equilibrium phase diagram (B). Annealing in the homogeneous austenite stability range (C) also results in an annealing temperature independent room temperature stacking fault energy for austenite. Intercritical annealing medium Mn steel in the two phase, ferrite and austenite, stability range (D) results in the Mn and C enrichment of the intercritical austenite, and a lowering of the Ms temperature during annealing (E). (F) The intercritical temperature can be selected to obtain a room temperature austenite stacking fault energy in the range of 15–20 mJ/m^2.

Mn steels have a Mn content in the range 5–12 wt.-%. The C content is usually less than 0.4 wt.-%. Intercritical annealing makes it possible to increase the Mn and C content of the intercritical austenite, which results in the lowering of the Ms temperature of the intercritical austenite. The intercritical temperature is selected to obtain a room temperature austenite stacking fault energy in the range of 15–20 mJ/m^2.

A key aspect of the alloy design is the partitioning of C and Mn to the austenite phase during intercritical annealing. The partitioning makes it possible to obtain a stable ultra-fine microstructure relatively easily by intercritical annealing at the appropriate temperature. Evidence for the partitioning of C and Mn has been obtained by three-dimensional atom probe tomography as illustrated in Fig. 11.29.

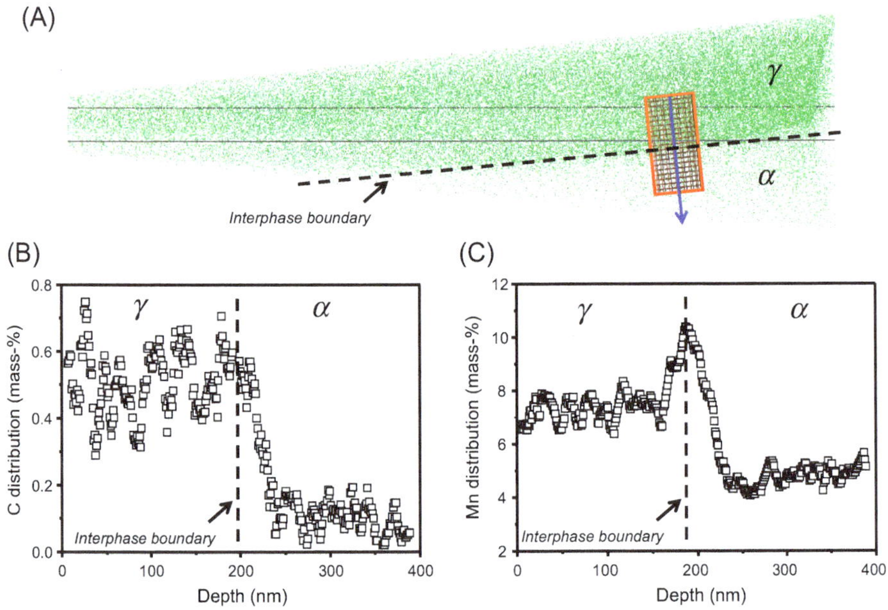

Figure 11.29 (A) Three-dimensional atom probe tomography showing the C distribution at a ferrite/austenite phase boundary in a medium Mn steel. Evidence for the partitioning of C (B) and Mn (C), for a measurement carried out within the rectangular area indicated in (A).

An important alloying addition is Al. It very effectively increases the stacking fault energy of the austenite, by approximately $+5\ \text{mJ/m}^2$ per wt.-% of Al. In addition, Al-addition makes it possible to obtain medium Mn steel microstructures containing δ-ferrite, which are characterized by a continuous elasto-plastic transition. The effect of the Al addition on the austenite phase fraction and composition is illustrated in the computational thermodynamics data shown in Fig. 11.30.

The microstructure of medium Mn steel is illustrated in Fig. 11.31. The figure also shows clear evidence for the activation of the TWIP + TRIP mechanism in medium Mn steel. In these ultra-fine grained ferrite + austenite steels, the control of the room temperature austenite stability and its stacking fault energy are achieved by the controlled partitioning of C and Mn to the intercritical austenite during the intercritical annealing treatment. If the composition-dependent stacking fault energy is in the range of $15-20\ \text{mJ/m}^2$, strain-induced twinning takes place in the austenite grains during room temperature deformation. This strain-induced twinning takes place at low strains. In addition, the strain-induced transformation to martensite nucleates at twin-twin intersections after the process of strain-induced twinning saturates. Both deformation-induced twinning and transformation are known to result in effective plasticity-enhancement mechanisms known as the TWIP and the TRIP effect, respectively. Although ultra-fine grained (UFG)

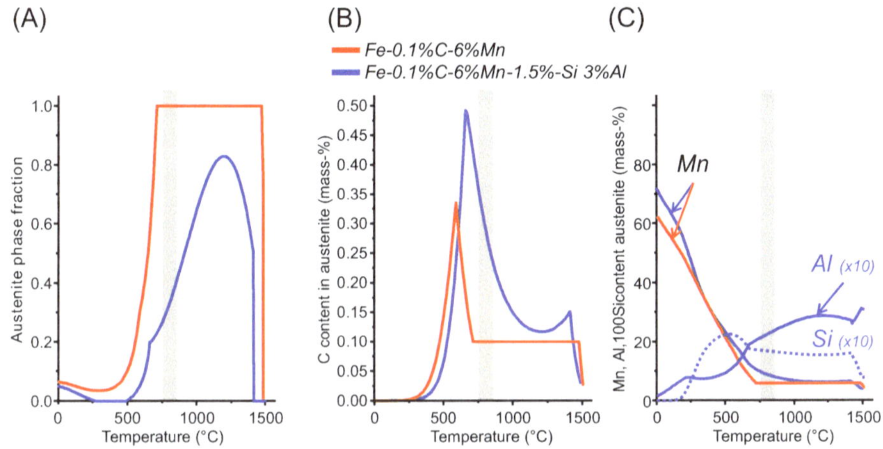

Figure 11.30 THERMOCALC computation showing the effect of Al-additions on (A) the austenite phase fraction, (B) the austenite C content, and the (C) Mn, Al, and Si content of the austenite. The grey band indicates the standard industrial annealing temperature range.

microstructures are usually characterized by a poor formability, the TWIP and TRIP effects occurring in succession result in the combination of an ultra-high strength (>1 GPa) and high ductility (>30%) of UFG medium Mn ferrite + austenite steels.

The key influence of the carbide precipitation and dissolution on the microstructure formation of intercritically annealed UFG medium Mn steel is illustrated in Fig. 11.32. The data reveals that the carbide precipitation reaction and the austenite formation mechanism are taking place during the heating of a Fe-10%Mn-0.3%C-3%Al-2%Si (in wt.-%) medium Mn steel during to the intercritical temperature range. The tempering of the initial cold-rolled high-C martensite results in the $\alpha' \rightarrow \alpha + \theta$ tempering reaction during heating. Cementite nucleates on the subgrain and lath boundaries of the martensite. The nucleation of austenite at the ferrite/cementite interface, $\alpha + \theta \rightarrow \gamma$, resulted in the formation of a carbide-free, UFG dual-phase ferrite-austenite microstructure. The austenite, nucleated at ferrite/cementite interfaces, has a high initial C content, which decreased with increasing austenite volume fraction at temperatures higher than 938 K (665°C).

11.3.2 Strain hardening control in medium Mn TWIP + TRIP steels

The tensile behavior of medium Mn steel is critically dependent on the intercritical annealing temperature and the phase fractions. The ferrite volume fraction controls both the grain size of the austenite and the composition of the austenite. Both factors determine the kinetics of the TWIP effect and the TRIP effect. The steels are characterized by two plasticity-enhancing mechanisms occurring in sequence during straining. At low strain, the TWIP effect is activated. At intermediate strains, when

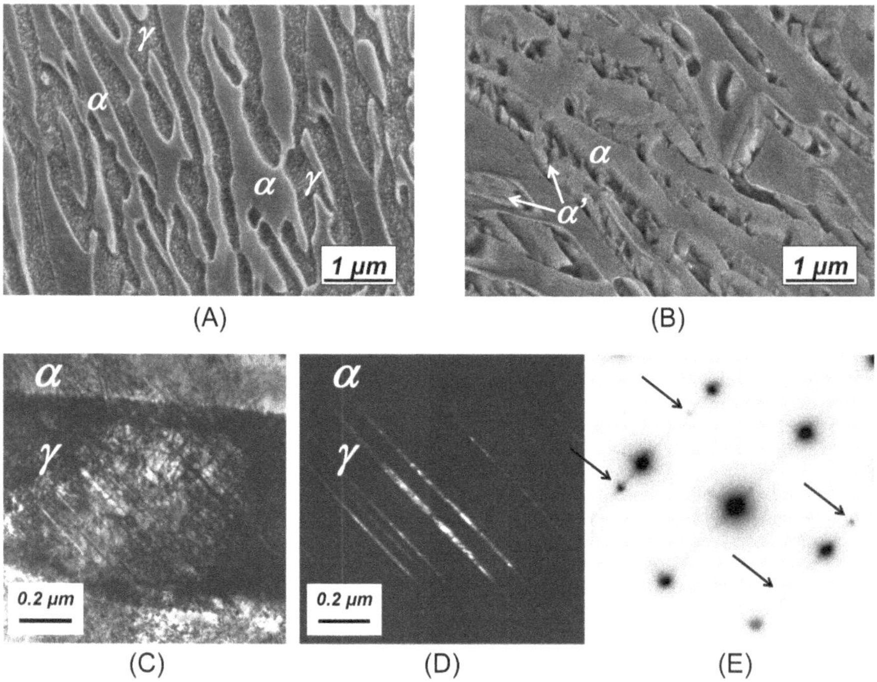

Figure 11.31 (A) As-annealed microstructure of Fe-8%Mn-0.4%C-3%Al-1%Si (in wt.-%) medium Mn steel intercritically annealed at 750°C. The austenite phase fraction is 0.57. (B) Deformed microstructure, showing that austenite has transformed to martensite during straining to an engineering strain of about 60%. (C) Bright-field and (D) dark-field transmission electron micrographs of an austenite grain showing the presence of twins formed prior to the strain-induced transformation to martensite. After saturation of the mechanical twinning, deformation-induced transformation, nucleated at twin-twin intersections, leads to the TRIP effect at higher strain. The TWIP + TRIP mechanism results in an ultra-high strength and a high ductility. (E) Electron diffraction pattern corresponding to (C) and (D). The arrows indicate the twin reflections.

the volume fraction of deformation twins has reached a saturation value, the TRIP effect is activated by nucleation at twin-twin intersections. Medium Mn TRIP steel exhibits the strain-induced martensite transformation $\gamma \rightarrow \alpha'$ at intersections of localized deformation bands, such as shear bands, stacking faults, twins, and ε-martensite plates. The phase transformation results in a high work hardening rate and delays the diffuse necking. The alloy composition and austenite grain size adjust the austenite stability and control the martensite transformation kinetics. The optimum austenite stability corresponds to the situation where strain-induced plasticity occurs over a wide strain range. Fig. 11.33 illustrates the annealing temperature dependence of the mechanical properties of a 6 wt.-% Mn steel. It can be seen that the optimum strength-ductility balance is obtained for an intercritical temperature of 780°C. Higher tensile strengths combined with a lower elongation are obtained at higher annealing

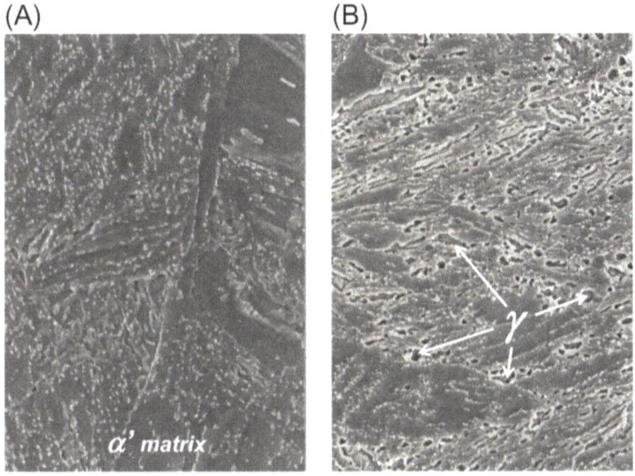

Figure 11.32 Scanning electron micrographs showing the microstructure evolution of Fe-0.30%C-10%Mn-3%Al-2%Si (in wt.-%) medium Mn steel during the interrupted heating to the intercritical annealing temperature. (A) At 650°C, martensite aging has occurred as evidenced by carbide precipitation. (B) At 750°C, austenite nucleation is initiated simultaneously with carbide dissolution.

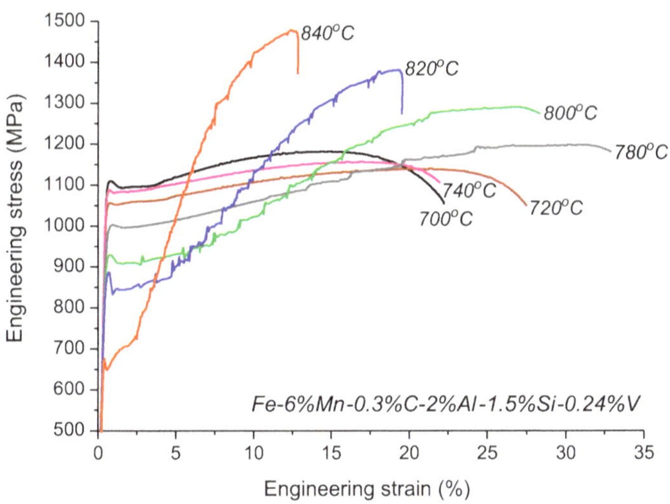

Figure 11.33 Example of the intercritical annealing temperature dependence of the mechanical properties of medium Mn steel.

temperature due to the rapid transformation of austenite to martensite. High yield strengths combined with a lower elongation are obtained at lower annealing temperature due to the suppression of the transformation of austenite to martensite.

TWIP steel exhibits deformation twinning resulting from the passage of a $a/2\langle 111\rangle\gamma$-type Shockley partial dislocation on every $\{111\}\gamma$-type close packed

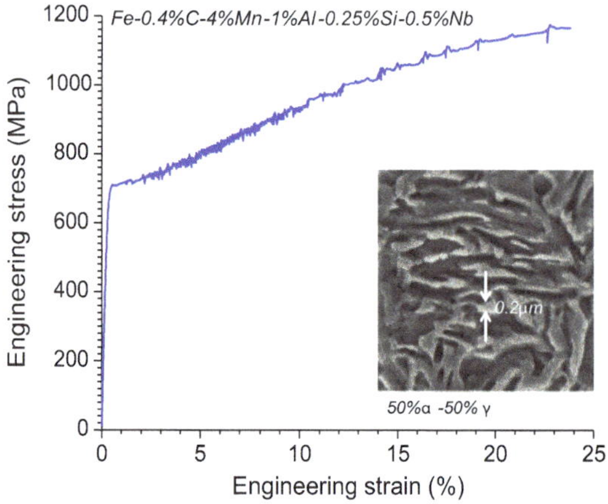

Figure 11.34 Ultra-fine grained Nb micro-alloyed medium Mn steel containing only 4 wt.-% Mn (YS: 700 MPa, UTS: 1180 MPa, TE: 24%) intercritically annealed at 700°C.

plane. The deformation twins act as effective obstacles to dislocation glide, thereby resulting in a pronounced work hardening rate. The kinetics of deformation twinning is dependent on the stacking fault energy of the austenite. The TWIP effect is activated when the stacking fault energy is in the range of 15–45 mJ/m^2. If the stacking fault energy is less than 15 mJ/m^2, strain-induced ε-martensite may be formed. The stacking fault energy is composition and temperature dependent. C and Al additions increase the stacking fault energy whereas Si and Cr additions decrease the stacking fault energy. The highest work hardening rate is obtained when the stacking fault energy is in the range of 15–20 mJ/m^2.

As the understanding of the physical metallurgy of medium Mn multi-phase steel exhibiting the TWIP + TRIP plasticity-enhancing mechanism is continuously improving, the properties of alloys with very low Mn contents are being explored. An example is shown in Fig. 11.34 for a Nb-alloyed 4 wt.-% Mn steel. The steel has a multiphase microstructure consisting of ferrite and austenite, obtained by intercritical annealing. The alloy composition and the intercritical annealing conditions are chosen to obtain an austenite stacking fault energy in the range of 15–20 mJ/m^2. Despite the very low alloy content, a very high tensile strength of 1.2 GPa is obtained after intercritical annealing. The succession of the TWIP and TRIP plasticity-enhancing mechanisms resulted in a high work hardening rate and a tensile elongation of 25%.

11.4 Outlook for high Mn TWIP steel and medium Mn steel

11.4.1 Industrial production

Fig. 11.35 illustrates the industrial production route for high Mn TWIP steel. In comparison to the processing of low C steel grades, precautionary measures have

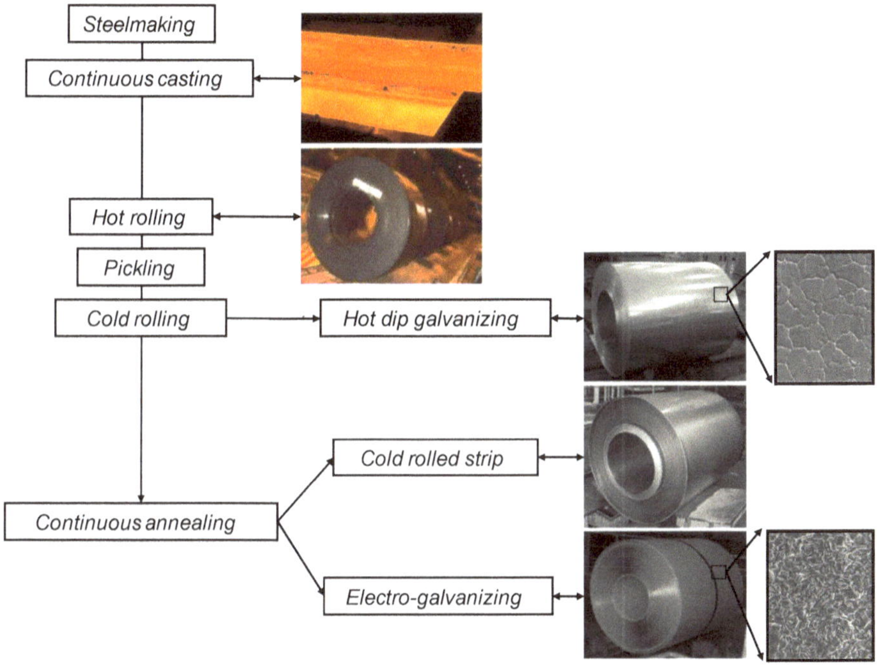

Figure 11.35 Production route industrial high Mn TWIP steel.

to be taken during secondary metallurgy, continuous casting, and rolling. Multiple production difficulties have been reported for the casting, the hot rolling, and the cold rolling of high Mn steels. Mn loss occurs due to the high Mn vapor pressure reducing the Mn yield during steelmaking. Temperature control of the steel melt is also challenging due to the large amount of alloying. The as-cast microstructure of TWIP steel is coarse with well-developed dendrites and coarse equiaxed grains. This is due to the wide solidification range and the low thermal conductivity of TWIP steel.

According to Scott et al. [139] the solidification range of Fe-22%Mn-0.6%C (in wt.-%) is 120°C, i.e., from 1407°C to 1287°C. The solidification range of Fe-18%Mn-0.6%C-1.5%Al (in wt.-%) is 110°C, i.e., from 1380°C to 1270°C. The hot ductility of austenitic steels with a Mn content in the range of 9–23 wt.-% and C content in the range of 0.6–0.9 wt.-% was studied by Bleck et al. [140]. They observed that a reduction of the Mn content resulted in an improvement of the hot ductility. The investigated steels solidified in a relatively wide temperature range. The fracture surfaces of Fe-16%Mn-0.8%C (in wt.-%) steel tested at 1201°C and 1000°C show dendrite branches, micro-shrinkage leading to the formation of interdendritic micro-porosities and interdendritic segregation [141]. This dendritic segregation is associated with an inhomogeneous Mn distribution in the steel (Fig. 11.36).

The grain boundary oxidation of continuously cast TWIP steel has been reported to result in the formation of surface and sub-surface defects during hot rolling.

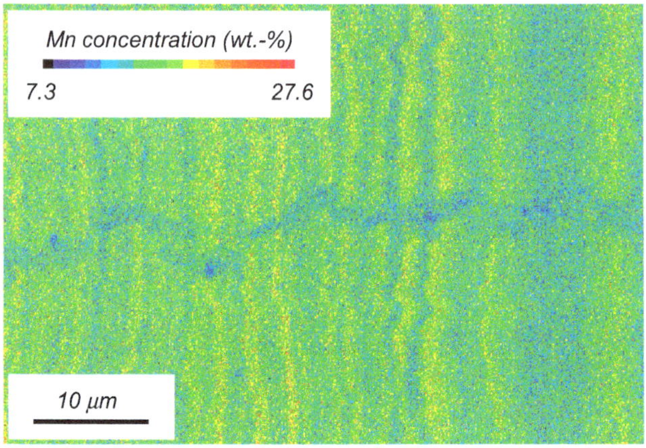

Figure 11.36 Electron probe micro-analysis of the Mn distribution in a hot rolled TWIP steel illustrating the inhomogeneous Mn distribution resulting from the inter-dendritic segregation during casting.

Direct strip casting has been proposed as a possible solution to this problem as it avoids the formation of bending-unbending cracks [142]. Edge cracks, poor surface quality, and rolling instabilities are typically reported in rolling [143].

11.4.2 Static and dynamic strain aging

Both static strain aging (SSA) and dynamic strain aging (DSA) occur in C-alloyed high Mn TWIP steel and medium Mn steel containing more that 0.2 wt.-% C. Allain et al. [144] carried out room temperature relaxation tests on a V-micro-alloyed Fe-22%Mn-0.6%C (in wt.-%) TWIP steel. They observed a yield point phenomenon and an associated yield plateau upon reloading. They ascribed this effect to the aging of dislocations by C atoms during the relaxation test. In addition, the relaxation experiment triggers the formation of serrations on the flow curve. Fig. 11.37 shows clear evidence for static strain aging at room temperature in both an Al-free and an Al-alloyed Fe-18%Mn-0.6%C (in wt.-%) TWIP steel. As the upper yield point and the yield point elongation are observed in conditions where twins have not formed, the static strain aging is due to an interaction between dislocations and point-defects in the initial as-annealed microstructure. Tsuji et al. also reported that TWIP steel with a grain size less than 2 μm exhibited a yield phenomenon, similar to ultra-fine grained (grain size <1 μm) fcc and bcc metals [145], but they did not provide an explanation for their observation.

DSA in TWIP steel has four main macroscopic effects: (1) the flow curve is serrated, (2) the strain is localized in Portevin-Le Châtelier (PLC) deformation bands, (3) the strain rate sensitivity is very small or negative, and (4) the post-uniform elongation, as measured in a uni-axial tensile test, is very limited.

In uni-axial tension, DSA manifests itself by jerky or serrated flow curves. The serrations are an indication of the non-homogeneous deformation of the

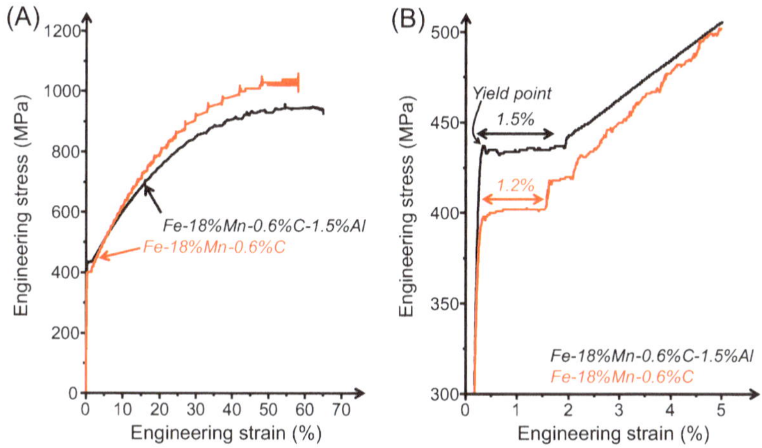

Figure 11.37 (A) Engineering stress-strain curve for Al-free and Al-alloyed Fe-18%Mn-0.6%C (in wt.-%) TWIP steel aged at room temperature. Serrations, due to DSA, are visible of the entire flow curve of the Al-free TWIP steel. (B) Magnification of the elasto-plastic transition region of (A), showing clearly that both steels are sensitive to static strain aging. A yield point elongation is visible on both stress-strain curves. A clear upper yield point is visible on the curve for the Al-alloyed Fe-18%Mn-0.6%C (in wt.-%) TWIP steel.

tensile samples. Strain localization takes place in propagating or static deformation bands. From a technological point of view, it is important to avoid DSA, to ensure stable material flow during sheet forming processes. DSA gives rise to non-homogeneous plastic flow and it is the cause of visible surface defects on press-formed parts. The negative strain rate sensitivity associated with DSA leads to the inability of C-alloyed TWIP steels to sustain large post-uniform strains after the initiation of necking. It has also been linked to the premature ductile fracture occurring during stretch-flanging of TWIP steels as DSA leads to rapid edge cracking during hole expansion [146,147]. The critical strain for the initiation of serration on the flow curve is temperature dependent, as illustrated in Fig. 11.38.

According to Chung et al. [20] the strain rate sensitivity of Fe-18%Mn-0.5%C-1.5%Al (in wt.-%) TWIP steel is −0.007. Shen et al. reported more negative values for the strain rate sensitivity of Fe-20%Mn-0.6%C (in wt.-%) TWIP steel [148], i.e., −0.021, for a steel with a grain size of 3.5 μm, and −0.029 for a steel with a grain size of 25 μm (Fig. 11.39).

Dynamic Strain Aging (DSA) is observed in C-added TWIP steel at room temperature, i.e., at a temperature at which the diffusivities of solutes, including that of solute interstitial C, is extremely low and diffusion of C to temporarily arrested dislocations is virtually impossible. Solute C in TWIP steel is known to have very low room temperature diffusivity and the room temperature interaction of solute C with dislocations during straining will therefore necessarily be of very short range nature. Even if local heating, e.g., due to the enhanced strain rate within in a PLC band, is taken into consideration, the C bulk diffusion at room temperature

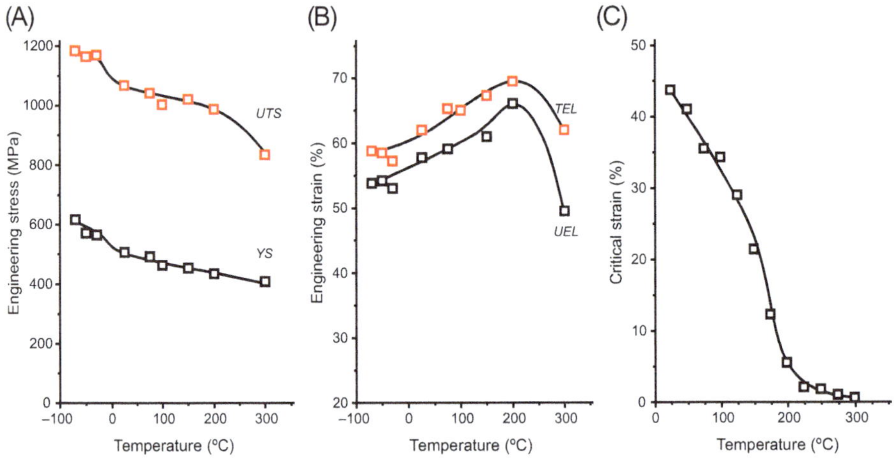

Figure 11.38 Temperature dependence of (A) the yield strength (YS) and the ultimate tensile strength (UTS), (B) the uniform and the total elongation (UEL and TEL), and (C) the critical stress for the initiation of serrations in the flow curve of Fe-18%Mn-0.6%C-1.5%Al (in wt.-%). The strain rate was 10^{-3} s^{-1}.

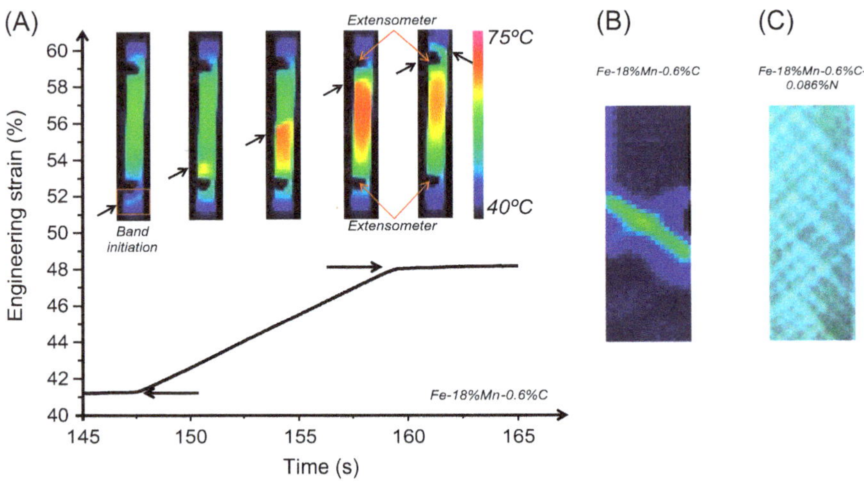

Figure 11.39 (A) Example of a solitary type A band which nucleates at one end of the tensile specimen and propagates across the entire sample length in Fe-18%Mn-0.6%C (in wt.-%) TWIP steel. The passage of the band is only detected when it is moving within the extensometer range. This gives rise to characteristic type A serrations on the stress-strain curves. (B) High resolution infrared image of a type A band, which consists of a main band and a secondary band. (C) Trace of multiple deformation bands which give rise to type B serrations on the stress-strain curve of Fe-18%Mn-0.6%C-0.086%N (in wt.-%) TWIP steel.

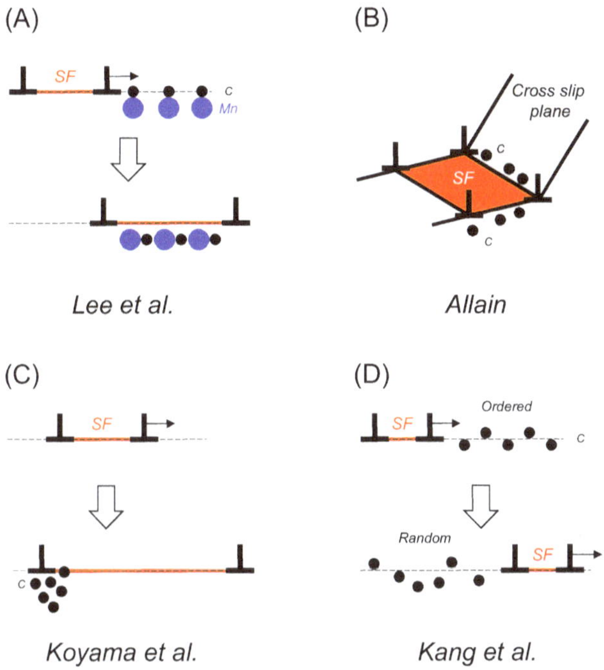

Figure 11.40 Schematics for the micro-mechanical mechanism of DSA in C-alloyed TWIP steel according to (A) Lee et al. [149], (B) Allain [52], (C) Koyama et al. [150], and (D) Kang et al. [53].

is too limited to allow for a DSA mechanism based on large diffusion distances. Several micro-mechanical models for DSA have therefore been proposed to relate DSA to the presence of solute C and the formation of twins. These models are reviewed in Fig. 11.40. Lee et al. [149] have suggested that the immobile C-Mn point defect complexes interact with the stacking faults of gliding dislocations. They argued that C atoms located on the stacking fault plane will be relocated to a high energy tetrahedral position by the shear from the leading partial dislocation. The C atoms then reorient themselves by a single diffusive hop to an octahedral interstitial site out of the stacking fault plane. This process, which is associated with a local reduction of the stacking fault energy, enhances planar glide. This leads to larger dislocation pile-ups and a reduction of the critical stress for deformation twinning, as implied by the Venables equation. Allain [52] has suggested that the flow instabilities and the negative strain rate sensitivity observed for C-alloyed TWIP steel were due to the dynamic interaction between solute C atoms and dislocations, which interacts with deformation twinning to stabilize the type A deformation bands. He proposed that the C-related DSA may have an important influence on the strain hardening mechanisms such as deformation twinning. Based on earlier work by Andrews et al. [151], Allain suggested that C, despite the fact that it increases γ_{isf}, may still increase the rate of deformation twinning.

As C increases the lattice friction, it inhibits dislocations glide and allows for the activation of deformation twinning as an alternative plasticity mechanism at higher stresses. He argued that the presence of solute C atoms at the core of partial dislocations also inhibits the formation of constrictions on screw dislocations. This process inhibits cross slip and promotes slip planarity. Koyama et al. [150] observed that deformation twinning in $<114>$-oriented grains was not influenced by an increase of the C content, despite the fact that carbon increases γ_{isf}. They also noted that deformation twins were formed in high-γ_{isf} Fe-18%Mn-1.2%C (in wt.-%) TWIP steel ($\gamma_{isf} = 55$ mJ/m^2), which had a serrated flow curve. They proposed that both effects were due to the trapping of the trailing partial dislocations by solute C atoms, when the velocity of the leading and trailing partial dislocations was significantly different. They argued that when this occurs, an infinite separation of the partials, and hence deformation twinning, can be initiated. Kang et al. computed the effect of short range ordering (SRO) on the yield strength of TWIP steel [53]. Short range ordering occurs when the number of unlike atom pairs in an alloy is larger than in the random solution. Based on arguments first developed by Gerold and Karnthaler [152], Kang et al. also related the occurrence of SRO with an enhancement of planar glide due to a glide plane softening phenomenon.

DSA associated with the interaction between point defects and arrested dislocations gives rise to an increase in the flow stress. Kim et al. have shown that this strength increase was small, at approximately 20 MPa [127]. The relation between DSA and strain hardening is complex. The room temperature data shown in Fig. 11.41 reveals that the addition of Al to a Fe-18%Mn-0.6%C (in wt.-%) TWIP steel results in a decrease of the strain hardening and a pronounced suppression of serrations. Both effects are related to the increase of γ_{isf} by the Al-addition, which has two effects: (1) a suppression of deformation twinning, and (2) a reduced interaction between point defects and arrested dislocations. A reduction of the stacking fault width will result in a reduced interaction, if this interaction involves the re-orientation of C-Mn point defect complexes in the stacking fault plane.

Reorientation of the C in a tetragonally distorted $CFe_{6-x}Mn_x$ point defect complex by the single diffusional hop to an energetically more favorable neighboring octahedral position relative to the stress field of a partial dislocation or the lattice contraction perpendicular to the stacking fault plane has been proposed as a DSA mechanism by Lee et al. [149,153]. They argued that DSA is due to a single diffusive hop of the C atom in the point defect complex in the stacking fault region. DSA is only observed when the C atom reorientation time is shorter than the residence time of the stacking fault at the location of the point defect complex (Fig. 11.42).

11.4.3 Crystallographic Texture Control

11.4.3.1 Rolling texture

All high γ_{isf} fcc metals and alloys are similar in that the deformation is by homogeneous $\{111\}<110>$-type dislocation slip and easy dislocation cross-slip.

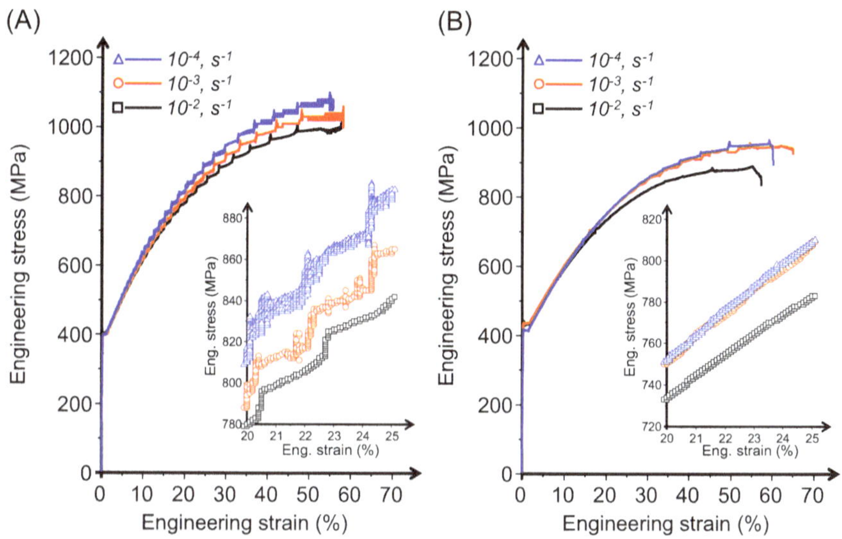

Figure 11.41 Stress-strain curves for (A) Al-free Fe-18%Mn-0.6%C (in wt.-%) TWIP steel, and (B) with Fe-18%Mn-0.6%C-1.5% (in wt.-%) Al TWIP steel. Both steels have a negative strain rate sensitivity, i.e., the flow stress increases with decreasing strain rate. Type A serrations, characterized by a sharp increase in flow stress followed by a plateau, are present on both flow curves. Whereas type A serrations appear on the stress-strain curve of the Al-free TWIP steel from the start of the plastic deformation, a large critical strain is needed for the initiation of the serrations in the case of the Al-added TWIP steel.

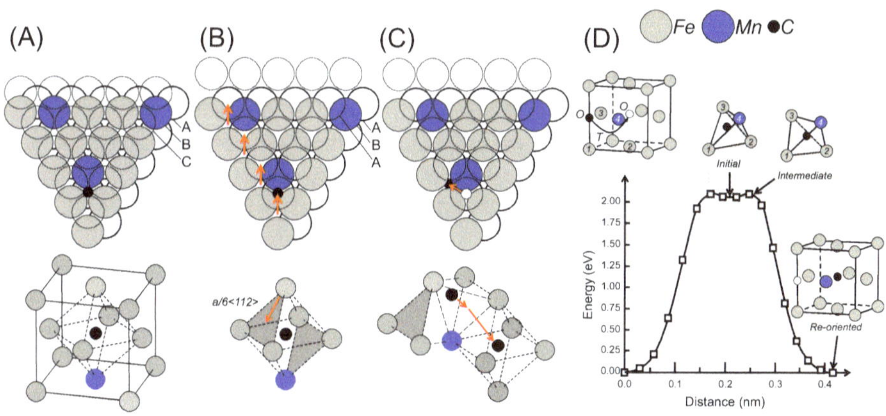

Figure 11.42 Schematic illustrating the reorientation of the carbon atom in a tetragonally distorted $CFe_{6-x}Mn_x$ point defect complex during the passage of a partial dislocation. (A) Starting situation for the $CFe_{6-x}Mn_x$ point defect complex. (B) Lattice shear by a partial dislocation causes the carbon atom to be transferred from an octahedral interstitial position to a high energy tetrahedral position. (C) Single diffusional hop of the carbon atom to an energetically more favorable neighboring octahedral position out of the stacking fault plane [154]. (D) Ab initio calculation of the energy pathway for C atom diffusion in γ-Fe at 0 K illustrating that when an interstitial C is transferred from an octahedral site to a near-tetrahedral site, when it is located in the stacking fault plane of a leading partial dislocation, no thermal energy needs to be supplied for the C atom to jump to a nearby octahedral site away from the stacking fault plane [153].

The {111}<110> slip and the grain rotations in rolling result in the development of a β-fiber orientation. The β-fiber is an ODF-skeleton line going from the Cu-{112}<111> component, via the S-{123}<634> component to the Brass-{110}<112> component (ODF = orientation distribution function). It usually consists of a relatively high fraction of S-{123}<634> component.

A Brass-{110}<112>-type texture is expected as the stable rolling texture in low-to-medium γ_{isf} TWIP steel as both octahedral slip and mechanical twinning are active deformation mechanisms. The limited cross-slip and deformation twinning result in a texture characterized by α-fiber orientations spreading from the Brass-{110}<112> to the Goss-{110}<100> orientation. When the Brass-texture is pronounced, the volume fraction of the Cu-{112}<111> and S-{123}<634> texture components is negligible. According to the classical model for the emergence of the Brass-texture in low γ_{isf} alloys deformation twinning plays a key role in the texture development. Grains with the Cu-{112}<111> orientation first twin into the CuT-{552}<115> orientation during straining. Dislocation glide then causes the CuT-{552}<115>-oriented grains to rotate around <110>//TD towards the Goss-{110}<001> orientation and the stable Brass-{110}<112> final orientation. Hirsch has proposed that the CuT-{552}<115> oriented lamellas formed by twinning rotated to an intermediate Brass-R-{111}<112> orientation first, and then rotated to the Goss orientation during the shear band formation [155,156]. Smallman and Green [157] and Dillamore [158] associated the texture transition to Brass-{110}<112> with the limited cross-slip in low γ_{isf} metals and alloys.

At high deformations a lamellar microstructure of matrix and twins aligns parallel to the rolling plane. According to Morii et al, slip cannot continue when the slip planes are almost parallel to the surface, and this leads to the formation of shear bands [159]. This alignment of the lamellar microstructure of matrix and twins parallel to the rolling plane does not appear to be necessary to form shear bands, and there is evidence for the fact that shear bands form early on when the rotation of the lamellae is initiated. Vercammen et al. [160] studied the evolution of microstructure and texture of a Fe-30%Mn-3%Al-3%Si (in wt.-%) TWIP-steel during cold rolling. They reported that a Brass-type texture was already present at small rolling strains. The Brass orientation {110}<112> was dominant at every strain level, and the Cu-orientation intensity remained low. The absence of a pronounced Cu-orientation texture was related to the early formation of deformation twins. The Goss and Cu-twin orientations developed due to an increase of the twin volume fraction with strain. Most of the grains were twinned at larger rolling reductions and their twinning planes were reoriented so as to be parallel to the rolling plane, making homogeneous deformation difficult. Shear banding was initiated because crystallographic slip was inhibited. At a reduction corresponding to an engineering strain of 65%, non-homogeneous shear band formation became the dominant deformation mechanism. The deformation microstructure consisted of bands approximately oriented at +35° and −35° to the RD. γ-fiber {111}//ND texture components, {111}<110> and {111}<112>, were formed due to the alignment of twin lamellae parallel to the rolling plane (Fig. 11.43).

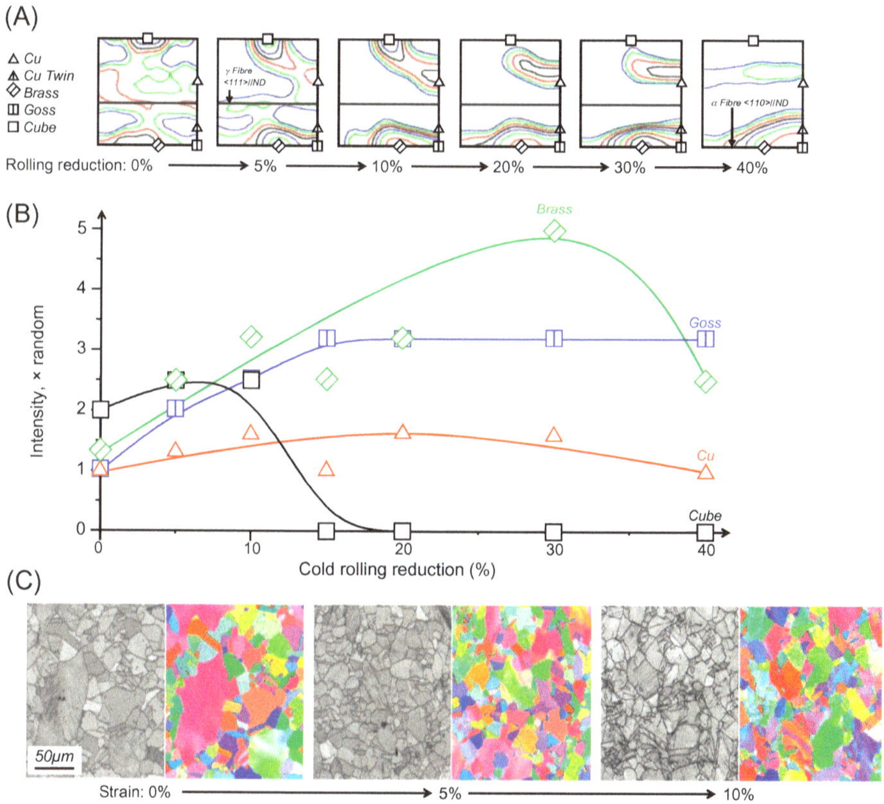

Figure 11.43 (A) Evolution of the ODF during cold rolling of a Fe-22%Mn-0.6%C-0.08%N (in wt.-%) TWIP steel. (B) Evolution of the volume fraction of the main texture components in (A). (C) Image quality (IQ) maps and inverse pole figure (IPF) maps showing that the main twinning activity is in the deformation range of 5–10%. Note the α-fiber texture from Brass-{110} < 112 > to Goss-{110} < 100 > and the weakening of the Cu- {112} < 111 > component at high strains. The rolling texture can be described as a Brass-type {110} < 112 > -texture with a spread towards the Goss-type texture {110} < 001 >.

11.4.3.2 Recrystallization texture

Lü et al. [161] studied the texture development in Fe-21.6%C-0.38%C (in wt.-%) TWIP steel cold rolled to a 50% rolling reduction and recrystallization annealed in the temperature range of 560°C to 700°C. They reported that the nucleation of new grains was site saturated, i.e., all nuclei were generated at the start of the recrystallization anneal. The nucleation sites were shear bands, grain boundaries, and triple junctions. The growth rate of the nuclei decreased with increasing annealing time, due to the recovery by pipe-diffusion controlled dislocation climb and the decrease of the driving force driving the grain boundary migration velocity. The grain structure after annealing was inhomogeneous due to the non-random nucleation site distribution. They reported that, in the initial stages of the annealing, i.e., prior to the onset of the recrystallization, the S {123}<634> component increased in intensity for a short period due to the transformation of the strain-induced ε-martensite to austenite [161]. The intensity of the dominant texture components after cold rolling, i.e., the S {123}<634>, Brass {110}<112>, and Goss {110}<100> components, gradually decreased during annealing. The texture after full recrystallization was similar to the deformation texture, but very weak. Texture was weak after annealing above 630°C. Un-recrystallized domains had a Goss-{110}<100> orientation. These grain are known to have a low stored energy after cold work [162]. After annealing at 560°C the intensity of the α-fiber <110>//RD and the <100>-fiber increased due to selective growth of grains with this orientation. Bracke et al. [88,163] studied the recrystallization texture of annealed cold rolled Fe-22%Mn-0.6%C (composition in wt.-%, $\gamma_{isf} = 15$ mJ/m^2). The hot rolled strip texture was weak with a low intensity Cube-100<100> component, which is a characteristic recrystallization texture for high γ_{isf} fcc alloys. For cold rolling, they reported that their TWIP steel had a characteristic low-γ_{isf} alloy behavior from the point of view of deformation texture evolution in rolling. This assessment is based on two points: (1) a decrease of the β-fiber component intensity, and (2) the development of an α-fiber with a high intensity between the Brass and Goss component. They also considered the presence of the CuT-552<115> texture component as a direct proof for twinning and the fact that twin formation does leave a trace in the texture. They also observed a new texture component which they referred to as the "X-component" which is twin related to the Brass component. They cold rolled and recrystallization annealed at a slightly higher temperature of 725°C, as compared to 700°C in the case of Lü et al. [164]. They found a recrystallization texture which was almost identical to the cold rolled texture, except for a marginal increase of the Goss-{110}<100> component in the recrystallized steel. This phenomenon was independent of the reheating rate. The high nucleation density which they observed at the start of the recrystallization was evidence for site-saturation nucleation. The nucleation was homogeneous throughout the microstructure. At high rolling strains, the recrystallization started in the shear bands. There was therefore no strong orientation selection occurring during the nucleation and growth of the recrystallized grains. Once recrystallized grains nucleated, their growth was sluggish, due to the impingement of the growing grains soon after their formation. This process resulted in small grain sizes. The results implied that the stored energy of deformation was homogeneously distributed. Grain growth and the formation of recrystallization twins did not have a statistically-relevant effect. As the nucleation process involved a random sampling of the orientations in the deformed microstructure, the recrystallization texture replicated the initial deformation texture.

In the case of TWIP steel, grain growth appears to be limited and oriented nucleation in subgrains in the matrix and in shear bands results in the retention of the deformation orientations during growth. Other orientations exist, but they do not change the overall texture significantly. Oriented growth appears to be also suppressed during the further recrystallization and grain growth. It could be that oriented growth is suppressed by the alloying as suggested by Lücke [165]. A solute drag effect resulting from the presence of solutes at the grain boundaries could homogenize the grain boundary mobilities and allow for the complete retention of the deformation texture.

11.4.4 Recovery annealing

Recovery annealing is an unusual microstructure optimization process, consisting of a partial recrystallization annealing treatment in the temperature range of 575–625°C. It is effective for the production of TWIP steel with a wide range of strength-ductility combinations. The retention of the deformation twins in cold deformed TWIP steel during recovery annealing is utilized to obtain a microstructure with both a high yield strength and an appreciable elongation. Suitable combinations of cold rolling and annealing temperature and times make it possible to tailor the mechanical properties of recovery annealed TWIP steel over a wide range. The retained twin boundaries result in a higher yield strength after recovery annealing as compared to fully recrystallized TWIP steel. The recovery process also significantly improves the ductility as compared to the cold-rolled state. While recovery annealing is somewhat similar to a grain refinement method, it does not result in a lower tensile ductility as in the case of a micro-structure refinement by smaller grain sizes. According to Dao et al. [166] the introduction of a large number of nanoscale twin boundaries leads to a high strength and the retention of the ductility. They explained this observation by the fact that, whereas dislocation activity is limited in small grain crystals, this is not the case in twinned crystals because dislocations can cross twin boundaries [166].

Kang et al. [167] have studied the recrystallization of 60% cold rolled Fe-18% Mn-0.6%C-1.5%Al (in wt.-%) TWIP steel in the temperature range of 550–1100°C. While the deformation twins were thermally stable during a recovery annealing at 550°C, the dislocation density was reduced. The static recrystallization started at 600°C and was completed at 700°C. The recrystallized grains were 2.38 μm in size. The increase of the yield strength and tensile strength observed after annealing in the temperature range of 700–800oC was due to M_3C carbide precipitation. At 840°C, grain coarsening occurred as a result of the dissolution of the M_3C carbides. During recovery annealing of a cold rolled TWIP steel below 600°C, a low dislocation density microstructure was obtained which still contained the nanoscale deformation twins.

Dini et al. [113] fabricated sub-micron grained Fe–31%Mn–3%Al–3%Si (in wt.-%) TWIP steel by a partial recrystallization annealing treatment in the temperature range of 575–625°C. They also confirmed that an excellent tensile strength-ductility balance could be obtained in the fine grained material. Their partially recrystallized 80%-cold rolled TWIP steel annealed at 575°C had a yield strength of about 1 GPa and a total elongation close to 30%. Similarly, Bouaziz et al. [168] fabricated a nano-twinned Fe-22%Mn-0.6%C (in wt.-%) TWIP steel by a recovery annealing treatment of 3600 s at 500°C aimed at reducing the dislocation density without changing the twinning nanostructure created during a prior 50% cold reduction. Their recovery annealed TWIP steel had a yield strength of 1200 MPa, a tensile strength of approximately 1700 MPa and a uniform elongation of 8%.

Haase et al. [102,107,169] used the evolution of the crystallographic texture during recovery annealing at 550°C and 630°C to optimize the recovery annealing process of a Fe-23%Mn-0.3%C-1.5%Al (in wt.-%) TWIP steel. The texture analysis revealed that the highest twin density was obtained after a 50% cold reduction. They observed a slight texture sharpening during recovery, which was indicated by the increased intensity of the deformation texture components, which was followed by the texture randomization during recrystallization, as indicated by an increase of the volume fraction of randomly oriented grains and a decrease of the intensity of the main texture components after cold rolling. The optimal annealing time was at the transition between the texture strengthening due to recovery and the texture randomization during primary recrystallization. The recrystallized volume fraction after recovery annealing was found to be less than 10%.

Haase et al. [170] also studied the mechanical properties of the recovery annealed Fe-17%Mn-0.6%C-1.5%Al and Fe-23%Mn-0.3%C-1.5%Al (in wt.-%) TWIP steels. They reported that the high yield strength retained after recovery annealing resulted in a significant increase of the energy-absorption capacity and superior crash test properties as compared the fully recrystallized state. They also demonstrated, by means of cross-shaped cup drawing experiments, that a considerable ductility was regained after a recovery annealing treatment.

11.4.5 Fracture

The failure mode of TWIP steel is ductile. The negative strain rate sensitivity makes diffuse necking strain small, and fast fracture is observed once necking is initiated. Hamada et al. [171] analyzed the room temperature fracture surface of Fe-25%Mn and Fe-25%Mn-3%Al (in wt.-%) TWIP steels after tensile straining. Despite the presence of strain-induced martensite in these steels, they reported that the appearance of the fracture surface was consistent with a ductile fracture by micro-void coalescence. They observed a dimpled fracture surface with nonmetallic inclusions visible inside some of the dimples. These inclusions deform less than austenite. The inclusions can fracture, or the strain gradients which develop at the matrix/inclusion interface can result in interfacial decohesion. The coalescence of cavities formed by these processes can result in fracture prior to necking. Lorthios et al. [172] used synchrotron X-ray micro-tomography to reveal the damage which develops in the material up to 1000 μm below the fracture surface of a Fe-33%Mn-0.6%C (in wt.-%) TWIP steel. They observed the formation of larger strings of voids along the rolling direction which had previously not been evidenced in standard light optical metallography. Fig. 11.44 shows the appearance of a fracture surface for Fe-18%Mn-0.6%C-1.5%Al (in wt.-%) TWIP steel after tensile straining. The image shows two types of dimples. Small equiaxed dimples and larger dimples often associated with a small particle, which may have been the microvoid initiation site in the course of ductile fracture.

Fracture of TWIP steel has been given considerable attention in connection to three phenomena (1) edge fracture in hole expansion, (2) liquid metal-induced

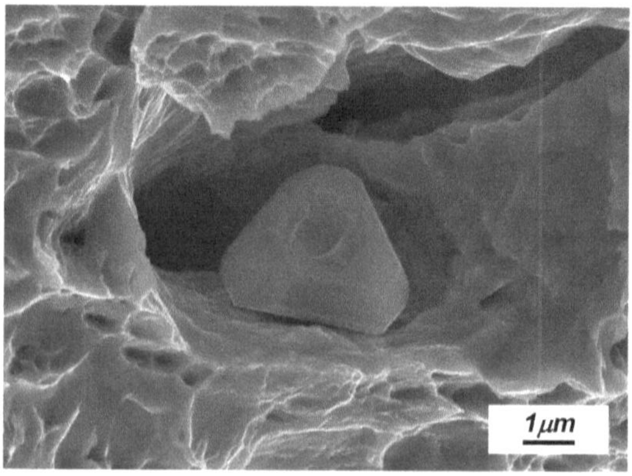

Figure 11.44 Fracture surface for Fe-18%Mn-0.6%C-1.5%Al (in wt.-%) TWIP steel after tensile straining, showing a large density of small dimples and a very low density of coarser cavities. The dimples are indicative of ductile fracture by micro-void coalescence. The AlN single crystal precipitate in the coarser cavity may have nucleated its formation by decohesion.

embrittlement (LMIE), and (3) the hydrogen-related delayed fracture. These aspects of fracture are reviewed in the following paragraphs.

11.4.6 Fracture in hole expansion

Cut edge stretch-ability is an important aspect of sheet formability. Whereas the press forming performance of TWIP steels is excellent due its high strain hardening, the lower normal anisotropy and the negative strain rate sensitivity of TWIP steel result in premature ductile fracture during stretch-flanging (i.e., hole expansion) when the initial hole is made using a method that leads to considerable prior deformation of the hole edge, such as hole punching [147]. The phenomenon has been analyzed in detail using advanced in situ measurement methods that allow for a local measurement of the principal strains and the temperature. The strain distribution analysis revealed that the hole edge deformation in a deep drawing deformation mode was similar to a tensile deformation. Fig. 11.45 compares the hole expansion performance of an interstitial free steel and a TWIP steel. Edge cracking was ductile in both cases. Very ductile fracture at the hole edge occurred after considerable edge thinning and localized necking in the case of the interstitial-free steel. In the case of the TWIP steel a single catastrophic crack occurred when the sample failed. In both cases the fracture surface consisted of shallow dimples, which are indicative of a ductile fracture by micro-void coalescence.

11.4.7 Hydrogen-related delayed fracture

Fig. 11.46 illustrates the spontaneous cracking of a deep drawn cup of Fe-22%Mn-0.6%C (in wt.-%) TWIP steel during a hydrogen-charging test. Both intergranular and transgranular cracks are present on the outer surface in addition to the single

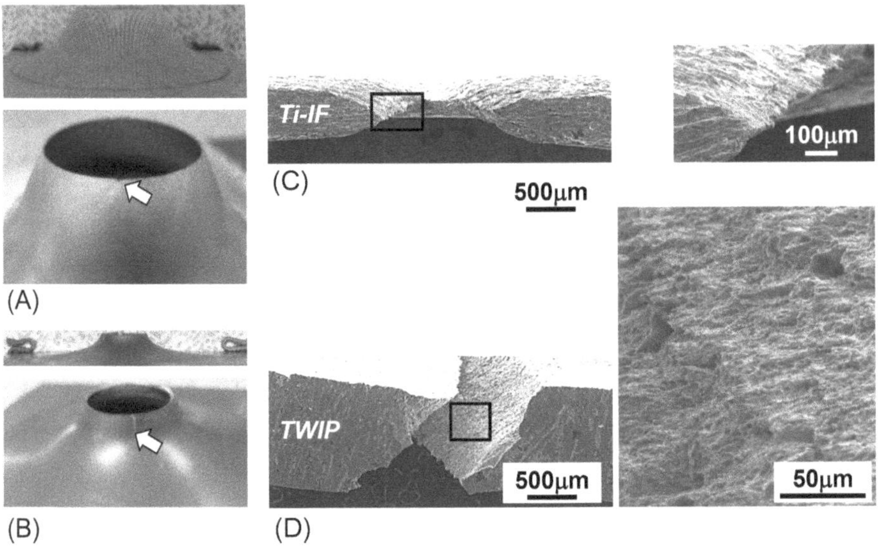

Figure 11.45 (A) Hole expansion test result for an interstitial-free steel. (B) Hole expansion test result for a TWIP steel. The arrows indicate the crack location. (C) Edge crack appearance for the interstitial-free steel, showing extensive thinning prior to rupture. (D) Edge crack appearance for the TWIP steel, showing absence of diffuse necking prior to shear rupture. The dimpled appearance of the fracture surface shows evidence of a ductile fracture mode by micro-void coalescence.

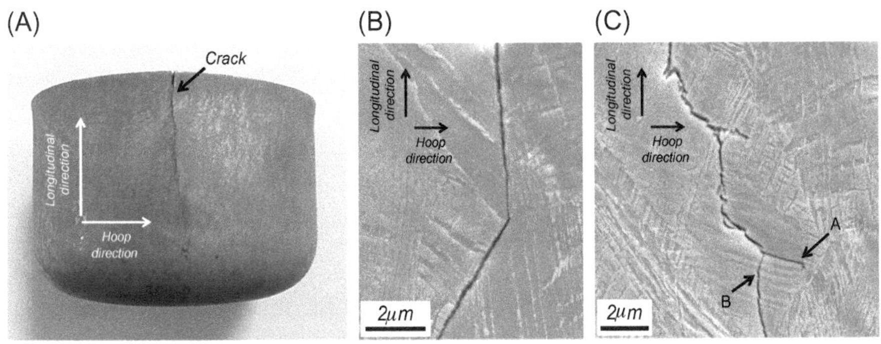

Figure 11.46 (A) Large longitudinal crack on a deep drawn cup of Fe-17%Mn-0.4%C-1.3% Al (in wt.-%) TWIP steel (grain size = 15 μm; γ_{isf} = 17 mJ/m^2) H-charged for 32 hours. (B) Intergranular crack. (C) Complex transgranular crack showing crack propagation (A) parallel to the twin/matrix interface and (B) across deformation twins. Note that the cracks shown in (B) and (C) are two examples of the much finer cracks present on the outer side of the sample shown in (A). TEM analysis of samples taken in the cup wall revealed the presence of ε-martensite, and magnetic measurements also showed the presence of 0.5–1.5 volume percent of α'-martensite.

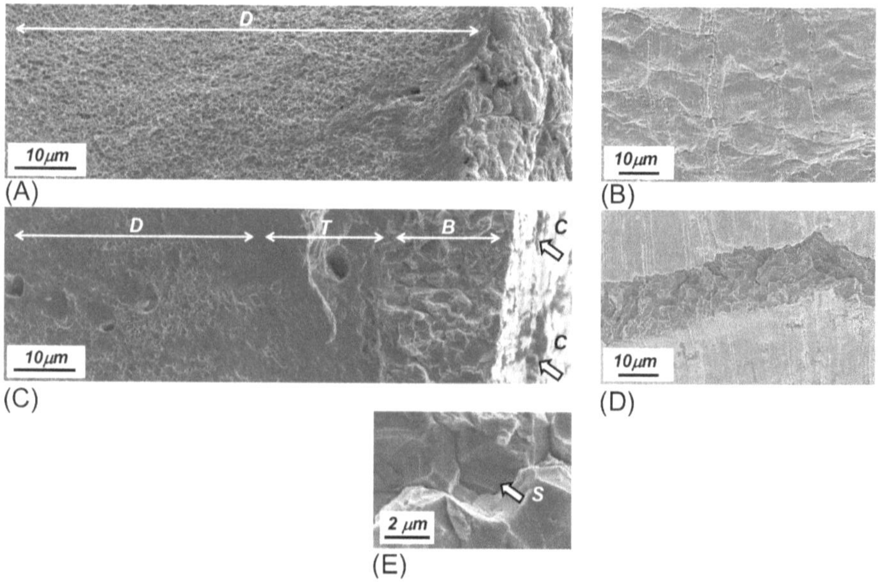

Figure 11.47 (A) Fracture surface and (B) sample surface of a Fe-18%Mn-0.6%C-1.5%Al (in wt.-%) TWIP steel. The fracture surface is fully ductile, consisting entirely of dimples, and no cracks appear at the sample surface. (C) Fracture surface and (D) sample surface of the same steel pre-strained 20% at 150°C, hydrogen-charged to 4.9 ppm of solute H, and tested to fracture at room temperature. The choice of the higher pre-straining temperature was made to avoid formation of strain-induced martensite. In (C) the fracture consists of a brittle intergranular fracture zone near the surface (area labeled "B"), a fracture-mode transition zone (area labeled "T") and a large area of ductile fracture (area labeled "D"). The micrograph of the surface (D) has a high density of wide cracks due to intergranular de-cohesion. Micrograph (E) shows a magnified view of a brittle fracture at the surface showing slip traces, labeled "S", at the surface of the grains in a zone where intergranular fracture has occurred.

major crack. The transgranular crack propagates through deformation twins and along twin/matrix interface boundaries. Although the H-delayed fracture of deep drawn cups is quite dramatic, the influence of H pre-charging on the tensile properties of TWIP steels is generally minor as the room temperature diffusion distance into the matrix is very limited. H diffusion in fcc Fe is considerably slower than in bcc Fe, by a factor of approximately 10^6 [173]. This low H diffusivity implies that the testing of the sensitivity to H delayed fracture of TWIP steel using H pre-charged tensile specimens results in an inhomogeneous fracture surface, with a brittle fracture close to the sample surface in the region of solute hydrogen penetration, and a ductile fracture in the central part of the sample. This is illustrated in Fig. 11.47, which shows that only the near-surface regions of H-charged TWIP steel are subject to a predominantly intergranular brittle fracture. The bulk remains very ductile. The surface of the grains in the intergranular fracture zone also show clear evidence of planar slip. This suggests that the H has an influence of the dislocations but that the H-penetration does not embrittle the matrix.

The delayed cracking phenomenon in deep drawn Erichsen cups made of TWIP steel shares many similarities with stress corrosion cracking. It is made possible by (1) the uptake of H in the material during steel production and/or processing, (2) the high tensile stresses at the edge of the deep drawn cup, and (3) crack-initiating features such as the intersection of twins, strain-induced ε or α' martensite, or the grain boundaries with an intrinsically low cohesion strength.

Al additions to TWIP steel improve their resistance to H-induced embrittlement. The Fe-(15-18)%Mn-1.5%Al-(0.15-0.60%)C (in wt.-%) TWIP steel concept, which proved to be resistant to H-delayed fracture, was first reported by Jung et al. [174]. Chun et al. [175] studied the influence of Al additions on the properties of H pre--charged notched tensile samples. The addition of Al was found to be a very effective way to suppress the sensitivity of TWIP steel to H-delayed cracking. Park et al. [176] also observed a decrease of solute H at shallow trapping sites. They proposed an alternative mechanism for the suppression of H-delayed fracture of TWIP steel by Al-addition. They argued that the improvement was due to the formation of a double oxide layer at the sample surface. This surface layer, which consisted of a $(Fe_{0.8}Mn_{0.2})O$ surface layer and a α-Al_2O_3 interface layer, acted as a H diffusion barrier preventing the H uptake during H charging.

The influence of Cu and Si additions on the hydrogen embrittling of TWIP steels has also been studied. According to Kwon et al. [177] Cu additions (0−2 wt.-%) to Fe-17%Mn-0.8%C (in wt.-%) TWIP steel improves the H embrittlement resistance. According to Lee et al. [178], who studied the effect of Si additions (0.01−2.95 wt.-%) to Fe-18%Mn-0.6%C (in wt.-%) TWIP steel, a double oxide layer was formed on the surface of the Si-added TWIP steel. This layer consisted of a surface layer, an oxide mixture of (Fe,Mn)O and $(Fe,Mn)_2SiO_4$, and a pure $(Fe,Mn)_2SiO_4$ interface layer. The layer reduced the hydrogen uptake during H charging of the TWIP steel. The $(Fe,Mn)_2SiO_4$ layer was less efficient than the α-Al_2O_3 layer on the Al-added TWIP steel in reducing the H uptake.

Ryu et al. [179] argued that, when solute H is present, intergranular and transgranular fracture of TWIP steel were due to the accumulation of dislocations at grain boundaries and twin boundaries, respectively. In the presence of strain-induced ε-martensite formed in H-charged TWIP steel, ε-martensite plates fracture in a brittle manner and the crack path is transgranular.

Soo et al. [180] tested the influence of the H content on the fracture stress, fracture strain and time to fracture of smooth and notched samples of Fe-18% Mn-0.6%C-1.5%Al (in wt.-%) TWIP steel. They reported that the pre-charged specimens were immune to H-induced cracking, as the properties were found to be relatively insensitive to the amount of H.

There are two H trapping sites, a site with a low binding energy of 35 mJ/mol and a site with a high binding energy of 62 mJ/mol. These trapping sites have been identified as dislocations (or grain boundaries), and twin boundaries, respectively. So et al. [181] have reported that H was trapped at dislocations and grain boundaries in TWIP steel. According to So et al, deformation twin boundaries could also act as H trap sites for H. Dislocations and grain boundaries are considered to be shallow or “diffusible” H trap sites. Twin boundaries act as deep or “non-diffusible”

H trap sites. The diffusible H is considered to be responsible for H-delayed fracture. Non-diffusing solute H, strongly bound at deep traps, does not contribute to H-delayed fracture.

In general, most authors report that whereas the flow stress and the work hardening behavior of tensile specimens are not influenced by the ex situ H pre-charging of recrystallization annealed TWIP steel, the elongation is slightly decreased [182,183]. In contrast to other workers, Koyama et al. [184,185] studied the change in mechanical properties during in situ H-charging of Fe-22%Mn-0.6%C (in wt.-%) TWIP steel. They also observed that fracture occurred during stress relaxation tests [186]. These effects are illustrated in Fig. 11.48. Their in-depth fractography revealed that the crack path was mostly intergranular. Transgranular fractures appeared to propagate along twin boundaries. According to Koyama et al, cracks started at the intersection between primary and secondary twins and propagated transgranularly along twin/matrix interfaces in a zigzag manner.

Lee et al. reported that longitudinal cracks in the wall of H-delayed fractured deep drawn cups were mostly intergranular, but could become transgranular and propagate along twin boundaries [187]. The longitudinal cracks, which appeared to be initiated at inclusions, propagated mainly along the grain boundaries. Some hoop cracks were also found to propagate perpendicular to twin boundaries. They reported that the H-induced delayed fracture of deep drawn TWIP steel was related to the magnitude of the residual hoop and longitudinal stresses. Van Tol et al. [188–190] have argued that the nucleation of the delayed H-fracture cracks in the cup wall may be due to the presence of a small volume fraction of strain-induced α'-martensite formed at intersections of shear band or twins.

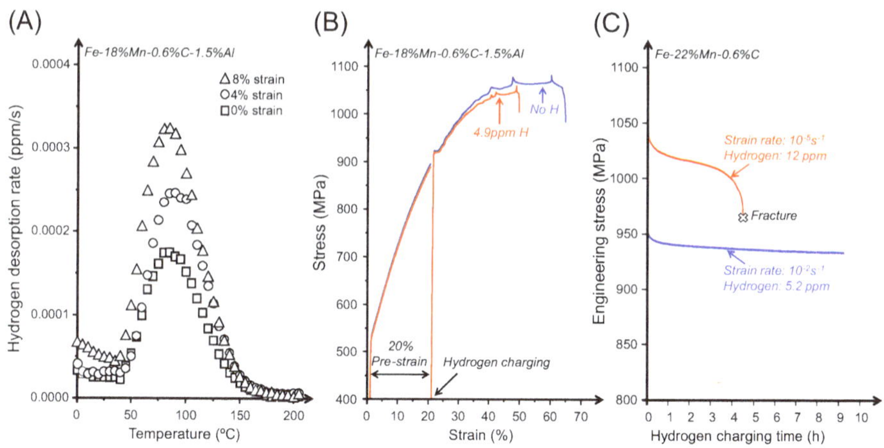

Figure 11.48 (A) H-desorption rate curves for TWIP steel for three different pre-strain levels: undeformed, 4% pre-strained and 8% pre-strained. The samples were H-charged for 24 h [183]. (B) Influence of pre-straining on the mechanical properties of hydrogen charged TWIP steel [183]. (C) Stress relaxation during the in situ charging of TWIP steel [186].

11.4.8 Liquid metal-induced embrittlement (LMIE)

The issue of Zn-LMIE of TWIP steel is mainly relevant to resistance spot welding, a key process in automotive body assembly. The sensitivity to Zn-LMIE of TWIP steel has been studied in detail [191–193]. Zn-LMIE of TWIP steel requires (1) the presence of liquid Zn, Zn_{liq}, or a Zn-rich liquid alloy, and (2) a high, uniformly distributed or locally concentrated tensile stress. LMIE is due to transgranular or intergranular decohesion of a metal or an alloy by the apparent rapid penetration of another liquid metal or alloy into its microstructure. The penetration is usually along grain boundaries. When tensile stresses are present, a liquid film appears to cause grain boundary decohesion, as the solid boundary is replaced by a liquid film. This results in a brittle fracture due to the rapid intergranular propagation of one or more fracture cracks formed in the region where the liquid metal has penetrated. The percolation of the liquid may be preceded by the rapid solid state diffusion of the embrittling atoms along grain boundaries. As TWIP steel is considered to be a potential material for structural applications in the automotive industry, it should be galvanized to protect it from perforation and cosmetic corrosion. The Zn and Zn-alloy coating of TWIP steel can result in Zn liquid metal embrittlement (Zn-LMIE) at high temperature (Fig. 11.49). The Zn-LMIE of TWIP steel has been evaluated by Kang et al. [194] by means of the microstructural analysis of sub-critical LMIE cracks in Zn-coated TWIP steel tested in uni-axial tension at high temperature. Zn was found to penetrate into the steel matrix along the grain boundaries. Both Zn grain boundary diffusion and Zn_{liq} percolation along grain boundaries was observed. Evidence of Zn_{liq} percolation was inferred from the presence of the Γ-$(Fe,Mn)_3Zn_{10}$ phase along the Zn penetration region. The Γ-$(Fe,Mn)_3Zn_{10}$ phase was also observed at the tip of the Zn penetration deep below the surface. Kang et al. proposed two mechanisms for the Zn- LMIE of Zn-coated TWIP steel, which are compatible with the features of the

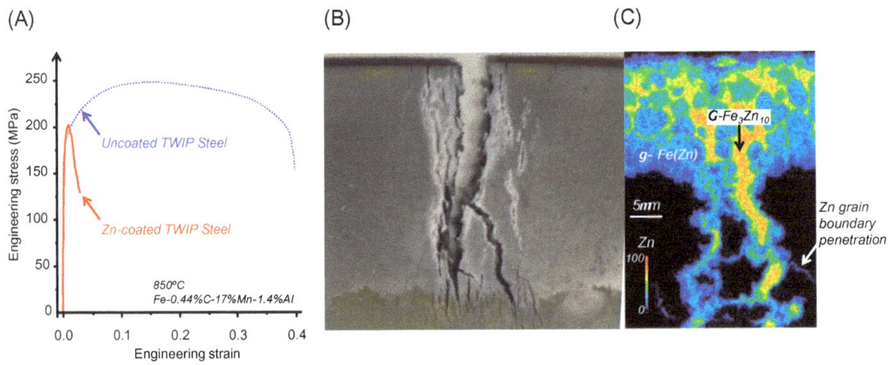

Figure 11.49 (A) Engineering stress-strain curves of Zn-coated TWIP steel tested at 850°C. (B) Fracture of a Zn-coated TWIP steel specimen after tensile testing at 850°C. The LMIE fracture has a main crack with multiple secondary branches. (C) Electron-probe microanalysis of the Zn distribution in an area of Zn penetration. Γ-$(Fe,Mn)_3Zn_{10}$ is present in the crack opening area of the Zn penetration. The Zn content at the tip of the Zn penetration is lower than the maximum solubility of Zn in γ-(Fe,Mn) at the test temperature.

Zn penetrations observed by micro characterization: (1) the Krishtal-Gordon-An model [195,196] for LMIE crack formation initiated by the rapid intergranular solid state diffusion of Zn, and (2) the Stoloff-Johnson-Westwood-Kamdar model [197,198] for LMIE crack formation due to intergranular liquid metal percolation.

11.5 Summary

The engineering of the strain hardening mechanisms in high Mn TWIP steel and ultra-fine grained medium Mn TWIP + TRIP steel has shown that the introduction of plasticity-enhancing mechanisms in the microstructure of steel can result in materials properties which combine high strength with formability. The study of the physical metallurgy of these two types of steel has advanced considerably, and the general consensus is that both alloy concepts achieve exceptional levels of strength and ductility, combining ultra-high strengths (UTS > 1 GPa) with a considerable plasticity (TEL > 20%). An intense effort is currently underway to optimize these two new steel grade concepts in terms of composition, industrial processing and application properties. Both are being developed industrially for use in the automotive industry, as their potential contribution to vehicle lightweighting is expected to be considerable.

Acknowledgments

The authors gratefully acknowledge the support and contributions form the following persons: Dr. Gwan Keun Chin (POSLAB), Dr. Jin Kyung Kim (POSTECH), Professor Wolfgang Bleck (RWTH-Aachen), Professor Young-Kook Lee (Yonsei University), Dr. Stephanie Sandlöbes (Max-Planck-Institut für Eisenforschung), Professor Dierk Raabe (Max-Planck-Institut für Eisenforschung), Professor Se Kyun Kwon (POSTECH), and Mr. Hojun Gwon (POSTECH).

List of abbreviations

$\mathbf{b_P}$	partial dislocation Burger's vector
$\Delta G_C^{\gamma\to\varepsilon}$	chemical driving force for the $\gamma\to\varepsilon$ transformation
D	grain diameter
DSA	dynamic strain ageing
ε_i	critical strain for deformation twin nucleation
E	Young's modulus
$\mathbf{F_{T,\,sat}}$	saturation twin volume fraction
$\mathbf{F_T}(\varepsilon)$	strain-dependent evolution of the twin volume fraction
$\mathbf{f_T}$	twin volume fraction
γ_{isf}	intrinsic stacking fault energy
γ_{usf}	unstable stacking fault energy
γ_{utf}	unstable twinning fault energy
G	shear modulus
k_{YS}^{HP}	dislocation unpinning parameter
$\mathbf{L_0}$	length of the sessile partial dislocation of the twin nucleus

LME	liquid metal-induced embrittlement
M	Taylor orientation factor
MBIP	micro band-induced plasticity
ν	Poisson's ratio
σ_0	lattice friction stress
$\sigma^{\gamma/\varepsilon}$	γ/ε interface energy
σ_T	twinning stress
σ_{YS}	yield stress
SBIP	shear band-induced plasticity
SFE	stacking fault energy
τ_{CRSS}	critical resolved shear stress
τ_T	twinning shear stress
θ	the orientation-dependent ratio of the Schmid factors for slip and twinning
T_N^{γ}	Néel temperature
TRIP	transformation-induced plasticity
TWIP	twinning-induced plasticity

References

[1] S.W. Lee, S.J. Lee, B.C. De Cooman, Acta Mater. 59 (2011) 7546–7553.
[2] T.-H. Lee, C.-S. Oh, S.-J. Kim, Scripta Mater. 58 (2008) 110–113.
[3] O. Bouaziz, S. Allain, C. Scott, Scripta Mater. 58 (2008) 484–487.
[4] B.C. De Cooman, O. Kwon, K.-G. Chin, Mater. Sci. Technol. 28 (5) (2012) 513–527.
[5] O. Bouaziz, S. Allain, C.P. Scott, P. Cugy, D. Barbier, Curr. Opin. Solid State Mater. Sci. 15 (2011) 141–168.
[6] R.L. Miller, Metall. Trans. 3 (1972) 905–912.
[7] A. Arlazarov, M. Goune, O. Bouaziz, A. Hazotte, G. Petitgand, P. Barges, Mater. Sci. Eng. A 542 (2012) 31–39.
[8] S. Lee, K. Lee, B.C. De Cooman, Metall. Mater. Trans. A 46 (2015) 2356–2363.
[9] G. Frommeyer, U. Brüx, Steel Res. Int. 77 (2006) 627–633.
[10] W.K. Choo, J.H. Kim, J.C. Yoon, Acta Mater. 45 (1997) 4877–4885.
[11] J.D. Yoo, K.-T. Park, Mater. Sci. Eng. A 496 (1–2) (2008) 417–424.
[12] O. Grässel, G. Frommeyer, C. Derder, H. Hofmann, J. Phys. IV 7 (1997) 383.
[13] O. Grässel, G. Frommeyer, Mater. Sci. Technol. 14 (1998) 1213.
[14] O. Grässel, L. Krüger, G. Frommeyer, L.W. Meyer, Int. J. Plasticity 16 (2000) 1391–1409.
[15] O. Grässel, G. Frommeyer, Stahl Eisen 122 (2002) 65–69.
[16] G. Frommeyer, U. Brüx, P. Neumann, ISIJ Intl. 43 (2003) 438–446.
[17] O. Bouaziz, S. Allain, C.P. Scott, P. Cugy, D. Barbier, Curr. Opin. Solid State Mater. Sci. 15 (4) (2011) 141–168.
[18] B.C. De Cooman, Ohjoon Kwon, Kwang-Geun Chin, Mater. Sci. Technol. 28 (5) (2012) 513.
[19] R.W. Neu, Mater. Performance Charact. 2 (2013) 244–284.
[20] K. Chung, K. Ahn, D.H. Yoo, K.H. Chung, M.H. Seo, S.H. Park, Int. J. Plasticity 27 (2011) 52–81.
[21] J.A. Venables, in: R.E. Reed-Hill, J.P. Hirth, H.C. Rogers (Eds.), Metallurgical Society Conference. Deformation Twinning, vol. 25, Gordon&Breach Science Publishers Inc, Metallurgical Society of the AIME, Philadephia, PA, 1964, pp. 77–111.
[22] L. Remy, A. Pineau, Mater. Sci. Eng. 26 (1976) 123–132.

[23] S. Allain, J.P. Chateau, O. Bouaziz, S. Migot, N. Guelton, Mater. Sci. Eng. 387–389 (2004) 143–147.
[24] A.S.Hamada, PhD thesis, University of Oulu, Linnanmaa, Finland, 2007.
[25] S. Vercammen, B.C. De Cooman, B. Blanpain, N. Akdut, P. Wollants, Steel Res. 74 (2003) 370–375.
[26] B.W. Oh, S.J. Chao, Y.G. Kim, Y.P. Kim, S.H. Hong, Mater. Sci. Eng. A 197 (1995) 147–156.
[27] L. Remy, A. Pineau, Mater. Sci. Eng. 28 (1977) 99–107.
[28] L. Remy, Metall. Trans. A 12 (1981) 387–408.
[29] S. Allain, J.-P. Chateau, O. Bouaziz, Mater. Sci. Eng. A 143 (2004) 387–389.
[30] T.-H. Lee, E. Shin, C.S. Oh, H.Y. Ha, S.J. Kim, Acta Mater. 58 (2010) 3173–3186.
[31] V.H.Schumann, N. Hütte, vol. 17, 1972, pp. 605–609.
[32] J. Nakano, P.J. Jacques, CALPHAD: Computer Coupling of Phase Diagrams and Thermochemistry 34 (2010) 167–175.
[33] A. Saeed-Akbari, J. Imlau, U. Prahl, W. Bleck, Metall. Mater. Trans. A 40A (2009) 3076–3090.
[34] G. Olson, M. Cohen, Metall. Trans. A 7 (1976) 1897.
[35] J.P. Hirth, Metall. Trans. A 1A (1970) 2367–2374.
[36] D. Pierce, J. Jiminez, J. Bentley, D. Raabe, C. Oskay, J. Wittig, Acta Mater. 68 (2014) 238.
[37] W.S. Wang, C.M. Wang, J. Mater. Sci. 25 (1990) 1821–1823.
[38] J.R. Rice, J. Mech. Phys. Solids 40 (1992) 239.
[39] E.B. Tadmor, S. Hai, J. Mech. Phys. Solids 51 (2003) 765.
[40] K. Oda, H. Fujimura, H. Ino, J.Phys.: Condens. Matter 6 (1994) 679–692.
[41] H.K.D.H. Bhadeshia, J. Mater. Sci. 39 (2004) 3949–3955.
[42] R.B. McLellan, C. Ko, Acta Metall. 36 (1988) 531–537.
[43] S.S. Babu, H.K.D.H. Bhadeshia, J. Mater. Sci. Letters 14 (1995) 314–316.
[44] A. Abassi, A. Dick, T. Hickel, J. Neugebauer, Acta Mater. 59 (2011) 3041.
[45] N.I. Medvedeva, M.S. Park, D.C. Van Aken, J.E. Medvedeva, J. Alloys Compounds 582 (2014) 475–482.
[46] K.R. Limmer, J.E. Medvedeva, D.C. Van Aken, N.I. Medvedeva, Comput. Mater. Sci. 99 (2015) 253–255.
[47] D.J. Hepburn, D. Ferguson, S. Gardner, G.J. Ackland, Phys. Rev. B 88 (2013) 024115.
[48] H.Gholizadeh, PhD thesis, Montan Universität, Leoben, Austria, 2013.
[49] K.R.Limmer, PhD thesis, Missouri University of Science and Technology, USA, 2014.
[50] T. Hickel, S. Sandlöbes, R.K.W. Marceau, A. Dick, I. Bleskov, J. Neugebauer, et al., Acta Mater. 75 (2014) 147–155.
[51] W.S. Choi, B.C. De Cooman, S. Sandloebes, D. Raabe, Acta Mater. 98 (2015) 391–404.
[52] S.Allain, Thèse d'Habilitation, Université de Lorraine, INPL, Grenoble, 2012.
[53] J.H. Kang, T. Ingendahl, J. von Appen, R. Dronskowski, W. Bleck, Mater. Sci. Eng. A 614 (2014) 122–128.
[54] T. Gebhardt, D. Music, M. Ekholm, I.A. Abrikosov, L. Vitos, A. Dick, et al., J. Phys.: Condens. Matter 23 (2011) 246003–246010.
[55] J.E. Jin, Y.K. Lee, Acta Mater. 60 (2012) 1680–1688.
[56] M. Kang, W. Woo, Y.K. Lee, B.S. Seong, Mater. Lett. 76 (2012) 93–95.
[57] T. Shun, C.M. Wan, J.G. Byrne, Acta Metall. Mater. 40 (1992) 3407–3412.
[58] K.T. Park, K.G. Jin, S.H. Ho, S.W. Hwang, K. Choi, C.S. Lee, Mater. Science Eng. A 527 (2010) 3651–3661.
[59] X. Tian, Y.S. Zhang, Mater. Sci. Eng. A 516 (2009) 73–77.
[60] K.-H. Jeong, J.-E. Jin, Y.-S. Jung, S. Kang, Y.-K. Lee, Acta Mater. 61 (2013) 3399–3410.

[61] S.J. Lee, Y.S. Jung, S.I. Baik, Y.W. Kim, M. Kang, W. Woo, et al., Scripta Mater. 92 (2014) 23–26.
[62] B.X. Huang, X.D. Wang, L. Wang, Y.H. Rong, Metall. Mater. Trans. A 39 (2008) 717–724.
[63] B.H. Jiang, X. Qi, S. Yang, W. Zhou, T.Y. Hsu, Acta Mater. 46 (1998) 501–510.
[64] S. Lee, J. Kim, S.J. Lee, B.C. De Cooman, Scripta Mater. 65 (2011) 1073–1076.
[65] M. Müller, J. Magn. Mater. 78 (1989) 337–346.
[66] G. Hausch, J. Magn. Mater. 10 (1979) 163–169.
[67] S. Reeh, D. Music, T. Gebhardt, M. Kasprzak, T. Jäpel, S. Zaefferer, et al., Acta Mater. 60 (2012) 6025–6032.
[68] I.C. Jung, B.C. De Cooman, Acta Mater. 61 (2013) 6724–6735.
[69] I.C.Jung, PhD thesis, POSTECH, South Korea, 2013.
[70] J.H. Kang, T. Ingendahl, W. Bleck, Mater. Des. 90 (2016) 340–349.
[71] W.S. Choi, B.C. De Cooman, S. Sandlöbes, D. Raabe, Acta Mater. 98 (2015) 391–404.
[72] S.Z. Wu, H.W. Yen, M.X. Huang, A.H.W. Ngan, Scripta Mater. 67 (2012) 641–644.
[73] A. Dumay, J.P. Chateau, S. Allain, S. Migot, O. Bouaziz, Mater. Sci. Eng. A 483–484 (2008) 184–187.
[74] J.P. Hirth, in: R.E. Reed-Hill, J.P. Hirth, H.C. Rogers (Eds.), Metallurgical Society Conference. Deformation Twinning, 25, Gordon&Breach Science Publishers Inc, Metallurgical Society of the AIME, Philadephia, PA, 1964, pp. 112–115.
[75] S.M. Copley, B.H. Kear, Acta Metall. 16 (1968) 227–231.
[76] T.S. Byun, Acta Mater. 51 (1993) 3060–3071.
[77] S. Mahajan, G.Y. Chin, Acta Metall. 21 (1975) 815.
[78] D.R. Steinmetz, T. Jäpel, B. Wietbrock, P. Eisenlohr, I. Gutierrez-Urrutia, A. Saeed-Akbari, et al., Acta Mater. 61 (2013) 494–510.
[79] B. Mahato, S.K. Shee, T. Sahu, S. Ghosh Chowdhury, P. Sahu, D.A. Porter, et al., Acta Mater. 86 (2015) 69–79.
[80] S. Kibey, J. Liu, D. Johnson, H. Sehitoglu, Acta Mater. 55 (2007) 6843.
[81] J.B. Liu, X.H. Liu, W. Liu, Y.W. Zeng, K.Y. Shu, Phil. Mag. 91 (2011) 4033–4044.
[82] H. Fujita, T. Mori, Scripta Metall. 9 (1975) 631.
[83] J.B. Cohen, J. Weertman, Acta Metall. 11 (1963) 996.
[84] H. Idrissi, K. Renard, L. Ryelandt, D. Schryvers, P.J. Jacques, Acta Mater. 58 (2010) 2464–2476.
[85] S.Miura, J.Takamura, N.Narita, Supplement Trans. Japan Inst. Metal. 9, 1968, 555–562.
[86] F. de las Cuevas, M. Reis, A. Ferraiuolo, G. Pratolongo, L.P. Karjalainen, J. Alkorta, et al., Key Eng. Mater. 423 (2010) 147–152.
[87] J.A. Venables, Phil. Mag. 30 (1974) 1165.
[88] L. Bracke, L. Kestens, J. Penning, Scripta Mater. 61 (2009) 220–222.
[89] K.M. Rahman, V.A. Vorontsov, D. Dye, Acta Mater. 89 (2015) 247–257.
[90] N. Narita, J. Takamura, Phil. Mag. 29 (1974) 1001.
[91] H. Suzuki, C.S. Barrett, Acta Metall. 6 (1958) 156–165.
[92] M.A. Meyers, O. Vöhringer, V.A. Lubarda, Acta Mater 49 (2001) 4025–4039.
[93] I. Gutierrez-Urrutia, S. Zaefferer, D. Raabe, Mater. Sci. Eng. A 527 (2010) 3552–3560.
[94] J.A. Venables, J. Phys. Chem. Solids 25 (1964) 693–700.
[95] S.A. Kibey, J.B. Liu, D.D. Johnson, H. Sehitoglu, Acta Mater. 55 (2007) 6843.
[96] S.A. Kibey, L.L. Wang, J.B. Liu, H. Johnson, H. Sehitoglu, D.D. Johnson, Phys. Rev. B 79 (2009) 214202.
[97] S. Ogata, J. Li, S. Yip, Phys. Rev. B 71 (2005) 224102.
[98] S.Allain, PhD thesis, INPL, France, 2004.
[99] D.Barbier, PhD thesis, Université Paul Verlaine de Metz, France.

[100] J.Kim, PhD thesis, POSTECH, South Korea.
[101] A. Soulami, K.S. Choio, Y.F. Chen, W.N. Liu, X. Sun, M.A. Khaleel, Mater. Sci. Eng. A 528 (2011) 1402–1408.
[102] C. Haase, L.A. Barrales-Mora, F. Roters, D.A. Molodov, G. Gottstein, Acta Mater. 80 (2014) 327–340.
[103] J. Kim, Y. Estrin, B.C. DeCooman, Metall. Mater. Trans. A 44 (2013) 4168–4182.
[104] K. Renard, P. Jacques, Mater. Sci. Eng. A 542 (2012) 8–14.
[105] H. Beladi, I.B. Timokina, Y. Estrin, J. Kim, B.C. De Cooman, S.K. Kim, Acta Mater. 59 (2011) 7787–7799.
[106] V. Shterner, A. Molotnikov, I. Timokhina, Y. Estrin, H. Beladi, Mater. Sci. Eng. A 613 (2014) 224–231.
[107] C. Haase, L.A. Barrales-Mora, D.A. Molodov, G. Gottstein, Metall. Mater. Trans. A 44 (2013) 4445–4449.
[108] O. Bouaziz, H. Zurob, B. Chehab, J.D. Embury, S. Allain, M. Huang, Mater. Sci. Technol. 27 (2011) 707–709.
[109] C.P. Scott, B. Remy, J.L. Collet, A. Cael, C. Bao, F. Danoix, et al., Int. J. Mater. Res. 102 (2011) 538.
[110] G.J. Sevillano, F. de las Cuevas, Scripta Mater. 66 (2012) 978–981.
[111] Y.F. Shen, N. Jia, R.D.K. Misra, L. Zuo, Acta Mater. 103 (2016) 229–242.
[112] S.Lee, B.C. De Cooman, unpublished research.
[113] G. Dini, A. Najafizadeh, R. Ueji, S.M. Monir-Vaghefi, Mater. Des. 31 (2010) 3395–3402.
[114] H.Gwon, B.C.De Cooman, unpublished research.
[115] T. Lee, M. Koyama, K. Tsuzaki, Y.H. Lee, C.S. Lee, Mater. Lett. 75 (2012) 169–171.
[116] S. Wang, Z. Liu, G. Wang, Acta Metall. Sinica 45 (2009) 1083–1090.
[117] J.H. Kang, S. Duan, S.J. Kim, W. Bleck, Metall. Mater. Trans. (2016). Available from: http://dx.doi.org/10.1007/s11661-016-3399-5.
[118] J. Li, Y. Chou, Metall. Mater. Trans. A 1 (1970) 1145–1159.
[119] J.P. Chateau, A. Dumay, S. Allain, A. Jacques, 15th International Conference on the Strength of Materials ICSMA-15, Journal of Physics: Conference Series 240, 2010. Available from: http://dx.doi.org/10.1088/1742-6596/240/1/012020.
[120] H.W. Yen, M. Huang, C.P. Scott, J.R. Yang, Scripta Mater. 66 (2012) 1018–1023.
[121] R. Ueji, N. Tsuchida, H. Fujii, D. Kondo, K. Kunishige, J. Jpn. Inst. Met. 71 (2007) 815–821.
[122] R. Ueji, N. Tsuchida, D. Terada, N. Tsuji, Y. Tanaka., A. Takemura, et al., Scripta Mater. 59 (2008) 963–966.
[123] I. Gutierrez-Urrutia, S. Zaefferer, D. Raabe, Acta Mater. 66 (2010) 3552.
[124] S. Asgari, E. El-Danaf, S.R. Kalidindi, R.D. Doherty, Metall. Trans. A 28 (1997) 1781.
[125] S.R. Kalidindi, Int. J. Plasticity 14 (1998) 1265–1277.
[126] J. Kim, Y. Estrin, H. Beladi, I. Timokina, K.G. Chin, S.K. Kim, et al., Metall. Mater. Trans. A 43 (2012) 479–490.
[127] J. Kim, Y. Estrin, B.C. De Cooman, Metall. Mater. Trans. A 44 (2013) 4168–4182.
[128] J.G. Sevillano, Scripta Mater. 60 (2009) 336–339.
[129] Y.N. Dastur, W.C. Leslie, Metall. Trans. A 12A (1981) 749.
[130] P.H. Adler, G.B. Olson, W.S. Owen, Metall. Trans. A 12A (1986) 1725.
[131] A. Butz, M. Zapara, A. Erhart, D. Croizet, N. Stenberg, J. Hagstrom, et al, IDDRG 2014 Conference Proceedings, June 1–4, Paris France, pp. 38–48.
[132] Y. Ha, Y. Kim, K.H. Kwon, S.G. Lee, S.H. Lee, N.J. Kim, Metall. Mater. Trans. A 46 (2015) 545–548.
[133] S. Curtze, V.T. Kuokkala, Acta Mater. 58 (2010) 5129–5141.
[134] S.W. Lee, B.C. De Cooman, Metall. Mater. Trans. A 45A (2014) 709–716.

[135] S.W. Lee, B.C. De Cooman, Metall. Mater. Trans. A 45A (2014) 5009–5016.
[136] S.W. Lee, B.C. De Cooman, Metall. Mater. Trans. A 45A (2014) 6039–6052.
[137] S.W. Lee, B.C. De Cooman, Steel Res. Int. 86 (2015). Available from: http://dx.doi.org/10.1002/srin.201500038.
[138] S.W. Lee, W.C. Woo, B.C. De Cooman, Metall. Mater. Trans. A 45A (2014) 5823–5828.
[139] C. Scott, S. Allain, M. Faral, N. Guelton, Rev. Metall. 103 (2006) 293.
[140] W. Bleck, K. Phiu-on, C. Heering, G. Hirt, Steel Res. Int. 78 (2007) 536–545.
[141] P. Lan, J. Zhang, Steel Res. Int. 86 (2015) 1–12.
[142] K.H. Spitzer, F. Ruppel, R. Viscorova, R. Scholz, J. Kroos, V. Flaxa, Steel Res. Int. 74 (2003) 724–731.
[143] H. Hoffmann, B. Engl, M. Menne, T. Heller, W. Zimmermann, US Patent, 2003, No. US 2003/0145911 A1; T.W. Kim, Y.G. Kim, S.H. Park, US Patent, 1997, No. 5647922.
[144] S. Allain, O. Bouaziz, T. Lebedkina, M. Lebyodkin, Scripta Mater. 64 (2011) 741–744.
[145] N. Tsuji, Y. Ito, Y. Saito, Y. Minamino, Scripta Mater. 47 (2002) 893–899.
[146] L. Chen, J.K. Kim, S.K. Kim, G.S. Kim, K.G. Kim, B.C. De Cooman, Steel Res. Int. 81 (2010) 552–568.
[147] L. Xu, L. Chen, B.C. De Cooman, D. Steglich, F. Barlat, Int. J. Mater. Form. 3 (1) (2010) 247–250.
[148] Y.F. Shen, N. Jia, R.D.K. Misra, L. Zuo, Acta Mater. 103 (2016) 229–242.
[149] S.J. Lee, J. Kim, S. Kane, B.C. De Cooman, Acta Mater. 59 (2011) 6809–6819.
[150] M. Koyama, T. Sawaguchi, K. Tsuzaki, ISIJ Intl. 55 (2015) 1754–1761.
[151] S.D. Andrews, H. Sehitoglu, I. Karaman, J. Appl. Phys. 87 (2000) 2194–2230.
[152] V. Gerold, H.P. Karnthaler, Acta Metall. 37 (1989) 2177–2183.
[153] D.E. Jiang, E.A. Carter, Phys. Rev.B 67 (2003) 214103.
[154] I.C. Jung, S.J. Lee, B.C. De Cooman, Scripta Mater. 66 (2012) 729–732.
[155] J. Hirsch, K. Lücke, M. Hatherly, Acta Metall. 36 (1988) 2863–3059.
[156] J. Hirsch, K. Lücke, M. Hatherly, Acta Metall. 36 (1988) 2905–2927.
[157] R.E. Smallman, D. Green, Acta Metall. 12 (1964) 145–153.
[158] I.L. Dillamore, Metall. Trans. 1 (1970) 2463–2470.
[159] K. Morii, H. Mecking, Y. Nakayama, Acta Metall. 33 (1985) 379–386.
[160] S. Vercammen, B. Blanpain, B.C. De Cooman, P. Wollants, Acta Mater. 52 (2004) 2005–2012.
[161] Y. Lü, B. Hutchinson, D.A. Molodov, G. Gottstein, Acta Mater. 58 (2010) 217.
[162] R.K. Ray, J.J. Jonas, R.E. Hook, Int. Mater. Rev. 39 (1994) 129.
[163] L. Bracke, K. Verbeken, L. Kestens, J. Penning, Acta Mater. 57 (2009) 1512–1524.
[164] Y. Lü, D.A. Molodov, G. Gottstein, Acta Mater. 59 (2011) 3229–3243.
[165] K. Lücke, R. Rixen, F.W. Rosenbaum, in: H. Hu (Ed.), Nature and Behaviour of Grain Boundaries, Plenum Press, New York, USA, 1972, p. 245.
[166] M. Dao, L. Lu, Y.F. Shen, S. Suresh, Acta Mater. 54 (2006) 5421–5432.
[167] S. Kang, Y.S. Jung, J.H. Jun, Y.K. Lee, Mater. Sci. Eng. A 527 (2010) 745–751.
[168] O. Bouaziz, C.P. Scott, G. Petitgand, Scripta Mater. 60 (2009) 714–716.
[169] C. Haase, L.A. Barrales-Mora, D.A. Molodov, G. Gottstein, Adv. Mater. Res. 922 (2014) 213–218.
[170] C. Haase, T. Ingendahl, O. Güvenç, M. Bambach, W. Bleck, D.A. Molodov, et al., Mater. Sci. Eng. A 649 (2016) 74–84.
[171] A.S. Hamada, L.P. Karjalainen, M.C. Somani, Mater. Sci. Eng. A 467 (2007) 114–124.
[172] J. Lorthios, M. Mazière, X. Lemoine, P. Cugy, J. Besson, A.F. Gourgeus-Lorenzon, Scripta Mater. 63 (2010) 1220–1223.
[173] B.J. Lee, J.W. Jang, Acta Mater. 55 (2007) 6779–6788.

[174] J.K. Jung, O.Y. Lee, Y.K. Park, D.E. Kim, K.G. Jin, S.K. Kim, et al., J. Korean Inst. Met. Mater. 46 (2008) 627–633.
[175] Y.S. Chun, K.T. Park, C.S. Lee, Scripta Mater. 66 (2012) 960–965.
[176] I.J. Park, K.H. Jeong, J.G. Jung, C.S. Lee, Y.K. Lee, Int. J. Hydrogen Energy 37 (2012) 9925–9932.
[177] Y.J. Kwon, T. Lee, J. Lee, Y.S. Chun, C.S. Lee, Int. J. Hydrogen Energy 40 (2015) 7409–7419.
[178] S.M. Lee, I.J. Park, J.G. Jung, Y.K. Lee, Acta Mater. 103 (2016) 264–272.
[179] J.H. Ryu, S.K. Kim, C.S. Lee, D.W. Suh, H.K.D.H. Bhadeshia, Proc. R. Soc. A 469 (2015) 20120458.
[180] K.H. Soo, J.S. Kim, Y.S. Chun, K.T. Park, Y.K. Lee, ISIJ Intl. 49 (2009) 1952–1959.
[181] K.H. So, J.S. Kim, Y.S. Chun, K.-T. Park, Y.-K. Lee, C.S. Lee, ISIJ Intl. 49 (2009) 1952.
[182] J.A. Ronevich, S.K. Kim, J.G. Speer, D.K. Matlock, Scripta Mater. 66 (2012) 956–959.
[183] L. Chen, S.J. Lee, B.C. De Cooman, ISIJ Intl. 52 (2012) 1670.
[184] M. Koyama, E. Akiyama, K. Tsuzaki, Corr. Sci. 54 (2012) 277–281.
[185] M. Koyama, E. Akiyama, T. Sawaguchi, D. Raabe, K. Tsuzaki, Scripta Mater. 66 (2012) 459–462.
[186] M. Koyama, E. Akiyama, K. Tsuzaki, Scripta Mater. 66 (2012) 947–950.
[187] S. Lee, B.C. De Cooman, X. Gou, A. Saeed-Akabari, W. Bleck, Proceeding of the 2nd International Conference on High Mn Steel HMnS2014, Aachen, Germany.
[188] R.T. Van Tol, PhD thesis, Technische Universiteit Delft, Delft, The Netherlands, 2014.
[189] R.T. Van Tol, J.K. Kim, L. Zhao, J. Sietsma, B.C. De Cooman, J. Mater. Sci. 47 (2012) 4845–4850.
[190] R.T. van Tol, L. Zhao, H. Schut, Mater. Sci. Technol. 28 (3) (2012) 348–353.
[191] C. Béal, X. Kleber, D. Fabregue, M. Bouzekri, Mater. Sci. Eng., A 543 (2012) 76–83.
[192] D. Fabrege, C. Béal, X. Kleber, M. Bouzekri, Proceedings of the 1st International Conference on High Mn Steels, Seoul, 2011, B-44.
[193] C. Béal, PhD thesis, L'Institut National des Sciences Appliquées de Lyon, Lyon, France, 2011.
[194] H. Kang, L. Cho, C.W. Lee, B.C. De Cooman, Metall. Mater. Trans. A (2016). Available from: http://dx.doi.org/10.1007/s11661-016-3475-x.
[195] M.A. Krishtal, Sov. Phys. Dokl. 15 (1970) 614–617.
[196] P. Gordon, H. An, Met. Trans. A 13A (1982) 457–472.
[197] N.S. Stoloff, T.L. Johnston, Acta Metall. 11 (1963) 251–256.
[198] A.R.C. Westwood, M.H. Kamdar, Phil. Mag. 8 (1963) 787–804.

Hot formed steels*,**

E. Billur[1,2]
[1]Billur Metal Form Ltd., Bursa, Turkey, [2]Atılım University, Ankara, Turkey

12

12.1 Introduction

The automotive industry has to reduce vehicle weight and costs, and yet meet the ever-challenging new crash standards. To fulfill all these goals at the same time, vehicle makers and their suppliers are continuously developing and using new materials to reduce weight and improve safety performance. These materials are typically advanced high strength steels (AHSS), hot formed steels, aluminum alloys, magnesium alloys, carbon fiber reinforced polymers (CFRP), etc. [1].

As seen in Fig. 12.1, most aluminum and magnesium alloys have very limited formability and strength as well. In steels, as the strength is increased it is difficult to form complicated parts. Only 2nd generation advanced high strength steels can be formed to complex shapes without fracture. These steels have high alloying elements, which make them expensive and prone to delayed cracking [5]. Furthermore, in all materials, higher press forces and harder to control springback are also observed with increasing strength.

12.1.1 What is hot forming?

Hot forming is a special process for forming Mn-B alloyed steels (typically 22MnB5) into complex shapes. There are two methods to achieve this: (1) Direct process: the unformed blank is heated in a furnace, formed while it is hot (❷ in Fig. 12.2), and quenched in the die, and (2) Indirect process: the blank is cold formed and trimmed (❶ in Fig. 12.2), heated, and quenched in the die. The final part typically has over 1500 MPa ultimate tensile strength (UTS), if quenched properly, and negligible springback and/or distortion [8,9]. For 22MnB5—the most typical hot forming steel—if the part is cooled over 27°C/s, almost 100% martensite and hardness over 470 HV can be achieved. Hardness drops significantly with cooling rates lower than this critical value [10]. Details of these transformations are given in Section 12.2.

* This chapter has been derived from 'Metallurgy of Steels and Blank Materials. In: Hot Stamping of Ultra High-Strength Steels, From a Technological and Business Perspective' Billur, E. 2017. With permission of Springer.

** Every effort has been made to trace copyright holders and to obtain their permission for the use of copyright material. The publisher apologizes for any errors or omissions in the acknowledgements printed in this book and would be grateful if notified of any corrections that should be incorporated in future reprints or editions.

Automotive Steels. DOI: http://dx.doi.org/10.1016/B978-0-08-100638-2.00012-2

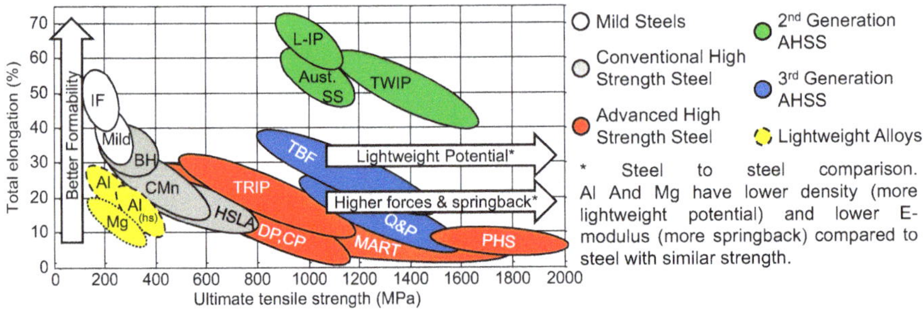

Figure 12.1 "Banana diagram" of strength and elongation showing different steel grades compared with lightweight alloys.
Source: Recreated after J.N. Hall, Evolution if advanced high strength steels in automotive applications, in: Great Designs in Steel, May 18th, Livonia, MI, USA, 2011; BaoShan Iron and Steel Co., Ltd., Automotive Advanced High Strength Steels, Catalogue, 2013; and Posco Technical Research Lab, 2016 Automotive Steel Data Book, 2016 [2–4].

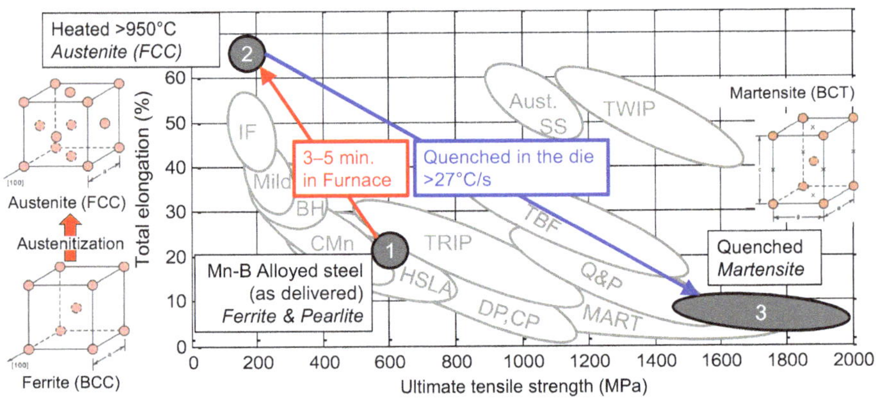

Figure 12.2 Summary of hot forming.
Source: Recreated after W.D. Callister, Fundamentals of Materials Science and Engineering, fifth ed., John Wiley & Sons, New York, NY, 2001, ISBN 0-471-39551-X; and E. Billur, C. Wang, C. Bloor, M. Holecek, H. Porzner, T. Altan, Advancements in tailored hot stamping simulations: cooling channel and distortion analyses, in: Proceedings of Numisheet 2014, AIP Conf. Proc. vol. 1567, 2014, pp. 1079–1084 [6,7].

12.1.2 A typical hot stamping line

A typical hot stamping line consists of (1) furnace/heating device, (2) material handling system, (3) press, (4) die set, (5) exit line, and (6) trimming/piercing systems.

Typically for heating, roller hearth furnaces are used. These are, depending on the type and production rate, furnaces that are 30 to 60 meters long. Roller hearth furnaces are typically gas and/or electric heated. To save space, "double decker furnaces" (as seen in Fig. 12.3) and "multi-chamber furnaces" have been designed and used in mass production [13–15]. Since 2013, conduction heating is also used by Lexus/Toyota to heat rectangular blanks for hot stamping [16,17]. In research

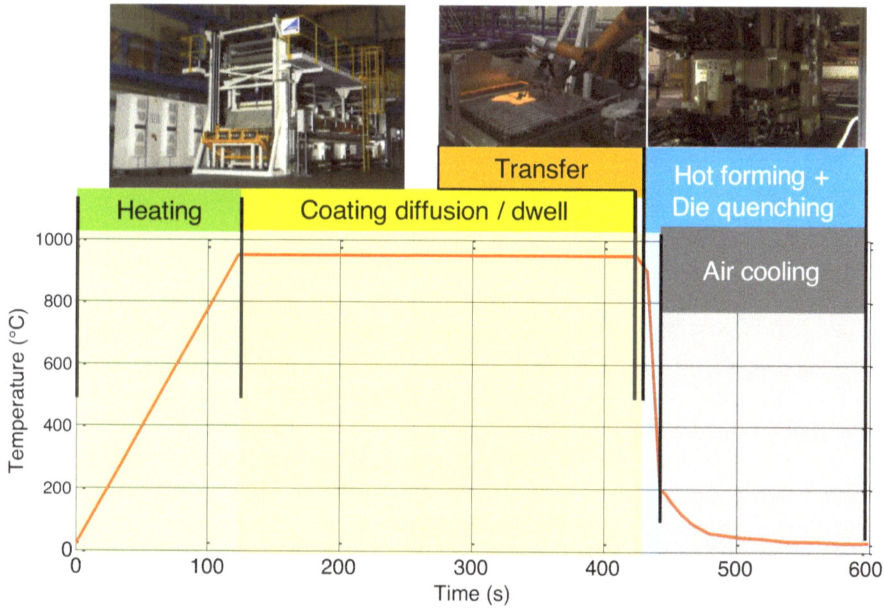

Figure 12.3 A typical hot stamping time-temperature profile.
Source: Recreated after A. Blaise, B. Bourouga, C. Dessain, Influence of the duration of transfer from the furnace to the press during the hot stamping process, ASME J. Thermal Sci. Eng. Appl. 7 (2015) 1–10; and M. Merklein, M. Wieland, M. Lechner, S. Bruschi, A. Ghiotti, Hot stamping of boron steel sheets with tailored properties: a review, J. Mater. Process. Technol. 228 (2016) 11–24[11,12], images are from: K. Eriksson, Bringing it together, in: AP&T Press Hardening, Next Step Seminar, Novi, MI, September 15th, 2010[13], [VW Media Services].

laboratories and for prototyping purposes (1) hot plate heating [18,19], (2) inductive heating [20], and (3) near infrared heating [21], were also used.

For material handling, manual feeding is not used in mass production, although could be used for prototyping purposes. During transfer, the blank would lose heat. Thus, in mass production robots or linear feeders are used to ensure the shortest (and most consistent) transfer times. To reduce the heat lost by radiation, new material transfer systems may have a reflective heat shield [22, 126]. A similar system is also used at the press exit.

In hot stamping, typically hydraulic presses have been used. During forming, the press loads are relatively low. However, high speed is required to ensure that the blank is still in austenite phase before the dies are closed. Several designs have been used to have high speed-low force at the beginning of the stroke and high force-low speed at the bottom:

1. Hydraulic press with accumulator drive: if the part has deep drawing, relatively high forces will be required for long distances. An accumulator drive may increase the speed of the press during forming [22,126].
2. Multi-cylinder hydraulic press: is a press with 5 hydraulic cylinders, but depending on the force requirement may employ only 1 or 3 of these. When less number of cylinders are used with the same pump, the press may speed up at low force conditions [126].

3. Hydraulic press with a flywheel: uses a flywheel instead of accumulators to store energy. The design eliminates the heat generation in the accumulator and may save more energy [23].
4. Servo mechanical presses: can control the motion and can dwell at the bottom dead center, which is required for quenching. According to Ormaetxea and Ibañez, a modified servo press would consume less energy than hydraulic press and have less maintenance problems [24].

In hot forming, the die set has two functions: (1) to form the part, and (2) to extract the heat from the blank. To achieve the critical cooling rate (27°C/s for 22MnB5), the dies should extract 17.5 kW/kg heat energy [25]. The industry standard for cooling the dies is to drill cooling channels to die pieces and assemble them together (Fig 12.4A). Another method is to use shell cooling, where the final part shape is machined on a shell die and cooling water is pumped between the shell and the core die (Fig. 12.4B). Lastly the die could be cast on cooling tubes (Fig 12.4C) [26,27,127].

Lastly, for trimming and piercing purposes, conventional die cutting is not commonly used in hot stamped components, as the die stresses would be extremely high [28]. Laser cutting is the industry standard for hot stamped components. In addition to laser cutting, the following are also used:

1. Blank development: where the initial blank shape is altered to reduce/eliminate the need for trimming.
2. In-die hot cutting: trimming/piercing the part while it is still hot and soft.
3. A combination of all three could be used [23,29].

12.1.3 Brief history

The hot forming technique, where the material is heated over its austenitization temperature, formed, and subsequently quenched, can be dated back to Japanese sword making. The earliest patent about "Preβhärten" (Press hardening) was

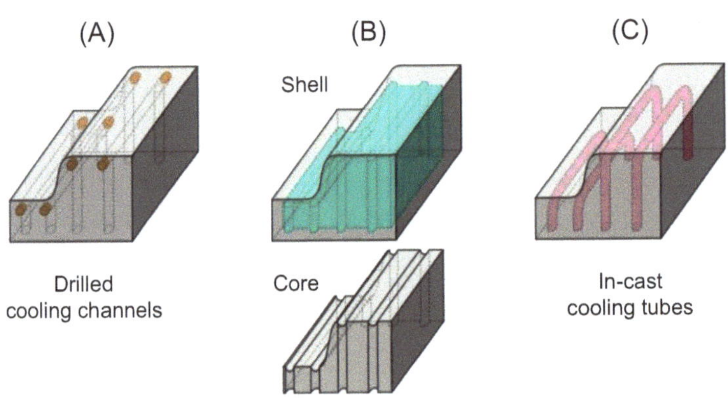

Figure 12.4 Die making strategies for hot stamping dies with cooling channels: (A) drilling the cooling channels, (B) shell and core design, and (C) in-cast cooling tubes [26].

granted in 1914 in Switzerland. The technique described in this patent has been used in agricultural products since 1930s [20,30,31].

Hot forming as we know it today started in 1970s in Sweden. In 1973, Norrbottens Järnverks (which later merged to form SSAB) applied for the first "modern hot forming" patent, which was granted in 1977 [32]. In 1975, a 6-year R&D project was started by Volvo Trucks, Norrbottens Järnverks, and Luleå University of Technology [33].

The first automotive application was side impact door beams of the SAAB 9000 which went to production in 1984. In 1986 Jaguar XJ and in 1993 Ford Mondeo also used hot formed door beams [34,35]. Ford Mondeo's North American versions Ford Contour and Mercury Mystique were the first American cars to have hot formed components [34,36]. In 1996, the first bumper beam was used by Renault [37]. In late 1990s the earliest hot formed B-pillar studies were done at SAAB [38]. The earliest hot formed A-pillar reinforcements were used in BMW E46 Cabrio (Production started in March 2000) and Citroën C5 (Production started in March 2001) [39,40]. Peugeot 307 (Production started in 2001) had hot formed B-pillar reinforcements and roof rail [41].

One of the biggest breakthroughs was the first generation Volvo XC90, which was introduced in 2002. There were a total of 10 hot formed parts, which accounted for 7% mass of the body-in-white. The second breakthrough was the 6th generation Volkswagen Passat (Production started in 2005). This vehicle had a total of 13 hot formed parts (parts shown in Fig. 12.5), which accounted for 19% of the body mass [42–44].

Since 2012, several vehicles have surpassed 20% barrier of hot formed body components. The first one was Volvo V40 (SOP 2012) with 20% hot stamped parts by mass. Later on several Volkswagen Group vehicles on MQB platform were introduced; these vehicles typically have 24–28% of hot formed body by mass. These vehicles include, but are not limited to, Audi A3 (3rd generation), VW Golf 7, VW Passat B8, Seat Leon Mk3 and Škoda Octavia Mk3 [45]. However, as of 2016, the highest usage of hot formed steels is in Volvo XC90 (2nd generation, production started in 2014), which accounts for 38% of the body mass [46].

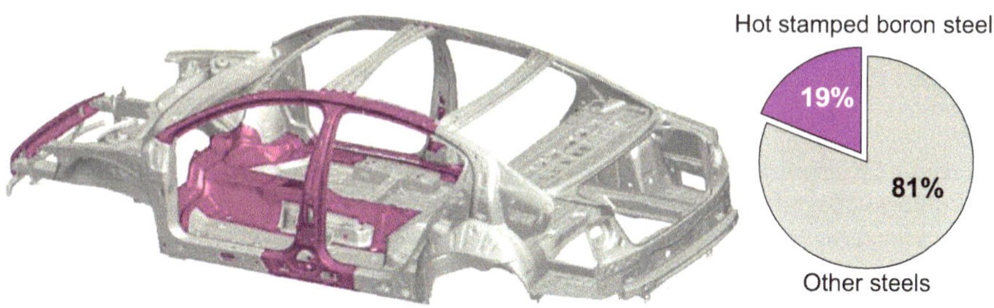

Figure 12.5 Hot stamped boron steel components in Passat B6 (SOP 2005) [42].

12.2 Physical metallurgy of hot forming steels

As mentioned, for hot stamping typically C-Mn-B alloyed steels are used. The alloy composition has to address several problems in industrial scale manufacturing; thus, the blanks have to:

1. Have high yield and ultimate tensile strength after quenching [10].
2. Undergo complete martensitic transformation only by using a water-cooled die [10].
3. Avoid diffusional transformation during blank transfer from the furnace to the press [47].
4. Be weldable (which requires low C and Mn) [48].

In hot stamped components, the high yield/tensile strength (and thus hardness) is achieved by solid solution hardening (by alloying elements) and transforming to almost 100% martensite [10]. The effect of main alloying elements can be summarized as:

Carbon (C): Carbon is the longest used and most well-known alloying element to achieve higher strength in steels. Strength increases with carbon content (up to 0.7–0.8 wt.%), and thus, the carbon content is chosen according to desired strength level. However, ductility and weldability are decreased with increasing carbon level [10,49,50]. Carbon also reduces the martensite start (shown as M_s in Fig. 12.6) significantly [51].
Manganese (Mn): Manganese is used as an alloying element in almost all steels, at least by 0.3 wt.%. Increased Mn content up to 2 wt.% would improve strength and toughness, however higher Mn contents cause reduced toughness and may cause cracking. However,

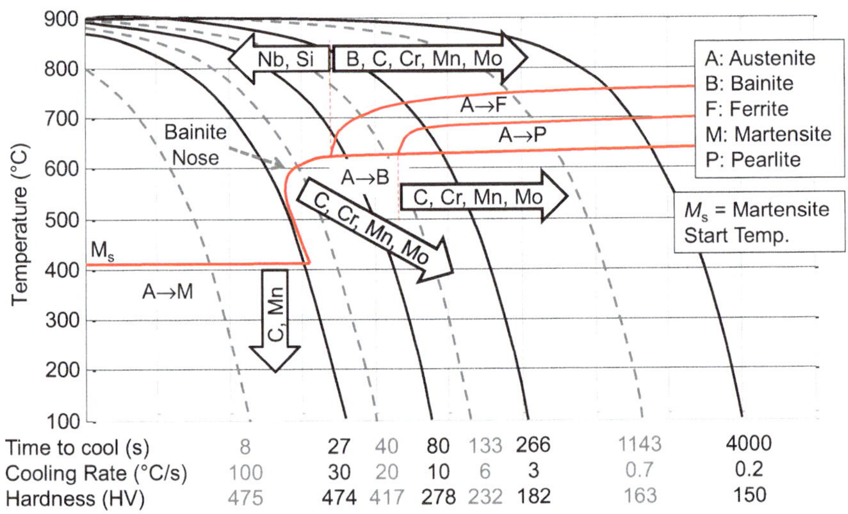

Figure 12.6 Continuous Cooling Transformation (CCT) curve of 22MnB5 steel and effects of additional alloying elements.
Source: Recreated after L. Vaissiere, J.P. Laurent, A. Reinhardt, Development of Pre-Coated Boron Steel for Applications on PSA Peugeot Citroën and RENAULT Bodies in White, SAE Technical Paper, 2002-01-2048, 2002; and H. Mohrbacher, Martensitic automotive steel sheet – fundamentals and metallurgical optimization strategies, Adv. Mater. Res. 1063 (2015) 130–142 [39,50].

above 5 wt.% Mn, a complex microstructure with austenite at room temperature may be present. The so-called Medium-Mn (typically 5–12 wt.% Mn) and High-Mn (TWIP, >15 wt.% Mn) steels have both high strength and elongation. There are several studies of using Medium-Mn steels in hot forming, for details see Section 12.3.3. Mn also lowers the austenitization temperature, which could reduce the furnace temperature in hot stamping, and thus save energy and reduce the carbon emissions during heating. [52–54]. In low carbon steels, Mn will retard bainite formation. In hot stamping, the main goal is to have almost 100% martensitic final part. So the actual cooling curve has to be on the left side of the bainite nose in the continuous cooling transformation (CCT) diagram (see Fig. 12.6). Mn also reduces martensite start temperature, but is not as effective as C [50,55].

Boron (B): A small amount of boron (0.001–0.003 wt.%) improves the hardenability of low carbon steels significantly. However, as the carbon level reaches 0.5–0.8 wt.%, adding boron may have no effect or may soften the steel. There are several explanations on how B increases the strength: (1) segregation of B atoms to austenite grain boundaries and therefore delaying the ferrite formation, (2) segregation of $Fe_{23}(CB)_6$ at austenite grain boundaries [10,52,56]. In any case, B delays the formation of ferrite and pearlite (see Fig. 12.6), thus it is possible to achieve martensitic structure with water-cooled dies. Naderi et al., have studied a number of C-Mn alloyed steels and have shown that without B-alloying, it was not possible to get 100% martensitic structure with water-cooled dies [57].

12.3 Hot forming steels

Hot forming, initially, was used for intrusion resistant parts only. The earliest parts produced were side impact door beams and bumper beams. Later, with developing technology, A- and B-pillars were hot stamped. Today, components with different local properties can be produced by using different materials and/or processes. By doing this, it is now possible to use hot formed components in energy absorbing areas, as well.

12.3.1 Intrusion resistant hot forming steels

The first commercially available hot forming steel, 22MnB5, is still the grade that is mostly used. As described in earlier sections, this steel typically has >1000 MPa yield strength and 1400–1600 MPa tensile strength [44]. There are several commercial names of this grade, including but not limited to:

1. BTR165: uncoated steel, Benteler [58].
2. USIBOR 1500: typically AlSi coated steel produced and patented by ArcelorMittal. Zn coated versions are also available by the name USIBOR 1500 GI or GA [59,60].
3. MBW1500: "Mangan-Bor-Stahl zum Warmumformhärten", literally meaning Manganese-Boron steel for hot form hardening, produced by ThyssenKrupp. Available uncoated or Al-Si coated (MBW1500 + AS) [61]. There was a ZnNi coated version [62] which was later discontinued [63].
4. HPF1470: Hot Press Forming 1470, uncoated/Al-Si coated/Zn-coated, by POSCO [64].
5. SQ1500: Sumiquench 1500 by Nippon-Sumitomo [65].
6. phs-ultraform: Zn coated steel by voestalpine [66].

Table 12.1 Chemical compositions of Manganese-Boron steels in wt.% (Fe balanced) [67,68]

Steel	Al	B	C	Cr	Mn	N	Ni	Si	Ti
20MnB5	0.04	0.001	0.16	0.23	1.05	–	0.01	0.40	0.034
22MnB5	0.03	0.002	0.23	0.16	1.18	0.005	0.12	0.22	0.040
27MnCrB5	0.03	0.002	0.25	0.34	1.24	0.004	0.01	0.21	0.042
28MnB5	–	0.005	0.28	–	1.30	–	–	0.40	–
34MnB5	–	0.005	0.34	–	1.30	–	–	0.40	–
37MnB4	0.03	0.001	0.37	0.19	0.81	0.006	0.02	0.31	0.046

Table 12.2 Yield and ultimate tensile strengths (YS and UTS) of several quenchable steels before and after quenching [10,68]

Steel	As delivered		Quenched	
	YS, MPa (ksi)	UTS, MPa (ksi)	YS, MPa (ksi)	UTS, MPa (ksi)
20MnB5	505 (73)	637 (92)	967 (140)	1354 (196)
22MnB5	457 (66)	608 (88)	1010 (146)	1478 (214)
27MnCrB5	478 (69)	638 (93)	1097 (159)	1611 (234)
28MnB5	–	–	1135 (165)	1740 (252)
34MnB5	–	–	1225 (178)	1919 (278)
37MnB4	580 (84)	810 (117)	1378 (200)	2040 (296)

Although some variations may occur, most of the grades listed above have very similar chemical compositions (see the second row in Table 12.1). There are also other Mn-B alloyed steels which could be used in hot forming process, as shown in Table 12.1. Depending on the thickness and the supplier, hot forming steels could be delivered hot or cold rolled, coated or uncoated.

These grades are delivered in soft ferritic-pearlitic condition. They have to be austenitized in an atmosphere controlled furnace. Once quenched, their strength levels are at least doubled, as listed in Table 12.2.

As seen in Table 12.2, some steel grades that contain higher carbon have higher strength values than the most common 22MnB5 grade. These materials may save even more weight, with equivalent intrusion resistance [69,70].

Problems with higher strength materials are (1) their lower toughness/energy absorption (i.e., typically even lower elongation than 22MnB5, see Fig. 12.7), (2) hydrogen embrittlement and (3) weldability [65]. Naderi [10] noted that, during tensile tests of 37MnB4 (UTS = 2040 MPa, 296 ksi), all of the hardened tensile samples were cracked outside the gage length.

Vickers Hardness (HV) values for conventional 22MnB5 steel are in the order of 470–510 HV after quenching. 37MnB4, on the other hand, has a Vickers Hardness

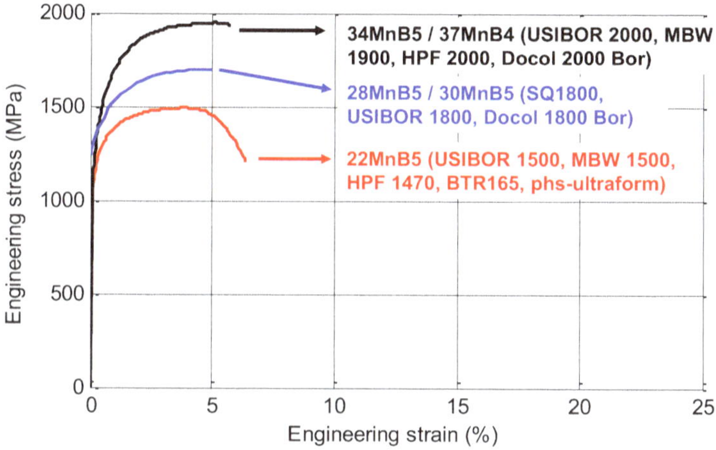

Figure 12.7 Engineering strain-stress diagram for 22MnB5 and higher strength boron steels. *Source*: Recreated after K. Hikita, et al., Properties of new TS1800 MPa grade hot staming steel and components, in: Materials in Car Body Engineering 2012, May 10–11, Bad Nauheim, Germany, 2012; S. Graff, T. Gerber, F.-J. Lenze, S. Sikora, About the simulation of microstructure evolution in the hot sheet stamping process and the correlation of resulting mechanical properties and crash-performance, Proceedings of chs2 2011, 2011, pp. 323–330; H. Lanzerath, et al., Simulation: enabler for the efficient design of lightweight boron intensive body structures, in: Insight Edition Conference, September 20–21, Gothenburg, Sweden, 2011; J. Overrath, F.-J. Lenze, S. Sikora, Aktuelle Entwicklung der Warmumformung im Automobilen Fahrzeugbau, (Current Developments of Hot Forming in Automobiles – in German), in 30. EFB-Kolloquium, March 2–3, Bad Boll, Germany, 2010; H.W. Lee, K.-H. Chung, A new body concept for electric vehicle, in: Materials in Car Body Engineering 2012, May 11, Bad Nauheim, Germany, 2012; H. Matsuoka, K. Fujihara, Mazda CX-5, in: EuroCarBody 2011, October 18–20, Bad Nauheim, Germany, 2011; and K. Fahlström, Laser welding of boron steels for light-weight vehicle applications, Licentiate Thesis, Högskolan Väst, 2015 [65,68,69,71–74].

of 600–610 HV [10]. Similarly, Overrath [71] found ~490 HV for MBW® 1500, ~530 and ~580 HV for MBW® 1900.

The first commercially available "higher-strength hot forming steel" (i.e., UTS > 1600 MPa) was Sumitomo Steel's SQ1800. This steel was used in the front and rear bumper beams of the Mazda CX-5 (production started in 2011). Studies have shown that 1.4 mm SQ1800 has the same intrusion resistance as 1.6 mm SQ1500 (22MnB5). Thus, by using SQ1800, Mazda could save 4.8 kg per vehicle [65,75].

In 2013, ThyssenKrupp started delivering MBW 1900 (34MnB5). Docol 1800 Bor from SSAB has 0.27 wt.% carbon. This steel has been available since 2013 and is similar to 28MnB5 shown in Table 12.1 and Fig. 12.7 [74]. SSAB is also developing a 37MnB4 grade with the commercial name Docol 2000 Bor [76]. POSCO is working on HPF1800 and HPF2000, two new grades already available for customer trials [4]. HPF2000 was already demonstrated in a number of prototype car bodies, including POSCO's own PBC-EV (POSCO Body Concept-Electric

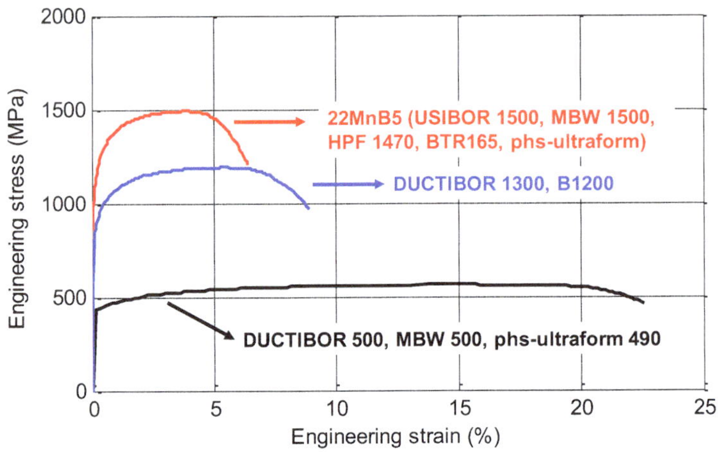

Figure 12.8 Engineering stress-strain curves of 22MnB5 in comparison with energy absorbing grades [3,59,80,81].

Vehicle) and Renault EOLAB concept car [77,78]. ArcelorMittal is developing USIBOR 1800 and USIBOR 2000 [59,73]. BaoSteel is also expected to roll an 1800 MPa hot forming steel [79].

12.3.2 Energy absorbing hot forming steels

For the automotive applications, the main purpose of hot stamped components is to improve crashworthiness and save weight. Therefore performance of these parts should be tested in crash conditions. Hot formed steels discussed in 12.2.1 have high strength but low ductility after quenching. This reduces the energy that can be absorbed during crash.

To improve energy absorption, several lower carbon steels were developed. Currently there are two separate areas of development: (1) 450–500 MPa strength level with >15% total elongation, and (2) 1000–1300 MPa strength level with >5% elongation. Fig. 12.8 shows approximate engineering stress-strain curves of energy absorbing steels compared with 22MnB5, all after quenching.

At present, several steelmakers offer 450–500 MPa level steels, including:

1. ArcelorMittal's DUCTIBOR 450 and DUCTIBOR 500, offered with Al-Si coating [60,82].
2. ThyssenKrupp's MBW500, uncoated or with Al-Si coating [63]. MBW500 is currently used in 6 components of the 2nd generation Volvo XC90 [46].
3. Voestalpine's phs-ultraform 490, offered with Zn coating [83].

Chemical compositions of these grades are listed in Table 12.3.

Table 12.3 Chemical compositions (wt.%, Fe balanced) of hot stamped steels with strength ~500 MPa [63,83,84,128]

Steel	C	Si	Mn	P	S	Al	Nb
Ductibor 500	0.05–0.08	<0.5	<1.4	<0.03	<0.01	0.02–0.04	0.03–0.07
MBW 500	<0.10	<0.35	<1.0	<0.03	<0.025	>0.015	<0.1
phs-ultraform 490	<0.11	<0.5	<1.4	<0.03	<0.025	<0.015	<0.1

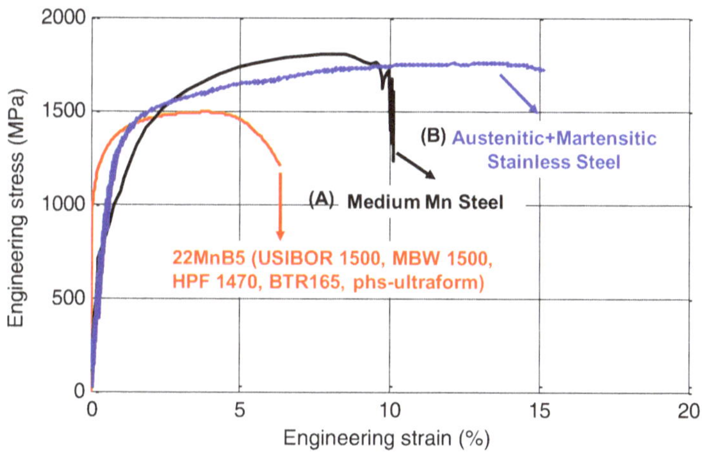

Figure 12.9 Comparison of 22MnB5 with potential new hot stamping materials: (A) Medium Mn Steel, and (B) Austenitic-Martensitic stainless steel. All tensile curves shown here are after hot stamping and quenching.
Source: Recreated after Han, Q., et al., Low temperature hot forming of medium-Mn steel, Proceedings of chs2 2015, 2015, pp. 381–390; and T. Fröhlich, Maximum safety and lightweight potential due to use of new high strength steels, in: Presentation at Outokumpu Experience 2013, May 22–23, London, UK, 2013[54,85].

12.3.3 Other steels considered for hot forming

In recent years, several other steels have also been considered for hot forming: (1) stainless steels, (2) Medium-Mn steels, and (3) conventional AHSS.

Although these are not currently in mass production, several companies and research institutes have studied the feasibility of hot forming and quenching these steels. Outokumpu, a stainless steel maker, has shown that a hot formed and quenched 1200 PH steel could have 1100–1300 MPa yield strength and over 1700 MPa tensile strength. This material is classified as duplex stainless steel, as it contains austenite and martensite. Its engineering stress—strain curve after quenching is compared with 22MnB5 in Fig. 12.9 [85].

Medium-Mn steels are being developed both for cold and hot forming applications. There are several advantages of Medium-Mn steels over 22MnB5:

1. Austenitization temperature is typically lower than 22MnB5 and decreases with increasing Mn content. This could reduce the energy requirement of the furnaces and save energy and cost.
2. Martensitic transformation can occur at very low cooling rates and thus simple dies could be built for hot forming of these grades. Martensite start/finish temperatures are also lower than 22MnB5. Some retained austenite may be present in the final part, which improves the elongation and thus energy absorption.
3. Medium-Mn steels, when properly hot stamped, may have high strength and high elongation. For example, tensile strength of 1800 MPa could be achieved with 10% total elongation, as seen in Fig. 12.9 [54]. Yi et al. achieved 1880 MPa tensile strength with 16% total elongation [86]. Rana et al. studied a number of heat treatment conditions with a 10 wt.% Mn steel and achieved 1330–1450 MPa tensile strength with 16–25% total elongation [87].

Lastly, there are studies about hot forming the conventional (designed for cold stamping) grades. Naderi [10] has shown that it was possible to achieve 100% martensitic structure by hot forming and die quenching several cold-formable dual-phase (DP) steel grades with various strength levels (DP800, DP1000, DP1400) and a martensitic steel grade (MS1200). In a transformation induced plasticity-aided steel (TRIP 800), some ferrite was observed when hot stamped with water-cooled dies [10]. There are also several studies on quenching and partitioning of hot formed steels to have retained austenite in the final part. These are explained in Section 12.5.3.

12.4 Blank coatings

In hot forming, blanks are heated over their austenitization temperature, which is typically around 900–950°C. At this temperature, if the surface of the steel is exposed to air, oxygen and other oxidizing gases react with the steel and cause scale formation on the surface – an oxide layer – typically composed of Fe_3O_4 [88]. The scale must be cleaned by sandblasting after the hot forming process [67]. Usually coated blanks form more oxide scales during reheating than uncoated ones [89]. Another phenomenon happening at high temperature is surface decarburization. Choi and De Cooman [90] have shown that up to a 30 μm layer could be decarburized. The hardness value of the base metal was measured as ~480 HV, whereas in the decarburization layer the average hardness was ~360 HV. Right at the surface, the hardness was as low as ~220 HV.

In addition to preventing scale formation and decarburization, the coating is desired to be:

1. Weldable.
2. Compatible with e-coating and paint bake cycles.
3. Corrosion protective [91].

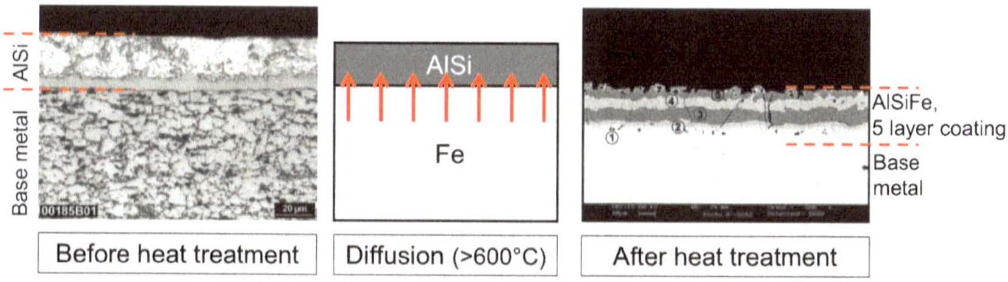

Figure 12.10 During heat treatment, Fe diffuses from the base steel to the coating and forms Al-Si-Fe coating.
Source: Recreated after T. Vietoris, Hot stamping with USIBOR1500®, in: AP&T Press Hardening, Next Step Seminar, Novi, MI, September 15th, 2010 [96].

There are different coatings available on the market, which could be classified under three main types [92]:

1. AlSi coating.
2. Zn based coatings.
3. Varnish coatings.

As of the time of writing, uncoated hot formed blanks are still used in automotive applications [93,94].

12.4.1 AlSi coating

AlSi coating was developed by Usinor, a French steelmaker that was later merged to Arcelor, and then to ArcelorMittal. The first pre-coated steel was used in Citroën C5 in 2001 [37,39,95].

The typical coating is approximately 25 μm thick and is composed of 10% Si + 90% Al as delivered. Once the pre-coated blank is heated, iron diffuses into coating and forms an Al-Si-Fe layer of 40 μm, as shown in Fig. 12.10. The diffusion is a time dependent process. If the diffusion is not completed (short heating time), or the blank is heated for long durations, the final quality may change [15,21,59,91]. The final properties of the coating are extremely important as they may affect:

1. Weldability of the final part.
2. Surface properties for e-coating and painting.

The total time[1] in furnace depends on three variables [96,97]:

1. The initial blank thickness; heating time to ensure austenitization.
2. The type of coating; AlSi coating requires <12°C/s heating rate, thus the heating time.
3. The initial coating thickness; dwell time to ensure coating diffusion.

[1] Total time in furnace is an addition of heating time + dwell time, see Fig. 12.3.

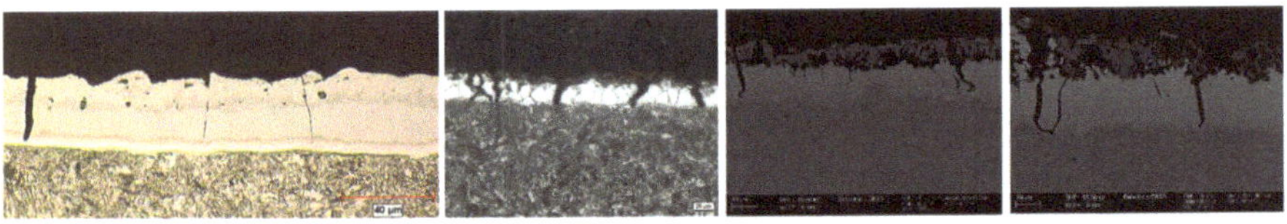

Figure 12.11 Coatings after hot stamping: (A) microcracks in AlSi coating does not penetrate to steel substrate [59], (B) GI coating melts and Zn may penetrate into base material, depending on coating thickness [97], (C–D) In GA and ZnAlMg coatings LME problem is reduced but still microcracks are formed at specific locations, predominantly at outside of the formed product [102].

AlSi coating prevents scale formation and decarburization even without an atmosphere controlled furnace. The coating has barrier corrosion protection. However, they cost more compared to uncoated blanks and require a longer time in the furnace. AlSi coated blanks cannot be cold formed (indirect hot stamping) because the coating would crack [91,98].

The typical AlSi coating weighs 150 g/m^2. In 2013, ArcelorMittal developed USIBOR 1500 P AS80 with 80 g/m^2 AlSi coating. There were two rationales behind this development:

1. Lower cost in order to be more competitive.
2. Reduced coating weight would reduce the heating/dwell time in furnace [99].

Fujita [100] has shown that hot formed steel with AS80 coating had approximately twice the blister width after cyclic corrosion test. Alden [101] has shown that the furnace dwell time could in practice be halved with AS80.

12.4.2 Zn based coatings

AlSi coated hot formed steels have limited corrosion protection. The protection comes from the so called "barrier protection," as the AlSi coating forms a barrier between the oxidizing environment and the bare steel. However, most of the car body components are already zinc coated. Thus, a similar level of corrosion protection is also required in hot formed steel areas [94].

Zn coated blanks, when hot formed with direct process, suffer from liquid metal embrittlement (LME). This phenomenon occurs when the coating is in liquid phase (melting point of zinc is around 420°C) and stress is applied to the base metal, which cannot be avoided in metal forming processes. When both conditions are met, the liquid coating may penetrate into the base metal, causing cracks on the surface, as seen in Fig. 12.11 [20,97,102].

These microcracks could reduce the fatigue life of the components and the bendability of the press hardened parts [97,103]. The bending angle at maximum load is long considered a good indicator of crashworthiness [104]. Kurz [97] has studied

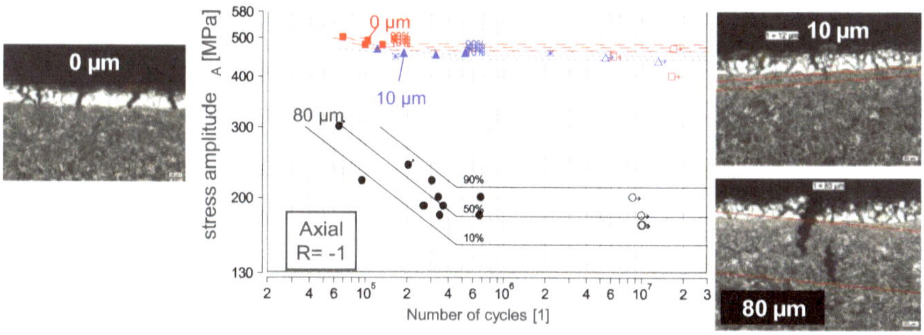

Figure 12.12 Effect of microcrack depth in base material on fatigue life of Zn coated 22MnB5 steel. Micrographs showing the cracks are also shown [97].

the crack depth and the fatigue stress. As seen in Fig. 12.12, if the cracks are deeper than 10 μm, the fatigue stress reduces drastically.

To solve the cracking issue, the first solution was to use indirect forming. This could be through two different methods [67,94]:

1. Cold deformation (most of the deformation is done in cold state) followed by hot calibration (very little deformation in hot state).
2. 100% of deformation, cutting, and piercing done at cold forming followed by a "form hardening" (where no deformation is done at hot state).

Another recent solution to crack formation is pre-cooling before plastic deformation. Ghanbari [105] found that microcrack formation occurs if forming is done over 782°C. Kurz et al. [94] developed a pre-cooling stage where the cooling rate is over 50°C/s but the cooling is interrupted before the martensite start temperature.

Hot dip galvanized (GI) steels may require an additional surface conditioning such as sand blasting to remove the top oxide layers. This is done to reduce the contact resistance in spot welding and to improve the phosphatability/paintability [92,99]. Galvannealing (GA) is a process where the galvanized steel is heated to 480–520°C. During this process iron diffuses into coating and the final coating may have 10–15% Fe & 85–90% Zn [106]. Fan and De Cooman [92] and Watkins [99] claimed GA coatings may be welded and painted without removing the oxide layer.

To reduce LME and/or the need for surface conditioning, alloying elements are added into coating. ZnNi coating was once commercialized by ThyssenKrupp with the name GammaProtect, but has been discontinued [63,107]. Other Zn-alloyed coatings could be ZnFe (Galvannealed) and ZnAlMg [92,108].

Currently ArcelorMittal, POSCO, and voestalpine offer Zn GI coatings. ArcelorMittal also offers GA coating. Tata Steel is developing MagiZinc (ZnAlMg) coating for press hardening, but this product is currently not in the market [4,60,94,109].

12.4.3 Varnish coatings

Another method to avoid scaling and decarburization is to apply varnish coatings. In this method, uncoated blanks are either coil coated or roll coated. The first such application for hot stamping process was developed by Nano-X company in 2005. The first generation x-tec was 6–7 μm thick Al + nano TiO_x. The coating had lower friction compared to uncoated and AlSi coated blanks, and was on a par with ZnNi coatings. The coating also allowed for faster heating compared to AlSi coating. However, the first generation coating had to be removed (sandblasted) before spot welding. To solve this problem 2nd generation x-tec was developed. This coating also allowed inductive, conductive, and near infrared heating and could be applied as thinly as 2–3 μm [88,110,111].

Recently, Henkel has also introduced Bonderite S-FN 7500 PH [91]. The product allowed fast heating, barrier corrosion protection, lubrication for hot stamping (friction coefficient is not published yet), can be e-coated and spot welded without removing/sandblasting the coating [91].

12.5 Typical automotive applications

Although the first application of hot formed steels in the automotive industry was side impact door beams, cold formed ultra high strength steels can replace (and have already started replacing in a few examples) hot formed steel in door and bumper beams [7]. The most typical application in the 2010s is the B-pillar. It is not wrong to state that, "If a car has hot formed component, then it most probably also has at least a hot formed B-pillar reinforcement". Typical hot formed automotive parts include A and B-pillar reinforcements, bumper and door beams, front and rear rails, roof rails, cantrails, and side-cross members, as shown in Fig. 12.13.

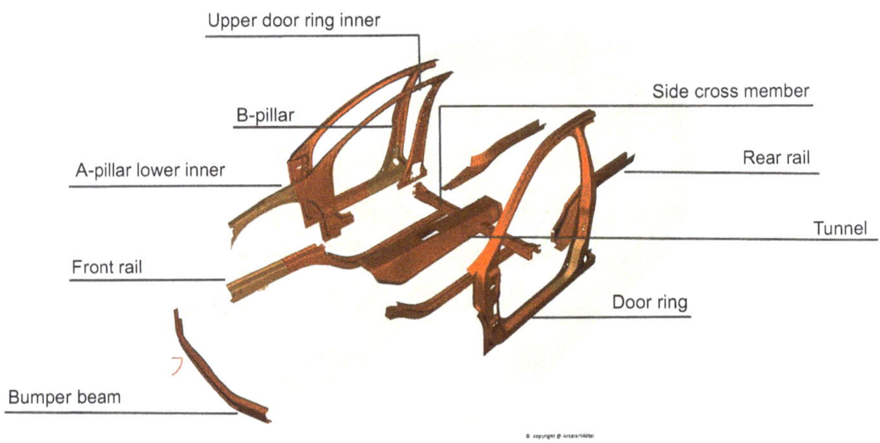

Figure 12.13 Examples of hot stamping applications.
Source: Courtesy of ArcelorMittal.

Several components, such as front and rear rails and B-pillar reinforcements, require both energy absorption and deformation resistance on the same component. In such components, a number of solutions are used. These are:

1. Welding energy absorbing materials (such as high strength low alloy (HSLA) steels or steels discussed in Section 12.3.2).
2. Reducing the thickness in energy absorbing areas to further reinforce with conventional high strength (HSS) or advanced high strength steels (AHSS).
3. Tailoring the properties of a hot formed part.

12.5.1 Tailor welded parts

Tailor welded blanks (TWBs) are made by butt welding two or more sheet blanks, and may have:

1. Different thicknesses.
2. Different materials.
3. A combination of (1) and (2).

Conventionally, HSLA steel grades were used in the energy absorbing areas. These have recently been replaced by energy absorbing hot forming steels, listed in Section 12.3.2. Munera [112] has shown the feasibility of welding USIBOR 1500 with DUCTIBOR 500 for applications such as door rings, front rails, and rear rails. The 2nd generation Volvo XC90 has a total of 6 TWB components with 22MnB5 and high elongation hot forming steel. These parts are: (1) B-pillars, (2) front rails, and (3) rear rails (3 parts x (left and right) = 6 parts) [46].

In 2013, Acura MDX was introduced, which had the first TWB hot formed door ring. The final part is as large as 1700 mm x 1600 mm, as seen in Fig. 12.14. The part has two different thicknesses, but the same material. By a nesting

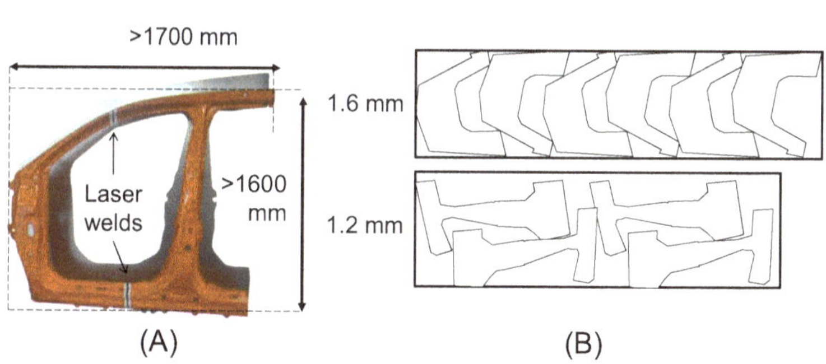

Figure 12.14 Door ring of Acura MDX: (A) final part with two thicknesses but one material (22MnB5), (B) sub-blanks nesting.
Source: Recreated after Altair Engineering, Industry First Hot Stamped Door Ring, Altair Enlighten Award Nominee, 2013; and R. Zum Mallen, J. Riggsby, Development of a global first SUV body construction, in: Great Designs in Steel, Livonia, MI, May 1st, 2013 [113,114].

optimization, the material utilization was improved from 53% to 63% (Fig. 12.14B). The design has also saved ~3 kg per vehicle [114].

Before welding sub-blanks, it is necessary to remove the Al-Si coating. This is typically done by "laser ablation." If Al enters the weld zone, the hardness of the weld zone may be reduced by ~40% [70].

12.5.2 Tailor rolled parts

Tailor rolled blanks (TRB) have been used in cold stamping since 2001 and in hot stamping since 2006. In tailored rolling, the thickness of the blank can be altered locally, as shown in Fig. 12.15A. The method has advantages of locally adjusting the stiffness and crash performance of the final part. It may also reduce the number of tools required (since reinforcements may be eliminated) and thus may lower the initial cost [116,117]. Since 2006, more than 50 million press-hardened TRB components have been manufactured by different Tier 1 companies [118]. The 2nd generation BMW X5 was one of the earliest examples of hot formed TRB, as shown in Fig. 12.15B [115]. Since then, a number of other vehicles also had hot formed TRB components, including, but not limited to: (1) 2010 Volvo S60, (2) 2011 Audi A6, (3) 2011 Ford Focus, (4) 2012 VW Golf VII, (5) 2013 Ford Kuga/Escape, (6) 2014 Peugeot 308, (7) 2014 Volvo XC90 [46, 119–121].

12.5.3 Tailor hardened parts

Hot formed steels generally have ultra high strength (>1400 MPa) but relatively poor elongation after quenching (~5–6% total elongation). Due to these two facts, in automotive industry several problems arise:

1. Low energy absorption (due to low elongation).
2. Heat affected zone in weld area.
3. Very high trimming/piercing loads.

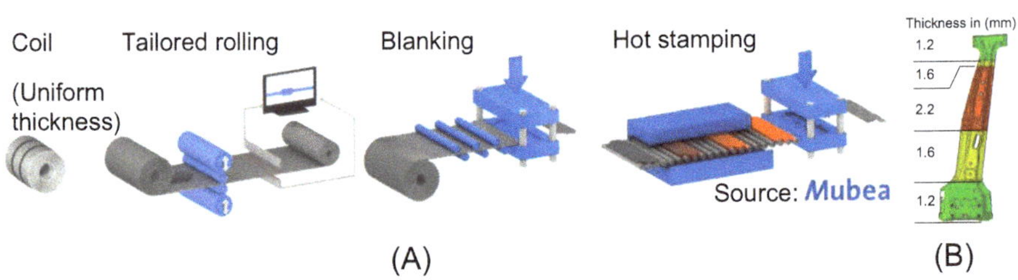

Figure 12.15 (A) Hot forming of Tailor Rolled Blanks, (B) B-pillar of 2nd generation BMW X5 [115].
Source: Courtesy of Mubea TRB GmbH.

To solve these problems, the hot forming process can be tailored so that some areas can be fully hardened (martensite) and some areas can be left soft (ferrite + pearlite or bainite) [67,122].

There are three main methods of obtaining tailored properties [7,12]:

1. By controlling the blank temperature during heating (the austenitized region during the blank heating).
2. By controlling the quenching rate.
3. By partially tempering a fully hardened part.

There are academic studies on tailoring the final properties of a press hardened steel to achieve quenched and partitioned (Q&P) or TRIP aided Bainitic Ferrite (TBF) steels. In Q&P steels, the quenching is stopped before martensite finish temperature (M_f) and held at constant temperature, or heated and held at constant temperature (called Partitioning Temperature) for some time (called Partitioning Time). The final microstructure is composed of carbon depleted martensite and carbon enriched austenite [123]. Liu et al. [124] have studied 22MnB5 hot stamping process. However, quenching was halted at 300°C and the blank was kept at this constant temperature for 10 seconds before subsequent quenching. The final microstructure was found to have retained austenite [124]. Seo et al. [125] have designed two new alloys for hot forming (HF) + Q&P process. These alloys with proper HF + Q&P process had at least 9.6% total elongation (at a level of 1396 MPa yield strength) and a minimum of 1098 MPa yield strength (with 17.1% total elongation) [125].

Chang et al. [129] have designed three different alloys (with higher C and Mn than 22MnB5). They interrupted the quenching stage before martensite start and reduced the cooling rate to achieve TBF properties in the final part [129]. In TBF steels, there are retained austenite, martenisite, and carbide free bainitic ferrite [56]. In the study by Chang et al., in some cases, carbides were formed. By using several simulated hot forming and interrupted quenching cycles, they were able to make steels of up to 10.6% total elongation with 1053 MPa yield strength. The strongest steel had 1458 MPa yield strength and 5.1% total elongation [129].

12.6 Summary and future outlook

Hot formed steels are formed at high temperature in soft form and quenched directly in the press die to have ultra high strength. In this way, complex shapes of automotive components are made possible with very negligible springback/distortion. These steels have been used significantly in the automotive industry and usage is expected to increase in the future. As of 2016, most of the industry is using 22MnB5 with approximately 1000 MPa yield strength and 1500 MPa tensile strength, with typically 5–7% total elongation. There are significant studies to improve strength, elongation, or both. To improve intrusion resistance and/or energy absorbing characteristics, researchers are working on changing the steel chemistry and/or the heat treatment cycle.

References

[1] H. So, D. Faßmann, H. Hoffmann, R. Golle, M. Schaper, An investigation of the blanking process of the quenchable boron alloyed steel 22MnB5 before and after hot stamping process, J. Mater. Process. Technol. 212 (2012) 437–449.

[2] J.N. Hall, Evolution if advanced high strength steels in automotive applications, in: Great Designs in Steel, May 18th, Livonia, MI, USA, 2011.

[3] BaoShan Iron and Steel Co., Ltd., Automotive Advanced High Strength Steels, Catalogue, 2013.

[4] Posco Technical Research Lab, 2016 Automotive Steel Data Book, 2016.

[5] B.C. De Cooman, O. Kwon, K.-G. Chin, State-of-the-knowledge on TWIP steel, Mater. Sci. Technol. 28 (5) (2012) 513–527.

[6] W.D. Callister, Fundamentals of Materials Science and Engineering, fifth ed., John Wiley & Sons, New York, NY, 2001, ISBN 0-471-39551-X.

[7] E. Billur, C. Wang, C. Bloor, M. Holecek, H. Porzner, T. Altan, Advancements in tailored hot stamping simulations: cooling channel and distortion analyses, Proceedings of Numisheet 2014, AIP Conf. Proc, vol. 1567, 2014, pp. 1079–1084.

[8] H. Engels, O. Schalmin, C. Müller-Bollenhagen, Controlling and monitoring of the hot-stamping process of boron-alloyed heat-treated steels, in: Proceedings of the International Conference on New Development in Sheet Metal Forming Technology, Stuttgart, Germany, 2006, pp. 135–150.

[9] H. Steinbeiss, H. So, T. Michelitsch, H. Hoffmann, Method for optimizing the cooling design of hot stamping tools, Prod. Eng. Res. Dev. 1 (2) (2007) 149–155.

[10] M. Naderi, Hot Stamping of Ultra High Strength Steels, Dissertation, RWTH Aachen, Aachen, Germany, 2007.

[11] A. Blaise, B. Bourouga, C. Dessain, "Influence of the duration of transfer from the furnace to the press during the hot stamping process", ASME J. Thermal Sci. Eng. Appl. 7 (2015) 1–10.

[12] M. Merklein, M. Wieland, M. Lechner, S. Bruschi, A. Ghiotti, Hot stamping of boron steel sheets with tailored properties: a review, J. Mater. Process. Technol. 228 (2016) 11–24.

[13] K. Eriksson, Bringing it together, in: AP&T Press Hardening, Next Step Seminar, Novi, MI, September 15th, 2010.

[14] F.J. Ebner, HotPHASE – press hardening automotive solutions by ebner, in: Proceedings of 3rd International Conference on Hot Sheet Metal Forming of High-Performance Steel, Kassel, Germany, July 13–17, 2011, pp. 247–253.

[15] K.S. Jhajj, Heat Transfer Modeling of Roller Hearth and Muffle Furnace, Master of Applied Science Thesis, University of Waterloo, Waterloo, ON, Canada, 2015.

[16] A. Mikuni, T. Sagisaka, T. Shiga, The New Lexus IS, in: EuroCarBody 2013, September 25–26, Bad Nauheim, Germany, 2013.

[17] Toyota Global Newsroom, accessed from <newsroom.toyota.co.jp>.

[18] V. Ploshikin, et al., New heating technology for the furnace-free press hardening process, Tools and Technologies for Processing Ultra High Strength Materials, September 19–21, Graz, Austria (2011).

[19] T. Barthel, R. Müller, D. Landgrebe, Alternative heating – press hardening – high speed impact cutting – energyefficient process chain for UHSS, in: Proceedings of IDDRG 2016, June 12–15, Linz, Austria, 2016, pp. 356–362.

[20] R. Kolleck, Press Hardening – Materials and Coatings, presented at Schuler Hot Stamping Workshop, May 14th, Dearborn, MI, USA, 2013.

[21] P. Siebert, M. Alsmann, H.-J. Watermeier, Influence of different heating technologies on the coating properties of hot-dip aluminized 22MnB5, in: Proceedings of chs2 2011, 2011, pp. 457–464.
[22] J. Karlsson, Press requirements, in: AP&T Press Hardening, Next Step Seminar, Novi, MI, September 15th, 2010.
[23] R. Hund, Continuous improvement of hot forming technology, in: Proceedings of 3rd International Conference on Hot Sheet Metal Forming of High-Performance Steel, Kassel, Germany, July 13–17, 2011, pp. 189–200.
[24] A. Ormaetxea, G. Ibañez, Servomechanical press makes foray into hot stamping, Stamping J. (2015) 18–21.
[25] U. Paar, I. Valls, Werkzeugstähle und Strategie für Warmumformung und Hartbeschneiden (Tool steels and strategy for hot forming and hard trimming – in German), in: Proceedings of 2. Erlangener Workshop Warmblechumformung (2. Erlangen Workshop of Hot Sheet Metal Forming), 2007, pp. 73–92.
[26] Ch. Escher, J.J. Wilzer, Tool steels for hot stamping of high strength automotive body parts, in: Proceedings of 3rd ICSCM, November 2–3, Bochum, Germany, 2013, pp. 219–228.
[27] R. Kolleck, W. Weiß, P. Mikoleizik, Cooling of tools for hot stamping applications, in: Proceedings of IDDRG 2010, May 31–June 02, Graz, Austria, 2010, pp. 111–119.
[28] T. Laumann, I. Picas, M. Grané, D. Casellas, M.D. Riera, I. Valls, Hard cutting of tailored hardened 22MnB5, in: Proceedings of IDDRG 2010, May 31–June 02, Graz, Austria, 2010, pp. 355–362.
[29] J. Friberg, Production process development, in: AP&T Press Hardening, Next Step Seminar, June 4th, Shanghai, China, 2012.
[30] H. Ebner, Verfahren und Vorrichtung zum Preßhärten von Sensen unter gleichzeitiger Herstellung der Höhlung, Richtung und Stellung des Blattes, (Method and apparatus for press hardening of scythes simultaneous preparation of the cavity, the direction and position of the sheet), Swiss Patent Nr. 69512, 1914.
[31] A.B. Shapiro, Using LS-dyna for hot stamping, in: 7th European LS-Dyna Conference, 2009.
[32] A.B. Norrbottens Järnverk, Manufacturing of a Hardened Steel Article, United Kingdom Patent, GB 1,490,535, Issued on November 2nd, 1977.
[33] G. Berglund, Boron Steel History, 2014.
[34] P. Fahlblöm, Analysis of the Choice of Packaging System, a Study Designed to Facilitate the Choice of Packaging System, Thesis, Department of Industrial Logistics, Luleå University of Technology, Luleå, Sweden, 1997.
[35] H. Lanzerath, A. Bach, G. Oberhofer, H. Gese, Failure Prediction of Boron Steels in Crash, SAE Technical Paper, 2007-01-0989, 2007.
[36] B.A. Behrens, Hot stamping, CIRP Encyclopedia of Production Engineering (2014). Available from: http://dx.doi.org/10.1007/978-3-642-35950-7_16722-2.
[37] X. Bano, J.P. Laurent, Heat treated boron steels in the automotive industry, in: 39th Mechanical Working and Steel Processing Conference Proceedings, Volume XXXV, Indianapolis, IN, pp. 673–677, 1998.
[38] P. Marklund, L. Nilsson, S. Rahmn, M. Jonsson, T. Svantesson, L. Hellgren, Optimization of a Press Hardened B-Pillar by Use of the Response Surface Method, SAE Technical Paper 1999-01-3236, 1999.
[39] L. Vaissiere, J.P. Laurent, A. Reinhardt, Development of Pre-Coated Boron Steel for Applications on PSA Peugeot Citroën and RENAULT Bodies in White, SAE Technical Paper, 2002-01-2048, 2002.

[40] J. Staeves, Höherfeste Stähle für die Karosserie (High strength steels for car body – in German), presentation at Technical University of Munich, June 28th, 2004.

[41] G. Plassart,G. Philip, Materials Criteria Selection and Certification Process for the Body in White in PSA PEUGEOT CITROËN, SAE Technical Paper 2002-01-2051, 2002.

[42] D. Holzkamp, Recent developments on UHSS welding and its simulation for prevention of heat distortion, in: Insight Edition Conference, September 20–21, Gothenburg, Sweden, 2011.

[43] P. Weigert, et al, Challenges in mass production of press hardened components focusing CO2 reduction, in: Insight Edition Conference, September 20–21, Gothenburg, Sweden, 2011.

[44] G. Georgiadis, A.E. Tekkaya, P. Weigert, J. Weiher, H. Kurz, Investigations on the manufacturability of thin press hardened steel components, Procedia CIRP 18 (2014) 74–79.

[45] Volkswagen Media Services, accessed from <www.volkswagen-media-services.com>.

[46] H. Ljungqvist, K. Amundsson, O. Lindblad, The all-new Volvo XC90 car body, in: EuroCarBody 2014, October 21–23, Bad Nauheim, Germany, 2014.

[47] S. Bruschi, G. Liu, Hot stamping, Chapter 3.03, ISBN 978-0080965338. Comprehensive Materials Processing, Elsevier, 2014.

[48] American Society of Metals, Welding and brazing, ASM Handbook 06, 2nd Edition, 1993.

[49] American Society of Metals, Handbook Volume 4: Heat Treating, 3rd printing, ASM International, Metals Park, OH, 1995.

[50] H. Mohrbacher, Martensitic automotive steel sheet – fundamentals and metallurgical optimization strategies, Adv. Mater. Res. 1063 (2015) 130–142.

[51] J. Bian, H. Mohrbacher, Novel alloying design for press hardening steels with better crash performance, Proceedings of the International Symposium on New Developments in Advanced High-Strength Sheet Steels, June 23–27, 2013, Vail, Colorado, USA.

[52] M. Maalekian, The Effects of Alloying Elements on Steels, Review Report, Technische Universität Graz, 2007.

[53] H. Aydin, E. Essadiqi, I.-H. Jung, S. Yue, Development of 3rd generation AHSS with medium Mn content alloying compositions, Mater. Sci. Eng. A 564 (2013) 501–508.

[54] Q. Han, et al., Low temperature hot forming of medium-Mn steel, Proceedings of chs2 2015 (2015) 381–390.

[55] W. Alqhadafi, Laser Welding of Boron and Bainitic Steels, Master of Science Thesis, Luleå University of Technology, Luleå, Sweden, 2012.

[56] N. Fonstein, ISBN 978-3-319-19164-5. Advanced High Strength Steels – Physical Metallurgy, Design, Processing, and Properties, Springer, 2015.

[57] M. Naderi, M. Ketabchi, M. Abbasi, W. Bleck, Analysis of microstructure and mechanical properties of different boron and non-boron alloyed steels after being hot stamped, Procedia Eng. 10 (2011) 460–465.

[58] B. Macek, Optimization side crash performance using a hot stamped B-pillar, in: Great Designs in Steel, Livonia, MI, March 8th, 2006.

[59] T. Vietoris, New developments in PHS: materials, coatings, production methods, in: AP&T Press Hardening, Next Step Seminar, Novi, MI, October, 2011.

[60] ArcelorMittal, Extract from the Product Catalogue, Last updated October 6th, 2015.

[61] J. Banik, Warmumformung im Automobilbau: Verfahren, Werkstoffe, Oberflächen, Volume 348 of Die Bibliothek der Technik, 2012.

[62] E. Hilfrich, Trends and potentials of new hotforming steels, in: Insight Edition Conference, September 19th, Neckarsulm, Germany, 2012.

[63] ThyssenKrupp Steel Europe AG, Mangan-Bor-Stähle Für die Warmumformung, Product Catalogue, November, 2014.
[64] Posco Technical Research Lab, 2015 Automotive Steel Data Book, 2015.
[65] K. Hikita, et al., Properties of new TS1800 Mpa grade hot staming steel and components, in: Materials in Car Body Engineering 2012, May 10–11, Bad Nauheim, Germany, 2012.
[66] voestalpine, "phs-ultraform®", Product catalogue, January, 2013.
[67] H. Karbasian, A.E. Tekkaya, A review on hot stamping, J. Mater. Process. Technol. 210 (2010) 2103–2118.
[68] S. Graff, T. Gerber, F.-J. Lenze, S. Sikora, About the simulation of microstructure evolution in the hot sheet stamping process and the correlation of resulting mechanical properties and crash-performance, Proceedings of chs2 2011 (2011) 323–330.
[69] H. Lanzerath, et al., Simulation: enabler for the efficient design of lightweight boron intensive body structures, in: Insight Edition Conference, September 20–21, Gothenburg, Sweden, 2011.
[70] D. Wenk, Global capability: hot stamping, presented at: Global Automotive Lightweight Materials Asia 2014, Shanghai, China, March 26–27, 2014.
[71] J. Overrath, F.-J. Lenze, S. Sikora, Aktuelle Entwicklung der Warmumformung im Automobilen Fahrzeugbau, (Current Developments of Hot Forming in Automobiles – in German), in 30. EFB-Kolloquium, March 2–3, Bad Boll, Germany, 2010.
[72] H.W. Lee, K.-H. Chung, A new body concept for electric vehicle, in: Materials in Car Body Engineering 2012, May 11, Bad Nauheim, Germany, 2012.
[73] M. Garcia, Remote laser welding, in: European Automotive Laser Applications (EALA) 2013, February 19–20, Bad Nauheim, Germany, 2013.
[74] K. Fahlström, Laser welding of boron steels for light-weight vehicle applications, Licentiate Thesis, Högskolan Väst, 2015.
[75] H. Matsuoka, K. Fujihara, Mazda CX-5, in: EuroCarBody 2011, October 18–20, Bad Nauheim, Germany, 2011.
[76] SSAB, Lets talk about Boron Steel, Product Catalogue, V2, 2015.
[77] G. Kim, POSCO's strategy for steel in car body lightweight design, in: Insight Edition Conference, September 19, Neckarsulm, Germany, 2012.
[78] Renault Media Services, accessed from <http://media.renault.com/>.
[79] J.-B. Nam, Development of new auto steels and application technology, presented in: China Automotive Steel Conference, 2013.
[80] H. Ferkel, O. Hoffmann, L. Keßler, Resource gentle light weight construction for today's and oncoming mobility, in: Proceedings of New Developments in Sheet Metal Forming, May 22–23, Stuttgart, Germany, 2012, pp. 1–16.
[81] voestalpine Media Services, accessed from <www.voestalpine.com/group/en/press>.
[82] R. Vierstraete, G. Black, Lasers Make Advances in Auto-body Door Rings, Industrial Laser Solutions, 2014.
[83] voestalpine, phs-ultraform 490 Z140, Data Sheet, February, 2013.
[84] Volvo Car Corporation, Aluminium/silicon-coated steels for hot forming, Standard VCS 5730, 3, Established: December, 2012.
[85] T. Fröhlich, Maximum safety and lightweight potential due to use of new high strength steels, in: Presentation at Outokumpu Experience 2013, May 22–23, London, UK, 2013.
[86] H.L. Yi, P.J. Du, B.G. Wang, A new invention of press-hardened steel achiving 1880 Mpa tensile strength combined with 16% elongation in hot-stamped, in: Proceedings of chs2 2015, 2015, pp. 725–734.

[87] R. Rana, C.H. Carson, J.G. Speer, Hot forming response of medium manganese transformation induced plasticity steels, in: Proceedings of chs2 2015, 2015, pp. 391–400.

[88] G. Frenzer, Nano-X GmbH x-tec® and Alsi® Coat Products against scale formation on steel, 2015.

[89] S. Sepeur, The company nano-X GmbH: products for the automotive industry, Presentation at Deutsche Börse, July 10th, Frankfurt, Germany, 2006.

[90] W.S. Choi, B.C. De Cooman, Characterization of the bendability of press-hardened 22MnB5 steel, Steel Res. Int. 85 (5) (2013) 824–835.

[91] W. Fristad, New coil-applied coating for press-hardening steel, in: Galvatech 2015 Proceedings, 2015, pp. 892–898.

[92] D.W. Fan, B.C. De Cooman, State-of-the-knowledge on coating systems for hot stamped parts, Steel Res. Int. 83 (5) (2012) 412–433.

[93] D. Copeland, M. Pfestorf, The body in white of the new BMW 5 series gran tourismo, in: Great Designs in Steel, Livonia, MI, May 5th, 2010.

[94] T. Kurz, G. Luckeneder, T. Manzenreiter, H. Schwinghammer, Zinc Coated Press-Hardening Steel – Challenges and Solutions, SAE Technical Paper 2015-01-0565, 2015.

[95] A. Reinhardt, Development of hot stamped ultra high strength steel parts on the Peugeot 307 and the Citroën C5, in: presented at EuroCarBody 2001 – 3rd Global Car Body Benchmarking Conference, Bad Nauheim, Germany, 2001.

[96] T. Vietoris, Hot stamping with USIBOR1500®, in: AP&T Press Hardening, Next Step Seminar, Novi, MI, September 15th, 2010.

[97] T. Kurz, New developments in zinc coated steel for press hardening, in: Insight Edition Conference, September 20–21, Gothenburg, Sweden, 2011.

[98] H. Lehmann, Furnaces for press hardening, in: AP&T Press Hardening, Next Step Seminar, Shanghai, China, March, 2010.

[99] E.J. Watkins, Hot stamping market, materials, coatings, and developments, in: Presented at Schuler Hot Stamping Workshop, Dearborn, MI, May 14th, 2013.

[100] S. Fujita, J. Maki, S. Yamanaka, H. Irikawa, M. Kurosak, Corrosion resistance after hot stamping of 22MnB5 steels aluminized with 80g/m^2 c.w. and ZnO coating, Proceedings of chs2 2015, 2015.

[101] R. Aldén, Metallurgical Investigation in Weldability of Aluminium Silicon Coated Boron Steel with Different Coating Thickness, Degree of bachelor thesis, KTH Royal Institute of Technology, Stockholm, Sweden, 2015.

[102] W.C. Verloop, Development of Zn-coated boron steel ZnX at tata steel, in: Insight Edition Conference, September 20–21, Gothenburg, Sweden, 2011.

[103] C.W. Lee, W.S. Choi, L. Cho, Y.R. Cho, B.C. De Cooman, Liquid-metal-induced embrittlement related microcrack propagation on Zn-coated press hardening steel, ISIJ Intl. 55 (1) (2015) 264–271.

[104] P. Larour, H. Pauli, T. Kurz, T. Hebesberger, Influence of post uniform tensile and bending properties on the crash behaviour of AHSS and press-hardening steel grades, Proceedings of IDDRG 2010, 2010.

[105] Z. Ghanbari, Zinc Coated Sheet Steel For Press Hardening, Master of science thesis, Colorado School of Mines, Golden, CO, USA, 2014.

[106] American Society of Metals, Corrosion, ASM Handbook 13, 4th Edition, 1992.

[107] O. Hoffmann, Environment oriented light weight design in steel, in: Presented at Hannover Messe, 2012.

[108] P.J. Belanger, J.N. Hall, J.J. Coryell, J.P. Singh, Automotive body press-hardened steel trends, Proceedings of International Symposium on New Developments in Advanced High-Strength Sheet Steels (2013) 239–250.

[109] Tata Steel, MagiZinc® Auto, Product Catalogue, 2015.
[110] J. Kondratiuk, P. Kuhn, Tribological Investigations on Friction and Wear Behaviour of Coatings for Hot Sheet Metal Forming, Wear 270, 2011, pp. 839–849.
[111] R. Neugebauer, et al., "Press hardening — an innovative and challenging technology", Arch. Civil Mech. Eng. 12 (2) (2012) 113–118.
[112] D.D. Munera, A. Pic, D. Abou-Khalil, F. Shmit, F. Pinard, Innovative Press Hardened Steel Based Laser Welded Blanks Solutions for Weight Savings and Crash Safety Improvements, SAE Technical Paper, 2008-01-1076, 2008.
[113] Altair Engineering, Industry First Hot Stamped Door Ring, Altair Enlighten Award Nominee, 2013.
[114] R. Zum Mallen, J. Riggsby, Development of a global first SUV body construction, in: Great Designs in Steel, Livonia, MI, May 1st, 2013.
[115] M. Pfestorf, J. van Rensburg, Functional properties of high strength steel in body in hite, in: Great Designs in Steel, Livonia, MI, March 8th, 2006.
[116] M. Rehse, Flexible rolling of tailor rolled blanks, in: Great Designs in Steel, Livonia, MI, March 8th, 2006.
[117] T. Ausmann, Hot stamping technologies, tailor rolled blanks, in: AP&T Press Hardening, Next Step Seminar, Novi, MI, September, 2010.
[118] H. Voswinckel, Private Communication, 2015.
[119] S. Morgans, 2013 Ford Escape, in: Great Designs in Steel, Livonia, MI, May 1st, 2013.
[120] R. Perez-Santiago, E. Billur, A. Ademaj, C. Sarmiento, R. Berlanga, T. Altan, Hot stamping of a B-pillar with tailored properties: experiments and preliminary simulation results, in: Proceedings of chs2 2013, 2013, pp. 83–90.
[121] C. Patois, M. Barbier, Body-in-white of the New Peugeot 308, in: Presented at EuroCarBody 2014 – 16th Global Car Body Benchmarking Conference, Bad Nauheim, Germany, 2014.
[122] O. Hedegärd, M. Åslund, Tempering of Hot-Formed Steel Using Induction Heating, Master of Science Thesis, Chalmers University of Technology, Gothenburg, Sweden, 2011.
[123] J.G. Speer, F.C.R. Assunção, D.K. Matlock, D.V. Edmonds, The "quenching and partitioning" process: background and recent progress, Mater. Res. 8 (4) (2005) 417–423.
[124] H. Liu, X. Jin, H. Dong, J. Shi, Martensitic microstructural transformations from the hot stamping, quenching and partitioning process, Mater. Character. 62 (2011) 223–227.
[125] E.J. Seo, L. Cho, B.C. De Cooman, Application of quenching and partitioning (Q&P) processing to press hardening steel, Metall. Mater. Trans. A 45A (2014) 4022–4037.
[126] J. Jonasson, E. Billur, A Hot Stamping Line, Chapter 5 in "Hot Stamping of Ultra High Strength Steels – An Overview of Technology and Business", Springer, in preparation.
[127] L. Etude, Press hardening seminar, Hot Forming Day, October 29th, St. Chamond, France.
[128] B.F. Rolfe, et al., Advances in tailored hot stamping - innovations in material and local patchwork topology, Proceedings of the 2nd International Conference of Advanced High Strength Steels and Press Hardening (ICHSU) (2015) 413–418.
[129] Y. Chang, G. Li, C. Wang, X. Li, H. Dong, "Effect of quenching and partitioning with hot stamping on martensite transformation and mechanical properties of AHSS, JMEPEG 24 (2015) 3194–3200. Available from: http://dx.doi.org/10.1007/s11665-015-1579-x.

Forging Grade Steels for Automotives*

13

O.N. Mohanty
RSB Group, Pune, India

13.1 Introduction

In passenger cars, trucks, and tractors (for agricultural applications), forged components are commonly used wherever one encounters load carrying and shock enduring points. Cars and trucks may contain more than 250 forgings, most of which are produced from carbon or alloy steel. Forged engine and powertrain components include connecting rods, crankshafts, transmission shafts and gears, differential gears, drive shafts, clutch hubs, and universal joint yokes and crosses. Forged components such as camshafts, pinions, gears, and rocker arms can develop a range of properties based on a variety of microstructures acquired through heat treatment. Wheel spindles, kingpins, axle beams and shafts, torsion bars, ball studs, idler arms, pitman arms, steering arms, and linkages for passenger cars, buses, and trucks exemplify applications requiring extreme conditions of strength and toughness.

Farm implements, in addition to engine and transmission components, utilize key forgings ranging from gears, shafts, levers, and spindles to tie-rod ends, spike harrow teeth, and cultivator shanks. Heavy tanks contain more than 550 separate forgings; armored personnel carriers employ more than 250.

Steel forgings are regularly specified where strength, resistance to shock and fatigue, reliability, and economy are vital considerations. Forged materials also offer the desired degree of high or low temperature performance, ductility, hardness, and machinability. Advances in forging technology have expanded the range of shapes, sizes, and properties available in forged products to meet an increasing variety of design and performance requirements.

Broadly, the steel forgings go through: (a) hot forging, (b) warm forging, or (c) cold forging. These are briefly described as follows:

a. *Hot forging of steel*: The forging temperatures are above the recrystallization temperature, and are typically between 950°C–1250°C. Usually, one experiences good formability (i.e., filling of die-cavity in the context of forging), low forming forces, and an almost uniform tensile strength of the work-piece.

* Every effort has been made to trace copyright holders and to obtain their permission for the use of copyright material. The publisher apologizes for any errors or omissions in the acknowledgements printed in this book and would be grateful if notified of any corrections that should be incorporated in future reprints or editions.

Automotive Steels. DOI: http://dx.doi.org/10.1016/B978-0-08-100638-2.00013-4

b. *Warm forging of steel*: Forging temperatures are still above the recrystallization temperature: between 750°C and 950°C. The scale-loss is lower at the work-piece surface and the tolerance is narrower compared to hot forging. One experiences limited formability and higher forming forces than in hot forging, but lower forming forces than in cold forming.

c. *Cold forging of steel*: Forging temperatures are around room conditions, adiabatic self-heating might bring the temperature up to 150°C. One experiences narrowest tolerances achievable and no scaling at work-piece surface. Further, an increase in strength and drop in ductility due to strain hardening might take place. The formability is rather low, and high forming forces are necessary.

In terms of the volume of industrial forgings, hot forging is the preferred process since a wide range of steels and components are amenable to this route. This chapter would therefore focus primarily on the hot forging of steels. Again, in the broad area of hot forging, the present chapter will be concentrating on closed-die forging (rather than on drop forging) which is used for producing finished products with close dimensional tolerance.

While looking at hot forging the important aspect that comes to notice is that the bar is reheated, typically in an induction furnace, to a temperature of ~1200°C, and then taken through the forging press. At the press, it may go through a multistage forging before being subjected to trimming. Thereafter, the formed component may be air-cooled or heat treated depending upon the target properties.

The majority of hot-forged steel forgings are made using plain carbon or low alloy steels with a carbon content selected to yield an acceptable combination of strength, toughness, and forgeability. High strength forgings conventionally contain carbon levels of about 0.2–0.5 weight percent, which allows forgings to be heat treated to the required strength following the forging operation. The heat treatment, primarily quenching and tempering ($Q + T$), consumes considerable energy (and hence is expensive) and adversely affects productivity. Additionally, quenching also induces the risk of high tensile residual stresses, distortion, and at times, cracks in the components. A straightening operation followed by stress relieving annealing is therefore required to minimize the tensile residual stresses. These operations obviously add to the total cost of processing. In many plants, the straightening operation is not followed by stress-relief annealing and therefore can lead to a lowering in fatigue life due to the presence of tensile residual stress. Consequently, there have been attempts to develop grades of steel (primarily, microalloyed, MA) that are free from some of these maladies. The schematic of both $Q + T$ treated and air-cooled MA forgings are shown in Fig. 13.1.

In the early 1970s, the use of medium carbon microalloyed steels for air-cooled hot forgings started in Europe, as well as the USA, in order to avoid quenching and tempering heat treatment and the concurrent problems. The strength requirement was met primarily through precipitation in the ferrite–pearlite matrix. This microstructure, however, leads to lowering of toughness and these steels suffer from poor weldability; therefore the use of such forgings was restricted to less critical components. A good account of these developments has been given by Hulka et al. [2].

The microalloyed, low carbon bainitic structures in the air-cooled conditions were found to yield an optimum combination of strength and toughness; new

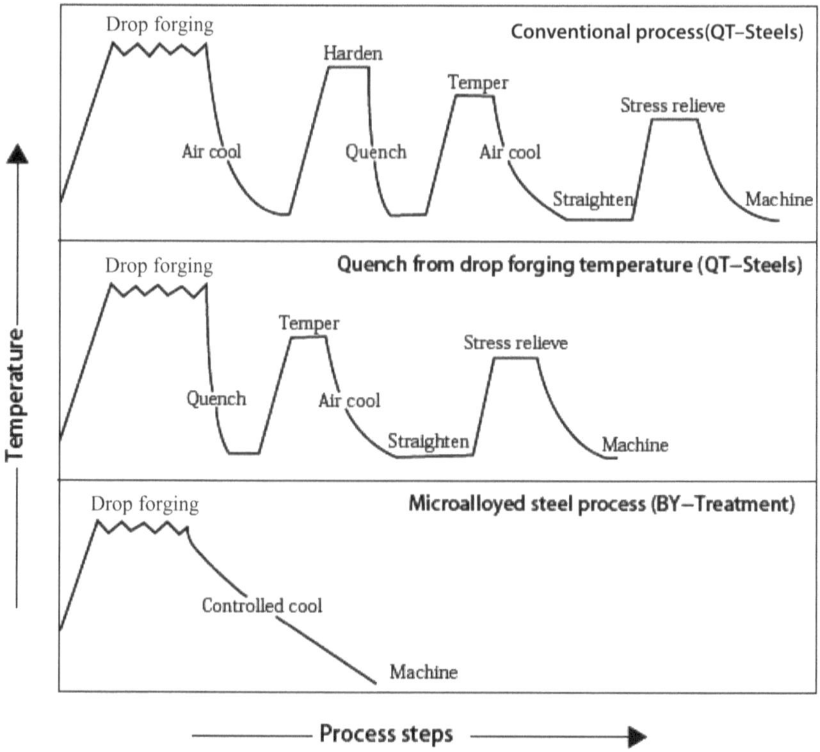

Figure 13.1 Comparison of processes involved in conventional quenched-tempered steels and in directly air-cooled MA-steels [1].

developments in this area have taken place in the last 15–20 years [e.g., 3]. In Japan for example, precipitation hardening through copper in the bainitic matrix has led to new air-cooled hot forgings, with high strength and reasonable toughness [4]. In the USA, there are patents to show the development of microalloyed medium carbon steels that can be used in forgings without heat treatment [5].

One other challenge has been bringing in a reasonable uniformity in strength properties at various depths of a large forging that would cool at varying rates. Several approaches have been employed; one attractive strategy is to use a chemistry that makes the resulting strength properties independent of cooling rates. This is basically possible only with very low carbon steels [4].

Further, fatigue resistance of the auto components is an important requirement. In general, the endurance limit (a common measure of fatigue resistance) is ~0.4–0.5 of UTS in most steels [e.g., 1,6].

As one can see from the data in Fig. 13.2, at the same tensile strength level, control rolled microalloyed (MA) steels demonstrate higher fatigue resistance.

More sophisticated approaches for developing a new generation of fatigue-resistant steels for automotives are also available [7].

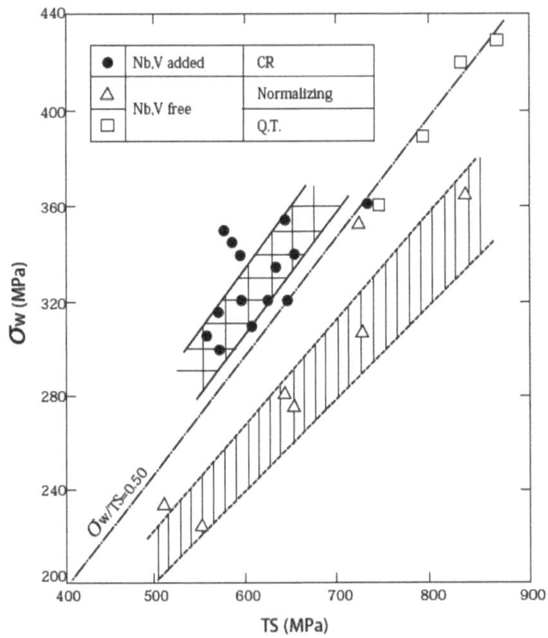

Figure 13.2 Inter-relation between endurance limit (σw) and tensile strength (TS) for microalloyed steels and carbon steels [1].

Another property that was not considered as critical earlier in the forging industry, but is increasingly being taken into consideration in modern designs, is weldability. Broadly, a simple measure of weldability is the carbon content in conjunction with the carbon-equivalent [C.E. = C + Mn/6 + (Ni + Cu)/15 + (Cr + Mo + V)/5]. Together, they show the under-bead (cold) cracking tendency after welding, as represented typically in a Graville diagram [8]. A carbon content below 0.1 wt.% is found to be safe and easy to weld. On the other hand, a combination of C > 0.15% with C.E. > 0.6 is considered difficult to weld, is vulnerable to cracking, and requires preweld and/or postweld treatments.

It is known that bars of the common forging grades of steel (e.g., 37 C 15; 40 Cr3 B and so on), when heated to ~1200°C, would show large austenite grain size and that the transformation products from such austenite would show low ductility. One of the challenges in forging therefore is to restrict the austenite grain size during forging. This has been attempted through microalloying and also by thermomechanical processing.

From the foregoing it would be evident that the steels for automotive forging ought to be able to combine strength, toughness, fatigue resistance and, in many instances, high weldability. With better road conditions, the demand for higher speed of automotive is constantly growing and with enhanced speed, the torque for transmission also increases, putting greater demand on many of the above properties. Additionally, with the concern for a reduced carbon footprint, the weight needs to be reduced without sacrificing any of the properties mentioned. Overall, the selection of steels, their forging process, and the posttreatment associated with automotive components, all play important roles in meeting the ever-growing challenges. This chapter essentially deals with microalloyed steels for automotive forging.

13.2 Basic physical metallurgy relevant to hot forging

As has been shown in Fig. 13.1, the steel bar meant for forging is heated to a high temperature of $\sim 1200°C$, typically in an induction furnace, then forged over a temperature range (down to $\sim 1050°C$, in a timeframe of $\sim 15-30$ seconds, depending upon the size and complexity of forging), trimmed to remove the flash (around a temperature of 950°C) and finally brought to use, either via the quench and temper route or through (a variety of) air-cooling routes. Consequently, the phase transformations involved during heating would include: conversion of the initial microstructure (usually ferrite and pearlite along with some precipitates, when microalloyed) into austenite; precipitate coarsening and dissolution; enrichment of austenite matrix by the (microalloying) solutes etc. Thereafter, the forging process would be associated with recrystallization and grain growth of austenite. Similarly, the cooling process after the forging would induce various phase transformations in the parent austenite and re-appearance of precipitates both in austenite and in ferrite. The basic principles governing some of these processes will be dealt with here.

13.2.1 Phase transformations associated with forging

All transformations that austenite might undergo have been captured in a succinct manner by Bhadeshia [9] and are shown in Fig. 13.3 below. Depending on the mechanism, the transformation products have been classified broadly into two categories: Reconstructive and Displacive.

Reconstructive are those that involve long-range diffusion, and are hence slow at low temperatures. Displacive transformations, on the other hand, are those where the change in crystal structure is achieved by a macroscopically homogeneous deformation, and the transformation is rather fast. Many of these are encountered in the forging process, depending on the cooling rates involved.

The formation of pearlite (and many others as shown) requires the diffusion of all elements including iron. The movement of iron or any substitutional solute occurs through a vacancy mechanism and hence can be slow at low temperatures. As mentioned earlier, prior to forging, the steel bar is heated into austenite region. It is important now to distinguish between the nature of the transformation taking place during heating and during cooling.

The transformation involving diffusion during heating is characterized by a monotonically increasing rate with temperature (since both the driving force and the diffusivity increase with temperature).

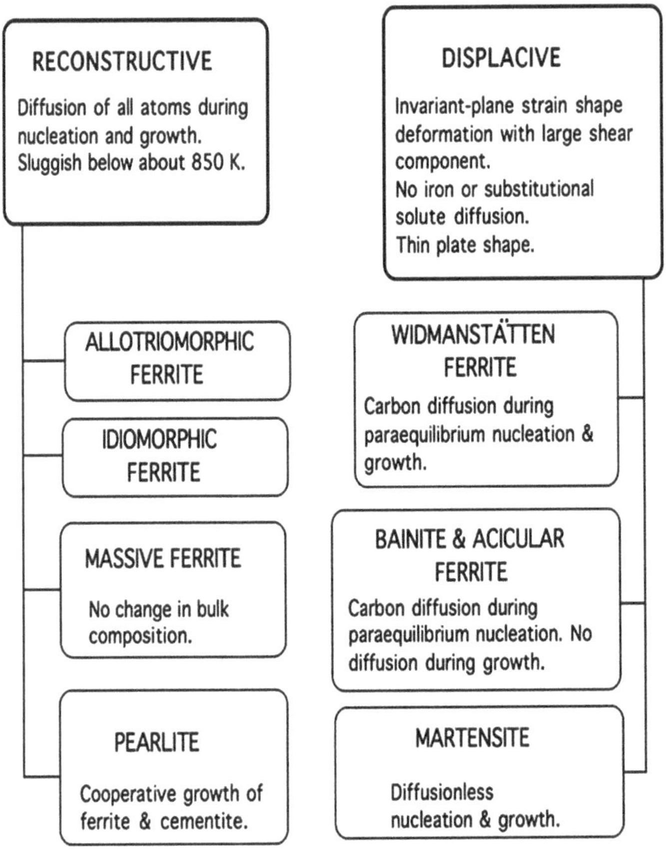

Figure 13.3 Classification of phase transformations in steels as a function of atomic mechanisms [9].

On the other hand, the transformations of austenite during cooling follow a C-curve kinetics since diffusion coefficients decrease but the magnitude of the driving force (ΔG) increases with fall in temperature. Thus, the overall rate of transformation goes through a maximum as a function of the undercooling below the equilibrium temperature. A schematic representation of the kinetics of transformation during heating/cooling is depicted in Fig. 13.4. Here, the Ae_1 and Ae_3 represent the temperatures at which austenite formation starts and completes respectively, under equilibrium heating conditions.

Over the past two decades, the field of predicting the C-curve for transformations (for isothermal as well as for continuous cooling conditions) of austenite using thermodynamic models has expanded appreciably [e.g., 11–15]. Currently, the Java-based Material Properties, JMATPRO has become a common commercial software that can provide information on evolution of phases (shown in Fig. 13.5 for example) from austenite at various temperatures and cooling rates, mechanical

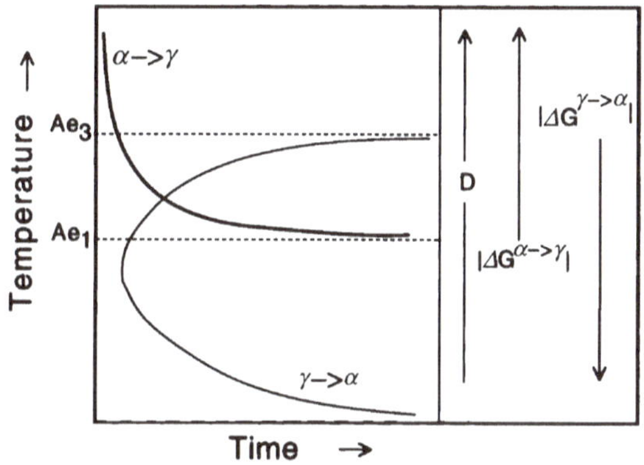

Figure 13.4 The TTT (time, temperature, transformation) curves for the $\gamma \rightarrow \alpha$ transformation and for the $\alpha \rightarrow \gamma$ reverse transformation [10].

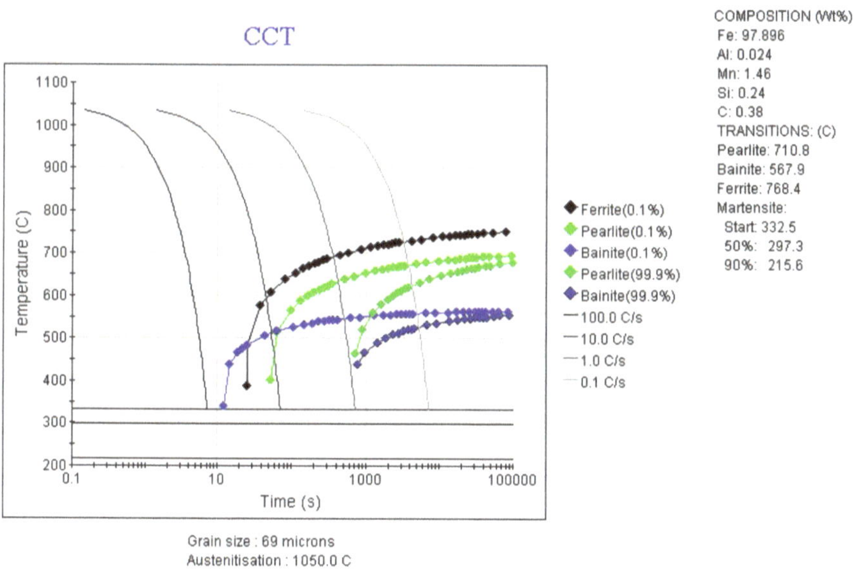

Figure 13.5 CCT-diagram showing the start/finish of pearlite/bainite transformations predicted using JMATPRO, for a typical 37 C 15 grade forging steel with superimposed cooling rates [16] (austenitizing temp. & grain size of austenite given at the bottom; composition and color codes at the side)

properties (primarily strength) of the steel upon transformation, thermo-physical properties, and so on.

Predicted CCT diagrams, such as the one in Fig. 13.5, were profitably used by the author in deciding on a much lower start-temperature for direct quenching following forging (without reheat to austenite, thus saving energy). In the process,

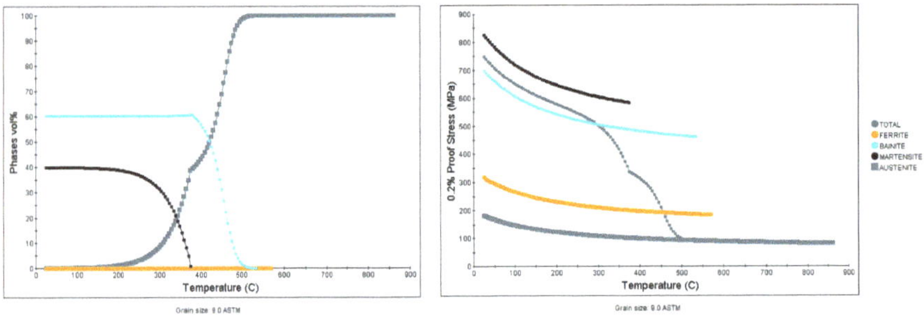

Figure 13.6 (A) The evolution of microstructure of a 3310 steel cooled from 900°C to various temperatures at a constant cooling rate of 10°K/s. (B) The corresponding proof stress data for each phase and also for the entire material as a function of temperature [15].

the risk of inducing quench-cracks is practically eliminated and yet the efficacy of quenching in terms of near-complete martensite formation is not compromised.

For calculation of yield or proof stress of steels, the standard Hall–Petch equation and solid solution hardening are considered. Phase transformation data are used to calculate the strength of the material as a function of microstructure, grain size, and cooling rate. One such example is given in Fig. 13.6 [15].

13.2.2 Stability of precipitates

The microalloying constituents in steel, as has been broadly mentioned above, can exert significant influence on the phase transformation taking place during heating and cooling, on the size of austenite grains/ferrite-grains, the occurrence of precipitates, and so on. Most of these effects are, again, inter-related. In microalloyed steels, precipitates, a result of interaction between solute atoms in iron, may exist in austenite and/or in ferrite phases. Their basic characteristics have been long studied [e.g., 17,18]. The stability of precipitates has been of prime concern and is the subject of investigation both from thermodynamic and kinetic standpoints.

i. *Thermodynamic stability: concept of solubility product and its application*

For a microalloying constituent M, reacting with carbon/nitrogen X (both in solution in austenite/ferrite), forming a precipitate of carbide/nitride/carbo-nitride etc. represented as MX, one can write a reaction as:

$$[M] + [X] = [MX] \ldots. \quad (13.1)$$

The equilibrium constant ($k_{s)}$ for the reaction (i) in terms of the activities of reactants and product:

$$k_s = a[MX]/a[M] \times a[X]) \ldots. \quad (13.2)$$

here,

$$a[MX] = 1; \;\; a[M] \cong [M] \times a[X] \cong [X]$$

where, $[M]$ and $[X]$ are mole fractions of concerned species in solution in austenite/ferrite matrix;

$$k_s = 1/[M][X] \ldots \ \ldots \tag{13.3}$$

Now, free energy change for the reaction:

$$\Delta F = \Delta H - T\Delta S = -RT\ln k_s \ \ldots . \tag{13.4}$$

$$\ln[M][X] = -\ln ks = -\Delta S/R + \Delta H/RT \ldots . \tag{13.5}$$

By convention the following form is used:

$$\mathbf{log}\boldsymbol{k_s} = \mathbf{log}[\boldsymbol{M}][\boldsymbol{X}] = \boldsymbol{A} - \boldsymbol{B}/\boldsymbol{T} \ldots . \tag{13.6}$$

The term $\boldsymbol{k_s}$ is known as *solubility product*. The values of A, B, the thermodynamic parameters, can be found in the literature for several precipitates [17–26] so that the solubility products at various temperatures can be evaluated.

For the NbC precipitate in austenite, the constants A & B reported are: 3.11 and 7530 respectively [19]; while they are 3.42 and 7900 in another reference [20]. One could still draw important conclusions about the thermodynamic stability of microalloy precipitates based on solubility products, notwithstanding these small variations.

It is to be noted that as the value of solubility product increases, the extent of solubility of the constituent elements of the precipitate in the matrix (here, austenite) also goes up; which is to say that the stability of the precipitate decreases. Thus, as shown in Fig. 13.7, TiN is the most stable and VC precipitate is the most easily dissolved (least stable), while NbC has intermediate solubility, and stability among the common precipitates used in industry, such as: AlN, NbC, NbN, VN, TiC, and TiN etc. Studies show the presence also of some nonstoichiometric species such as $Nb\ C_{0.83}\ N_{0.14}$ [24,25].

Indeed, DeArdo provides [25] a comprehensive list of techniques adopted by authors for the measurement of solubility products for NbC, NbN, and NbCN and demonstrates the variation in their values.

Further, mutual solubility between precipitate species in steels containing more than one microalloying element has also been reported [18,25,26].

The application of solubility product for NbC, is shown in Fig. 13.8 as example. The solubility product isotherm at 1100°c is based on the equation by Narita [20].

$$\log[Nb][C] = 3.42 - 7900/T(^\circ\mathrm{K}) \ldots . \tag{13.7}$$

Superimposed is the slope representing the stoichiometric composition of NbC. One may observe that the stability of NbC precipitate increases with carbon content in austenite. It is also to be expected from Eq. (13.7) that as the temperature is raised, more precipitates would be taken into solution (i.e., the isotherm is now raised) and one needs a higher amount of Nb in order for undissolved precipitates of NbC to be available in the matrix of austenite.

The solubility product of microalloy precipitates varies with temperature differently for austenite and for ferrite. It is useful to know in designing alloys if a precipitate is likely to dissolve or remain in solution as one moves from ferrite to austenite during heating. This is depicted in Fig. 13.9, for a set of carbides and nitrides.

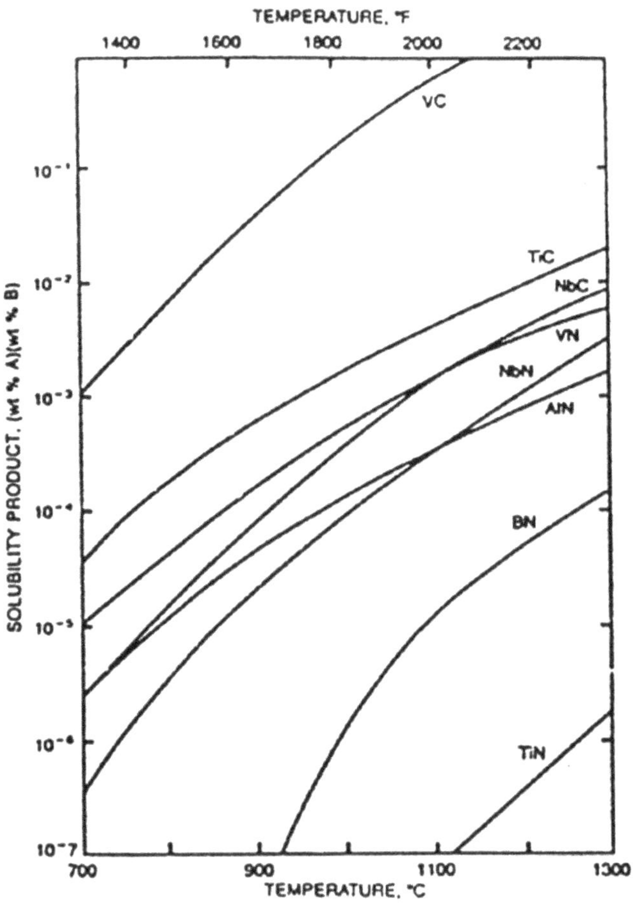

Figure 13.7 Solubility products of selected carbides and nitrides in austenite [23].

From the figure, one observes the following: all precipitates are more soluble in austenite than in ferrite; TiN has high stability even in austenite; precipitates which are soluble in austenite would be available in principle for precipitation in ferrite and thus can contribute to strengthening.

According to Fig. 13.10, at 1100°C for example, up to 0.1% V would go into solution in austenite, at 200 ppm nitrogen. Thus, the whole amount of V and nitrogen present can precipitate out in ferrite during cooling, providing precipitation hardening. By the same argument, there would be no VN precipitate available to restrict austenite grain growth at this temperature. With Nb, the picture reverses: even at 1200°C, in a 0.3% C steel, roughly 0.02% Nb would have gone into solution. If the steel contains 0.04% Nb (which is typical), one is still left with sufficient NbC precipitate; and during rapid cooling very little precipitation would take place in ferrite.

From the foregoing, it would be clear that the concept of solubility product is quite useful in designing the composition of alloys where one could predict if the precipitates form or dissolve in austenite/ferrite at temperatures of interest.

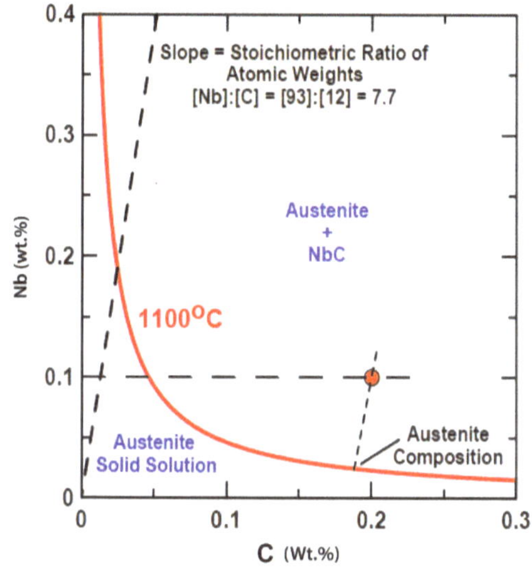

Figure 13.8 Solubility product of NbC in austenite at 1100°C [27].

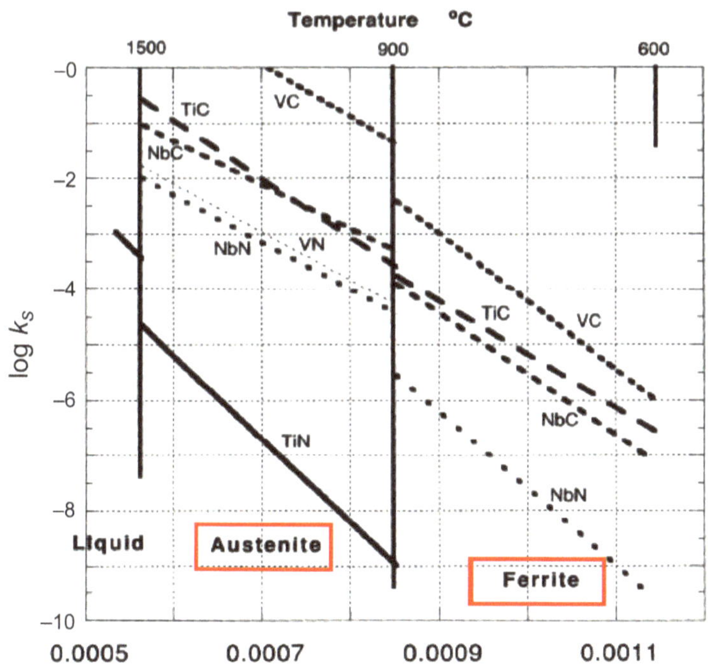

Figure 13.9 Solubility product of precipitates in austenite and in ferrite [27].

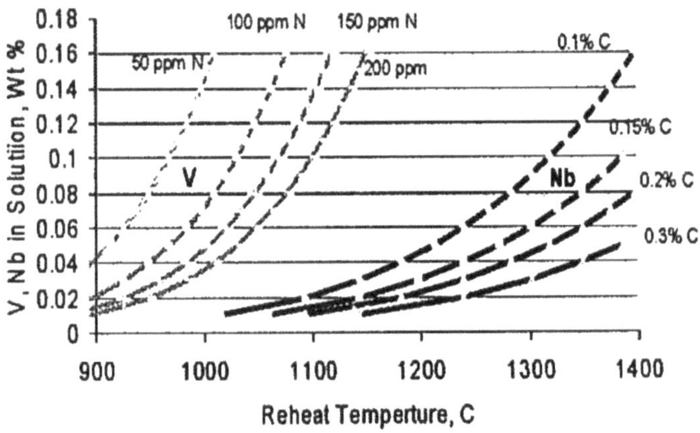

Figure 13.10 Effect of reheat temperature on maximum solubility of VN/NbC in Austenite [28].

ii. *Kinetic aspect of precipitate stability and dissolution*

As shown earlier, the thermodynamic parameter (i.e., solubility product) is conventionally invoked in order to judge whether a specific precipitate would or would not dissolve in austenite as the temperature is raised. Lee et al, on the other hand, have recently explored the kinetic aspects of dissolution of NbC precipitates [29].

Although the work essentially examines the kinetics of dissolution of NbC precipitates during slab reheating, it is extendable to the case of forging as well. In both cases, the temperature range is comparable and there is a size distribution of NbC particles that are present prior to subjecting the material to heating to austenite region. The steel composition chosen was 0.10C-0.89Mn-0.20Si-0.018Nb wt.%, Lee et al calculated (a) number density, (b) phase fraction, and (c) equilibrium phase fractions of NbC precipitates (using Thermo Calc with TCFE7 database) before the heating process started. The kinetics of NbC dissolution during heating was simulated with the Mat Calc software (ver.5.52). The dissolution behavior of NbC in slab during reheating process was investigated both experimentally and using simulations that have, as an input, the experimentally measured initial NbC precipitate size distribution.

The average radius and the fraction of NbC precipitates predicted as a function of temperature, for various heating rates are shown in Fig. 13.11 (A) and (B) respectively. These were also validated by experiments [29].

The kinetic calculation of the work reliably predicted the dissolution kinetics, which was confirmed by microstructural observations, using interrupted quenching. One observes from Fig. 13.11(A) that at a very slow rate (i.e., 1.3×10^{-3}°C/s) there is excessive growth of the surviving NbC precipitate. From Fig. 13.11(B) it may be noticed that at 1.3°C/s heating rate, the dissolution temperature could be raised by close to 100°C. The work indicates that kinetics could limit the amount of NbC that dissolves to values less than those expected from equilibrium (primarily, solubility product) during heating, and these considerations could play a critical role in the size and amount of NbC precipitate that in turn would affect the characteristics of austenite.

Extending the arguments to the case of the dissolution of NbC precipitates during forging, it would appear that since a range of precipitate sizes exists in the starting bar, and the

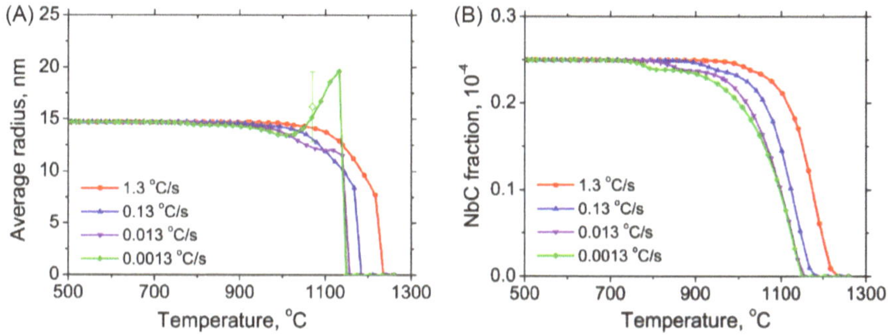

Figure 13.11 Average radius (A) and fraction (B) of NbC precipitates [29].

rate of heating in induction furnace is much faster than those adopted by Lee et al, the entire picture of dissolution of NbC in microalloyed forgings needs to be reconsidered. It cannot be arrived at from the solubility product information alone.

13.2.3 Precipitates, solute atoms and grains: mechanism of their interaction and consequence

During the heating of a bar, well into the austenite region prior to forging, the initial microstructure gets transformed into austenite that may grow in size with time and temperature. Subsequently, during forging, which is primarily a hot working process, static and dynamic recrystallization takes place when a high-energy deformed structure is replaced by a strain-free low energy structure, by the migration of a high angle grain boundary [30]. The recrystallized austenite grain would also grow under favorable conditions. This prior austenite grain size largely controls the resultant final grain size of ferrite, as well as the size of the transformation products. A fine grain size of ferrite and other microstructural constituents is associated with a better combination of strength and toughness. The microalloyed precipitates are known to play a key role in controlling the prior austenite grain size as well as final ferritic grain size [e.g., 17,18].

i. *Mechanism of controlling grain size of austenite and ferrite*

Two effective mechanisms have been advanced explaining the slowing down of the grain boundary movement: Zener pinning by precipitates [17,18,31,32] and drag by solute(s) in solution, known as solute drag [33,34].

The simple model on which the Zener pinning mechanism is based essentially describes a situation in which a precipitate lies on the plane of a grain boundary. If such a grain were to grow, the missing segment of the grain boundary needs to be created which, however, requires extra energy. In other words, there is a threshold force (Zener pressure) that opposes the motion of the boundary and has to be overcome for the grain boundary movement.

The solute drag effect, on the other hand, reduces the mobility of the boundary in a nonlinear manner; it depends on velocity and a number of parameters such as the binding energy of solute to the grain boundary (E_b), the diffusivity of solute across the boundary (D_{trans}), and the mobility of the boundary in the absence of solute (M_o).

An in-depth analysis for the case of Nb microalloyed steel has been made by Hutchinson et al [35] to examine the relative effectiveness of solute Nb and NbC precipitate in restricting the grain size in austenite as well as in ferrite. Only the broad findings are discussed here.

Hutchinson et al have developed a complex equation, (number 10 in their paper), the sign of which decides which is more effective in impeding the grain boundary: the precipitate or the solute. Again, the sign of the equation is decided by the sign of a parameter 'Ω in their equation number 10. If '$\Omega < 0$ (negative), then precipitates are more effective; and if '$\Omega > 0$ (positive), then solute in solution is more effective.

The parameter 'Ω is evaluated by Hutchinson et al. as a function of temperature, for a range of mean precipitate sizes (R), in both ferrite and austenite. A bulk Nb content of 0.05 at% (0.1 wt.%) in steel and a driving force (DF) for recrystallization of 0.5 MPa are assumed. In both ferrite and austenite, a grain boundary energy, Υ, of 0.5 J m^{-2} is used. All other parameters necessary for the solute drag calculation for ferrite grains are taken from the work of Sinclair et al. [36] while those for austenite are taken from previous works of Zurob et al. [37,38].

Fig. 13.12(A) shows that solute Nb is more effective than NbC (of mean radii of precipitate greater than 1.5 nm) at retarding recrystallization in ferrite. Whilst the precipitation of NbC in ferrite is desirable for increasing the yield strength, Nb in solid solution would be more effective at retarding recrystallization, at least as far as growth is concerned. In fact, the precipitation of NbC in ferrite could even lead to faster recrystallization kinetics due to a decrease of solute drag. To utilize the Nb-solute effect in ferrite, it is commented [35] that ultra-low carbon matrix would be preferred for thermomechanical treatment where Nb exists more in solution, and hence would be effective in restricting the ferrite grain size.

In the case of austenite, Fig. 13.12(B), the comparative retarding effects of NbC and solute Nb are strongly influenced by temperature and the precipitate size (R). The precipitation of fine NbC ($R = 1-5$ nm) is more effective than solute drag at retarding grain growth over most of the traditional hot working temperatures.

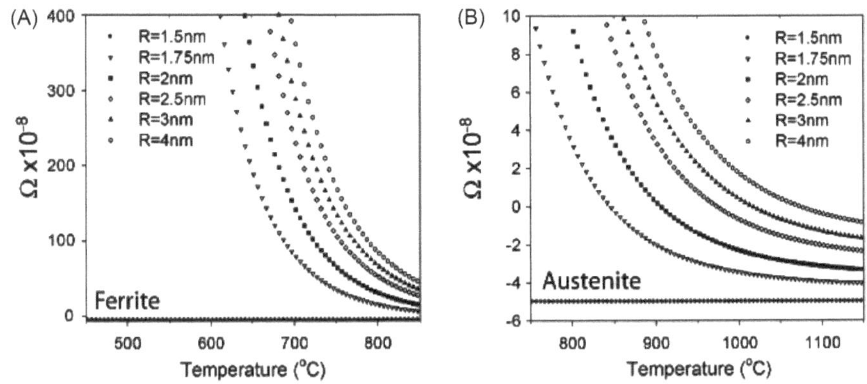

Figure 13.12 Interaction parameter 'Ω of Hutchinson et al. as a function of temperature and mean precipitate size (R) for ferrite (A) and for austenite (B) [35].

The above analysis provides a comparison between the Nb-solute and NbC with regard to their grain growth restriction potential and does not relate to the actual possibility of their existence at a given temperature. Further, it is also to be noted that the treatment here relies on a number of parameters that have uncertainties. All these difficulties notwithstanding, it must be appreciated that the work provides an insight that would help design microalloyed steels and their processing routes with greater confidence.

ii. *Normal and abnormal grain growth—basics*

Zener (31,32) arrived at the following simple expression for the equilibrium grain size (Deq.) from the basic pinning effect:

$$\text{Deq.} = 4R/3F \quad \ldots \tag{13.7}$$

where, $R \rightarrow$ precipitate size; $F \rightarrow$ volume fraction of precipitates

The equilibrium grain size decreases as the precipitate size decreases and its volume fraction increases.

Gladman [39] however pointed out that Zener's assumption of an isolated spherical grain overestimates the driving force for grain growth. In a polycrystalline material, as one grain shrinks, the grain boundary area of others must increase. On the assumption of space filled by tetrakaihedra, there would be a net reduction of grain boundary area if the ratio of the radius of the growing grains to that of the neighbors is greater than 1.33. He therefore postulated that heterogeneity in grain size is essential for grain growth and such heterogeneity is available in commercial alloys. Gladman arrived at the maximum precipitate size (Rm) for restricting grain growth as:

$$\mathbf{Rm} = 6\mathbf{Do}F/\pi(3/2-2/\mathbf{Z})^{-1} \ldots. \tag{13.8}$$

where, Do $\rightarrow$ size of matrix grains; $z \rightarrow$ ratio of the radii of the growing grains and matrix grains.

The Eq. (13.8) by Gladman [39] indicates that complete dissolution of pinning precipitates is not a precondition for grains to grow; the growth of some grains would be allowed once the pinning precipitates reach a critical size. Now, the pinning precipitates can grow in size through growth (solutes coming from the surrounding matrix) or coarsening (solutes coming from smaller particles that are gradually shrinking in the process reducing the overall interfacial energy). The theory of precipitate coarsening, or its more common term, Ostwald Ripening, was initially laid out by Lifshitz and Slyozov [40]; it is valuable in explaining the changing size of pinning precipitates with temperature and time.

The Gladman analysis (Eq. 13.8) has been examined for HSLA steels. The characteristic feature of normal grain growth (NGG) in which there is a gradual increase in grain size with temperature was observed in steels without grain refining particles [41,42]. In the presence of pinning particles, however, the grain growth takes place with the growth of a few selected grains to abnormally large dimensions while the other grains are unable to grow until further particle coarsening has occurred. This behavior of grain growth has been known in literature as abnormal grain growth (AGG) or secondary recrystallization [43,44].

AGG takes place only when NGG is impeded; it is particularly likely to occur when the temperature is raised and the particle dispersion becomes unstable [30]. The other condition of AGG is found to be associated with local heterogeneity; that is, when the ratio of the particle size to the volume fraction of particles exceeds the mean value of

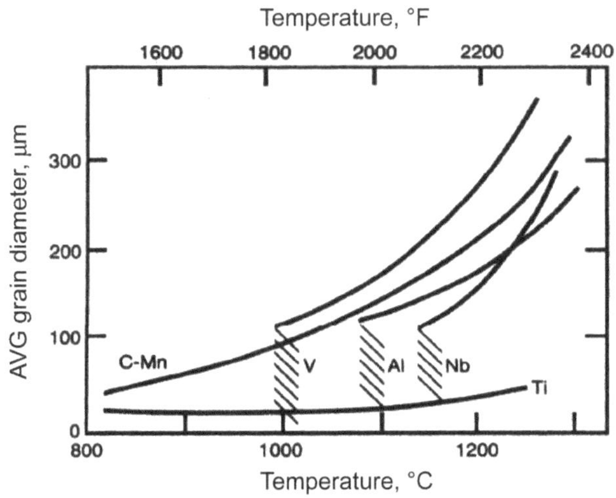

Figure 13.13 Austenite grain size as a function of temperature/presence of abnormal growth [47].

this ratio. This situation is caused by a high Z value [refer to Eq. (13.8), given by Gladman]; this has also been observed subsequently by Eist et al. [45]. Austenite grain growth behavior in Al−V−N and Al−V−Ti−N steels with low carbon (~0.12 wt.%) was studied by Gao and Baker [46]; they confirmed the presence of AGG in some cases.

iii. *Grain growth in microalloyed steels—examples*

Investigations on austenite grain size measurements have been made (e.g., 47) in microalloyed steels. Some results are shown in Fig. 13.13. As can be observed, in the plain C-Mn steel, the grains grow continuously with temperature; while microalloy additions of Ti, Nb, V, Al, etc. help restrict austenite grain size up to at least ~1000°C. Beyond this temperature, V in steel cannot prevent grain growth. Similarly, beyond ~1150°C, Nb-bearing steel also displays grain growth; in between, austenite grains in the Al containing steel start growing. Indeed, grain growth in these microalloyed steels appears to take place discontinuously. The grain growth in the V-grade goes beyond that displayed by the plain C-Mn steel. The investigation further reports that with the onset of AGG, isolated grains appear that are as large as >6X the mode grain size in the matrix.

A case of normal and AGG in austenite has been reported in a Ti (0.03 wt.%) and varying Nb-containing carburizing grade 8620 steel which was primarily taken up for vacuum carburizing [48,49]. The aim here was to adopt a higher than conventional carburizing temperature (~930°C) in order to take advantage of higher diffusional rate, thereby reducing time and enhancing productivity. The challenge was to restrict the grain size of austenite at a higher temperature and hence Nb was added.

The mean grain diameter and microstructures corresponding to various temperatures for one steel (with 0.02% Nb) are shown in Fig. 13.14 [48]. At the lowest and highest temperatures, NGG occurs, while at the two intermediate temperatures one encounters AGG. The AGG shows isolated single grains that are much larger than

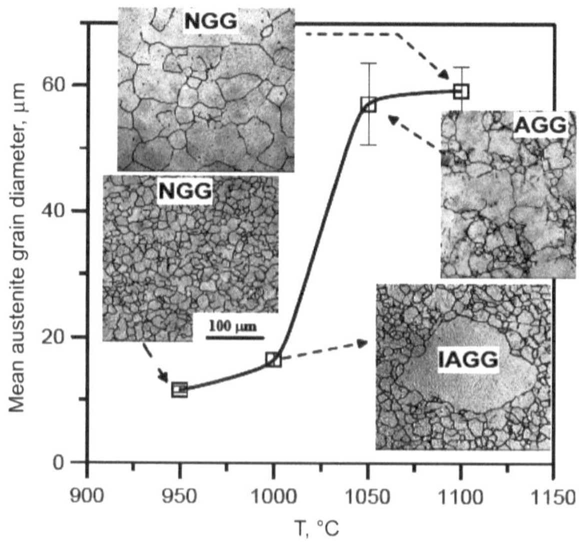

Figure 13.14 Description of austenite grain growth in a Nb-bearing steel [48].

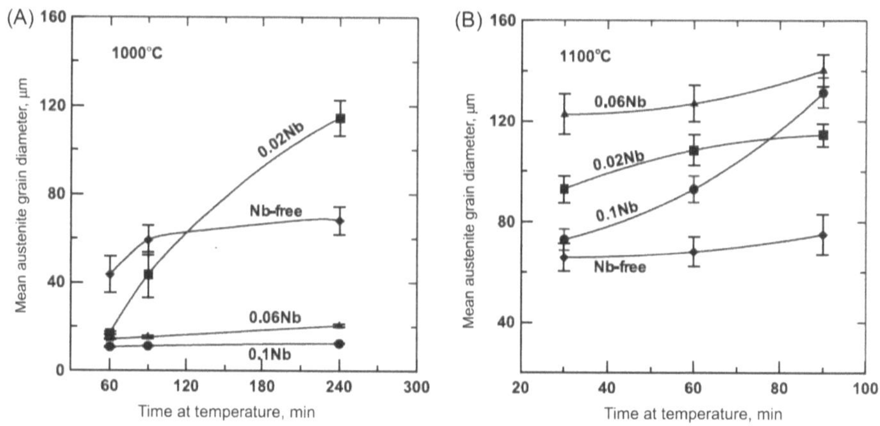

Figure 13.15 Carburized steel 8620, containing various levels of Nb, heated to notional carburizing temperatures: (A) 1000°C and (B) 1100°C showing austenite grain size [49].

the average grains. In Fig. 13.15 (A) and (B), prior austenite grain sizes for steels with different Nb-contents heated to 1000°C and 1100°C respectively for varying holding times have been shown [49]. At 1000°C the 02% Nb steel shows AGG, while at 1100°C, all the Nb-containing steels, in particular 0.1 Nb steel, show higher size than in the Nb-free steel.

The above results are to be reconciled in the light of the Gladman analysis for grain growth (Eq. 13.8).

The nomenclature used in Fig. 13.14 are described below:

NGG: Normal Grain Growth
AGG: Abnormal Grain Growth
IAGG: Initial Abnormal Grain Growth

A study on the characteristics of austenite grain growth in Nb microalloyed steels in homogeneous and in segregated situations were reported by Davis et al. [50]. An AGG, where a large grain is surrounded by many smaller grains, is seen here. The heating condition here coincides with the dissolution of Al-nitride pinning particles, according to the authors.

Broadly, the work found that during solidification, segregation results in an inhomogeneous distribution of microalloying elements, on the scale of the secondary dendrite arm spacing (SDAS). The resulting nonuniform spatial composition leads to different local microalloy precipitate distributions and therefore, on reheating, different local precipitate dissolution behavior. This leads to varying grain boundary pinning forces, which in turn would decide whether abnormal, bimodal, and NGG would occur during reheating. Strictly, such a situation is not expected in the bar reheated prior to forging.

In addition to the references made in the earlier sections, further investigations are also available in literature [51–53] throwing light on the austenite grain growth characteristics during heating. Overall, one finds a great deal of attention [e.g., 45–53] paid to the grain growth aspect of microalloyed steels. Broadly, it is observed that in the Nb-bearing steels the grain growth is suppressed until about 1050°C once the temperature exceeds ~1100°C, austenite grains grow continuously and rapidly, as the pinning precipitates coarsen. Under specific situations, one would also encounter AGG in which only a few selected grains grow to very large size. The role of solutes, in particular Nb, in restricting the grain growth through solute drag effect needs to be elucidated further in order to use the same for designing steel chemistry and for selecting optimal industrial processing.

13.2.4 Strengthening mechanisms operative in microalloyed forging steels

The general mechanisms governing the strengthening of steels have been covered in several text books and numerous research papers [e.g., 17,18,25]. In the case of forgings, particularly those for automotives, the accent currently is on adopting non-hear treated routes for energy saving, eliminating distortion, and enhancing productivity. Consequently, the major strengthening principles that appear attractive for microalloyed (MA) steel forgings include: Grain size Refinement, Precipitation Hardening, and Transformation Strengthening.

Before embarking on a somewhat detailed analysis of the above aspects, one might invoke an often-used figure to provide a broad insight into the influence that the three common microalloying elements exert on steels. In this case, a 0.08% C, 0.90% Mn steel:

It is known in the literature that grain refinement, while raising the strength, also contributes to improved ductility/toughness [e.g., 17,18]. Precipitation hardening, on the other hand, raises the transition temperature (i.e., lowering of toughness) while contributing to strength enhancement. Thus, the net effect is an important consideration. While Nb at ~0.04 wt.% level leads to an excellent combination of strength and toughness, V addition (~0.08% at comparable strength level) results in a net increase in transition temperature. The effect of Ti addition (~0.12 wt.% for comparable strength) lies in-between that of Nb and V. Further, earlier work also shows that grain refinement enhances resistance to environmental embrittlement and cleavage fracture as well [29]. Some details will follow in subsequent sections.

i. *Influence of grain size refinement and precipitation*

Connecting strength with grain size refinement through a Hall–Petch type relation is commonly accepted. However, for the usual microalloying elements it is rather difficult to isolate some of the individual effects. It has therefore been considered more rational to combine many of the factors contributing to strength into one expression [cited 25] that has the form of an expanded Hall–Petch equation for the observed yield strength (*YSobs*).

$$\boldsymbol{YSobs.} = [\boldsymbol{YSpn} + \Delta \boldsymbol{YSss} + \Delta \boldsymbol{YStexture} + \Delta \boldsymbol{YSdisl.}] + \Delta \boldsymbol{YSpptn.} + \mathbf{Ky}\mathrm{D}^{-1/2} \quad \ldots. \tag{13.9}$$

Where,

YSobs. → observed yield strength; *YSpn* → lattice friction stress;

Δ*YSss* → stress increment caused by solid solution; Δ*YStexture* → stress increment by texture

Δ*YSdisl.* → stress increment caused by dislocations; Δ*YSpptn* → stress increment by precipitation

$\mathrm{KyD}^{-1/2}$ → contribution by grain size

The above linear additive expression comprising various strength components has been analyzed in the context of NbC precipitate in ferrite [25]. The linear summation law (Eq. 13.9) permits the evaluation of the contribution of the precipitation hardening by subtracting the contribution from all other components from the observed yield stress.

Now, there are some issues as well in using the Eq. (13.9) for strength prediction in ferrite. It is usually observed that in the presence of Nb, the transformation from austenite to ferrite takes place at a temperature lower than the equilibrium value and acicular substructures may form with low angle grain boundaries. Therefore, whether the Hall–Petch type relation that is applied conventionally to high angle grain boundaries is also extendable to the case of low angle grain boundaries is not established beyond doubt.

An example of precipitates examined under TEM is shown in Fig. 13.16 [cited 27].

The pinning of dislocations (shown in Fig. 13.16) by fine precipitates is inversely proportional to the separation distance between precipitates [18]. The finer dispersion would lead to higher strength.

The Fig. 13.17 brings out the fact that as the reheat temperature is raised, more and more of Nb would be taken into solution in austenite, and the NbC precipitation in ferrite increases, contributing to strengthening of a 0.25 C,0.5 Si.1.2 Mn steel (base steel), modified with 0.05 wt.%Nb. Superimposed is the increment due to grain size [56]. The base steel showed yield strength of between 400–480 MPa in the temperature range shown. (Fig. 13.17).

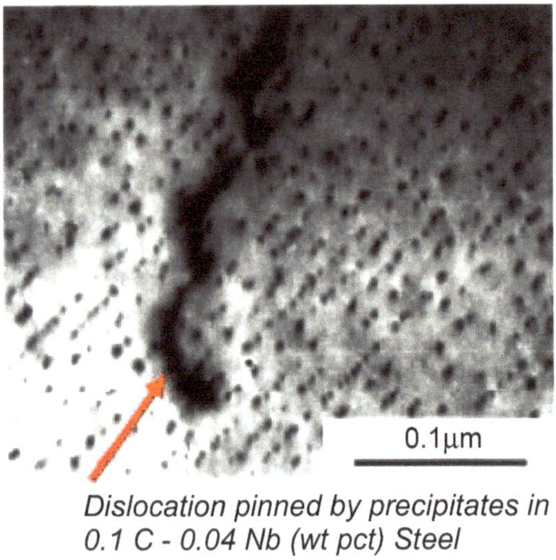

Figure 13.16 Precipitations in a Nb-bearing steel, also showing pinning of dislocations [27].

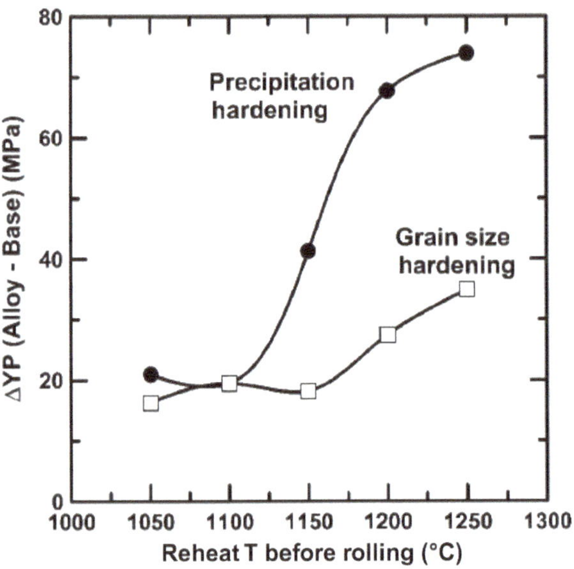

Figure 13.17 Incremental contribution to yield strength due to precipitation and grain size [56,57].

The shaded regions in Fig. 13.18 show the inter phase (IP) precipitates in the TTT space [cited 25]. As is seen, the IP precipitate forms at high (>700°C) temperatures, unlike the common precipitates encountered in ferrite, at lower temperatures. (Fig. 13.18).

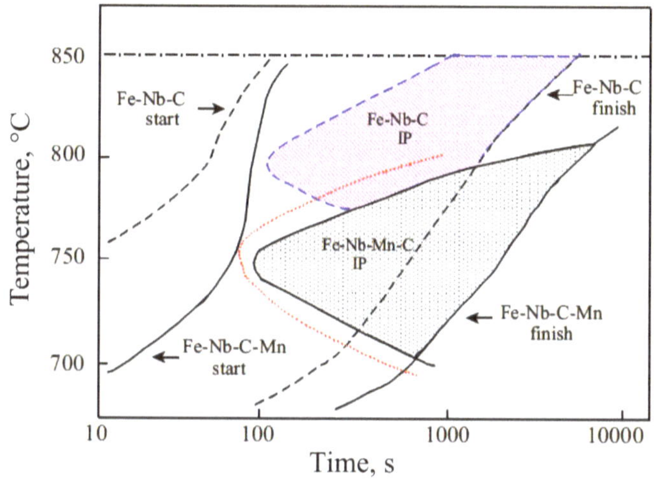

Figure 13.18 Interphase precipitation in 0.036 Nb, 0.09 C and in 0.036 Nb, 0.09 C, 1.07 Mn steel and in a 0,28 Nb, 0.07 C, 1.1 Mn steel (*red dotted line*) [cited in 25].

ii. *Transformation strengthening and thermo-mechanical processing*

Most of the present-day high strength forgings are based on medium carbon Mn steels that are quenched and tempered. Tensile strength and toughness are controlled by the tempering process (temperature and time) adopted following quenching. The packet size of as-quenched martensite is finer with smaller prior austenite grain size [27]. The smaller packet size, in turn, leads to improved yield strength and toughness of the tempered steel.

Microalloying additions to the medium carbon grades would permit air-cooling, avoiding the quench-and-temper heat treatment with appreciable advantages [e.g., 1,58,59]. Broadly, forged microalloy based components, when properly designed, could develop fine pearlite with tiny precipitates in the air-cooled condition. It is known that if a pearlitic structure is produced with fine inter lamellar spacing the strength is high. On the other hand, the thickness of cementite primarily controls the toughness in terms of the transition temperature. A correlation between MA-content and inter lamellar spacing of pearlite/ thickness of cementite as a function of microalloy content is given in Fig. 13.19.

One observes that the microalloying constituents, Nb, V, and Ti lower the transformation temperature, and through this effect lower the inter lamellar spacing in pearlite. Interestingly, the effect of Nb seems to get reversed beyond ~0.05 wt.%. In such a situation, a combination of microalloying elements is recommended. As for the thickness of cementite, only Nb beyond ~0.06% seems to improve the situation. One needs more data on these fronts, to draw conclusively on the aspect.

In order to improve the combination of strength and toughness (and other characteristics such as weldability) it is desirable to lower the carbon and deal with the strength aspect, through nonpearlitic lower temperature transformation products, even under moderate cooling. Results from one such attempt is given in Figs. 13.20 and 13.21. In Fig. 13.20, a TTT diagram (using JMATPRO) is shown [60]. The bainitic portion may be noticed clearly.

It may be observed that the TTT diagram (Fig. 13.20) and the dilatometric data (Fig. 13.21 'A') indicate the formation of a nonpearlitic microstructure upon cooling; this

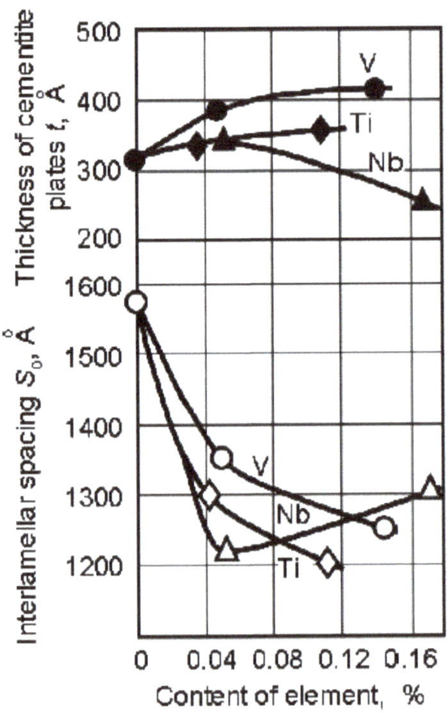

Figure 13.19 Influence of content of microalloy elements on the inter lamellar spacing and thickness of cementite in pearlite, in a 30CMn1Cr steel [59].

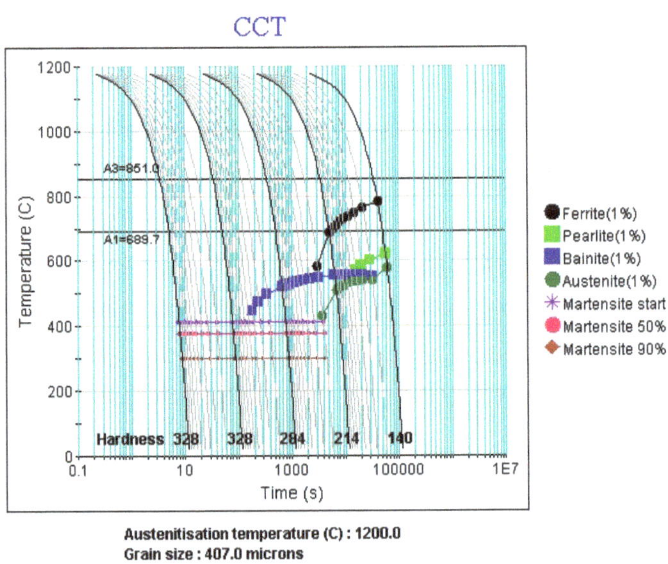

Figure 13.20 TTT diagram (using JMATPRO) of a low (~0.1) carbon MA steel [60].

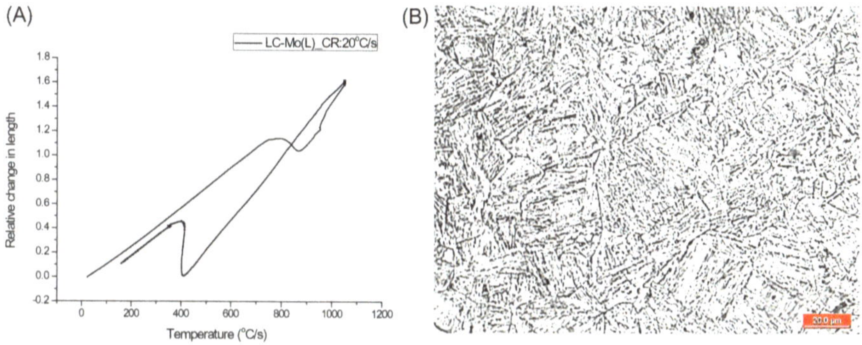

Figure 13.21 Dilatometric data (A) and corresponding microstructure (B) of a low (~0.1 wt.%) carbon MA-steel controlled-cooled at 20°C/s from austenite region at 1050°C in the dilatometer.

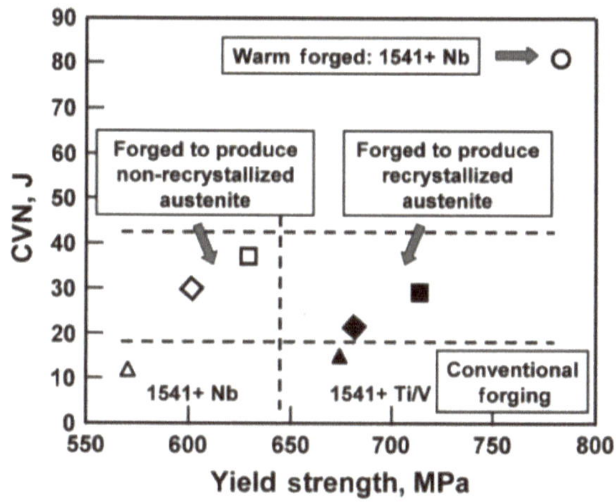

Figure 13.22 Toughness and yield strength combination for conventional forging of 1541 steel with those in thermo-mechanically processed (TMP) material [56].

is confirmed from the microstructure and hardness (~360 VHN). The microstructure is close to bainite/acicular ferrite. Indeed, this steel is amenable auto-tempering upon quenching in water, thus combining very high strength (YS ~900 MPa) with good (>50 J at room temperature) toughness [54,60].

Thermo-mechanical processing (TMP) has long been adopted in flat products and has resulted in vastly improved performance of products. With microalloying and superior process control, TMP is being increasingly explored for forgings as well. One example is shown by Boyd and Jhao on 1541 grade steel (0.41 C) modified with 0.014 Ti and 0.114 V or 0.039 Nb [61]. Both hot forming (in austenite) or warm forming (in two-phase region of austenite + ferrite) was done. Interrelationships between precipitation and recrystallization of austenite, formation of sub-structures, precipitation following transformation to ferrite, etc. were investigated. Consolidated results, as adapted by Matlock and Speer [56], are in Fig. 13.22.

In Fig. 13.22, one finds the 1541 steel (0.41 C) possessing poor toughness in the conventionally forged condition. With the addition of Nb along with warm forging (open circle), the combination of strength-toughness is excellent. Other alternatives seem to be inferior.

It may be mentioned that a good deal of studies would be required to understand the processes involved during TMP, in particular, the warm forging. The forging industry would take some more time to consider adopting such processes, particularly warm forging, that would also warrant enhancing mill capacity.

13.3 Evolution of microalloyed forging steels

The principles of the development of and the fundamental aspects associated with forging steels, in particular microalloyed (MA) grades, have been provided in the preceding section. An attempt will be made here to deal with the evolution of the MA forging steels in the application world, making use of the basics already discussed. As has been shown (cf. Fig. 13.1), the MA additions have the potential to avoid heat treatment cycles for forgings, making the process less expensive.

One finds mention of the influence of MA constituents on steel, specifically of Nb, both in open literature as well as in the patent sites, quite early [62–65]. In the 1980s, 1990s, and also thereafter, the interest in MA-steels continued both in the research as well as the industry communities [2,25,56,66–70].

13.3.1 Microalloying in medium carbon steel forgings

The use of Nb in forgings was dealt with by Hulka and Heisterkamp [2], primarily in medium carbon steels that are the most widely used current grades. For closed-die forgings the steel grade 49MnVS3 (0.50% C-0.25% Si- 0.70% Mn-0.040% S- 0.10% V by wt.) was established in the mid-1970s [71]. The strength in this steel could be raised by increasing the V-content, however the toughness fell at higher strength levels. For improved toughness, a 0.38% C-1.00% Mn-0.07% V-0.03% Nb steel was developed and applied in the as-forged condition for automobile parts with improved safety [72]. The microstructure of Nb-modified AISI 1141 steel was compared with that of standard AISI 1141 steel (0.37–0.45% C; 1.35–1.65% Mn) used in forged connecting rods. The refinement in microstructure contributed to enhancing the hardness of the Nb-bearing steel to 97 HRB vs. 92 HRB for the standard grade, and concurrently improved the toughness in connecting rods, as shown by Bucher [73]. The modified steel is also used for weld yoke or universal joint couplings [74].

Sampei et al. [75] confirmed that Nb-added medium carbon steel bars (0.25–0.45% C) had a good strength-toughness combination, in the controlled rolled condition. Hulka [2] referred to the idea of controlled forging, which had been conceptualized by Pawelski et al in 1978 [76], and demonstrated that finishing forging temperature in the range 950–1100°K led to a better balance of tensile strength and toughness in Nb-bearing base steels (0.32% C, 1.47% Mn by wt.).

Table 13.1 Average chemical composition (wt.%) of the major METASAFE grades [79]

	C	Si	Mn	Cu	Ni	Nb + V
METASAFE 800	0.22	0.15	1.5	–	–	0.19
METASAFE 1000	0.43	0.15	1.5	–	–	0.16
METASAFE 1200	0.21	0.55	1.5	1.4	1.5	0.13

However, a high load during forging at such low temperatures shortened the life of the forming dies and therefore limits the use of controlled forging commercially. Producing ultra-fine (<1 micron) grains of ferrite can improve the properties (better strength-toughness combination) appreciably [77] and Nb-addition has the potential to achieve the same.

Improved toughness even at a higher strength soon became a major requirement for microalloyed forgings. This led to the development of Nb-V steels in France [78–80], Germany [81], and Italy [82], and at the later stage V-Ti steels in Germany [83]. It was shown [83] that the steel design could produce useful strength-toughness combinations, provided a minimum V-level of 0.10 wt.% is coupled with a high nitrogen level, and a stoichiometric Ti:N ratio is maintained.

The development of Nb-V forging steels took advantage of the triple role of niobium to refine grain, reduce pearlite inter lamellar spacing and cause precipitation strengthening [Tither, 68]. The "METASAFE" steels in France became the main family of Nb-V microalloyed forging steels [78–80]; the chemical compositions of the principal grades are given in Table 13.1.

The designation number indicates the tensile strength [MPa] of the steels. In these steels the Nb-content is ~0.04 wt.% and the balance is V-content in the combined values shown in the Table 13.1. Depending on the grade, the carbon content is maintained between 0.15% and 0.45% by wt.; the lower end of which leads to improved toughness and good weldability. Two of the METASAFE grades therefore do not strictly come under the medium carbon category. These steels are mainly used for suspension arms, antisway bars, swivel axle spindles, and connecting rods, etc., in the automotive industry requiring higher toughness. An example of a component is shown in Fig. 13.23.

A number of steels based on carbon levels between 0.35 and 0.55 wt.%, along with Nb up to 0.05 wt.% (with or without V) have been commercially developed by Gerlach-Werke in Germany, Krupp & Volkswagen in Brazil, and FIAT Auto and Deltasider in Italy [68]. Of these, the modified C-38 developed by Gerlach with Nb 0.05 wt.%, V 0.05–0.12 wt.% and with Y.S. of 600 MPa minutes; T.S. of 900 (minutes); and R.A. 12% (minutes) became popular.

13.3.2 Low carbon microalloyed forging steels (without copper)

There has been substantial developments and investigations in this area [68]. The "multiphase" family of steels (primarily, nonpearlitic) is based on alloying a low

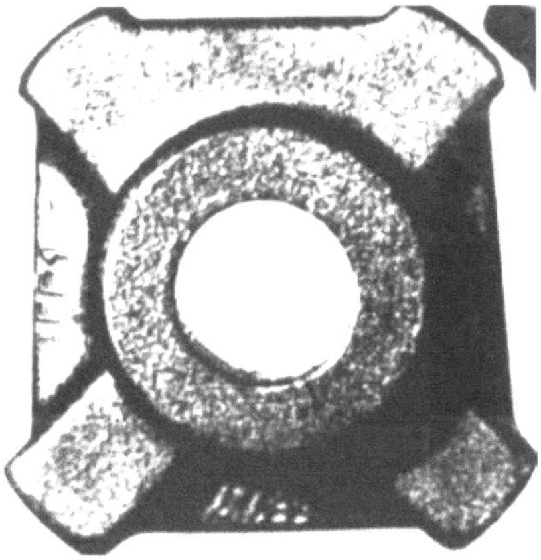

Figure 13.23 Swivel axle spindle carrier using METASAFE 700 hot-forged [79].

carbon (~0.10 wt.%) steel with Mn, Mo, and Nb. Two major steel grades here developed in North America; they are: BHS-1 [3,84–86] and FreeForm™ [87].

Niobium is added for both austenite conditioning during controlled hot processing and to manipulate transformation characteristics during cooling. The addition of Mn (1.4%–2.0%) and Mo (0.10%–0.50%) has been done for transformation control. Commercial trials undertaken using the BHS-1 steel have involved the manufacture of connecting rods, idler arm brackets (steering brackets), and lower control arms. The last mentioned component was directly quenched after hot forging and did not undergo any further heat treatment. The mechanical properties exhibited by the components made from conventional steels (BC 1038 grade, containing 0.34–0.42 wt.% C, 0.60–0.90 wt.% Mn, in QT condition and 1541 grade, with compositions given earlier, in QT condition) and from BHS-1 (with Mn–Mo–Nb) are given in Table 13.2 for comparison [3].

The mechanical properties of BHS-1, in particular strength and fatigue resistance, in direct water quenched (DWQ) condition, are quite appreciable.

The Free Form grade, is leaner (in terms of Mo) and is a trademark of Ispat Inland Steel Company. This steel is known to be amenable to cold reduction and yet would retain a good amount of ductility, and also displays high fatigue resistance after cold reduction.

A direct quenching microalloyed forging grade was commercialized in the 1990s in the USA [88,89]. This was designated as MicrotuffR (a trade mark of the Chaparral Steel Company). The steel contained 0.10%–0.20% wt.% C with ~0.10% Nb. A fine dispersion of undissolved Nb (C, N) particles retards the recrystallization and prevents austenite grain growth during forging, trimming, and during any entry delay into the quenchant that might be experienced. The grain size remains fine with better than ASTM 5 at forging temperatures up to 1290°C. Direct quenching produces a martensitic microstructure which is allowed to auto-temper. The fine-grained auto-tempered martensitic forging exhibits high yield strength (945–1225 MPa)

Table 13.2 **Mechanical properties (including Charpy CVN) of components made from BHS-1 and two conventional steels (BC 1038QT and 1541 QT) [3]**

(a) Idler Arm Bracket							
Steel	YS MPa	UTS MPa	RA (%)	CVN@RT (Joules)		Average kilocycles to failure	
BC 1038 (QT)	607	697	59	86		134.79 ± 36.5	
BHS-1	828	1049	43	96		261.85 ± 46.9	
(b) Lower Control Arm							
Steel/Processing	YS MPa	UTS MPa	RA (%)	CVN (Joules)		Average kilocycles to failure	
				25°C–40°C		Smooth	Notched
1541 (QT)	820	930	60	60	22	105.7	58.6
BHS-1 (DWQ)	935	1197	63	50	32	>1000	>500

Table 13.3 **Composition (wt.%) and strength/toughness of MicrotuffR**

C	**Mn**	**P**	**S**	**Si**	**Cu**	**Ni**	**Cr**	**Mo**	**Nb**	**N**
$\frac{0.10}{0.15}$ or $\frac{0.15}{0.20}$	$\frac{1.65}{2.00}$	0.03[a]	0.03[a]	$\frac{0.50}{0.70}$	0.35[a]	0.20[a]	0.25	$\frac{0.15}{0.20}$	$\frac{0.09}{0.12}$	$\frac{0.012}{0.020}$
Y.S. MPa	T.S. MPa	RA%	EI%	Hr_c	CVN Energy at 0°C Joules					
$\frac{945}{1225}$	$\frac{1190}{1540}$	25 min	8 min	$\frac{38}{45}$	40 min					

[a]Denotes a maximum value.
Top portion of table originally from W.A. Szilva et al., Proceedings of the International Symposium on Microalloyed Bar and Forging Steels, Ontario, 26-29th, The Iron and Steel Section of the Metallurgical Society of CIM, Hamilton, August (1990), p. 227. Reproduced with permission from the Canadian Institute of Mining, Metallurgy and Petroleum.

coupled with an impact transition temperature well below zero. The MicrotuffR compositions and mechanical properties are given in Table 13.3. The mechanical properties achieved in MicrotuffR are comparable to those attained by the conventional reheated, quenched, and tempered low alloy AISI 4140 steel [89].

Table 13.4 **Direct quenched steels from US, Europe, and Japan; composition (wt.%) and properties**

Property measured	BHS (US) [84]	IMAFORM (European) [90]	KOBE (Japan) [91]
Composition [mass-%]	0.10%C 1.0%Mn 0.6%Mo 0.3–0.5% Cr 0.05%Nb –	0.06%C 0.9%Mn – 1.1%Cr (0.04%Nb) 50 ppm B	0.12%C 2.0%Mn 0.25%Mo 1.0%Cr 0.12%Nb 20 ppm B
Tensile strength CVN impact energy	950 MPa 90 J	950 MPa 60 J	1250 MPa 75 J

Table 13.5 **Composition (wt.%) and properties of MPC steel [92]**

C	Mn	Si	S	P	Al	Cr	Mo	Nb	N
0.1	1.00	0.30	0.02	0.01	0.04	0.50	0.70	0.05	0.007

Mechanical properties

Steel & treatment	YS, MPa	UTS, MPa	RA, %	CVN at −18°C, Joule	CVN at, 24°C Joule
MPC/WQ	786	950	63	45	72
MPC/AC	427	668	61	62	92

A comparison of three direct quenched, low carbon MA-steels introduced in the USA [84], Europe [90], and Japan [91] are shown in Table 13.4. These steels (two of them, IMAFORM and KOBE steels, also contain some Boron).

Mention may be made of a low carbon variety of steel developed and patented in the USA by DeArdo [92] that can be used for both cold forming and hot forging (noninduction hardened) applications without any heat treatment. The composition and mechanical properties of this steel, termed MPC steel, are given in Table 13.5 (A) and (B).

From the foregoing details, it would be clear that the low carbon, nonhear treated forging steels have already seen applications in a number of industrial components. However, extensive use of different grades is yet to take place.

13.3.3 Microalloyed copper bearing forging steels

Copper has been known for some time to induce strengthening in steels through precipitation hardening [93,94]. However, the understanding of details of the precipitation mechanism has improved considerably in recent years, with the use of more sophisticated instruments. These efforts have also opened up new opportunities in copper bearing steel products. For instance, a number of grades with low carbon, based on copper addition, have been developed at North Western University over the past years [95,96]. There is also a unique challenge in the form of hot shortness in these steels that causes embrittlement in extreme cases [97]. These aspects, viz. precipitation hardening and hot shortness, both due to copper, will be dealt with before taking up some details on copper bearing industrial forgings.

i. *Precipitation behavior of copper in steel*

Copper precipitation in steel takes place on a very fine scale (<5 nm level) and therefore one needs the use of special techniques such as high resolution transmission electron microscopy (HRTEM), atom-probe field ion microscopy (APFIM), small angle neutron and x-ray scattering (SANS and SAXS), extended x-ray absorption fluorescence spectroscopy (EXAFS), x-ray absorption near edge spectroscopy (XANES), and so on. It has been established that useful copper precipitation occurs in the temperature range of 450°C–600°C and that the precipitation follows the following sequence [95,98–101]:

Super-saturated solution of Cu in α*-iron* → *bcc cluster* → *9 R Cu* → *3 R Cu* →*fcc* ε*-Cu*

At diameters of less than 5 nm, the copper is in meta-stable bcc cluster structure; in the 5–10 nm size the bcc clusters martensitically transform to the twinned and faulted fcc-like 9 R structure. At ~17 nm size, the 9 R spherical precipitates get transformed to stacking fault-free fcc equilibrium structure,ε-Cu [100,101]. It has further been observed that upon reaching a size of about 30 nm, strain energy minimization [102,103] induces a change in shape from a spherical to a rod-like morphology.

Early research by Goodman et al [103,104] indicated that the bcc Cu precipitates comprised about 50 at.% iron and 50 at.% copper for sizes up to about 2.5 nm; however, more recent atom-probe field ion microscopy and small angle neutron scattering studies claimed that the bcc copper clusters contain only small amounts of iron, ~2 at.% [105]. In the peak-aged condition, the coherent, bcc clusters generally have an average diameter of between 1 and 5 nm [99,106], with about 50% of the copper still remaining in solid solution [107].

Investigation using atom-probe tomography on the bcc Cu precipitates in a NUCu-100 grade steel specimen, solutionized at 1100°C and directly aged at 490°C, demonstrated the existence of chemically complex Cu-rich precipitates containing Fe, Ni, Mn, and Al. The precipitate/α-Fe matrix hetero-phase interfaces are enriched in Ni and Mn, forming a spherical shell-like structure, whereas the Al enhancement is found toward the inner region of the Cu precipitates [108–112].

Studies on the kinetics of coprecipitation of bcc copper and NbC have been done in some detail by Gagliano & Fine, in a copper and niobium containing steel. The nucleation rate and the growth rate of these two varieties of precipitates have been computed using Thermocalc and compared with experiments [113]. It was found that the presence of Nb contributes to a resistance against softening associated with over ageing of Cu precipitates, even at very long times of ageing (>15 hours.). Further, the two types of precipitates do not attain a peak at the same time during ageing, as their kinetics are very different. The nucleation rate for bcc copper precipitate is order of magnitudes higher than the nucleation rate for NbC in ferrite, under similar austenitizing and ageing conditions [113]. This area of work however necessitates further research.

ii. *Copper-induced surface hot shortness (SHS) in steels*

It is long since known [97,114] that the presence of copper (and other elements such as Sn and Sb) in steel may cause cracks to form on the surface during hot working; this phenomenon is known as surface hot shortness (SHS).

It is accepted that the phenomenon occurs due to copper enrichment at the scale–metal interface during heating. The enrichment of copper takes place as a result of selective oxidation of iron and because the solubility of copper in the iron oxide (scale) is very low. The Cu-enriched phase liquefies and penetrates into austenite grain boundaries during the hot working and causes cracking. Tin is known to encourage this phenomenon [115,116]. In view of the importance of the phenomenon of hot shortness in steel, ISIJ International devoted a full issue, vol. 37 (1997), to this topic. A number of articles on the subject could be found in this issue [115–118]. In spite of the antiquity of the subject of SHS in steels, the field is still active and new key information is being generated in recent years as well [119–122].

There are interesting approaches made by Shibata et al [120], where the investigation was targeted towards enhancing the permissible limit of Cu for various applications, from the standpoint of SHS (in sheet steel from existing limit 0.06→0.12 and in bars from the existing 0.4→0.5) by resisting the liquid Cu penetration of the grain boundaries with preventive measures. The amount of liquid Cu-enriched phase can be reduced by suppression of oxidation, occlusion of liquid phase into scale [97], back-diffusion of Cu into steel matrix, suppression of wettability or fluidity of liquid phase, etc. [120]. Shibata's work also shows the influence of alloying elements, temperature, deformation etc. [120].

The most recommended alloying element for combating SHS due to copper is nickel. The solubility of Cu in austenite as well as occlusion of Cu in the scale are both enhanced by nickel, hence the beneficial effect with regard to SHS. The presence of nickel also raises the melting point of the Cu-rich phase formed on the steel surface that otherwise leads to SHS [115,116]. The earlier recommendation was to add the same level of nickel as that of copper for resisting SHS; it is now believed that nickel to the extent of only 50% of the Cu-level should suffice.

Nickel addition is in any case an expensive solution, and other alternatives have been explored. Silicon has been found to be effective in suppressing SHS; the amount of Ni required for resisting SHS is reduced appreciably in the presence of Si, as Shibata's work has shown [120]. Indeed, the work has also demonstrated that a good quantity of the liquid Cu-enriched phase is occluded into the scale in the presence of Si. Similarly, Mn and B also work in the same direction; however, the effectiveness is lower.

Heating to a high temperature is found to reduce the risk of SHS. The beneficial effect of a soaking temperature range of 1200°C–1300°C, with or without nickel, is now confirmed [120,122]. The main reason for the absence of SHS in this high temperature region apparently stems from the rapid diffusion of copper in austenite. Indeed, there appears to be a critical temperature range at which the steel is very much prone to SHS [120,122].

Shibata's work further showed that very slow deformation rate reduces the effect of SHS due to the reduction of stress level through dynamic recrystallization; similarly at very high deformation rates one finds reduced tendency for SHS as the critical stress for liquid embrittlement also increases. Thus, again, there is a window that needs to be avoided for preventing SHS. It was further observed that a large grain size of austenite and presence of even a small amount of moisture (H_2O) enhances the tendency for SHS [120].

The work of Garza & van Tyne [121] dealt with the issue of SHS in a variety of 1045 forging steels, with the amount of copper varying between 0.09% and 0.39%. This work also supported the observation that when heated to 1200°C, the steels did not show any cracking due to SHS. Again, most of the steels had a high vulnerability to SHS in the temperature range of 1120°C−1140°C. The work showed that even a low level of copper, as mentioned above, could cause SHS in the medium carbon grades such as 1045.

It may be observed from the above discussion that SHS continues to raise uncertainties when the use of precipitation hardened grades based on copper is considered; innovative methods need to be developed that depend very little on costly nickel.

iii. *Copper bearing bar steel grades: examples from industry*

In section 2.4 (Figs. 13.20, 13.21), as well as in section 3.2, it was pointed out that as one moved into lower carbon levels in steel, one would encounter lower temperature, nonpearlitic transformations. One of the main nonpearlitic transformation products (if one did not use quenching to get martensite), is bainite or some of its variants [details in ref. 6]. An interesting development of an extremely low carbon (0.009 wt.%), non-heat treated, bainitic steel, along with its upgraded versions (improved strength characteristics), with copper precipitation, was done by the Kawasaki group in Japan [4].

The early attempt towards developing non-heat treated steel grades in Japan contained 0.2−0.5 mass %C ferrite−pearlite steels, strengthened by V-carbonitride. These steels, however, suffered from low toughness and their applications were consequently limited. Further, non-heat treated bainitic steels that could provide higher strength and toughness than ferrite-pearlite steels were also developed. Their mass effect was however large and the size of machine components based on them had to be limited. Therefore, these non-heat treated bainitic steels were of little practical use.

In order to overcome the above disadvantages, earlier investigations by Okatsu et al. [123] on extremely low carbon bainitic steel were revisited [4]. Additionally, a new microstructure control technique, called the TPCP (Thermo-mechanical Precipitation Control Process), was developed by Amano et al. [124]. The basic concept of the TPCP is to select a steel structure in which the cooling rate dependence of the microstructure is extremely low and the steel is strengthened by precipitation, and not by phases forming under high rate of cooling.

The chemical compositions (wt.%) of the chosen very low carbon steel and of a conventional low alloy steel (AISI 4137) are given in Table 13.6. One may also observe the high ($\sim$2 wt.%) amount of Mn in the extremely low carbon bainitic steel.

The CCT diagrams of the two steels are given below in Figs. 13.24 and 13.25 respectively. For the extremely low carbon bainitic steel (Fig.13.24) the hardness changes very little (from VHN 225 to VHN 174) as one moves from the highest to the lowest cooling rate . On the other hand, for a more conventional AISI 4137 steel (Fig.13.25), an appreciable variation in hardness (VHN 587 to VHN 219) is observed with cooling rate.

It is desirable to obtain a uniform microstructure through the cross-section of a large component. This is possible if the cooling rate dependence after hot deformation is minimal. By lowering the carbon content of steel to less than 0.02% (mass) at which carbon partition does not occur during austenite to ferrite transformation and by optimizing other alloying elements, it is possible to

Table 13.6 Chemical composition of TPCP steel of Kawasaki steel [4]

Wt.% of elements in steel	Extremely low carbon bainitic steel	Conventional low alloy steel (AISI 4137)
C	0.009	0.34
Si	0.26	0.22
Mn	1.99	0.80
P	0.015	0.016
S	0.015	0.014
Al	0.034	0.026
Cr	tr	1.08
Mo	tr	0.21
Others (Nb, Ti, B)	Not specified	Not specified

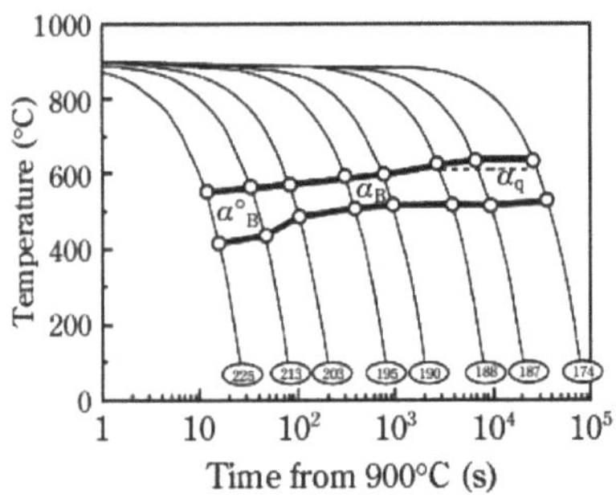

Figure 13.24 CCT diagram of extremely low carbon bainitic steel. [4]
$\alpha_B^0 \rightarrow$ Bainitic ferrite,
$\alpha_B \rightarrow$ Granular bainitic ferrite,
$\alpha_q \rightarrow$ Quasi-polygonal ferrite

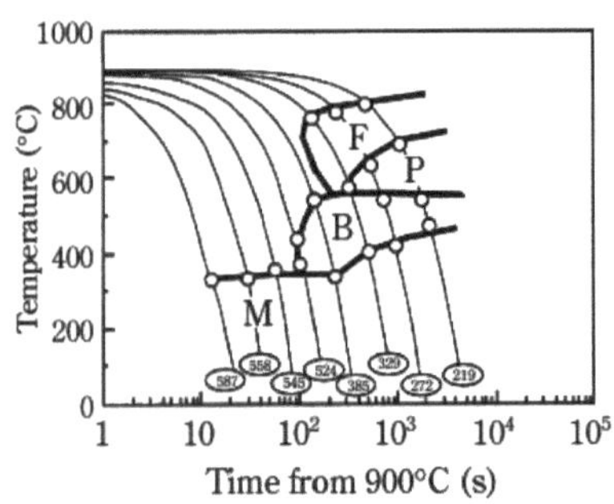

Figure 13.25 CCT-diagram of conventional low alloy (AISI 4137) steel. [4]
M→Martensite, B→Bainite, F→Ferrite, P→Pearlite

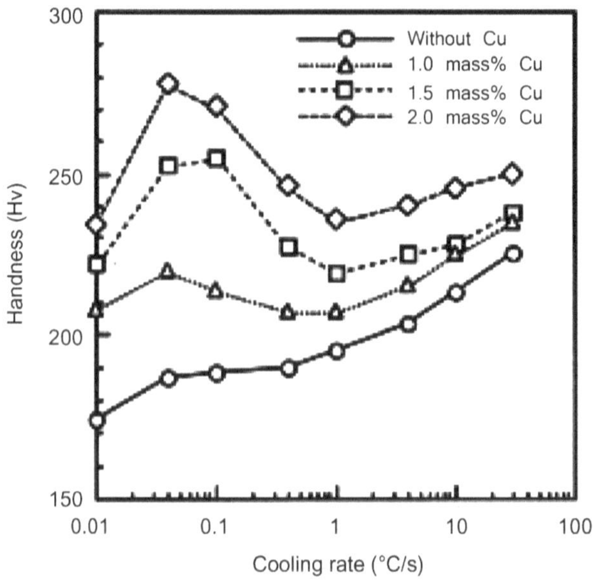

Figure 13.26 Hardness of TPCP Cu-precipitated steels as a function of cooling rate [4].

obtain a uniform, extremely low carbon bainitic microstructure that has only a small cooling rate dependence. The various phases that form here as a function of cooling rate (nomenclature of various given in Fig. 13.24) belong to the same broad category ferrites in the bainite. For the AISI-4137 steel, on the other hand, the transformation products vary a great deal with the cooling rates, hence the wide difference in hardness between slowest cooled and fastest cooled samples.

Now, the extremely low carbon steel was alloyed with copper for precipitation hardening. However, unlike the case of flat products where Cu is precipitated out after quenching and holding at various temperatures, in the TPCP steels, the precipitation was allowed after the hot deformation of a bar. The typical changes in hardness can be seen in Fig. 13.26 [4].

One may observe that while the no-Cu steel shows a decreasing hardness with lowering cooling rate the Cu-containing steels show an enhancement of hardness in a window of cooling rate due to precipitation and some 'over ageing' effect as a result of too-slow cooling rates. One also finds a solid solution hardening effect due to copper.

The fine precipitation of copper (and cluster formation) was studied by a three-dimensional atom-probe. Based on the above considerations, a new TPCP steel was developed, composition given in Table 13.7.

The TPCP steels were evaluated with regard to mechanical properties (strength, toughness, fatigue, etc.) as well as weldability, machinability, etc.

The mechanical properties in the TPCP and in the AISI 4137 Q&T steel have been shown in Table 13.8.

Table 13.7 Composition of TPCP steel along with that of conventional steel for Q&T [4]

(mass)									
Steel	**C**	**Si**	**Mn**	**P**	**S**	**Al**	**Cr**	**Mo**	**Others**
TPCP steel	0.007	0.24	2.01	0.015	0.016	0.036	Tr.	Tr.	Cu, Ni, Nb, Ti, B
AISI 4137	0.34	0.22	0.80	0.016	0.014	0.026	1.08	0.21	–

Table 13.8 Mechanical properties of TPCP and AISI steels [4]

Steel	Position	0.2% PS (MPa)	TS (MPa)	YR (%)	E1 (%)	RA (%)
TPCP steel	Surface[a]	730	840	87	26	74
	1/4*D*	708	818	87	25	72
	1/2*D*	689	813	85	22	66
Quench-tempered AISI4137	Surface[a]	644	820	79	23	62
	1/4*D*	638	811	79	22	59
	1/2*D*	636	807	79	20	52

[a]15 mm inside from surface.

The strength properties in the TPCP Grade are good from surface to interior, at a higher ductility, and at the same strength level (UTS ~820 MPa and Proof Strength ~700 MPa) as the Q + T AISI 4137 steel.

Remarkable is the yield ratio (Proof Strength/UTS) of ~85% in the TPCP steel, in the nonheat treated condition. The fracture transition temperature, notch-toughness, and fatigue resistance, etc., of the new steel are shown to be better than the $Q + T$ conventional variety of steel [4].

The Fig. 13.27 shows machinability, as determined through tool life (test conditions given in Ref.), is found to be better for TPCP, compared to the conventional steels.

It would appear that all mechanical properties including toughness and machinability are excellent for the TPCP steel. However, since the exact composition in terms of Cu, Nb, Ni, etc., are not given (only indicated in the Table 13.7), the cost etc. of such steels would be difficult to estimate.

One other observation may be made relating to the Cu-precipitation carbon content of the TPCP steel. Most of the Cu-addition in flat steels have been made to higher carbon (~0.05) matrix. The kinetics of Cu-precipitation, as done by Morris Fine and coworkers [e.g., 95,113] need to be revisited. The Cu-precipitation in extremely low carbon IF-steels (Typical: C-0.003, Cu-0.8, Nb- 0.011, Ti-0.05, Mn-0.12]) was investigated by Rana et al. [125–128]. It would be interesting to compare the kinetics of copper precipitation in the matrices where the c-contents are different from the one studied in the Kawasaki work.

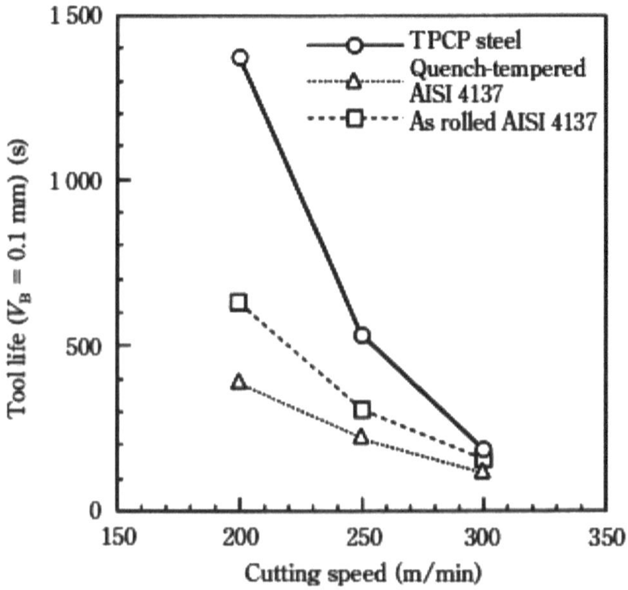

Figure 13.27 Machinability of TPCP steel in comparison with AISI in $Q + T$ and as-rolled conditions [4].

13.4 Steels for automotive forging—the way forward

The scenario of auto forging steel has changed over the past two to three decades and is poised for more radical changes now. Newer materials, some of them superior in terms of strength-to-weight ratio and with comparable mechanical properties, have made their entry. One thus finds composite materials going for propeller shafts, crank shafts, and so on. From within the field of ferrous materials itself, tough competition has started coming from the casting area, particularly after the advent of austempered ductile iron (ADI). Several auto majors (Toyota in particular) have listed out a number of traditional steel forged components that have the potential to switch to ADI for cost competitiveness. Near net shape castings and possibilities of semi-fluid castings for auto components are not far away, In short, weight reduction and cost are the two key elements that would decide the future of the automotive steel forging sector [e.g., 129–133].

Viewed in the context as described, it is clear that the forged steel components of the future could not afford to continue with the quench-and-temper process for long. The medium carbon, high manganese grades which are the dominant material today ought to be made amenable to air-cooling/direct quenching. Consequently, as a first step, microalloying of medium carbon grades would have to be worked out extensively. For this to happen, all the issues discussed in this chapter specifically, the state in which the microalloying element remains (in solution or as precipitate) during forging; rate of cooling that would induce the right phases and their desirable amounts etc. would have to be looked into, in view of the role that kinetic factors play.

Another major issue is the variation in properties across the cross-section due to cooling rate. One way to deal with this problem is to select very low carbon compositions (similar to those as at Kawasaki) that would render the component insensitive to the cooling rate. The use of copper in the grade and the possibility of freedom from resorting to high cooling rate after forging needs to be reconsidered, since hot shortness is an important issue although right temperatures and strain rate are known ways of dealing with the problem. One interesting approach to steel grade selection for forging is that propagated by Prof. Bhadeshia of Cambridge; here one could go in for steels transforming over very long period [e.g., 9,10], bringing in strong and tough steels.

One aspect not covered in the present article is the close working needed between people in the material selection area and those in die-designing. Further, the processing-map approach alongside the currently propagated phenomenological approach to forging simulation would have to be pursued as an interdisciplinary area.

Demand for higher weldability in forgings (which is not the case right now) would push the steel compositions to lower carbon content. In such a case, the challenge would be to obtain the desired strength level by resorting to non-conventional strengthening mechanisms some of which have been discussed in the present paper.

References

[1] M. Katsumata, Steel wire rods and bars and special steels (Ch. 5), in: H. Takechi, et al. (Eds.), How to Use Niobium for Improvement of Steel Properties, Pub. CBMM Asia Co. Ltd, 2005.

[2] K. Hulka, F. Heisterkamp, in: G. Krauss, S.K. Banerji (Eds.), Role and Implication of Niobium in As-Forged Steels in Fundamentals of Microalloying Forging Steels, 255, AIME, 1986.

[3] C.I. Garcia et al., SAE Technical Papers 910143 and 910145, SAE International Congress and Exposition, 25 February–1 March, Detroit, MI, USA (1991).

[4] K. Hase et al., New extremely low carbon bainitic high strength steel bar having excellent machinability and toughness produced by TPCP Technology, Kawasaki Steel technical report No. 47 (2002), p. 35.

[5] J.F. Held et al., As Hot Rolled Bar Steel; US Patent No. 4,806,177, Feb. (1989).

[6] H.K.D.H. Bhadeshia, Bainite in Steels (2nd and 3rd Eds.), Pub. the Institute of Materials (2001).

[7] Lars Elek et al., New generation of Forging Steels for Cyclic Loaded Safety Components with Improved Fatigue Properties, in SAE International Conference on Automotive Materials & Manufacturing (2014), p. 1

[8] B.A. Graville, Cold cracking in welds in HSLA steels, welding of HSLA (Microalloyed) structural steels, Proc. Intl. Conf. ASM, 1976.

[9] H.K.D.H. Bhadeshia, Phase transformations contributing to the properties of modern steels, Bull. Polish Acad. Sci. 58 (2010) 255.

[10] H.K.D.H. Bhadeshia, Very short & long heat treatments in the processing of steel, Proceedings of International Conference on, SIMPRO, Ranchi, 2008, p. 3.
[11] J.S. Kirkaldy, D. Venugopalan, in: A.R. Marder, J.I. Goldstein (Eds.), Phase Transformation in Ferrous Alloys, AIME, Warrandale, 1984, p. 125.
[12] N. Saunders, A.P. Miodownik, in: R.W. Cahn (Ed.), CALPHAD- Calculation of Phase Diagrams, Pergamon Materials Series, Vol. 1, Oxford Elsevier Science, 1998.
[13] N. Saunders, et al., in: J.C. Zhao, et al. (Eds.), Materials Design Approaches and Experiences, Warrandate, 2001, p. 185.
[14] N. Saunders, et al., J. Metals 55 (12) (2003) 60.
[15] W. Diekmann, Calculation of Steel Data using JMATPRO. Proceedings of International Conference on COMAT (2012).
[16] CCT diagram using JMATPRO package at IIT Kharagpur (courtesy: Prof. S.B. Singh, 2015)
[17] W.C. Leslie, The Physical Metallurgy of Steels, McGraw-Hill International Book Co, New York, NY, 1982.
[18] T. Gladman, The Physical Metallugry of Microalloyed Steels; The Institute of Metallurgy, London (1997); New Ed. Pub.: Maney for Institute Materials London (2002).
[19] H. Nordberg, B. Aronsson, J. Iron Steel Inst. 206 (1968) 1263.
[20] K. Narita, Trans. ISIJ 15 (1975) 145.
[21] H. Adrian, Thermodynamic calculations of carbonitride precipitations as a guide for alloy design of microalloyed steels, Proceedings of the International Conference Microalloyed'95, Iron and Steel Soc., Pittsburg, PA, (1995), p. 285.
[22] E.J. Palmiere, Precipitation phenomena in microalloyed steels, Proceedings of the International Conference Microalloyed'95, Iron and Steel Soc., Pittsburg, PA, (1995), p. 307.
[23] E. Turkdogan, Causes and effects of nitride and carbonitride precipatation during continuous casting, Iron & Steel Maker 16 No.5 (1989) 61.
[24] P.R. Rios, Expressions for solubility product in niobium carbonitride in austenite, Mater. Sci. Eng. A 4 (1998) 324.
[25] A.J. DeArdo, et al., in: S. Hashimoto, et al. (Eds.), Basic Metallurgy of Modern Niobium Steels, in International Symposium on Niobium Microalloyed Sheet Steel for Automotive Application, TMS, 2006, p. 499.
[26] B. Loberg, et al., The role of alloy compositions on the stability of nitrides in Ti-microalloyed steels during weld thermal cycles, Met Trans. A 15 (1984) 33.
[27] D. Matlock et al., Microalloyed Forging Steels for Automoive Applications. Presentation at AIST Bar Conference, Winter Park, Colorado (2006).
[28] R.J. Glodowski, A review of microalloying in hot rolled steel sheet products. International Seminar on Application Technologies of Vanadium in Flat Rolled Sheets (2005).
[29] H.Y. Lee, et al., Dissolution behavior of niobium carbide during slab reheating, ISIJ Int. 54 (2014) 1671.
[30] F.J. Humphreys, Recrystallisation and Related Annealing Phenomena, Elsevier, New York, NY, 2004.
[31] C.S. Smith, Trans. AIME 175 (1948) 15.
[32] V.P. Paranjpe, Trans. Indian Inst. Metals 6 (1952) 203.
[33] J.W. Cahn, Acta Metal. 10 (1962) 789.
[34] M. Hillert, B. Sundman, Acta Metal. 24 (1976) 731.
[35] C.R. Hutchinson, et al., The comparative effectiveness of Nb solute and NbC precipitates at impeding grain boundary motion in Nb Steels, Scripta Mater. 59 (2008) 635.

[36] C.W. Sinclair, et al., Metal. Mater. Trans. A 38A (2007) 821.
[37] H.S. Zurob, et al., Acta Mater. 50 (2002) 3075.
[38] H.S. Zurob, et al., ISIJ Int. 45 (2005) 713.
[39] T. Gladman, Proc. R. Soc. (London) 294 A (1966) 298.
[40] I.M. Lifshitz, V.V. Slyozov, Kinetics of precipitation from supersaturated solutions, J. Phys. Chem. Solids 9 (1961) 35.
[41] M. Hillert, Acta Met. 13 (1965) 227.
[42] K.J. Irvine, et al., J. Iron Steel Inst. 205 (1967) 161.
[43] T. Gladman, F.B. Pickering, J. Iron Steel Inst. 205 (1967) 653.
[44] T. Gladman, JOM 9 (1992) 21.
[45] R. Eist, et al., Acta Metal. 36 (1988) 1723.
[46] N. Gao, T.N. Baker, Austenitic Grain Growth Behavior of Microalloyed Al-V-N and Al-V-Ti-N Steels, ISIJ Intl. 38 (7) (1988) 744.
[47] A.J. DeArdo, An overview of microalloyed steels, in: J. Hartman, et al. (Eds.), Proceedings of 8th Processessing and Technology Forum on The Effect of Microalloying on the Hot Working of Ferrous Alloys, Iron & Steel Soc., Warrandale, 1988, p. 67.
[48] K.A. AlOgab, Ph.D. Dissertation, Metallurgical & Materials Engg.CSM (2007).
[49] K.A. AlOgab, et al., ISIJ Int. 47 (2007) 307.
[50] C. Davis, Grain size development in microalloyed steels during reheating – effect of segregation during solidification. Proceedings of 4th International Conference on SIMPRO'16 (2016), p. 50.
[51] Y. Jhao, et al., Kinetics of austenitic grain growth in medium carbon niobium bearing steel, J. Jhejang Univ. Sci. A (Appl. Phys. Eng.) 12 (3) (2011) 171.
[52] D.S. Martin, et al., Austenite grain coarsening under influence of niobium carbonitride, Mater. Trans 45 (# 9) (2004) 2797.
[53] O. Flores, L. Martinez, Abnormal grain growth of austenite in a V-Nb microalloyed steel, J. Mater. Sci. 32 (1997) 5985.
[54] C. Klinkenberg, Niobium in microalloyed structural & engineering steel, Mater. Sci. Forum 539–543 (2007) 4261.
[55] T. Gladman et al., Structure property relationship in high strength microalloy steels. Proceedings of International Conference Microalloying on HSLA; University College (1977) p. 32.
[56] D.K. Matlock, J.G. Speer, Microalloying Concepts and Application in Long Products, Mater. Sci. Technol. 25 (9) (2009) 1118.
[57] S. Hashimoto, M. Nakamura, ISIJ Int 46 (2006) 1510.
[58] G. Krauss, S.K. Banerjee (Eds.), Fundamentals of Micro alloying Forging Grades, TMS, Warrendale, 1987.
[59] N. Fonstein, in: S.G. Jansto, Jitendra Patel (Eds.), Effects of Nb,V and Ti on the Evolution of Structures in Medium Carbon Steels during Various Hot Forging Steps;in Niobium Bearing Structural Steels, TMS, Warrendale, 2010.
[60] O.N. Mohanty and S.B. Singh, Developing low carbon forging steels (private communication, 2016).
[61] J.D. Boyd, P. Jhao, in: J.G. Speer, et al. (Eds.), Proceedings of International Conference on New Developments in Long and Forged Products: Metallurgy and Applications, AIST, Warrendale, 2006, p. 169.
[62] F.M. Becket and R. Franks, Steel, U.S. Patent No. 2,158,651, (1939).
[63] F.M. Becket and R. Franks, Steel, U.S. Patent No. 2,264,355, December 1941.

[64] A.G. Barkow, Columbium steel in high-pressure line pipe service, AISI Regional Technical Meeting No. 59, Buffalo, N.Y., (1960); See also U.S. Patent No.3,010,822, (1961).
[65] B. Morrison, The influence of small niobium additions on the properties of carbon–manganese steels, JISI 201 (1963) 317.
[66] A.J. DeArdo, J.M. Gray, L. Meyer, in: H. Stuart (Ed.), "Fundamental Metallurgy of Niobium in Steel", Niobium, AIME, 1981, p. 685.
[67] C.J. Van Tyne, G. Krauss, D.K. Matlock (Eds.), Fundamentals and Applications of Microalloying Forging Steels, TMS, Warrendale, PA, 1996.
[68] G. Tither, Niobium Science & Technology. Proceedings of International Symposium Niobium 2001, Orlando, Florida, December(2001), p. 845.
[69] D.K. Matlock, J.G. Speer, C.J. Van Tyne, "Microalloyed Forging Steel", Colorado School of Mines, IFC (2008), p. 189
[70] C.J. Van Tyne, D.K. Matlock, and J.G. Speer, Microalloyed forging steels for automotive applications. Proceedings of International Conference on Automotive Materials & Manufacturing, TMS (2010), p. 51
[71] A. Von den Steinen, et al., Investigations of Steels with about 0.5% C and small addition of vanadium or niobium, Stahl und Eisen 95 (6) (1975) 209.
[72] J. Stoeter, J. Kneller, Metal Progress 3 (1985) 61.
[73] J.H. Buche, in: H. Stuart (Ed.), Niobium in Engineering Bar Steels, Niobium, AIME, 1981, p. 989.
[74] J.H. Bucher and J.F. Held, Microalloyed cold finished and hot rolled bars, SAE Technical Paper Series, SAE Inc., No.811003 (1981).
[75] T. Sampei, T. Abe, H. Osuzu and I. Kozasu, Microalloyed bar for machine structural use, AMS Metals Mater. Technol. Ser., (1983), 8306.
[76] O. Pawelski, et al., The hot deformation simulator of the Max–Planck-Institut fuer Eisenforschung – a new concept for the investigation of fast hot Deformation processes, Stahl und Eisen 98 (5) (1978) 181.
[77] T. Maki, New Metallurgy for creating ultrafine-grained steel and its future prospect, Metal Technol. 71 (8) (2001) 59.
[78] L. Backer, P. Charlier, and R. Hechelski, Seminar on modification of quality requirements in the demand for steel, ONV, Turin, June, (1982).
[79] P. Charlier, A. Guimier, Proceedings of the 7th International Conference on the Strength of Metals and Alloys, Montreal, 2, Pergamon Press, 1986, p. 1019.
[80] P. Charlier, A. Guimier, Proceedings of an International Seminar on Automotive Steels, Moscow, Pub, CBMM, Brazil, 1988, p. 167.
[81] J. Kneller et al., Proceedings of an International Symposium, "Fundamentals of Microalloying Forging Steels", Colorado, TMS, (1986), p. 425.
[82] A. Blarasi, et al., Metal. Sci. Technol. 7 (2) (1989) 104.
[83] S. Engineer, et al., Entwicklung von Mikrolegierten Perlitischen Edelbaustaehlen mit verbesserten Zahigkeitseigenschaften, Information Day on Microalloyed Engineering Steels, Commission of the European Communities, VDI-Dusseldorf, 1998.
[84] C.I. Garcia, et al., Proceedings of the 8th Process Technology Conference, ISS-AIME, (Warrendale, PA, USA, (1987), p. 79.
[85] C.I. Garcia, et al. Proceedings of the 8th Process Technology Conference, Vol. 8, ISS-AIME, (Warrendale, PA, USA, (1988), p. 59.
[86] C.I. Garcia, et al. Proceedings of "Cold FormingTechnology Clinic", SME, April (Southfield, USA, Paper MF92 (1992), p. 139.

[87] W.E. Heitmann and L. Vea, Inland fre form™ microalloy steel for cold forging inland steel bar company.
[88] P.H. Wright, Advanced materials and processes, Metals Progress 12 (1988) 29–34.
[89] W.A. Szilva et al., Proceedings of the International Symposium on Microalloyed Bar and Forging Steels, Ontario, 26–29th, The Iron and Steel Section of the Metallurgical Society of CIM, Hamilton, August (1990), p. 227.
[90] V. Ollilainen, J. Kohopåå, Microalloyed Bar and Forging Steels, CIM, Montreal (Canada), 1990, pp. 237–248.
[91] M. Katsumata, et al., Materials Trans. JIM 8 (1991) 715–728.
[92] A.J. DeArdo, et al., Multiphase microalloyed steel and method thereof, U.S.Patent No. 5213634 (1993), p. 93.
[93] C.S. Smith, E.W. Palmer, Trans. AIME 105 (1933) 133.
[94] E. Hornbogen, Trans. ASM 57 (1964) 120.
[95] M.S. Gagliano, M.E. Fine, Characterization of the nucleation & growth behavior of copper precipitation in low carbon steels, Metall. Mater. Trans. A 35 (2004) 2323.
[96] S. Vaynman, et al., High-strength low-carbon ferritic steel containing Cu–Fe–Ni–Al–Mn Precipitates, Metall. Mater. Trans. A 39 (2008) 363.
[97] A. Nicholson, J.D. Murray, Surface hot shortness in low carbon steels, JISI 204 (1965) 1007.
[98] F. Maury, et al., J. Phys. Condn. Matter. 6 (1994) 569.
[99] N. Maruyama, et al., Mater. Trans, JIM 40 (No.4) (1999) 268.
[100] P.J. Othen, et al., Phil. Mag. Lett 64 (1991) 383.
[101] R. Monzen, et al., Phil. Mag. A 80 (2000) 711.
[102] U. Dahmen, et al., Acta Metall. 32 (1984) 803.
[103] S.R. Goodman, S.S. Brenner, J.R. Low Jr, Metall. Trans. 4 (1973) 2363.
[104] S.R. Goodman, S.S. Brenner, J.R. Low Jr, Metall. Trans. 4 (1973) 2371.
[105] G.M. Worrall, et al., J. Nucl. Mater. 148 (1987) 107.
[106] R.P. Foley and M.E. Fine, Proceedings of International. Conference on Processing, microstructure and properties of microalloy and other high strength low alloy steels, A.J. DeArdo, ed., Pittsburgh, (1991), p. 315.
[107] S.W. Thompson, et al., J. Mater. Sci. Lett. 17 (1998) 2075.
[108] M.K. Miller, Atom Probe Tomography, Kluwer Academic/PlenumPublishers, New York, 2000.
[109] T.F. Kelly, M.K. Miller, Rev. Sci. Instrum. 78 (2007) 31.
[110] D.N. Seidman, Annual. Rev. Mater. Res. 37 (2007) 127.
[111] D. Isheim, D.N. Seidman, Surf. Interface Anal. 36 (2004) 569.
[112] D. Isheim, et al., ActaMater. 54 (2006) 841.
[113] M.S. Gagliano, M.E. Fine, Precipitation kinetics of niobium carbide and copper in a low carbon, chromium-free steel, Calphad 25 (2001) 207–216.
[114] D.A. Melford, The influence of Sn and As on surface hot shortness copper containing steels, JISI (1966) 495.
[115] N. Imai, et al., Effect of Cu, Sn and Ni on hot workability of hot rolled mild steel, ISIJ Int. 37 (1997) 224.
[116] N. Imai, et al., Effect of Cu, Sn and Ni on hot workability of hot rolled mild steel, ISIJ Int. 37 (1997) 523.
[117] H. Ohtani, et al., Solid/liquid equilibria in Fe–Cu based ternary systems, ISIJ Int. 37 (1997) 207.
[118] S.J. Seo, et al., Effects of 0.4% Si and 0.02% P additions on surface hot shortness in 0.1% C–0.5% Mn steels containing 0.5% Cu, ISIJ Int. 37 (1997) 240.

[119] H. Fredriksson, On the mechanism of liquid copper penetration into iron grain boundaries, Scand. J. Metall. 30 (2001) 41.
[120] K. Shibata, et al., Suppression of surface hot shortness due to Cu in recycled steels, Mater. Trans. 43 (3) (2002) 292.
[121] L.G. Garza, C.J. Van Tyne, Surface hot-shortness of 1045 forging steel with residual copper, J. Mater. Process. Technol. 159 (2) (2005) 169.
[122] M. Torkar, Effect of trace and residual elements on the hot brittleness, hot shortness and properties of 0.15–0.3% C Al-killed steel with a solidification microstructure, Mater. Technol. (in English translation) 44 (6) (2010) 327.
[123] M. Okatsu, et al. Kawasaki Steel Giho, 30 (1998) 3, 131 (in Japanese), [ref. 8 in 124].
[124] K. Amano, et al., Materia 39 (184) (2000) (in Japanese) [ref. 9 in 124]
[125] R. Rana, et al., Laboratory investigations on copper-alloyed interstitial free steel part I: effect of annealing, Steel Res. Int. 78 (8) (2007) 612.
[126] R. Rana, et al., Laboratory investigations on copper-alloyed interstitial free steel – part II: effect of coiling temperature on annealing, Steel Res. Int. 78 (8) (2007) 622.
[127] R. Rana, et al., Laboratory investigations on copper-alloyed interstitial free steel – part III: ageing behavior, Steel Res. Int. 78 (8) (2007) 631.
[128] R. Rana, et al., Development of high strength interstitial free steel by copper precipitation hardening, Mater. Lett. 61 (14–15) (2007) 2919.
[129] J. Yang, S.K. Putatunda, Influence of a novel two step austempering process on the strainhardening behavior of austempered ductile cast iron (ADI), Mater. Sci. Eng. A382 (2004) 265.
[130] T. Dorn, et al. The current state of worldwide standards for ductile iron, 2003 Keith D. Mills Symposium on Ductile Iron, HiltonHead, SC, USA, 20–23 Oct. (2003).
[131] K.L. Hayrynen, J.R. Keough, Is it time for a grade 800 ADI in North America? AFS Trans. 113 (Pape4) (2005) 851.
[132] J.R. Keough and K.L. Hayrynen, Automotive applications of austempered ductile iron (ADI): a critical review in Designing and Achieving Lightweight Vehicles (SP1684). SAE Technical Paper 010.0764. (2000).
[133] A. Nofal, Advances in the metallurgy and applications of ADI, J. Metal. Eng. (ME) 2 (1) (2013) 1.

Index

Note: Page numbers followed by "*f*" and "*t*" refer to figures and tables, respectively.

FSC
www.fsc.org
MIX
Papier | Fördert gute Waldnutzung
FSC® C083411

Kolibri 360 GmbH
Ferdinand-Jühlke-Straße 7
99095 Erfurt, Deutschland
produktsicherheit@kolibri360.de